PHYSICAL FOUNDATIONS OF COSMOLOGY

Inflationary cosmology has been developed over the last 20 years to remedy serious shortcomings in the standard hot big bang model of the universe. Taking an original approach, this textbook explains the basis of modern cosmology and shows where the theoretical results come from.

The book is divided into two parts: the first deals with the homogeneous and isotropic model of the universe, while the second part discusses how initial inhomogeneities can explain the observed structure of the universe. Analytical treatments of traditionally highly numerical topics – such as primordial nucleosynthesis, recombination and cosmic microwave background anisotropy – are provided, and inflation and quantum cosmological perturbation theory are covered in great detail. The reader is brought to the frontiers of current cosmological research by the discussion of more speculative ideas.

This is an ideal textbook both for advanced students of physics and astrophysics and for those with a particular interest in theoretical cosmology. Nearly every formula in the book is derived from basic physical principles covered in undergraduate courses. Each chapter includes all necessary background material and no prior knowledge of general relativity and quantum field theory is assumed.

VIATCHESLAV MUKHANOV is Professor of Physics and Head of the Astroparticle Physics and Cosmology Group at the Department of Physics, Ludwig-Maximilians-Universität München, Germany. Following his Ph.D. at the Moscow Physical-Technical Institute, he conducted research at the Institute for Nuclear Research, Moscow, between 1982 and 1991. From 1992, he was a lecturer at Eidgenössische Technische Hochschule (ETH) in Zürich, Switzerland, until his appointment at LMU in 1997. His current research interests include cosmic microwave background fluctuations, inflationary models, string cosmology, the cosmological constant problem, dark energy, quantum and classical black holes, and quantum cosmology. He also serves on the editorial boards of leading research journals in these areas.

In 1980–81, Professor Mukhanov and G. Chibisov discovered that quantum fluctuations could be responsible for the large-scale structure of the universe. They calculated the spectrum of fluctuations in a model with a quasi-exponential stage of expansion, later known as inflation. The predicted perturbation spectrum is in very good agreement with measurements of the cosmic microwave background fluctuations. Subsequently, Professor Mukhanov developed the quantum theory of cosmological perturbations for calculating perturbations in generic inflationary models. In 1988, he was awarded the Gold Medal of the Academy of Sciences of the USSR for his work on this theory.

PHYSICAL FOUNDATIONS OF COSMOLOGY

VIATCHESLAV MUKHANOV

Ludwig-Maximilians-Universität München

CAMBRIDGE
UNIVERSITY PRESS

University Printing House, Cambridge CB2 8BS, United Kingdom

Published in the United States of America by Cambridge University Press, New York

Cambridge University Press is part of the University of Cambridge.

It furthers the University's mission by disseminating knowledge in the pursuit of education, learning and research at the highest international levels of excellence.

www.cambridge.org
Information on this title: www.cambridge.org/9780521563987

© V. Mukhanov 2005

This publication is in copyright. Subject to statutory exception and to the provisions of relevant collective licensing agreements, no reproduction of any part may take place without the written permission of Cambridge University Press.

First published 2005
3rd printing 2012

A catalogue record for this publication is available from the British Library

ISBN 978-0-521-56398-7 Hardback

Cambridge University Press has no responsibility for the persistence or accuracy of URLs for external or third-party internet websites referred to in this publication, and does not guarantee that any content on such websites is, or will remain, accurate or appropriate.

Contents

Foreword by Professor Andrei Linde		*page* xi
Preface		xiv
Acknowledgements		xvi
Units and conventions		xvii
Part I	**Homogeneous isotropic universe**	1
1	Kinematics and dynamics of an expanding universe	3
1.1	Hubble law	4
1.2	Dynamics of dust in Newtonian cosmology	8
	1.2.1 Continuity equation	9
	1.2.2 Acceleration equation	9
	1.2.3 Newtonian solutions	10
1.3	From Newtonian to relativistic cosmology	13
	1.3.1 Geometry of an homogeneous, isotropic space	14
	1.3.2 The Einstein equations and cosmic evolution	19
	1.3.3 Friedmann equations	22
	1.3.4 Conformal time and relativistic solutions	24
	1.3.5 Milne universe	27
	1.3.6 De Sitter universe	29
2	Propagation of light and horizons	37
2.1	Light geodesics	37
2.2	Horizons	38
2.3	Conformal diagrams	41
2.4	Redshift	55
	2.4.1 Redshift as a measure of time and distance	58
2.5	Kinematic tests	60
	2.5.1 Angular diameter–redshift relation	60
	2.5.2 Luminosity–redshift relation	64

v

		2.5.3	Number counts	66
		2.5.4	Redshift evolution	67
3	The hot universe			69
	3.1	The composition of the universe		69
	3.2	Brief thermal history		72
	3.3	Rudiments of thermodynamics		74
		3.3.1	Maximal entropy state, thermal spectrum, conservation laws and chemical potentials	75
		3.3.2	Energy density, pressure and the equation of state	79
		3.3.3	Calculating integrals	82
		3.3.4	Ultra-relativistic particles	85
		3.3.5	Nonrelativistic particles	88
	3.4	Lepton era		89
		3.4.1	Chemical potentials	92
		3.4.2	Neutrino decoupling and electron–positron annihilation	94
	3.5	Nucleosynthesis		97
		3.5.1	Freeze-out of neutrons	98
		3.5.2	"Deuterium bottleneck"	104
		3.5.3	Helium-4	108
		3.5.4	Deuterium	112
		3.5.5	The other light elements	117
	3.6	Recombination		120
		3.6.1	Helium recombination	120
		3.6.2	Hydrogen recombination: equilibrium consideration	122
		3.6.3	Hydrogen recombination: the kinetic approach	123
4	The very early universe			131
	4.1	Basics		132
		4.1.1	Local gauge invariance	133
		4.1.2	Non-Abelian gauge theories	135
	4.2	Quantum chromodynamics and quark–gluon plasma		138
		4.2.1	Running coupling constant and asymptotic freedom	141
		4.2.2	Cosmological quark–gluon phase transition	146
	4.3	Electroweak theory		150
		4.3.1	Fermion content	151
		4.3.2	"Spontaneous breaking" of $U(1)$ symmetry	153
		4.3.3	Gauge bosons	154
		4.3.4	Fermion interactions	158
		4.3.5	Fermion masses	160
		4.3.6	CP violation	162

	4.4	"Symmetry restoration" and phase transitions	165
		4.4.1 Effective potential	165
		4.4.2 $U(1)$ model	170
		4.4.3 Symmetry restoration at high temperature	173
		4.4.4 Phase transitions	174
		4.4.5 Electroweak phase transition	176
	4.5	Instantons, sphalerons and the early universe	180
		4.5.1 Particle escape from a potential well	180
		4.5.2 Decay of the metastable vacuum	184
		4.5.3 The vacuum structure of gauge theories	190
		4.5.4 Chiral anomaly and nonconservation of the fermion number	196
	4.6	Beyond the Standard Model	199
		4.6.1 Dark matter candidates	203
		4.6.2 Baryogenesis	210
		4.6.3 Topological defects	216
5		Inflation I: homogeneous limit	226
	5.1	Problem of initial conditions	226
	5.2	Inflation: main idea	229
	5.3	How can gravity become "repulsive"?	233
	5.4	How to realize the equation of state $p \approx -\varepsilon$	235
		5.4.1 Simple example: $V = \frac{1}{2}m^2\varphi^2$.	236
		5.4.2 General potential: slow-roll approximation	241
	5.5	Preheating and reheating	243
		5.5.1 Elementary theory	244
		5.5.2 Narrow resonance	245
		5.5.3 Broad resonance	249
		5.5.4 Implications	255
	5.6	"Menu" of scenarios	256
Part II		**Inhomogeneous universe**	263
6		Gravitational instability in Newtonian theory	265
	6.1	Basic equations	266
	6.2	Jeans theory	267
		6.2.1 Adiabatic perturbations	269
		6.2.2 Vector perturbations	270
		6.2.3 Entropy perturbations	270
	6.3	Instability in an expanding universe	271
		6.3.1 Adiabatic perturbations	273
		6.3.2 Vector perturbations	275

		6.3.3	Self-similar solution	275
		6.3.4	Cold matter in the presence of radiation or dark energy	276
	6.4		Beyond linear approximation	279
		6.4.1	Tolman solution	281
		6.4.2	Zel'dovich solution	283
		6.4.3	Cosmic web	286
7	Gravitational instability in General Relativity			289
	7.1		Perturbations and gauge-invariant variables	290
		7.1.1	Classification of perturbations	291
		7.1.2	Gauge transformations and gauge-invariant variables	292
		7.1.3	Coordinate systems	295
	7.2		Equations for cosmological perturbations	297
	7.3		Hydrodynamical perturbations	299
		7.3.1	Scalar perturbations	299
		7.3.2	Vector and tensor perturbations	309
	7.4		Baryon–radiation plasma and cold dark matter	310
		7.4.1	Equations	311
		7.4.2	Evolution of perturbations and transfer functions	314
8	Inflation II: origin of the primordial inhomogeneities			322
	8.1		Characterizing perturbations	323
	8.2		Perturbations on inflation (slow-roll approximation)	325
		8.2.1	Inside the Hubble scale	327
		8.2.2	The spectrum of generated perturbations	329
		8.2.3	Why do we need inflation?	332
	8.3		Quantum cosmological perturbations	334
		8.3.1	Equations	335
		8.3.2	Classical solutions	337
		8.3.3	Quantizing perturbations	340
	8.4		Gravitational waves from inflation	348
	8.5		Self-reproduction of the universe	352
	8.6		Inflation as a theory with predictive power	354
9	Cosmic microwave background anisotropies			356
	9.1		Basics	357
	9.2		Sachs–Wolfe effect	360
	9.3		Initial conditions	363
	9.4		Correlation function and multipoles	365
	9.5		Anisotropies on large angular scales	367
	9.6		Delayed recombination and the finite thickness effect	369
	9.7		Anisotropies on small angular scales	374
		9.7.1	Transfer functions	374

		9.7.2 Multipole moments	377
		9.7.3 Parameters	379
		9.7.4 Calculating the spectrum	382
9.8	Determining cosmic parameters		385
9.9	Gravitational waves		391
9.10	Polarization of the cosmic microwave background		395
		9.10.1 Polarization tensor	396
		9.10.2 Thomson scattering and polarization	398
		9.10.3 Delayed recombination and polarization	400
		9.10.4 E and B polarization modes and correlation functions	402
9.11	Reionization		407

Bibliography	410
Expanding universe (Chapters 1 and 2)	410
Hot universe and nucleosynthesis (Chapter 3)	411
Particle physics and early universe (Chapter 4)	412
Inflation (Chapters 5 and 8)	414
Gravitational instability (Chapters 6 and 7)	416
CMB fluctuations (Chapter 9)	417
Index	419

Foreword by Professor Andrei Linde

Since the beginning of the 1970s, we have witnessed spectacular progress in the development of cosmology, which started with a breakthrough in the theoretical understanding of the physical processes in the early universe and culminated in a series of observational discoveries. The time is ripe for a textbook which summarizes the new knowledge in a rigorous and yet accessible form.

The beginning of the new era in theoretical cosmology can be associated with the development of the gauge theories of weak, electromagnetic and strong interactions. Until that time, we had no idea of properties of matter at densities much greater than nuclear density $\sim 10^{14}$ g/cm^3, and everybody thought that the main thing we need to know about the early universe is the equation of state of superdense matter. In the beginning of the 1970s we learned that not only the size and the temperature of our universe, but also the properties of elementary particles in the early universe were quite different from what we see now. According to the theory of the cosmological phase transitions, during the first 10^{-10} seconds after the big bang there was not much difference between weak and electromagnetic interactions. The discovery of the asymptotic freedom for the first time allowed us to investigate the properties of matter even closer to the big bang, at densities almost 80 orders of magnitude higher than the nuclear density. Development of grand unified theories demonstrated that baryon number may not be conserved, which cleared the way towards the theoretical description of the creation of matter in the universe. This in its turn opened the doors towards inflationary cosmology, which can describe our universe only if the observed excess of baryons over antibaryons can appear after inflation.

Inflationary theory allowed us to understand why our universe is so large and flat, why it is homogeneous and isotropic, why its different parts started their expansion simultaneously. According to this theory, the universe at the very early stages of its evolution rapidly expanded (inflated) in a slowly changing vacuum-like state, which is usually associated with a scalar field with a large energy density. In the simplest version of this theory, called 'chaotic inflation,' the whole universe could emerge from a tiny speck of space of a Planckian size 10^{-33} cm, with a total mass smaller than 1 milligram. All elementary particles surrounding us were produced as a result of the decay of this vacuum-like state at the end of inflation. Galaxies emerged due to the growth of density perturbations, which were produced from quantum fluctuations generated and amplified during inflation. In certain cases, these quantum fluctuations may accumulate and become so large that they can be responsible not only for the formation of galaxies, but also for the formation of new exponentially large parts of the universe with different laws of low-energy physics operating in each of them. Thus, instead of being spherically symmetric and uniform, our universe becomes a multiverse, an eternally growing fractal consisting of different exponentially large parts which look homogeneous only locally.

One of the most powerful tools which can be used for testing the predictions of various versions of inflationary theory is the investigation of anisotropy of the cosmic microwave background (CMB) radiation coming to us from all directions. By studying this radiation, one can use the whole sky as a giant photographic plate with the amplified image of inflationary quantum fluctuations imprinted on it. The results of this investigation, in combination with the study of supernova and of the large-scale structure of the universe, have already confirmed many of the predictions of the new cosmological theory.

From this quick sketch of the evolution of our picture of the universe during the last 30 years one can easily see how challenging it may be to write a book serving as a guide in this vast and rapidly growing area of physics. That is why it gives me a special pleasure to introduce the book *Physical Foundations of Cosmology* by Viatcheslav Mukhanov.

In the first part of the book the author considers a homogeneous universe. One can find there not only the description of the basic cosmological models, but also an excellent introduction to the theory of physical processes in the early universe, such as the theory of nucleosynthesis, the theory of cosmological phase transitions, baryogenesis and inflationary cosmology. All of the necessary concepts from the general theory of relativity and particle physics are introduced and explained in an accurate and intuitively clear way. This part alone could be considered a good textbook in modern cosmology; it may serve as a basis for a separate course of lectures on this subject.

But if you are preparing for active research in modern cosmology, you may particularly appreciate the second part of the book, where the author discusses the formation and evolution of the large-scale structure of our universe. In order to understand this process, one must learn the theory of production of metric perturbations during inflation.

In 1981 Mukhanov and Chibisov discovered, in the context of the Starobinsky model, that the accelerated expansion can amplify the initial quantum perturbations of metric up to the values sufficient for explaining the large-scale structure of the universe. In 1982, a combined effort of many participants of the Nuffield Symposium in Cambridge allowed them to come to a similar conclusion with respect to the new inflationary universe scenario. A few years later, Mukhanov developed the general theory of inflationary perturbations of metric, valid for a broad class of inflationary models, including chaotic inflation. Since that time, his approach has become the standard method of investigation of inflationary perturbations.

A detailed description of this method is one of the most important features of this book. The theory of inflationary perturbations is quite complicated not only because it requires working knowledge of General Relativity and quantum field theory, but also because one should learn how to represent the results of the calculations in terms of variables that do not depend on the arbitrary choice of coordinates. It is very important to have a real master guiding you through this difficult subject, and Mukhanov does it brilliantly. He begins with a reminder of the simple Newtonian approach to the theory of density perturbations in an expanding universe, then extends this investigation to the general theory of relativity, and finishes with the full quantum theory of production and subsequent evolution of inflationary perturbations of metric.

The last chapter of the book provides the necessary link between this theory and the observations of the CMB anisotropy. Everyone who has studied this subject knows the famous figures of the spectrum of the CMB anisotropy, with several different peaks predicted by inflationary cosmology. The shape of the spectrum depends on various cosmological parameters, such as the total density of matter in the universe, the Hubble constant, etc. By measuring the spectrum one can determine these parameters experimentally. The standard approach is based on the numerical analysis using the CMBFAST code. Mukhanov made one further step and derived an analytic expression for the CMB spectrum, which can help the readers to obtain a much better understanding of the origin of the peaks, of their position and their height as a function of the cosmological parameters.

As in a good painting, this book consists of many layers. It can serve as an introduction to cosmology for the new generation of researchers, but it also contains a lot of information which can be very useful even for the best experts in this subject.

We live at a very unusual time. According to the observational data, the universe is approximately 14 billion years old. A hundred years ago we did not even know that it is expanding. A few decades from now we will have a detailed map of the observable part of the universe, and this map is not going to change much for the next billion years. We live at the time of the great cosmological discoveries, and I hope that this book will help us in our quest.

Preface

This textbook is designed both for serious students of physics and astrophysics and for those with a particular interest in learning about theoretical cosmology. There are already many books that survey current observations and describe theoretical results; my goal is to complement the existing literature and to show where the theoretical results come from. Cosmology uses methods from nearly all fields of theoretical physics, among which are General Relativity, thermodynamics and statistical physics, nuclear physics, atomic physics, kinetic theory, particle physics and field theory. I wanted to make the book useful for undergraduate students and, therefore, decided not to assume preliminary knowledge in any specialized field. With very few exceptions, the derivation of every formula in the book begins with basic physical principles from undergraduate courses. Every chapter starts with a general elementary introduction. For example, I have tried to make such a geometrical topic as conformal diagrams understandable even to those who have only a vague idea about General Relativity. The derivations of the renormalization group equation, the effective potential, the non-conservation of fermion number, and quantum cosmological perturbations should also, in principle, require no prior knowledge of quantum field theory. All elements of the Standard Model of particle physics needed in cosmological applications are derived from the initial idea of gauge invariance of the electromagnetic field. Of course, some knowledge of general relativity and particle physics would be helpful, but this is not a necessary condition for understanding the book. It is my hope that a student who has not previously taken the corresponding courses will be able to follow all the derivations.

This book is meant to be neither encyclopedic nor a sourcebook for the most recent observational data. In fact, I avoid altogether the presentation of data; after all the data change very quickly and are easily accessible from numerous available monographs as well as on the Internet. Furthermore, I have intentionally restricted the discussion in this book to results that have a solid basis. I believe it is premature to present detailed mathematical consideration of controversial topics in a book on the foundations of cosmology and, therefore, such topics are covered only at a very elementary level.

Inflationary theory and the generation of primordial cosmological perturbations, which I count among the solid results, are discussed in great detail. Here, I have tried to delineate carefully the robust features of inflation which do not depend on the particular inflationary scenario. Among the other novel features of the book is the analytical treatment of some topics which are traditionally considered as highly numerical, for example, primordial nucleosynthesis, recombination and the cosmic microwave background anisotropy.

Some words must be said about my decision to imbed problems in the main text rather than gathering them at the end. I have tried to make the derivations as transparent as possible so that the reader should be able to proceed from one equation to the next without making calculations on the way. In cases where this strategy failed, I have included problems, which thus constitute an integral part of the main text. Therefore, even the casual reader who is not solving the problems is encouraged to read them.

Acknowledgements

I have benefited very much from a great number of discussions with my colleagues and friends while planning and writing this book. The text of the first two chapters was substantially improved as a result of the numerous interactions I had with Paul Steinhardt during my sabbatical at Princeton University in 2002. It is a great pleasure to express to Paul and the physics faculty and students at Princeton my gratitude for their gracious hospitality.

I have benefited enormously from endless discussions with Andrei Linde and Lev Kofman and I am very grateful to them both.

I am indebted to Gerhard Buchalla, Mikhail Shaposhnikov, Andreas Ringwald and Georg Raffelt for broadening my understanding of the Standard Model, phase transitions in the early universe, sphalerons, instantons and axions.

Discussions with Uros Seljak, Sergei Bashinsky, Dick Bond, Steven Weinberg and Lyman Page were extremely helpful in writing the chapter on CMB fluctuations. My special thanks to Alexey Makarov, who assisted me with numerical calculations of the transfer function T_o and Carlo Contaldi who provided Figures 9.3 and 9.7.

It is a pleasure to extend my thanks to Andrei Barvinsky, Wilfried Buchmuller, Lars Bergstrom, Ivo Sachs, Sergei Shandarin, Alex Vilenkin and Hector Rubinstein, who read different parts of the manuscript and made valuable comments.

I am very much obliged to the members of our group in Munich: Matthew Parry, Serge Winitzki, Dorothea Deeg, Alex Vikman and Sebastian Pichler for their valuable advice on improving the presentation of different topics and for technical assistance in preparing the figures and index.

Last but not least, I would like to thank Vanessa Manhire and Matthew Parry for their heroic and hopefully successful attempt to convert my "Russian English" into English.

Units and conventions

Planckian (natural) units Gravity, quantum theory and thermodynamics play an important role in cosmology. It is not surprising, therefore, that all fundamental physical constants, such as the gravitational constant G, Planck's constant $\hbar$, the speed of light c and Boltzmann's constant k_B, enter the main formulae describing the universe. These formulae look much nicer if one uses (Planckian) natural units by setting $G = \hbar = c = k_B = 1$. In this case, all constants drop from the formulae and, after the calculations are completed, they can easily be restored in the final result if needed. For this reason, nearly all the calculations in this book are made using natural units, though the gravitational constant and Planck's constant are kept in some formulae in order to stress the relevance of gravitational and quantum physics for describing the corresponding phenomena.

After the formula for some physical quantity is derived in Planckian units, one can immediately calculate its numerical value in usual units simply by using the values of the elementary Planckian units:

$$l_{Pl} = \left(\frac{G\hbar}{c^3}\right)^{1/2} = 1.616 \times 10^{-33} \text{ cm},$$

$$t_{Pl} = \frac{l_{Pl}}{c} = 5.391 \times 10^{-44} \text{ s},$$

$$m_{Pl} = \left(\frac{\hbar c}{G}\right)^{1/2} = 2.177 \times 10^{-5} \text{ g},$$

$$T_{Pl} = \frac{m_{Pl} c^2}{k_B} = 1.416 \times 10^{32} \text{ K} = 1.221 \times 10^{19} \text{ GeV}.$$

Planckian units with other dimensions can easily be built out of these quantities. For example, the Planckian density and the Planckian area are $\varepsilon_{Pl} = m_{Pl}/l_{Pl}^3 = 5.157 \times 10^{93}$ g cm^{-3} and $S_{Pl} = l_{Pl}^2 = 2.611 \times 10^{-66}$ cm^2, respectively.

Two examples below show how to make calculations using Planckian units.

Example 1 *Calculate the number density of photons in the background radiation today.* In usual units, the temperature of the background radiation is $T \simeq 2.73$ K. In dimensionless Planckian units, this temperature is equal to

$$T \simeq \frac{2.73 \text{ K}}{1.416 \times 10^{32} \text{ K}} \simeq 1.93 \times 10^{-32}.$$

The number density of photons in natural units is

$$n_\gamma = \frac{3\zeta(3)}{2\pi^2} T^3 \simeq \frac{3 \times 1.202}{2\pi^2} \left(1.93 \times 10^{-32}\right)^3 \simeq 1.31 \times 10^{-96}.$$

To determine the number density of photons per cubic centimeter, we must multiply the *dimensionless* density obtained by the Planckian quantity with the corresponding dimension cm^{-3}, namely l_{Pl}^{-3}:

$$n_\gamma \simeq 1.31 \times 10^{-96} \times \left(1.616 \times 10^{-33} \text{ cm}\right)^{-3} \simeq 310 \text{ cm}^{-3}.$$

Example 2 *Determine the energy density of the universe 1 s after the big bang and estimate the temperature at this time.* The early universe is dominated by ultra-relativistic matter, and in natural units the energy density ε is related to the time t via

$$\varepsilon = \frac{3}{32\pi t^2}.$$

The time 1 s expressed in dimensionless units is

$$t \simeq \frac{1 \text{ s}}{5.391 \times 10^{-44} \text{ s}} \simeq 1.86 \times 10^{43};$$

hence the energy density at this time is equal to

$$\varepsilon = \frac{3}{32\pi \left(1.86 \times 10^{43}\right)^2} \simeq 8.63 \times 10^{-89}$$

Planckian units. To express the energy density in usual units, we have to multiply this number by the Planckian density, $\varepsilon_{Pl} = 5.157 \times 10^{93}$ g cm^{-3}. Thus we obtain

$$\varepsilon \simeq \left(8.63 \times 10^{-89}\right) \varepsilon_{Pl} \simeq 4.45 \times 10^{5} \text{ g cm}^{-3}.$$

To make a rough estimate of the temperature, we note that in natural units $\varepsilon \sim T^4$, hence $T \sim \varepsilon^{1/4} \sim \left(10^{-88}\right)^{1/4} = 10^{-22}$ Planckian units. In usual units,

$$T \sim 10^{-22} T_{Pl} \simeq 10^{10} \text{ K} \simeq 1 \text{ MeV}.$$

From this follows the useful relation between the temperature in the early Universe, measured in MeV, and the time, measured in seconds: $T_{\text{MeV}} = O(1) \, t_{\text{sec}}^{-1/2}$.

Astronomical units In astronomy, distances are usually measured in parsecs and megaparsecs instead of centimeters. They are related to centimeters via

$$1 \text{ pc} = 3.26 \text{ light years} = 3.086 \times 10^{18} \text{ cm}, \quad 1 \text{ Mpc} = 10^6 \text{ pc}.$$

The masses of galaxies and clusters of galaxies are expressed in terms of the mass of the Sun,

$$M_\odot \simeq 1.989 \times 10^{33} \text{ g}.$$

Charge units We use the Heaviside–Lorentz system for normalization of the elementary electric charge e. This system is adopted in most books on particle physics and in these units the Coulomb force between two electrons separated by a distance r is

$$F = \frac{e^2}{4\pi r^2}.$$

The dimensionless fine structure constant is $\alpha \equiv e^2/4\pi \simeq 1/137$.

Signature Throughout the book, we will always use the signature $(+, -, -, -)$ for the metric, so that the Minkowski metric takes the form $ds^2 = dt^2 - dx^2 - dy^2 - dz^2$.

Part I

Homogeneous isotropic universe

1

Kinematics and dynamics of an expanding universe

The most important feature of our universe is its large scale homogeneity and isotropy. This feature ensures that observations made from our single vantage point are representative of the universe as a whole and can therefore be legitimately used to test cosmological models.

For most of the twentieth century, the homogeneity and isotropy of the universe had to be taken as an assumption, known as the "Cosmological Principle." Physicists often use the word "principle" to designate what are at the time wild, intuitive guesses in contrast to "laws," which refer to experimentally established facts.

The Cosmological Principle remained an intelligent guess until firm empirical data, confirming large scale homogeneity and isotropy, were finally obtained at the end of the twentieth century. The nature of the homogeneity is certainly curious. The observable patch of the universe is of order 3000 Mpc (1 Mpc $\simeq 3.26 \times 10^6$ light years $\simeq 3.08 \times 10^{24}$ cm). Redshift surveys suggest that the universe is homogeneous and isotropic only when coarse grained on 100 Mpc scales; on smaller scales there exist large inhomogeneities, such as galaxies, clusters and superclusters. Hence, the Cosmological Principle is only valid within a limited range of scales, spanning a few orders of magnitude.

Moreover, theory suggests that this may not be the end of the story. According to inflationary theory, the universe continues to be homogeneous and isotropic over distances larger than 3000 Mpc, but it becomes highly inhomogeneous when viewed on scales *much much* larger than the observable patch. This dampens, to some degree, our hope of comprehending the entire universe. We would like to answer such questions as: What portion of the entire universe is like the part we find ourselves in? What fraction has a predominance of matter over antimatter? Or is spatially flat? Or is accelerating or decelerating? These questions are not only difficult to answer, but they are also hard to pose in a mathematically precise way. And, even if a suitable mathematical definition can be found, it is difficult to imagine how we could verify empirically any theoretical predictions concerning scales greatly exceeding the observable universe. The subject is too seductive to avoid speculations altogether, but we will, nevertheless, try to focus on the salient, empirically testable features of the observable universe.

It is firmly established by observations that our universe:

- *is homogeneous and isotropic on scales larger than* 100 Mpc *and has well developed inhomogeneous structure on smaller scales;*
- *expands according to the Hubble law.*

Concerning the matter composition of the universe, we know that:

- *it is pervaded by thermal microwave background radiation with temperature* $T \simeq 2.73$ K;
- *there is baryonic matter, roughly one baryon per* 10^9 *photons, but no substantial amount of antimatter;*
- *the chemical composition of baryonic matter is about 75% hydrogen, 25% helium, plus trace amounts of heavier elements;*
- *baryons contribute only a small percentage of the total energy density; the rest is a dark component, which appears to be composed of cold dark matter with negligible pressure* ($\sim 25\%$) *and dark energy with negative pressure* ($\sim 70\%$).

Observations of the fluctuations in the cosmic microwave background radiation suggest that:

- *there were only small fluctuations of order* 10^{-5} *in the energy density distribution when the universe was a thousand times smaller than now.*

For a review of the observational evidence the reader is encouraged to refer to recent papers and reviews. In this book we concentrate mostly on theoretical understanding of these basic observational facts.

Any cosmological model worthy of consideration must be consistent with established facts. While the standard big bang model accommodates most known facts, a physical theory is also judged by its predictive power. At present, inflationary theory, naturally incorporating the success of the standard big bang, has no competitor in this regard. Therefore, we will build upon the standard big bang model, which will be our starting point, until we reach contemporary ideas of inflation.

1.1 Hubble law

In a nutshell, the standard big bang model proposes that the universe emerged about 15 billion years ago with a homogeneous and isotropic distribution of matter at very high temperature and density, and has been expanding and cooling since then. We begin our account with the Newtonian theory of gravity, which captures many of the essential aspects of the universe's dynamics and gives us an intuitive grasp of

1.1 Hubble law

what happens. After we have reached the limits of validity of Newtonian theory, we turn to a proper relativistic treatment.

In an expanding, homogeneous and isotropic universe, the relative velocities of observers obey the *Hubble law*: the velocity of observer B with respect to A is

$$\mathbf{v}_{B(A)} = H(t)\mathbf{r}_{BA}, \qquad (1.1)$$

where the Hubble parameter $H(t)$ depends only on the time t, and $\mathbf{r}_{BA}$ is the vector pointing from A to B. Some refer to H as the Hubble "constant" to stress its independence of the spatial coordinates, but it is important to recognize that H is, in general, time-varying.

In a homogeneous, isotropic universe there are no *privileged* vantage points and the expansion appears the same to all observers wherever they are located. The Hubble law is in complete agreement with this. Let us consider how two observers A and B view a third observer C (Figure 1.1). The Hubble law specifies the velocities of the other two observers relative to A:

$$\mathbf{v}_{B(A)} = H(t)\mathbf{r}_{BA}, \quad \mathbf{v}_{C(A)} = H(t)\mathbf{r}_{CA}. \qquad (1.2)$$

From these relations, we can find the relative velocity of observer C with respect to observer B:

$$\mathbf{v}_{C(B)} = \mathbf{v}_{C(A)} - \mathbf{v}_{B(A)} = H(t)(\mathbf{r}_{CA} - \mathbf{r}_{BA}) = H\mathbf{r}_{CB}. \qquad (1.3)$$

The result is that observer B sees precisely the same expansion law as observer A. In fact, the Hubble law is the *unique* expansion law compatible with homogeneity and isotropy.

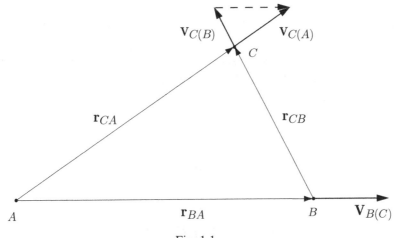

Fig. 1.1.

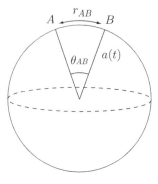

Fig. 1.2.

Problem 1.1 In order for a general expansion law, $\mathbf{v} = \mathbf{f}(\mathbf{r},t)$, to be the same for all observers, the function $\mathbf{f}$ must satisfy the relation

$$\mathbf{f}(\mathbf{r}_{CA} - \mathbf{r}_{BA}, t) = \mathbf{f}(\mathbf{r}_{CA}, t) - \mathbf{f}(\mathbf{r}_{BA}, t). \tag{1.4}$$

Show that the only solution of this equation is given by (1.1).

A useful analogy for envisioning Hubble expansion is the two-dimensional surface of an expanding sphere (Figure 1.2). The angle θ_{AB} between any two points A and B on the surface of the sphere remains unchanged as its radius $a(t)$ increases. Therefore the distance between the points, measured along the surface, grows as

$$r_{AB}(t) = a(t)\theta_{AB}, \tag{1.5}$$

implying a relative velocity

$$v_{AB} = \dot{r}_{AB} = \dot{a}(t)\theta_{AB} = \frac{\dot{a}}{a} r_{AB}, \tag{1.6}$$

where dot denotes a derivative with respect to time t. Thus, the Hubble law emerges here with $H(t) \equiv \dot{a}/a$.

The distance between any two observers A and B in a homogeneous and isotropic universe can be also rewritten in a form similar to (1.5). Integrating the equation

$$\dot{\mathbf{r}}_{BA} = H(t)\mathbf{r}_{BA}, \tag{1.7}$$

we obtain

$$\mathbf{r}_{BA}(t) = a(t)\boldsymbol{\chi}_{BA}, \tag{1.8}$$

where

$$a(t) = \exp\left(\int H(t)\, dt\right) \tag{1.9}$$

is called the scale factor and is the analogue of the radius of the 2-sphere. The integration constant, χ_{BA}, is the analogue of θ_{BA} and can be interpreted as the distance between points A and B at some particular moment of time. It is called the Lagrangian or *comoving* coordinate of B, assuming a coordinate system centered at A.

In the 2-sphere analogy, $a(t)$ has a precise geometrical interpretation as the radius of the sphere and, consequently, has a fixed normalization. In Newtonian theory, however, the value of the scale factor $a(t)$ itself has no geometrical meaning and its normalization can be chosen arbitrarily. Once the normalization is fixed, the scale factor $a(t)$ describes the distance between observers as a function of time. For example, when the scale factor increases by a factor of 3, the distance between any two observers increases threefold. Therefore, when we say the size of the universe was, for instance, 1000 times smaller, this means that the distance between any two comoving objects was 1000 times smaller — a statement which makes sense even in an infinitely large universe. The Hubble parameter, which is equal to

$$H(t) = \frac{\dot{a}}{a}, \qquad (1.10)$$

measures the expansion rate.

In this description, we are assuming a perfectly homogeneous and isotropic universe in which all observers are comoving in the sense that their coordinates χ remain unchanged. In the real universe, wherever matter is concentrated, the motion of nearby objects is dominated by the inhomogeneities in the gravitational field, which lead, for example, to virial orbital motion rather than Hubble expansion. Similarly, objects held together by other, stronger forces resist Hubble expansion. The velocity of these objects relative to comoving observers is referred to as the "peculiar" velocity. Hence, the Hubble law is valid only on the scales of homogeneity.

Problem 1.2 Typical peculiar velocities of galaxies are about a few hundred kilometers per second. The mean distance between large galaxies is about 1 Mpc. How distant must a galaxy be from us for its peculiar velocity to be small compared to its comoving (Hubble) velocity, if the Hubble parameter is 75 km s^{-1} Mpc^{-1}?

The current value of the Hubble parameter, H_0, can be determined by measuring the ratio of the recession velocity to the distance for an object whose peculiar velocity is small compared to its comoving velocity. The recessional velocity can be accurately measured because it induces a Doppler shift in spectral lines. The challenge is to find a reliable measure of the distance. Two methods used are based on the concepts of "standard candles" and "standard rulers." A class of objects is called a standard candle if the objects have about the same luminosity. Usually, they possess a set of characteristics that can be used to identify them even when they are far away. For example, Cepheid variable stars pulse at a periodic rate, and Type IA supernovae are bright, exploding stars with a characteristic spectral pattern. The distances to nearby objects in the class are measured directly (for example, by parallax) or by comparing them to another standard candle whose distance has already been calibrated. Once the distance to a subset of a given standard candle class has been measured, the distance to further members of that class can be determined: the inverse square law relates the apparent luminosity of the distant objects to that of the nearby objects whose distance is already determined. The standard ruler method is exactly like the standard candle method except that it relies on identifying a class of objects of the same size rather than the same luminosity. It is clear, however, that only if the variation in luminosity or size of objects within the same class is small can they be useful for measuring the Hubble parameter. Cepheid variable stars have been studied for nearly a century and appear to be good standard candles. Type IA supernovae are promising candidates which are potentially important because they can be observed at much greater distances than Cepheids. Because of systematic uncertainties, the value of the measured Hubble constant is known today with only modest accuracy and is about 65–80 km s^{-1} Mpc^{-1}.

Knowing the value of the Hubble constant, we can obtain a rough estimate for the age of the universe. If we neglect gravity and consider the velocity to be constant in time, then two points separated by $|\mathbf{r}|$ today, coincided in the past, $t_0 \simeq |\mathbf{r}|/|\mathbf{v}| = 1/H_0$ ago. For the measured value of the Hubble constant, t_0 is about 15 billion years. We will show later that the exact value for the age of the universe differs from this rough estimate by a factor of order unity, depending on the composition and curvature of the universe.

Because the Hubble law has a kinematical origin and its form is dictated by the requirement of homogeneity and isotropy, it has to be valid in both Newtonian theory and General Relativity. In fact, rewritten in the form (1.8), it can be immediately applied in Einstein's theory. This remark may be disconcerting since, according to the Hubble law, the relative velocity can exceed the speed of light for two objects separated by a distance larger than $1/H$. How can this be consistent with Special Relativity? The resolution of the paradox is that, in General Relativity, the relative velocity has no invariant meaning for objects whose separation exceeds $1/H$, which represents the curvature scale. We will explore this point further in context of the Milne universe (Section 1.3.5), following the discussion of Newtonian cosmology.

1.2 Dynamics of dust in Newtonian cosmology

We first consider an infinite, expanding, homogeneous and isotropic universe filled with "dust," a euphemism for matter whose pressure p is negligible compared

1.2 Dynamics of dust in Newtonian cosmology

to its energy density ε. (In cosmology the terms "dust" and "matter" are used interchangeably to represent nonrelativistic particles.) Let us choose some arbitrary point as the origin and consider an expanding sphere about that origin with radius $R(t) = a(t)\chi_{com}$. Provided that gravity is weak and the radius is small enough that the speed of the particles within the sphere relative to the origin is much less than the speed of light, the expansion can be described by Newtonian gravity. (Actually, General Relativity is involved here in an indirect way. We assume the net effect on a particle within the sphere due to the matter outside the sphere is zero, a premise that is ultimately justified by Birkhoff's theorem in General Relativity.)

1.2.1 Continuity equation

The total mass M within the sphere is conserved. Therefore, the energy density due to the mass of the particles is

$$\varepsilon(t) = \frac{M}{(4\pi/3)R^3(t)} = \varepsilon_0 \left(\frac{a_0}{a(t)}\right)^3, \tag{1.11}$$

where ε_0 is the energy density at the moment when the scale factor is equal a_0. It is convenient to rewrite this conservation law in differential form. Taking the time derivative of (1.11), we obtain

$$\dot{\varepsilon}(t) = -3\varepsilon_0 \left(\frac{a_0}{a(t)}\right)^3 \frac{\dot{a}}{a} = -3H\varepsilon(t). \tag{1.12}$$

This equation is a particular case of the nonrelativistic continuity equation,

$$\frac{\partial \varepsilon}{\partial t} = -\nabla(\varepsilon \mathbf{v}), \tag{1.13}$$

if we take $\varepsilon(\mathbf{x}, t) = \varepsilon(t)$ and $\mathbf{v} = H(t)\mathbf{r}$. Beginning with the continuity equation and assuming homogeneous initial conditions, it is straightforward to show that the unique velocity distribution which maintains homogeneity evolving in time is the Hubble law: $\mathbf{v} = H(t)\mathbf{r}$.

1.2.2 Acceleration equation

Matter is gravitationally self-attractive and this causes the expansion of the universe to decelerate. To derive the equation of motion for the scale factor, consider a probe particle of mass m on the surface of the sphere, a distance $R(t)$ from the origin. Assuming matter outside the sphere does not exert a gravitational force on the particle, the only force acting is due to the mass M of all particles within the sphere. The equation of motion, therefore, is

$$m\ddot{R} = -\frac{GmM}{R^2} = -\frac{4\pi}{3}Gm\frac{M}{(4\pi/3)R^3}R. \tag{1.14}$$

Using the expression for the energy density in (1.11) and substituting $R(t) = a(t)\chi_{com}$, we obtain

$$\ddot{a} = -\frac{4\pi}{3}G\varepsilon a. \tag{1.15}$$

The mass of the probe particle and the comoving size of the sphere χ_{com} drop out of the final equation.

Equations (1.12) and (1.15) are the two master equations that determine the evolution of $a(t)$ and $\varepsilon(t)$. They *exactly* coincide with the corresponding equations for dust ($p = 0$) in General Relativity. This is not as surprising as it may seem at first. The equations derived do not depend on the size of the auxiliary sphere and, therefore, are exactly the same for an infinitesimally small sphere where all the particles move with infinitesimal velocities and create a negligible gravitational field. In this limit, General Relativity *exactly* reduces to Newtonian theory and, hence, relativistic corrections should not arise.

1.2.3 Newtonian solutions

The closed form equation for the scale factor is obtained by substituting the expression for the energy density (1.11) into the acceleration equation (1.15):

$$\ddot{a} = -\frac{4\pi}{3}G\varepsilon_0\frac{a_0^3}{a^2}. \tag{1.16}$$

Multiplying this equation by $\dot{a}$ and integrating, we find

$$\frac{1}{2}\dot{a}^2 + V(a) = E, \tag{1.17}$$

where E is a constant of integration and

$$V(a) = -\frac{4\pi G\varepsilon_0 a_0^3}{3a}.$$

Equation (1.17) is identical to the energy conservation equation for a rocket launched from the surface of the Earth with unit mass and speed $\dot{a}$. The integration constant E represents the total energy of the rocket. Escape from the Earth occurs if the positive kinetic energy overcomes the negative gravitational potential or, equivalently, if E is positive. If the kinetic energy is too small, the total energy E is negative and the rocket falls back to Earth. Similarly, the fate of the dust-dominated universe – whether it expands forever or eventually recollapses – depends on the sign of E. As pointed out above, the normalization of a has no invariant meaning in Newtonian gravity and it can be rescaled by an arbitrary factor. Hence, only the sign of E is physically relevant. Rewriting (1.17) as

$$H^2 - \frac{2E}{a^2} = \frac{8\pi G}{3}\varepsilon, \qquad (1.18)$$

we see that the sign of E is determined by the relation between the Hubble parameter, which determines the kinetic energy of expansion, and the mass density, which defines the gravitational potential energy.

In the rocket problem, the mass of the Earth is given and the student is asked to compute the minimal escape velocity by setting $E = 0$ and solving for the velocity v. In cosmology, the expansion velocity, as set by the Hubble parameter, has been reasonably well measured while the mass density was very poorly determined for most of the twentieth century. For this historical reason, the boundary between escape and gravitational entrapment is traditionally characterized by a critical density, rather than critical velocity. Setting $E = 0$ in (1.18), we obtain

$$\varepsilon^{cr} = \frac{3H^2}{8\pi G}. \qquad (1.19)$$

The critical density decreases with time since H is decreasing, though the term "critical density" is often used to refer to its current value. Expressing E in terms of the energy density $\varepsilon(t)$ and the Hubble constant $H(t)$, we find

$$E = \frac{4\pi G}{3}a^2\varepsilon^{cr}\left(1 - \frac{\varepsilon}{\varepsilon^{cr}}\right) = \frac{4\pi G}{3}a^2\varepsilon^{cr}\left[1 - \Omega(t)\right], \qquad (1.20)$$

where

$$\Omega(t) \equiv \frac{\varepsilon(t)}{\varepsilon^{cr}(t)} \qquad (1.21)$$

is called the cosmological parameter. Generally, $\Omega(t)$ varies with time, but because the sign of E is fixed, the difference $1 - \Omega(t)$ does not change sign. Therefore, by measuring the current value of the cosmological parameter, $\Omega_0 \equiv \Omega(t_0)$, we can determine the sign of E.

We shall see that the sign of E determines the spatial geometry of the universe in General Relativity. In particular, the spatial curvature has the opposite sign to E. Hence, *in a dust-dominated universe*, there is a direct link between the ratio of the energy density to the critical density, the spatial geometry and the future evolution of the universe. If $\Omega_0 = \varepsilon_0/\varepsilon_0^{cr} > 1$, then $E < 0$ and the spatial curvature is positive (closed universe). In this case the scale factor reaches some maximal value and the universe recollapses, as shown in Figure 1.3. When $\Omega_0 < 1$, E is positive, the spatial curvature is negative (open universe), and the universe expands hyperbolically. The

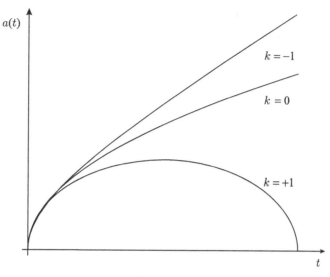

Fig. 1.3.

special case of $\Omega = 1$, or $E = 0$, corresponds to parabolic expansion and flat spatial geometry (flat universe). For both flat and open cases, the universe expands forever at an ever-decreasing rate (Figure 1.3). In all three cases, extrapolating back to a "beginning," we face an "initial singularity," where the scale factor approaches zero and the expansion rate and energy density diverge.

The reader should be aware that the connection between Ω_0 and the future evolution of the universe discussed above is not universal, but depends on the matter content of the universe. We will see later that it is possible to have a closed universe that never recollapses.

Problem 1.3 Show that $\dot{a} \to \infty$, $H \to \infty$ and $\varepsilon \to \infty$ when $a \to 0$.

Problem 1.4 Show that, for the expanding sphere of dust, $\Omega(t)$ is equal to the absolute value of the ratio of the gravitational potential energy to the kinetic energy. Since dust is gravitationally self-attractive, it decelerates the expansion rate. Therefore, in the past, the kinetic energy was much larger than at present. To satisfy the energy conservation law, the increase in kinetic energy should be accompanied by an increase in the magnitude of the negative potential energy. Show that, irrespective of its current value, $\Omega(t) \longrightarrow 1$ as $a \longrightarrow 0$.

Problem 1.5 Another convenient dimensionless parameter that characterizes the expansion is the "deceleration parameter":

$$q = -\frac{\ddot{a}}{aH^2}. \tag{1.22}$$

The sign of q determines whether the expansion is slowing down or speeding up. Find a general expression for q in terms of Ω and verify that $q = 1/2$ in a flat dust-dominated universe.

To conclude this section we derive an explicit solution for the scale factor in a flat matter-dominated universe. Because $E = 0$, (1.17) can be rewritten as

$$a \cdot \dot{a}^2 = \frac{4}{9}\left(\frac{da^{3/2}}{dt}\right)^2 = \text{const}, \tag{1.23}$$

and, hence, its solution is

$$a \propto t^{2/3}. \tag{1.24}$$

For the Hubble parameter, we obtain

$$H = \frac{2}{3t}. \tag{1.25}$$

Thus, the current age of a flat ($E = 0$) dust-dominated universe is

$$t_0 = \frac{2}{3H_0}, \tag{1.26}$$

where H_0 is the present value of the Hubble parameter. We see that the result is not very different from the rough estimate obtained by neglecting gravity. The energy density of matter as a function of cosmic time can be found by substituting the Hubble parameter (1.25) into (1.18):

$$\varepsilon(t) = \frac{1}{6\pi G t^2}. \tag{1.27}$$

Problem 1.6 Estimate the energy density at $t = 10^{-43}$ s, 1 s and 1 year after the big bang.

Problem 1.7 Solve (1.18) in the limit $t \to \infty$ for an open universe and discuss the properties of the solution.

1.3 From Newtonian to relativistic cosmology

General Relativity leads to a mathematically consistent theory of the universe, whereas Newtonian theory does not. For example, we pointed out that the Newtonian picture of an expanding, dust-filled universe relies on Birkhoff's theorem, which is proven in General Relativity. In addition, General Relativity introduces key changes to the Newtonian description. First, Einstein's theory proposes that geometry is dynamical and is determined by the matter composition of the universe. Second, General Relativity can describe matter moving with relativistic velocities and having arbitrary pressure. We know that radiation, which has a pressure equal to one third of its energy density, dominated the universe for the first 100 000 years after the big bang. Additionally, evidence suggests that most of the energy density today has negative pressure. To understand these important epochs in cosmic history, we are forced to go beyond Newtonian gravity and turn to a fully relativistic theory. We begin by considering what kind of three-dimensional spaces can be used to describe a homogeneous and isotropic universe.

1.3.1 Geometry of an homogeneous, isotropic space

The assumption that our universe is homogeneous and isotropic means that its evolution can be represented as a time-ordered sequence of three-dimensional space-like hypersurfaces, each of which is homogeneous and isotropic. These hypersurfaces are the natural choice for surfaces of constant time.

Homogeneity means that the physical conditions are the same at every point of any given hypersurface. Isotropy means that the physical conditions are identical in all directions when viewed from a given point on the hypersurface. Isotropy *at every point* automatically enforces homogeneity. However, homogeneity does not necessarily imply isotropy. One can imagine, for example, a homogeneous yet anisotropic universe which contracts in one direction and expands in the other two directions.

Homogeneous and isotropic spaces have the largest possible symmetry group; in three dimensions there are three independent translations and three rotations. These symmetries strongly restrict the admissible geometry for such spaces. There exist only three types of homogeneous and isotropic spaces with simple topology: (a) flat space, (b) a three-dimensional sphere of constant positive curvature, and (c) a three-dimensional hyperbolic space of constant negative curvature.

To help visualize these spaces, we consider the analogous two-dimensional homogeneous, isotropic surfaces. The generalization to three dimensions is straightforward. Two well known cases of homogeneous, isotropic surfaces are the plane and the 2-sphere. They both can be embedded in three-dimensional Euclidean space with the usual Cartesian coordinates x, y, z. The equation describing the embedding of a two-dimensional sphere (Figure 1.4) is

$$x^2 + y^2 + z^2 = a^2, \tag{1.28}$$

where a is the radius of the sphere. Differentiating this equation, we see that, for two infinitesimally close points on the sphere,

$$dz = -\frac{xdx + ydy}{z} = \pm\frac{xdx + ydy}{\sqrt{a^2 - x^2 - y^2}}.$$

1.3 From Newtonian to relativistic cosmology

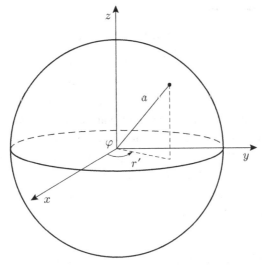

Fig. 1.4.

Substituting this expression into the three-dimensional Euclidean metric,

$$dl^2 = dx^2 + dy^2 + dz^2, \tag{1.29}$$

gives

$$dl^2 = dx^2 + dy^2 + \frac{(xdx + ydy)^2}{a^2 - x^2 - y^2}. \tag{1.30}$$

In this way, the distance between a pair of points located on the 2-sphere is expressed entirely in terms of two independent coordinates x and y, which are bounded, $x^2 + y^2 \leq a^2$. These coordinates, however, are degenerate in the sense that to every given (x, y) there correspond two different points on the sphere located in the northern and southern hemispheres. It is convenient to introduce instead of x and y the angular coordinates r', φ defined in the standard way:

$$x = r' \cos \varphi, \ y = r' \sin \varphi. \tag{1.31}$$

Differentiating the relation $x^2 + y^2 = r'^2$, we have

$$xdx + ydy = r'dr'.$$

Combining this with

$$dx^2 + dy^2 = dr'^2 + r'^2 d\varphi^2,$$

the metric in (1.30) becomes

$$dl^2 = \frac{dr'^2}{1-(r'^2/a^2)} + r'^2 d\varphi^2. \quad (1.32)$$

The limit $a^2 \to \infty$ corresponds to a (flat) plane. We can also formally take a^2 to be negative and then metric (1.32) describes a homogeneous, isotropic two-dimensional space with constant negative curvature, known as Lobachevski space. Unlike the flat plane or the two-dimensional sphere, Lobachevski space cannot be embedded in Euclidean three-dimensional space because the radius of the "sphere" a is imaginary (this is why this space is called a pseudo-sphere or hyperbolic space). Of course, this does not mean that this space cannot exist. Any curved space can be described entirely in terms of its internal geometry without referring to its embedding.

Problem 1.8 Lobachevski space can be visualized as a hyperboloid in Lorentzian three-dimensional space (Figure 1.5). Verify that the embedding of the surface $x^2 + y^2 - z^2 = -a^2$, where a^2 is positive, in the space with metric $dl^2 = dx^2 + dy^2 - dz^2$ gives a Lobachevski space.

Introducing the rescaled coordinate $r = r'/\sqrt{|a^2|}$, we can recast metric (1.32) as

$$dl^2 = |a^2| \left(\frac{dr^2}{1-kr^2} + r^2 d\varphi^2 \right), \quad (1.33)$$

where $k = +1$ for the sphere ($a^2 > 0$), $k = -1$ for the pseudo-sphere ($a^2 < 0$) and $k = 0$ for the plane (two-dimensional flat space). In curved space, $|a^2|$ characterizes the radius of curvature. In flat space, however, the normalization of $|a^2|$ does not have any physical meaning and this factor can be absorbed by redefinition of the coordinates. The generalization of the above consideration to three dimensions is straightforward.

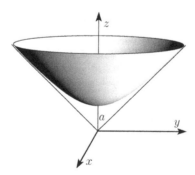

Fig. 1.5.

Problem 1.9 By embedding a three-dimensional sphere (pseudo-sphere) in a four-dimensional Euclidean (Lorentzian) space, verify that the metric of a three-dimensional space of constant curvature can be written as

$$dl_{3d}^2 = a^2 \left(\frac{dr^2}{1 - kr^2} + r^2(d\theta^2 + \sin^2\theta d\varphi^2) \right), \quad (1.34)$$

where a^2 is positive and $k = 0, \pm 1$. Introduce the rescaled radial coordinate $\bar{r}$, defined by

$$r = \frac{\bar{r}}{1 + k\bar{r}^2/4}, \quad (1.35)$$

and show that this metric can then be rewritten in explicitly isotropic form:

$$dl_{3d}^2 = a^2 \frac{(d\bar{x}^2 + d\bar{y}^2 + d\bar{z}^2)}{(1 + k\bar{r}^2/4)^2}, \quad (1.36)$$

where

$$\bar{x} = \bar{r}\sin\theta\cos\varphi, \quad \bar{y} = \bar{r}\sin\theta\sin\varphi, \quad \bar{z} = \bar{r}\cos\theta.$$

In many applications, instead of the radial coordinate r, it is convenient to use coordinate χ defined via the relation

$$d\chi^2 = \frac{dr^2}{1 - kr^2}. \quad (1.37)$$

It follows that

$$\chi = \begin{cases} \operatorname{arcsinh} r, & k = -1; \\ r, & k = 0; \\ \arcsin r & k = +1. \end{cases} \quad (1.38)$$

The coordinate χ varies between 0 and $+\infty$ in flat and hyperbolic spaces, while $\pi \geq \chi \geq 0$ in spaces with positive curvature $(k = +1)$. In this last case, to every particular r correspond two different χ. Thus, introducing χ removes the coordinate degeneracy mentioned above. In terms of χ, metric (1.34) takes the form

$$dl_{3d}^2 = a^2(d\chi^2 + \Phi^2(\chi)d\Omega^2) \equiv a^2 \left[d\chi^2 + \begin{pmatrix} \sinh^2\chi \\ \chi^2 \\ \sin^2\chi \end{pmatrix} d\Omega^2 \right] \begin{array}{l} k = -1; \\ k = 0; \\ k = +1, \end{array} \quad (1.39)$$

where

$$d\Omega^2 = (d\theta^2 + \sin^2\theta d\varphi^2). \quad (1.40)$$

Let us now take a closer look at the properties of the constant curvature spaces.

Three-dimensional sphere ($k = +1$) It follows from (1.39) that in a three-dimensional space with positive curvature, the distance element on the surface of a 2-sphere of radius χ is

$$dl^2 = a^2 \sin^2 \chi (d\theta^2 + \sin^2 \theta d\varphi^2). \tag{1.41}$$

This expression is the same as for a sphere of radius $R = a \sin \chi$ in flat three-dimensional space, and hence we can immediately find the total surface area:

$$S_{2d}(\chi) = 4\pi R^2 = 4\pi a^2 \sin^2 \chi. \tag{1.42}$$

As the radius χ increases, the surface area first grows, reaches its maximal value at $\chi = \pi/2$, and then decreases, vanishing at $\chi = \pi$ (Figure 1.6).

To understand such unusual behavior of a surface area, it is useful to turn to a low-dimensional analogy. In this analogy, the surface of the globe plays the role of three-dimensional space with constant curvature and the two-dimensional surfaces correspond to circles of constant latitude on the globe. Beginning from the north pole, corresponding to $\theta = 0$, the circumferences of the circles grow as we move southward, reach a maximum at the equator, where $\theta = \pi/2$, then decrease below the equator and vanish at the south pole, $\theta = \pi$. As θ runs from 0 to π, it covers the whole surface of the globe. Similarly, as χ changes from 0 to π, it sweeps out the whole three-dimensional space of constant positive curvature. Because the total area of the globe is finite, we expect that the total volume of the three-dimensional space with positive curvature is also finite.

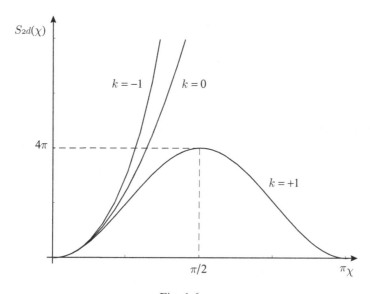

Fig. 1.6.

1.3 From Newtonian to relativistic cosmology 19

In fact, since the physical width of an infinitesimal shell is $dl = ad\chi$, the volume element between two spheres with radii χ and $\chi + d\chi$ is

$$dV = S_{2d}ad\chi = 4\pi a^3 \sin^2 \chi d\chi.$$

Therefore, the volume within the sphere of radius χ_0 is

$$V(\chi_0) = 4\pi a^3 \int_0^{\chi_0} \sin^2 \chi d\chi = 2\pi a^3 \left(\chi_0 - \tfrac{1}{2} \sin 2\chi_0\right). \tag{1.43}$$

For $\chi_0 \ll 1$, the volume,

$$V(\chi_0) = 4\pi (a\chi_0)^3/3 + \cdots,$$

grows in the same way as in Euclidean space. The total volume, obtained by substituting $\chi_0 = \pi$ in (1.43), is finite and equal to

$$V = 2\pi^2 a^3. \tag{1.44}$$

The other distinguishing property of a space of constant positive curvature is that the sum of the angles of a triangle constructed from geodesics (curves of minimal length) is larger than 180 degrees.

Three-dimensional pseudo-sphere ($k = -1$) The metric on the surface of a 2-sphere of radius χ in a three-dimensional space of constant negative curvature is

$$dl^2 = a^2 \sinh^2 \chi (d\theta^2 + \sin^2 \theta d\varphi^2), \tag{1.45}$$

and the area of the sphere,

$$S_{2d}(\chi) = 4\pi a^2 \sinh^2 \chi, \tag{1.46}$$

increases exponentially for $\chi \gg 1$. Since the coordinate χ varies from 0 to $+\infty$, the total volume of the hyperbolic space is infinite. The sum of angles of a triangle is less than 180 degrees.

Problem 1.10 Calculate the volume of a sphere with radius χ_0 in a space with constant negative curvature.

1.3.2 The Einstein equations and cosmic evolution

The only way to preserve the homogeneity and isotropy of space and yet incorporate time evolution is to allow the curvature scale, characterized by a, to be time-dependent. The scale factor $a(t)$ thus completely describes the time evolution of a homogeneous, isotropic universe. In relativistic theory, there is no absolute time and spatial distances are not invariant with respect to coordinate transformations. Instead, the infinitesimal spacetime interval between events is invariant. There exist, however, preferred coordinate systems in which the symmetries of the universe are clearly manifest. In one of the most convenient of such coordinate systems, the interval takes the form

$$ds^2 = dt^2 - dl_{3d}^2 = dt^2 - a^2(t)\left(\frac{dr^2}{1-kr^2} + r^2 d\Omega^2\right) \equiv g_{\alpha\beta} dx^\alpha dx^\beta, \quad (1.47)$$

where $g_{\alpha\beta}$ is the metric of the spacetime and $x^\alpha \equiv (t, r, \theta, \varphi)$ are the coordinates of events. We will use the Einstein convention for summation over repeated indices:

$$g_{\alpha\beta} dx^\alpha dx^\beta \equiv \sum_{\alpha,\beta} g_{\alpha\beta} dx^\alpha dx^\beta.$$

Additionally, we will always choose Greek indices to run from 0 to 3 with 0 reserved for the time-like coordinate. Latin indices run only over spatial coordinates: $i, l, \ldots = 1, 2, 3$. The spatial coordinates introduced above are comoving; that is, every object with zero peculiar velocity has constant coordinates r, θ, φ. Furthermore, the coordinate t is the proper time measured by a comoving observer. The distance between two comoving observers at a particular moment of time is

$$\int \sqrt{-ds_{t=const}^2} \propto a(t)$$

and, therefore, increases or decreases in proportion to the scale factor.

In General Relativity, the dynamical variables characterizing the gravitational field are the components of the metric $g_{\alpha\beta}(x^\gamma)$ and they obey the Einstein equations:

$$G_\beta^\alpha \equiv R_\beta^\alpha - \frac{1}{2}\delta_\beta^\alpha R - \Lambda \delta_\beta^\alpha = 8\pi G T_\beta^\alpha. \quad (1.48)$$

Here

$$R_\beta^\alpha = g^{\alpha\gamma}\left(\frac{\partial \Gamma_{\gamma\beta}^\delta}{\partial x^\delta} - \frac{\partial \Gamma_{\gamma\delta}^\delta}{\partial x^\beta} + \Gamma_{\gamma\beta}^\delta \Gamma_{\delta\sigma}^\sigma - \Gamma_{\gamma\delta}^\sigma \Gamma_{\beta\sigma}^\delta\right) \quad (1.49)$$

is the Ricci tensor expressed in terms of the inverse metric $g^{\alpha\gamma}$, defined via $g^{\alpha\gamma} g_{\gamma\beta} = \delta_\beta^\alpha$, and the Christoffel symbols

$$\Gamma_{\gamma\beta}^\alpha = \frac{1}{2} g^{\alpha\delta}\left(\frac{\partial g_{\gamma\delta}}{\partial x^\beta} + \frac{\partial g_{\delta\beta}}{\partial x^\gamma} - \frac{\partial g_{\gamma\beta}}{\partial x^\delta}\right). \quad (1.50)$$

The symbol δ_β^α denotes the unit tensor, equal to 1 when $\alpha = \beta$ and 0 otherwise; $R = R_\alpha^\alpha$ is the scalar curvature; and $\Lambda = $ const is the cosmological term. Matter is incorporated in Einstein's equations through the energy–momentum tensor, T^α_β. (In General Relativity the term "matter" is used for anything not the gravitational field.) This tensor is symmetric,

$$T^{\alpha\beta} \equiv g^{\beta\delta} T^\alpha_\delta = T^{\beta\alpha}, \tag{1.51}$$

and is (almost unambiguously) determined by the condition that the equations

$$\partial T^{\alpha\beta}/\partial x^\beta = 0 \tag{1.52}$$

must coincide with the equations of motion for matter in Minkowski spacetime. To generalize to curved spacetime, the equations of motion are modified:

$$T^{\alpha\beta}{}_{;\beta} \equiv \frac{\partial T^{\alpha\beta}}{\partial x^\beta} + \Gamma^\alpha_{\gamma\beta} T^{\gamma\beta} + \Gamma^\beta_{\gamma\beta} T^{\alpha\gamma} = 0, \tag{1.53}$$

where the terms proportional to Γ account for the gravitational field. Note that in General Relativity these equations do not need to be postulated separately. They follow from the Einstein equations as a consequence of the Bianchi identities satisfied by the Einstein tensor:

$$G^\alpha_{\beta;\alpha} = 0. \tag{1.54}$$

On large scales, matter can be approximated as a perfect fluid characterized by energy density ε, pressure p and 4-velocity u^α. Its energy–momentum tensor is

$$T^\alpha_\beta = (\varepsilon + p) u^\alpha u_\beta - p \delta^\alpha_\beta, \tag{1.55}$$

where the equation of state $p = p(\varepsilon)$ depends on the properties of matter and must be specified. For example, if the universe is composed of ultra-relativistic gas, the equation of state is $p = \varepsilon/3$. In many cosmologically interesting cases $p = w\varepsilon$, where w is constant.

Problem 1.11 Consider a nonrelativistic, dust-like perfect fluid ($u^0 \approx 1$, $u^i \ll 1$, $p \ll \varepsilon$) in a flat spacetime. Verify that the equations $T^{\alpha\beta}{}_{,\beta} = 0$ are equivalent to the mass conservation law plus the Euler equations of motion.

Another important example of matter is a classical scalar field φ with potential $V(\varphi)$. In this case, the energy–momentum tensor is given by the expression

$$T^\alpha_\beta = \varphi^{,\alpha} \varphi_{,\beta} - \left(\frac{1}{2} \varphi^{,\gamma} \varphi_{,\gamma} - V(\varphi) \right) \delta^\alpha_\beta, \tag{1.56}$$

where

$$\varphi_{,\beta} \equiv \frac{\partial \varphi}{\partial x^\beta}, \quad \varphi^{,\alpha} \equiv g^{\alpha\gamma} \varphi_{,\gamma}.$$

Problem 1.12 Show that the equations of motion for the scalar field,

$$\varphi^{;\alpha}{}_{;\alpha} + \frac{\partial V}{\partial \varphi} = 0, \tag{1.57}$$

follow from $T^{\alpha}_{\beta\,;\alpha} = 0$.

If $\varphi^{,\gamma}\varphi_{,\gamma} > 0$, then the energy–momentum tensor for a scalar field can be rewritten in the form of a perfect fluid (1.55) by defining

$$\varepsilon \equiv \tfrac{1}{2}\varphi^{,\gamma}\varphi_{,\gamma} + V(\varphi), \quad p \equiv \tfrac{1}{2}\varphi^{,\gamma}\varphi_{,\gamma} - V(\varphi), \quad u^{\alpha} \equiv \varphi^{,\alpha}/\sqrt{\varphi^{,\gamma}\varphi_{,\gamma}}. \tag{1.58}$$

In particular, assuming that the field is homogeneous ($\partial \varphi/\partial x^i = 0$), we have

$$\varepsilon \equiv \tfrac{1}{2}\dot{\varphi}^2 + V(\varphi),\ p \equiv \tfrac{1}{2}\dot{\varphi}^2 - V(\varphi). \tag{1.59}$$

For a scalar field, the ratio $w = p/\varepsilon$ is, in general, time-dependent. Additionally, w is bounded from below by -1 for any positive potential V and the weak energy dominance condition, $\varepsilon + p \geq 0$, is satisfied. However, the strong energy dominance condition, $\varepsilon + 3p \geq 0$, can easily be violated by a scalar field. For example, if a potential $V(\varphi)$ has a local minimum at some point φ_0, then $\varphi(t) = \varphi_0$ is a solution of the scalar field equations, for which

$$p = -\varepsilon = -V(\varphi_0). \tag{1.60}$$

As far as Einstein's equations are concerned, the corresponding energy–momentum tensor,

$$T^{\alpha}_{\beta} = V(\varphi_0)\delta^{\alpha}_{\beta}, \tag{1.61}$$

imitates a cosmological term

$$\Lambda = 8\pi G V(\varphi_0). \tag{1.62}$$

The cosmological term can therefore always be interpreted as the contribution of vacuum energy to the Einstein equations and from now on we include it in the energy–momentum tensor of matter and set $\Lambda = 0$ in (1.48).

1.3.3 Friedmann equations

How are the Newtonian equations of cosmological evolution (1.12), (1.15) and (1.18) modified when matter is relativistic? In principle, to answer this question we must simply substitute the metric (1.47) and energy–momentum tensor (1.55) into the Einstein equations (1.48). The resulting equations are the Friedmann equations and they determine the two unknown functions $a(t)$ and $\varepsilon(t)$. However, rather than starting with this formal derivation, it is instructive to explain how the nonrelativistic equations (1.12) and (1.15) must be modified.

1.3 From Newtonian to relativistic cosmology

If the pressure p within an expanding sphere of volume V is significant, then the total energy, $E = \varepsilon V$, is no longer conserved because the pressure does work, $-pdV$. According to the first law of thermodynamics, this work must be equal to the change in the total energy:

$$dE = -pdV. \quad (1.63)$$

Since $V \propto a^3$, we can rewrite this conservation law as

$$d\varepsilon = -3(\varepsilon + p)d\ln a \quad (1.64)$$

or, equivalently,

$$\dot{\varepsilon} = -3H(\varepsilon + p). \quad (1.65)$$

This relation is the new version of (1.12) and it turns out to be the energy conservation equation, $T^{\alpha}_{0\,;\alpha} = 0$, in an isotropic, homogeneous universe.

The acceleration equation is also modified for matter with nonnegligible pressure since, according to General Relativity, the strength of the gravitational field depends not only on the energy density but also on the pressure. Equation (1.15) becomes the first Friedmann equation:

$$\ddot{a} = -\frac{4\pi}{3}G(\varepsilon + 3p)a. \quad (1.66)$$

The real justification for the form of the pressure contribution is that the acceleration equation (1.66) follows from any diagonal spatial component of the Einstein equations. Multiplying (1.66) by $\dot{a}$, using (1.65) to express p in terms of ε, $\dot{\varepsilon}$ and H, and integrating, we obtain the second Friedmann equation:

$$H^2 + \frac{k}{a^2} = \frac{8\pi G}{3}\varepsilon. \quad (1.67)$$

This looks like the Newtonian equation (1.18) with $k = -2E$, though (1.67) applies for an arbitrary equation of state. However, k is not simply a constant of integration: the $0-0$ Einstein equation tells us that it is exactly the curvature introduced before, that is, $k = \pm 1$ or 0. For $k = \pm 1$, the magnitude of the scale factor a has a geometrical interpretation as the radius of curvature.

Thus, in General Relativity, the value of cosmological parameter, $\Omega \equiv \varepsilon/\varepsilon^{cr}$, determines the geometry. If $\Omega > 1$, the universe is closed and has the geometry of a three-dimensional sphere ($k = +1$); $\Omega = 1$ corresponds to a flat universe ($k = 0$); and in the case of $\Omega < 1$, the universe is open and has hyperbolic geometry ($k = -1$).

The combination of (1.67) and either the conservation law (1.65) or the acceleration equation (1.66), supplemented by the equation of state $p = p(\varepsilon)$, forms a complete system of equations that determines the two unknown functions $a(t)$ and

$\varepsilon(t)$. The solutions, and hence the future of the universe, depend not only on the geometry but also on the equation of state.

Problem 1.13 From (1.65) and (1.67), derive the following useful relation:

$$\dot{H} = -4\pi G(\varepsilon + p) + \frac{k}{a^2}. \tag{1.68}$$

Problem 1.14 Show that, for $p > -\varepsilon/3$, a closed universe recollapses after reaching a maximal radius while flat and open universes continue to expand forever. Verify that the spatial curvature term in (1.67), k/a^2, can be neglected as $a \to 0$ and give a physical interpretation of this result. Analyze the behavior of the scale factor for the case $-\varepsilon/3 \geq p \geq -\varepsilon$.

To conclude this section, let us reiterate the most important distinctions between the Newtonian and relativistic treatments of a homogeneous, isotropic universe. First, the Newtonian approach is incomplete: it is only valid (with justification from General Relativity) for nearly pressureless matter on small scales, where the relative velocities due to expansion are small compared to the speed of light. In Newtonian cosmology, the spatial geometry is always flat and, consequently, the scale factor has no geometrical interpretation. General Relativity, by contrast, provides a complete, self-consistent theory which allows us to describe relativistic matter with any equation of state. This theory is applicable on arbitrarily large scales. The matter content determines the geometry of the universe and, if $k = \pm 1$, the scale factor has a geometrical interpretation as the radius of curvature.

1.3.4 Conformal time and relativistic solutions

To find particular solutions of the Friedmann equations it is often convenient to replace the physical time t with the conformal time η, defined as

$$\eta \equiv \int \frac{dt}{a(t)}, \tag{1.69}$$

so that $dt = a(\eta)d\eta$. Equation (1.67) can then be rewritten as

$$a'^2 + ka^2 = \frac{8\pi G}{3}\varepsilon a^4, \tag{1.70}$$

where prime denotes the derivative with respect to η. Differentiating with respect to η and using (1.64), we obtain

$$a'' + ka = \frac{4\pi G}{3}(\varepsilon - 3p)a^3. \tag{1.71}$$

1.3 From Newtonian to relativistic cosmology

This last equation, which corresponds to the trace of the Einstein equations, is useful for finding analytic solutions for a universe filled by dust and radiation.

In the case of radiation, $p = \varepsilon/3$, the expression on the right hand side of (1.71) vanishes and the equation reduces to

$$a'' + ka = 0. \tag{1.72}$$

This is easily integrated and the result is

$$a(\eta) = a_m \cdot \begin{cases} \sinh \eta, & k = -1; \\ \eta, & k = 0; \\ \sin \eta, & k = +1. \end{cases} \tag{1.73}$$

Here a_m is one constant of integration and the other has been fixed by requiring $a(\eta = 0) = 0$. The physical time t is expressed in terms of η by integrating the relation $dt = a\, d\eta$:

$$t = a_m \cdot \begin{cases} (\cosh \eta - 1), & k = -1; \\ \eta^2/2, & k = 0; \\ (1 - \cos \eta), & k = +1. \end{cases} \tag{1.74}$$

It follows that in the most interesting case of a flat radiation-dominated universe, the scale factor is proportional to the square root of the physical time, $a \propto \sqrt{t}$, and hence $H = 1/2t$. Substituting this into (1.67), we obtain

$$\varepsilon_r = \frac{3}{32\pi G t^2} \propto a^{-4}. \tag{1.75}$$

Alternatively, the energy conservation equation (1.64) for radiation takes the form

$$d\varepsilon_r = -4\varepsilon_r d \ln a, \tag{1.76}$$

also implying that $\varepsilon_r \propto a^{-4}$.

Problem 1.15 Find $H(\eta)$ and $\Omega(\eta)$ in open and closed radiation-dominated universes and express the current age of the universe t_0 in terms of H_0 and Ω_0. Analyze the result for $\Omega_0 \ll 1$ and give its physical interpretation.

Problem 1.16 For dust, $p = 0$, the expression on the right hand side of (1.71) is constant and solutions of this equation can easily be found. Verify that

$$a(\eta) = a_m \cdot \begin{cases} (\cosh \eta - 1), & k = -1; \\ \eta^2, & k = 0; \\ (1 - \cos \eta), & k = +1. \end{cases} \tag{1.77}$$

For each case, compute $H(\eta)$ and $\Omega(\eta)$ and express the age of the universe in terms of H_0 and Ω_0. Show that in the limit $\Omega_0 \to 0$, we have $t_0 = 1/H_0$, in agreement with the Newtonian estimate obtained by ignoring gravity. (*Hint* Use (1.70) to fix one of the constants of integration.)

The range of conformal time η in flat and open universes is semi-infinite, $+\infty > \eta > 0$, regardless of whether the universe is dominated by radiation or matter. For a closed universe, η is bounded: $\pi > \eta > 0$ and $2\pi > \eta > 0$ in the radiation- and matter-dominated universes respectively.

Finally, we consider the important case of a flat universe with a mixture of matter (dust) and radiation. The energy density of matter decreases as $1/a^3$ while that of radiation decays as $1/a^4$. Therefore, we have

$$\varepsilon = \varepsilon_m + \varepsilon_r = \frac{\varepsilon_{eq}}{2}\left(\left(\frac{a_{eq}}{a}\right)^3 + \left(\frac{a_{eq}}{a}\right)^4\right), \quad (1.78)$$

where a_{eq} is the value of the scale factor at matter–radiation equality, when $\varepsilon_m = \varepsilon_r$. Equation (1.71) now becomes

$$a'' = \frac{2\pi G}{3}\varepsilon_{eq} a_{eq}^3 \quad (1.79)$$

and has a simple solution:

$$a(\eta) = \frac{\pi G}{3}\varepsilon_{eq} a_{eq}^3 \eta^2 + C\eta. \quad (1.80)$$

Again, we have fixed one of the two constants of integration by imposing the condition $a(\eta = 0) = 0$. Substituting (1.78) and (1.80) into (1.70), we find the other constant of integration:

$$C = \left(4\pi G \varepsilon_{eq} a_{eq}^4/3\right)^{1/2}.$$

Solution (1.80) is then

$$a(\eta) = a_{eq}\left(\left(\frac{\eta}{\eta_\star}\right)^2 + 2\left(\frac{\eta}{\eta_\star}\right)\right), \quad (1.81)$$

where

$$\eta_\star = \left(\pi G \varepsilon_{eq} a_{eq}^2/3\right)^{-1/2} = \eta_{eq}/(\sqrt{2} - 1) \quad (1.82)$$

has been introduced to simplify the expression. (The relation between $\eta_\star$ and η_{eq} immediately follows from $a(\eta_{eq}) = a_{eq}$.) For $\eta \ll \eta_{eq}$, radiation dominates and $a \propto \eta$. As the universe expands, the energy density of radiation decreases faster than that of dust. Hence, for $\eta \gg \eta_{eq}$, dust takes over and we have $a \propto \eta^2$.

Problem 1.17 Verify that, for a nonflat universe with a mixture of matter and radiation, one has

$$a(\eta) = a_m \cdot \begin{cases} (\eta_\star \sinh \eta + \cosh \eta - 1), & k = -1; \\ (\eta_\star \sin \eta + 1 - \cos \eta), & k = +1. \end{cases} \quad (1.83)$$

Problem 1.18 Consider a closed universe filled with matter whose equation of state is $w = p/\varepsilon$, where w is constant. Verify that the scale factor is then

$$a(\eta) = a_m \left(\sin\left(\frac{1+3w}{2} \eta + C \right) \right)^{2/(1+3w)}, \quad (1.84)$$

where C is a constant of integration. Analyze the behavior of the scale factor for $w = -1, -1/2, -1/3, 0$ and $+1/3$. Find the corresponding solutions for flat and open universes.

1.3.5 Milne universe

Let us consider an open universe with $k = -1$ in the limit of vanishing energy density, $\varepsilon \to 0$. In this case, (1.67) simplifies to

$$\dot{a}^2 = 1$$

and has a solution, $a = t$. The metric then takes the form

$$ds^2 = dt^2 - t^2(d\chi^2 + \sinh^2 \chi \, d\Omega^2), \quad (1.85)$$

and describes a spacetime known as a Milne universe. One might naturally expect that the solution of the Einstein equations for an isotropic space without matter must be Minkowski spacetime. Indeed, the Milne universe is simply a piece of Minkowski spacetime described in expanding coordinates. To prove this, we begin with the Minkowski metric,

$$ds^2 = d\tau^2 - dr^2 - r^2 d\Omega^2. \quad (1.86)$$

Replacing the Minkowski coordinates τ and r by the new coordinates t and χ, defined via

$$\tau = t \cosh \chi, \quad r = t \sinh \chi, \quad (1.87)$$

we find that

$$d\tau^2 - dr^2 = dt^2 - t^2 d\chi^2,$$

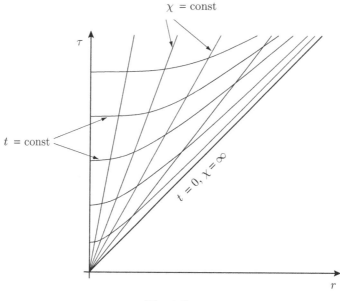

Fig. 1.7.

and hence the Minkowski metric reduces to (1.85). A particle with a given comoving coordinate χ moves with constant velocity

$$|\mathbf{v}| \equiv r/\tau = \tanh \chi < 1 \qquad (1.88)$$

in Minkowski space and its proper time, $\sqrt{1 - |\mathbf{v}|^2}\tau$, is equal to the cosmological time t. To find the hypersurfaces of constant proper time t, we note that

$$t^2 = \tau^2 - r^2. \qquad (1.89)$$

The hypersurface $t = 0$ coincides with the forward light cone; the surfaces of constant $t > 0$ are hyperboloids in Minkowski coordinates, all located within the forward light cone. Hence, the Milne coordinates cover only one quarter of Minkowski spacetime (Figure 1.7).

Despite its obvious deficiencies as a practical model, the Milne universe does illustrate some useful points. First, it shows the similarities and differences between an explosion (the popular misconception of the "big bang") and Hubble expansion. The Milne universe has a center. It is apparent from the fact that the Milne coordinates cover only one particular quarter of Minkowski spacetime. The curved Friedmann universe has no center. Second, the Milne universe reveals the subtleties in the physical interpretation of recessional velocity. If the recessional velocity of a particle were defined as $|\mathbf{u}| \equiv r/t = \sinh \chi$, it would exceed the speed of light for $\chi > 1$. Of course, there can be no contradiction with the principles of Special Relativity and we know that the particle is traveling on a physically allowed, time-like world-line. Special Relativity says that the speed measured using rulers and clocks of the *same* inertial coordinate system never exceeds the speed of light. In the definition of $|\mathbf{u}|$, however, we used the distance measured in the Minkowski coordinate system and the proper time of the moving particle. This corresponds to the spatial part of the 4-velocity, which can be arbitrarily large. The Hubble velocity in a Milne universe is also not bounded when defined in the usual way: $v_H = \dot{a}\chi = \chi$. Only $|\mathbf{v}| = \tanh \chi$ is well defined. Although both $|\mathbf{u}|$ and v_H are approximately equal to $|\mathbf{v}|$ for $\chi \ll 1$, for $\chi \geq 1$ they are very different and can have no invariant meaning. In curved spacetime, the situation is even more complicated. The inertial coordinate system can be introduced only locally, on scales much smaller than the four-dimensional curvature scale, roughly $1/H$. Hence, the relative Minkowski velocity, the quantity which can never exceed the speed of light, is only defined for particles whose separation is much less than $1/H$. Any definition of relative velocities at distances larger than the curvature scale, where the Hubble law predicts velocities which exceed unity, cannot have an invariant meaning. These remarks may be helpful in clarifying the notion of "superluminal expansion," a confusing term sometimes used in the literature to describe inflationary expansion.

The Milne solution is also useful as an illustration of the difference between 3-curvature and 4-curvature. A "spatially flat" universe ($k = 0$) generically has nonzero 4-curvature. For example, in the case of a dust-dominated universe with $\Omega = 1$, space is nonempty and the Riemann tensor is nonzero. The Milne universe is a complementary example with nonzero spatial curvature ($k = -1$) but zero 4-curvature. Milne coordinates correspond to foliating the locally flat spacetime with spatially curved homogeneous three-dimensional hypersurfaces. Hence, whenever the term "flat" is used in cosmology, it is important to distinguish between 3-curvature and 4-curvature.

Generally, one does not have a choice of foliation if it is to respect the homogeneity and isotropy of space. In particular, if the energy density is changing with time, the appropriate foliation is hypersurfaces of constant energy density. This choice is unique and has invariant physical meaning. Empty space, however, possesses extra time-translational invariance, so any space-like hypersurface has uniform "energy density" equal to zero. The other example of a homogeneous and isotropic spacetime with extra time-translational invariance is de Sitter space. In the next section we will see that de Sitter space can be covered by three-dimensional hypersurfaces of constant curvature with open, flat and closed geometry.

1.3.6 De Sitter universe

The de Sitter universe is a *spacetime* with positive constant 4-curvature that is homogeneous and isotropic in both space and time. Hence, it possesses the largest possible symmetry group, as large as the symmetry group of Minkowski spacetime (ten parameters in the four-dimensional case). In this book, we pay special attention to the de Sitter universe because it plays a central role in understanding the basic properties of inflation. In fact, in most scenarios, inflation is nothing more than a de Sitter stage with slightly broken time-translational symmetry.

To find the metric of the de Sitter universe, we use three different approaches which illustrate different mathematical aspects of this spacetime. First, we obtain the de Sitter metric in a way similar to that discussed in Section 1.3.1, namely, as a result of embedding a constant curvature surface in a higher-dimensional flat spacetime. For the sake of simplicity, we perform all calculations for two-dimensional surfaces. The generalization to higher dimensions is straightforward. As a second approach, we analytically continue metric (1.39), describing a homogeneous, isotropic three-dimensional space of constant positive curvature with Euclidean signature, to obtain a constant curvature space with Lorentzian signature. Finally, we obtain de Sitter spacetime as a solution to the Friedmann equations with positive cosmological constant.

De Sitter universe as a constant curvature surface embedded in Minkowski spacetime (two-dimensional case) Let us consider a hyperboloid

$$-z^2 + x^2 + y^2 = H_\Lambda^{-2}, \tag{1.90}$$

embedded in three-dimensional Minkowski space with the metric

$$ds^2 = dz^2 - dx^2 - dy^2. \tag{1.91}$$

This hyperboloid has positive curvature and lies entirely outside the light cone (Figure 1.8). Therefore, the induced metric has Lorentzian signature. (We noted in Problem 1.8 that Lobachevski space can also be embedded in a space with Lorentzian signature. However, Lobachevski space corresponds to a hyperbolic surface lying within the light cone and has an induced metric with Euclidean signature.) To parameterize the surface of the hyperboloid, we can use x and y coordinates. The metric of the hyperboloid can then be written as

$$ds^2 = \frac{(xdx + ydy)^2}{x^2 + y^2 - H_\Lambda^{-2}} - dx^2 - dy^2, \tag{1.92}$$

where $x^2 + y^2 > H_\Lambda^{-2}$. This is the metric of a two-dimensional de Sitter *spacetime* in x, y coordinates. As with the cases considered in Section 1.3.1, it is more convenient to use coordinates in which the symmetries of the spacetime are more explicit. The first choice is t, χ coordinates related to x, y via

$$x = H_\Lambda^{-1} \cosh(H_\Lambda t) \cos \chi, \quad y = H_\Lambda^{-1} \cosh(H_\Lambda t) \sin \chi. \tag{1.93}$$

1.3 From Newtonian to relativistic cosmology

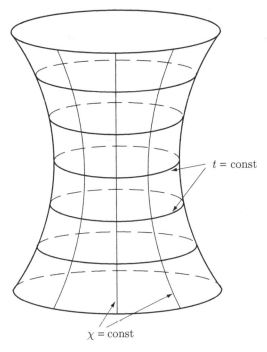

Fig. 1.8.

These coordinates cover the entire hyperboloid for $+\infty > t > -\infty$ and $2\pi \geq \chi \geq 0$ (Figure 1.8), and metric (1.92) becomes

$$ds^2 = dt^2 - H_\Lambda^{-2} \cosh^2(H_\Lambda t) d\chi^2. \qquad (1.94)$$

In the four-dimensional case, this form of the metric corresponds to a closed universe with $k = +1$.

Another choice of coordinates, namely,

$$x = H_\Lambda^{-1} \cosh(H_\Lambda \tilde{t}), \quad y = H_\Lambda^{-1} \sinh(H_\Lambda \tilde{t}) \sinh \tilde{\chi}, \qquad (1.95)$$

reduces (1.92) to a form corresponding to an open de Sitter universe:

$$ds^2 = d\tilde{t}^2 - H_\Lambda^{-2} \sinh^2(H_\Lambda \tilde{t}) d\tilde{\chi}^2. \qquad (1.96)$$

The range of these coordinates is $+\infty > \tilde{t} \geq 0$ and $+\infty > \tilde{\chi} > -\infty$, covering only the part of de Sitter spacetime where $x \geq H_\Lambda^{-1}$ and $z > 0$ (Figure 1.9). Moreover, the coordinates are singular at $\tilde{t} = 0$.

Finally, we consider the coordinate system defined via

$$x = H_\Lambda^{-1} \left[\cosh(H_\Lambda \bar{t}) - \frac{1}{2} \exp(H_\Lambda \bar{t}) \bar{\chi}^2 \right], \quad y = H_\Lambda^{-1} \exp(H_\Lambda \bar{t}) \bar{\chi}, \qquad (1.97)$$

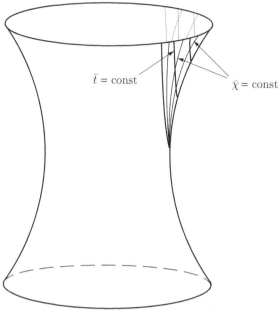

Fig. 1.9.

where $+\infty > \bar{t} > -\infty$ and $+\infty > \bar{\chi} > -\infty$. Expressing z in terms of $\bar{t}, \bar{\chi}$, one finds that only the half of the hyperboloid located at $x + z \geq 0$ is covered by these "flat" coordinates (Figure 1.10). The metric becomes

$$ds^2 = d\bar{t}^2 - H_\Lambda^{-2} \exp(2H_\Lambda \bar{t}) d\bar{\chi}^2. \tag{1.98}$$

The relation between the different coordinate systems in the regions where they overlap can be obtained by comparing (1.93), (1.95) and (1.97):

$$\begin{aligned} \cosh(H_\Lambda t) \cos \chi &= \cosh(H_\Lambda \tilde{t}) = \cosh(H_\Lambda \bar{t}) - \tfrac{1}{2} \exp(H_\Lambda \bar{t}) \bar{\chi}^2, \\ \cosh(H_\Lambda t) \sin \chi &= \sinh(H_\Lambda \tilde{t}) \sinh \tilde{\chi} = \exp(H_\Lambda \bar{t}) \bar{\chi}. \end{aligned} \tag{1.99}$$

De Sitter spacetime via analytical continuation (three-dimensional case) Since a de Sitter universe is a spacetime of constant *positive* curvature with Lorentzian signature, it can be obtained by analytical continuation of a metric describing a positive curvature space with Euclidean signature. To see how analytical continuation changes the signature of the metric let us consider (1.39) describing a closed universe $(k = +1)$. After the change of variables,

$$a \to H_\Lambda^{-1}, \chi \to H_\Lambda \tau, \theta \to \chi, \varphi \to \theta,$$

1.3 From Newtonian to relativistic cosmology

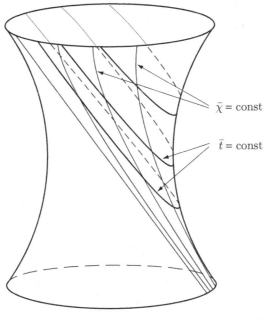

Fig. 1.10.

it is recast as

$$ds^2 = -dl_{3d}^2 = -d\tau^2 - H_\Lambda^{-2} \sin^2(H_\Lambda \tau)(d\chi^2 + \sin^2 \chi d\theta^2). \tag{1.100}$$

Then, analytically continuing $\tau \to it + \pi/2$, we obtain a three-dimensional de Sitter spacetime in the form of a closed Friedmann universe:

$$ds^2 = dt^2 - H_\Lambda^{-2} \cosh^2(H_\Lambda t)(d\chi^2 + \sin^2 \chi d\theta^2). \tag{1.101}$$

Note that the coordinate χ varies only from 0 to π, covering the entire space. The same construction works for a four-dimensional closed de Sitter universe.

To obtain an open de Sitter metric we must analytically continue *two* coordinates in (1.100) simultaneously: $\tau \to i\tilde{t}$ and $\chi \to i\tilde{\chi}$, giving

$$ds^2 = d\tilde{t}^2 - H_\Lambda^{-2} \sinh^2(H_\Lambda \tilde{t})(d\tilde{\chi}^2 + \sinh^2 \tilde{\chi} d\theta^2). \tag{1.102}$$

Generalizing the procedure to four dimensions is again straightforward.

De Sitter universe as a solution of Friedmann equations with cosmological constant (four-dimensional case) A cosmological constant is equivalent to a "perfect fluid" with equation of state $p_\Lambda = -\varepsilon_\Lambda$. It follows from (1.64) that

$$d\varepsilon_V = -3(\varepsilon_V + p_V) d\ln a = 0,$$

and hence the energy density stays constant during expansion. Substituting $\varepsilon_\Lambda =$ const into (1.66), we obtain

$$\ddot{a} - H_\Lambda^2 a = 0, \tag{1.103}$$

where

$$H_\Lambda = (8\pi G \varepsilon_\Lambda/3)^{1/2}.$$

A general solution of this equation is

$$a = C_1 \exp(H_\Lambda t) + C_2 \exp(-H_\Lambda t), \tag{1.104}$$

where C_1 and C_2 are constants of integration. These constants are constrained by Friedmann equation (1.67):

$$4H_\Lambda^2 C_1 C_2 = k. \tag{1.105}$$

Hence, in a flat universe ($k = 0$), one of the constants must be equal to zero. If $C_1 \neq 0$ and $C_2 = 0$, then (1.104) describes a flat expanding de Sitter universe and we can choose $C_1 = H_\Lambda^{-1}$. If both C_1 and C_2 are nonzero, the time $t = 0$ can be chosen so that $|C_1| = |C_2|$. For a closed universe ($k = +1$), we have

$$C_1 = C_2 = \frac{1}{2H_\Lambda},$$

while for an open universe ($k = -1$),

$$C_1 = -C_2 = \frac{1}{2H_\Lambda}.$$

The three solutions can be summarized as

$$ds^2 = dt^2 - H_\Lambda^{-2} \begin{pmatrix} \sinh^2(H_\Lambda t) \\ \exp(2H_\Lambda t) \\ \cosh^2(H_\Lambda t) \end{pmatrix} \left[d\chi^2 + \begin{pmatrix} \sinh^2 \chi \\ \chi^2 \\ \sin^2 \chi \end{pmatrix} d\Omega^2 \right] \begin{matrix} k = -1; \\ k = 0; \\ k = +1, \end{matrix} \tag{1.106}$$

where the radial coordinate χ changes from zero to infinity in flat and open universes. In contrast to a matter-dominated universe, where the spatial curvature is determined by the energy density, here all three types of solutions exist for any given value of ε_V. They all describe the same physical spacetime in different coordinate systems. One should not be surprised that it is possible to cover the same spacetime using homogeneous and isotropic hypersurfaces with different curvatures, since de Sitter spacetime is translational invariant in time. *Any* space-like hypersurface is a constant density hypersurface.

The behavior of the scale factor $a(t)$, shown in Figure 1.11, depends on the coordinate system. In a closed coordinate system, the scale factor first decreases,

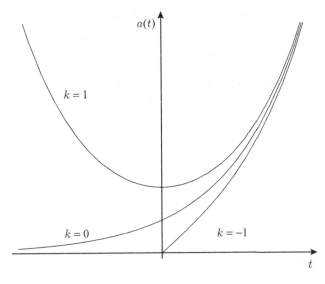

Fig. 1.11.

then reaches its minimum value, and subsequently increases. In flat and open coordinates, $a(t)$ always increases as t grows but vanishes as $t \to -\infty$ and $t = 0$, respectively. However, the vanishing of the scale factor does not represent a real physical singularity but simply signals that the coordinates become singular. For $t \gg H_\Lambda^{-1}$, the expansion is nearly the same in all coordinate systems, namely, exponential with $a \propto \exp(H_\Lambda t)$.

Problem 1.19 Calculate $H(t)$ and $\Omega(t)$ in open and closed de Sitter universes. Verify that $H(t) \to H_\Lambda$ and $\Omega(t) \to 1$ as $t \to \infty$ in both cases.

In a pure de Sitter universe, there is no real evolution. In this sense, de Sitter spacetime is similar to Minkowski spacetime. As in the case of the Milne universe, the apparent expansion reflects the nonstatic character of the chosen coordinate systems. However, unlike Minkowski spacetime, there exists no *static* coordinate system which can cover de Sitter spacetime on scales exceeding H_Λ^{-1}. We will see later that only a de Sitter solution with *slightly broken* time-translational symmetry plays an important role in physical applications. The notion of de Sitter expansion is still useful in the presence of perturbations that break the exact symmetry, and the coordinate systems (1.106) are well suited to study the behavior of these perturbations and the subsequent exit from the de Sitter stage.

Problem 1.20 Verify that for a flat universe filled with radiation and cosmological constant, the scale factor grows as

$$a(t) = a_0(\sinh 2H_\Lambda t)^{1/2}, \tag{1.107}$$

where H_Λ is defined in (1.103). Analyze and discuss the behavior of this solution in the limits $t \to 0$ and $t \to \infty$. Derive the corresponding solutions for $k = \pm 1$. (*Hint* Use (1.71), replacing the conformal time with the physical time.)

Problem 1.21 Show that the solution of (1.67) and (1.68) for a flat universe with cold matter (dust) and cosmological constant is

$$a(t) = a_0 \left(\sinh \frac{3}{2} H_\Lambda t \right)^{2/3}. \tag{1.108}$$

Verify that in this case the age of the universe is given by

$$t_0 = \frac{2}{3H_0} \frac{1}{\sqrt{1 - \Omega_m}} \ln \frac{1 + \sqrt{1 - \Omega_m}}{\sqrt{\Omega_m}}, \tag{1.109}$$

where H_0 is the current value of the Hubble constant and Ω_m is the cold matter contribution to the cosmological parameter today.

Problem 1.22 Given a nonvanishing cosmological constant, find the static solution for a closed universe filled with cold matter (*Einsteins's universe*). Why is this solution unstable?

Problem 1.23 Find the solutions for an energy component with equation of state $p = -\varepsilon/3$ in the presence of a cosmological term. Discuss the properties of these solutions.

2
Propagation of light and horizons

We obtain most of the information about the universe from light. Over the last century, the development of x-ray, radio and infrared detectors has given us new windows on the universe. Understanding the propagation of light in an expanding universe is therefore critical to the interpretation of observations.

Problem 2.1 Estimate the total amount of energy received by all optical telescopes over the course of the last century and compare this energy to that needed to return this book to your bookshelf.

There is a fundamental limit to how far we can see, since no particles can travel faster than light. The finite speed of light leads to "horizons" and sets an absolute constraint on our ability to comprehend the entire universe. The term "horizon" is used in different contexts in the literature, often without clear definition, and one of the purposes of this chapter is to carefully delineate the various usages. We will study in detail conformal diagrams, which are a useful pictorial way of representing horizons and the causal global structure of spacetime. Finally, we discuss the basic kinematical tests which aim to measure the distance, angular size, speed and acceleration of distant objects. Using these tests, one can obtain information about the expansion rate and deceleration parameter at earlier times, and thus probe the evolutionary history of the universe.

2.1 Light geodesics

In Special Relativity, the spacetime interval along the trajectory of a massless particle propagating with the speed of light is equal to zero:

$$ds^2 = 0. \tag{2.1}$$

In General Relativity, the same must be true in every local inertial coordinate frame. Then, since the interval is invariant, the condition $ds^2 = 0$ should be valid along the light geodesic in any curved spacetime.

We consider mainly the radial propagation of light in an isotropic universe in a coordinate system where the observer is located at the origin. The light trajectories look especially simple if, instead of physical time t, we use the conformal time

$$\eta \equiv \int \frac{dt}{a(t)}.$$

The metric (1.47) in η, χ coordinates is

$$ds^2 = a^2(\eta)(d\eta^2 - d\chi^2 - \Phi^2(\chi)(d\theta^2 + \sin^2\theta d\varphi^2)), \tag{2.2}$$

where

$$\Phi^2(\chi) = \begin{cases} \sinh^2\chi, & k = -1; \\ \chi^2, & k = 0; \\ \sin^2\chi, & k = +1. \end{cases} \tag{2.3}$$

By symmetry, it is clear that the radial trajectory $\theta, \varphi = $ const is a geodesic. The function $\chi(\eta)$ along the trajectory is then entirely determined by the condition $ds^2 = 0$, or

$$d\eta^2 - d\chi^2 = 0. \tag{2.4}$$

Hence, radial light geodesics are described by

$$\chi(\eta) = \pm\eta + \text{const}, \tag{2.5}$$

and correspond to straight lines at angles $\pm 45°$ in the η–χ plane.

2.2 Horizons

Particle horizon If the universe has a finite age, then light travels only a finite distance in that time and the volume of space from which we can receive information at a given moment of time is limited. The boundary of this volume is called the *particle horizon*. Today, the universe is roughly 15 billion years old, so a naive estimate for the particle horizon scale is 15 billion light years.

According to (2.5), the maximum comoving distance light can propagate is

$$\chi_p(\eta) = \eta - \eta_i = \int_{t_i}^{t} \frac{dt}{a}, \tag{2.6}$$

2.2 Horizons

where η_i (or t_i) corresponds to the beginning of the universe. At time η, the information about events at $\chi > \chi_p(\eta)$ is inaccessible to an observer located at $\chi = 0$. In a universe with an initial singularity, we can always set $\eta_i = t_i = 0$, but in some nonsingular spacetimes, for example, the de Sitter universe, it is more convenient to take $\eta_i \neq 0$. Multiplying χ_p by the scale factor, we obtain the physical size of the particle horizon:

$$d_p(t) = a(t)\chi_p = a(t) \int_{t_i}^{t} \frac{dt}{a}. \tag{2.7}$$

Until hydrogen recombination (see Section 3.6), which occurred when the universe was 1000 times smaller than now, the universe was opaque to photons. Therefore, in practice, our view is limited to the maximum distance light can travel since recombination. This is called the "optical" horizon:

$$d_{opt} = a(\eta)(\eta - \eta_r) = a(t) \int_{t_r}^{t} \frac{dt}{a}. \tag{2.8}$$

Problem 2.2 Calculate η_r/η_0 in a dust-dominated universe and verify that the present optical horizon is less than the particle horizon by only a small percentage.

Although the optical horizon is not very different from the particle horizon, it unfortunately obscures information about the most interesting stages of the evolution of the early universe. Primordial neutrinos and gravitational waves decouple from matter before photons, and so could, in principle, bring us this information. Sadly, the short-term prospects of detecting primordial neutrinos or cosmological gravitational waves are not very promising.

Let us calculate the size of the particle horizon in flat matter-dominated and radiation-dominated universes. Substituting $a(t) \propto t^{2/3}$ into (2.7), we find that in a matter-dominated universe $d_p(t) = 3t$ ($c = 1$). If the universe is dominated by radiation, then $a(t) \propto t^{1/2}$ and, correspondingly, $d_p(t) = 2t$.

Problem 2.3 Calculate the size of the particle horizon in a dust-dominated universe with an arbitrary value of the current cosmological parameter Ω_0 and show that

$$\Phi(\chi_p) = \frac{2}{a_0 H_0 \Omega_0}, \tag{2.9}$$

where the function Φ is defined in (2.3).

Curvature scale ("Hubble horizon") vs. particle horizon When matter satisfies the strong energy dominance condition, $\varepsilon + 3p > 0$, the particle horizon is usually of order the Hubble scale, $1/H$. Consequently, the terms "Hubble scale" and "particle horizon" are sometimes used interchangeably. Some authors even conjoin the terms and refer to a "Hubble horizon." However, the Hubble scale, H^{-1}, is conceptually distinct from a horizon. Whereas the particle horizon is a scale set by kinematical considerations, the curvature scale is a dynamical scale that characterizes the rate of expansion and enters the equations describing, for instance, the evolution of cosmological perturbations. Because H^{-1} is of order the 4-curvature scale, it also characterizes the "size" of the local inertial frame.

Although the Hubble scale and particle horizon are of similar magnitudes for some models, they can differ by a large factor when the strong energy condition is violated, $\varepsilon + 3p < 0$. In this case, from (1.66), $\ddot{a} > 0$, that is, the expansion is accelerating. Then, the integral in the expression

$$d_p(t) = a(t) \int^t \frac{dt}{a} = a(t) \int^a \frac{da}{a\dot{a}} \qquad (2.10)$$

converges as $t \to \infty$ and $a \to \infty$. At large t, the particle horizon is proportional to $a(t)$, but the curvature scale, $H^{-1} = a/\dot{a}$, grows more slowly since $\dot{a}$ also increases during accelerated expansion. For instance, the particle horizon in a flat de Sitter universe, where $a(t) \propto \exp(H_\Lambda t)$, is

$$d_p(t) = \exp(H_\Lambda t) \int_{t_i}^t \exp(-H_\Lambda t)dt = H_\Lambda^{-1}(\exp(H_\Lambda(t-t_i)) - 1). \qquad (2.11)$$

For $t - t_i \gg H_\Lambda^{-1}$, the size of the causally connected region grows exponentially fast, whereas the curvature scale, H_Λ^{-1}, is constant. Formally, as $t_i \to -\infty$, the particle horizon diverges, and hence all points were in causal contact. However, this has limited significance since the flat slicing of de Sitter spacetime is geodesically incomplete (see next section). Moreover, when applied as an approximation for inflation, we use only a part of the whole de Sitter spacetime. The beginning of inflation corresponds to a finite initial time t_i and, consequently, the particle horizon is finite.

Despite the fact that the curvature scale is not, if properly considered, a horizon, the use of the term "Hubble horizon" has become so widespread that we will occasionally follow the "traditional terminology." However, the reader is strongly advised to keep in mind the distinction between the *dynamical* curvature scale and the *kinematical* horizon.

Event horizon The *event horizon* is the complement of the particle horizon. The event horizon encloses the set of points from which signals sent *at a given moment of time* η will never be received by an observer in the *future*. These points have

2.3 Conformal diagrams

comoving coordinates

$$\chi > \chi_e(\eta) = \int_\eta^{\eta_{max}} d\eta = \eta_{max} - \eta. \tag{2.12}$$

Hence, the physical size of the event horizon at time t is

$$d_e(t) = a(t) \int_t^{t_{max}} \frac{dt}{a}, \tag{2.13}$$

where "max" refers to the final moment of time. If the universe expands forever, then t_{max} is infinite. However, the value of η_{max}, and hence d_e, can be either infinite or finite depending on the rate of expansion. In flat and open decelerating universes, t_{max} and η_{max} are both infinite, χ_e and d_e diverge, and so there is no event horizon. However, if the universe undergoes accelerated expansion, then the integral in (2.13) converges and the radius of the event horizon is finite, even if the universe is flat or open. In this case, η approaches a finite limit η_{max}, as $t_{max} \to \infty$.

An important example is a flat de Sitter universe, where

$$d_e(t) = \exp(H_\Lambda t) \int_t^\infty \exp(-H_\Lambda t) \, dt = H_\Lambda^{-1}, \tag{2.14}$$

that is, the size of the event horizon is equal to the curvature scale. Every event that occurs at a given moment of time at a distance greater than H_Λ^{-1} will never be seen by an observer and cannot influence his future because the intervening space is expanding too rapidly. For this reason, the situation is sometimes characterized as "superluminal expansion."

In a closed decelerating universe, the time available for future observations is finite since the universe ultimately collapses. Therefore, there is both an event horizon and a particle horizon.

Problem 2.4 Verify that, in a closed, radiation-dominated universe, the curvature scale H^{-1} is roughly equal to the *particle horizon* size at the beginning of expansion but roughly coincides with the radius of the *event horizon* during the final stages of collapse.

2.3 Conformal diagrams

The homogeneous, isotropic universe is a particular case of a spherically symmetric space. The most general form of metric respecting spherical symmetry is

$$ds^2 = g_{ab}(x^c) \, dx^a dx^b - R^2(x^c) \, d\Omega^2, \tag{2.15}$$

where the indices a, b and c run over only two values, 0 and 1, corresponding to the time and radial coordinates respectively. The angular part of the metric is rather simple. It is proportional to

$$d\Omega^2 \equiv d\theta^2 + \sin^2\theta \, d\varphi^2 \tag{2.16}$$

and describes a 2-sphere of radius $R(x^c)$. The only nontrivial piece of the metric is the temporal–radial part, which can describe spaces with different causal structure. The causal structure can be represented by a two-dimensional *conformal diagram*, in which every point corresponds to a 2-sphere.

The global properties of the spacetime can be completely explored by considering the radial geodesics of light. As we showed in Section 2.1, in a coordinate system where metric (2.15) takes the form

$$ds^2 = a^2(\eta, \chi)\left[d\eta^2 - d\chi^2 - \Phi^2(\eta, \chi) \, d\Omega^2\right], \tag{2.17}$$

the radial propagation of light is described by the equation

$$\chi(\eta) = \pm \eta + \text{const}, \tag{2.18}$$

or in other words, by straight lines at ± 45 degree angles in the η–χ plane.

In principle, it is always possible to find a coordinate system that allows us to write (2.15) as (2.17). In the coordinate transformation

$$x^a \to \tilde{x}^a \equiv \left(\eta\left(x^a\right), \chi\left(x^a\right)\right),$$

the freedom to choose the two functions η and χ means we can impose the two conditions

$$\tilde{g}_{01} = 0, \quad \tilde{g}_{00} = \tilde{g}_{11} \equiv a^2(\eta, \chi).$$

Solving the equations for η and χ can be difficult in general, but in cosmologically interesting cases the metric is already in the required form.

Typically, η and χ may extend over infinite or semi-infinite intervals. Since our goal is to visualize the causal structure of the full spacetime, in these cases we perform a further coordinate transformation that preserves the form of metric (2.17) but maps unbounded coordinates into coordinates which vary over a finite interval. We shall see that it is always possible to find such transformations. In this section, we reserve the symbols η and χ to refer only to bounded coordinates.

A conformal diagram is a picture of a spacetime plotted in terms of η and χ. Hence, *a conformal diagram always has finite size and light geodesics (null lines) are always represented by straight lines at ± 45 degree angles*. These are the defining features of a conformal diagram. Although the finite ranges spanned by the coordinates and the size of the diagram can be altered, its shape is uniquely determined.

2.3 Conformal diagrams

Note that the diagrams of different spacetimes are exactly the same if their metrics are related by a *nonsingular* conformal transformation: $\tilde{g}_{\mu\nu} = a^2(x) g_{\mu\nu}$.

In addition to the shape of the diagram, we must pay attention to the location of singularities. Singularities, as well as the boundaries of the diagram, are determined by the behavior of the scale factor $a(\eta, \chi)$ and the function $\Phi(\eta, \chi)$ in (2.17). We will see that it is possible to have two spacetimes whose conformal diagrams have the same shape but different singular boundaries.

Closed radiation- and dust-dominated universes For a closed universe filled with radiation or dust, the conformal diagram can be immediately drawn based on the solutions for $a(\eta)$ found in Section 1.3.4. Metric (2.2) becomes

$$ds^2 = a^2(\eta)(d\eta^2 - d\chi^2 - \sin^2\chi \, d\Omega^2), \tag{2.19}$$

where

$$a = a_m \sin \eta \tag{2.20}$$

in a radiation-dominated universe and

$$a = a_m(1 - \cos \eta) \tag{2.21}$$

in a dust-dominated universe (see (1.73) and (1.77)). In both cases, χ and η have finite ranges and cover the whole spacetime:

$$\pi \geq \chi \geq 0, \quad \pi > \eta > 0, \tag{2.22}$$

for a radiation-dominated universe and

$$\pi \geq \chi \geq 0, \quad 2\pi > \eta > 0, \tag{2.23}$$

for a dust-dominated universe. The conformal diagrams are a square and rectangle respectively, and are shown in Figures 2.1 and 2.2. Horizontal and vertical lines represent hypersurfaces of constant η and χ. The lower and upper boundaries correspond to physical singularities where the scale factor vanishes and the energy density and curvature diverge. In both cases, the lower half of the diagram describes an expanding universe and the upper half corresponds to a contracting phase. The scale factor reaches its maximum value at $\eta = \pi/2$ in the radiation-dominated universe and at $\eta = \pi$ in the dust-dominated universe.

The essential difference between the diagrams is the comparative ranges of η and χ: for the dust-dominated universe η has twice the range of χ, while η and χ have the same range for the radiation-dominated universe. This has important consequences for the particle and event horizons. In both cases, we can set $\eta_i = 0$ at the lower boundary of the diagram. Then the particle horizon for the observer at

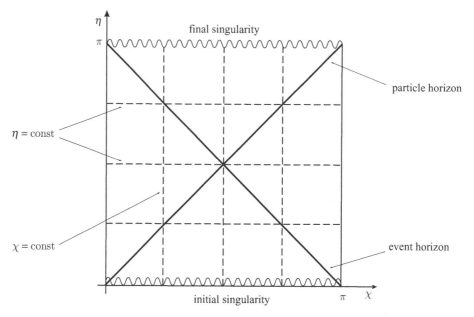

Fig. 2.1.

$\chi = 0$ is given by

$$\chi_p(\eta) = \eta - \eta_i = \eta. \tag{2.24}$$

In the radiation-dominated universe, the particle horizon spans the whole space when $\eta \to \pi$, that is, just as the universe recollapses. At this last moment of time, all points in space become visible. The light that reaches an observer from the most remote point, $\chi = \pi$, reveals information about the state of the universe at the beginning of expansion. In the dust-dominated universe, the whole universe also becomes entirely visible at $\eta = \pi$. However, here this corresponds to the moment of maximum expansion. There remains enough time for light to make a second trip across the whole space before the universe recollapses.

The event horizon is given by

$$\chi_e(\eta) = \eta_{\max} - \eta. \tag{2.25}$$

In the radiation-dominated universe, there exists an event horizon for any η since $\eta_{\max} = \pi$. In contrast, for the dust-dominated universe, where $\eta_{\max} = 2\pi$, the event horizon exists only during the contraction phase when $\eta > \pi$. All events that occur at $\eta < \pi$, no matter how far away, can be seen before the universe recollapses.

In summary, as shown in Figures 2.1 and 2.2, particle and event horizons exist at each η in the closed radiation-dominated universe. In the matter-dominated

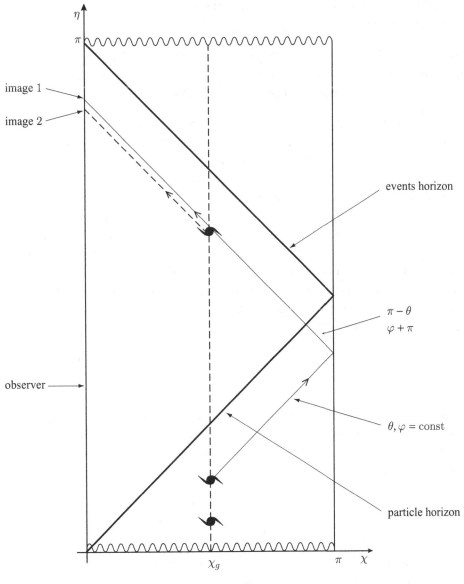

Fig. 2.2.

universe, the particle horizon exists only during the expansion phase, and the event horizon exists only during the contraction phase.

The points $\chi = 0, \pi$ are the opposite poles of the three-dimensional sphere describing the spatial geometry at any given moment of time. Light propagating with constant θ, φ away from an observer located at $\chi = 0$ reaches the opposite pole at $\chi = \pi$. Because the coordinate system we are using is singular at the poles, to clarify what happens to light after passing through the pole, one has to use another coordinate system which is regular near $\chi = \pi$.

Problem 2.5 Show that light propagating away from an observer at $\chi = 0$ in the direction (θ, φ) begins to propagate back towards the observer along the direction $(\tilde{\theta} = \pi - \theta, \tilde{\varphi} = \varphi + \pi)$ after it passes through the pole at $\chi = \pi$.

Thus, a light geodesic is "reflected" from the boundary at $\chi = \pi$ and its angular coordinates θ and φ change. This change of the angular coordinates is not apparent from the conformal diagram because they are suppressed there.

Let us use a conformal diagram to infer how a galaxy located at $\chi = \chi_g =$ const appears to an observer at $\chi = 0$ in a dust-dominated universe. As is clear from Figure 2.2, at $\eta > 2\pi - \chi_g$, when the universe is contracting, there are two geodesics along which light emitted by the galaxy can reach the observer. Hence, the observer simultaneously sees two images of the same galaxy in opposite directions in the sky. One image is older than the other by $\Delta \eta = 2(\pi - \chi_g)$. In a radiation-dominated universe, only one image of the galaxy can be seen because light does not have enough time to travel around the pole at $\chi = \pi$ and reach an observer before the universe recollapses.

Problem 2.6 Using (1.83), draw the conformal diagram for a closed universe filled with a mixture of dust and radiation.

De Sitter universe De Sitter spacetime is an example of how different coordinate systems used for the same spacetime can lead to different conformal diagrams. We begin by rewriting metric (1.106) in terms of conformal time instead of physical time t. For a *closed universe*, the relation is

$$\eta = \int_{\infty}^{t} \frac{dt}{H_\Lambda^{-1} \cosh(H_\Lambda t)} = \arcsin[\tanh(H_\Lambda t)] - \frac{\pi}{2}. \quad (2.26)$$

The conformal time η is always negative and ranges from $-\pi$ to 0 as t varies from $-\infty$ to $+\infty$. It follows from (2.26) that

$$\cosh(H_\Lambda t) = -(\sin \eta)^{-1}, \quad (2.27)$$

which allows us to write the metric of the *closed* de Sitter universe as

$$ds^2 = \frac{1}{H_\Lambda^2 \sin^2 \eta} \left(d\eta^2 - d\chi^2 - \sin^2 \chi \, d\Omega^2 \right). \quad (2.28)$$

Since the spatial coordinate χ varies from 0 to π and the temporal coordinate η changes from $-\pi$ to 0, the conformal diagram for a closed de Sitter universe is

2.3 Conformal diagrams

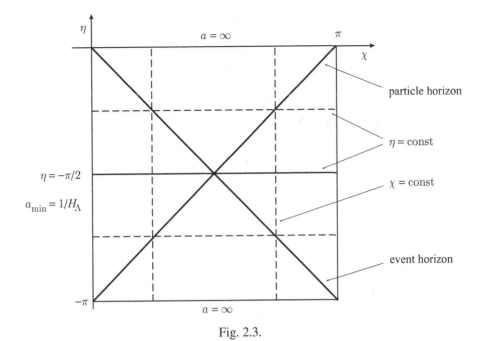

Fig. 2.3.

a square. In fact, it has the same shape as the diagram for a closed, radiation-dominated universe, with the difference that there are no singularities at $\eta_i = -\pi$ and $\eta_{\max} = 0$ – see Figure 2.3. Moreover, in a de Sitter universe, the scale factor, $a(\eta) = -1/H_\Lambda \sin \eta$, is infinite at the lower boundary of the diagram where $\eta \to -\pi$, decreases as η changes from $-\pi$ to $-\pi/2$, reaches its minimum value $1/H_\Lambda$, and then grows to infinity again as $\eta \to 0$. The blowing up of the scale factor does not signify a singularity. We have seen that all curvature invariants are constant in de Sitter spacetime, and hence, the infinite growth of the scale factor is entirely a coordinate effect.

As with the closed radiation-dominated universe, de Sitter spacetime has both a particle horizon,

$$\chi_p(\eta) = (\eta - \eta_i) = \eta + \pi, \tag{2.29}$$

and an event horizon,

$$\chi_e(\eta) = (\eta_{\max} - \eta) = -\eta, \tag{2.30}$$

which exist at any time η. In both the closed de Sitter and radiation-dominated universes, the physical size of the event horizon $d_e(t)$ approaches the curvature scale H^{-1} near the upper boundary of the conformal diagram. However, in de Sitter spacetime, H, and consequently the size of the event horizon, remain constant; in a radiation-dominated universe, H increases and the size of the event horizon shrinks.

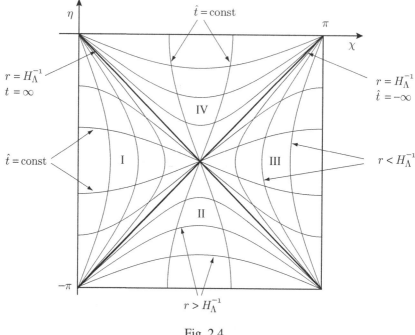

Fig. 2.4.

Problem 2.7 One can utilize for de Sitter spacetime the so called "static coordinates" $\hat{t}, r$, related to η, χ via

$$\tanh(H_\Lambda \hat{t}) = \frac{\cos \eta}{\cos \chi}, \qquad H_\Lambda r = \frac{\sin \chi}{\sin \eta}. \qquad (2.31)$$

Verify that in these coordinates the metric takes the following form:

$$ds^2 = \left[1 - (H_\Lambda r)^2\right] d\hat{t}^2 - \frac{dr^2}{\left[1 - (H_\Lambda r)^2\right]} - r^2 d\Omega^2. \qquad (2.32)$$

The hypersurfaces of constant r and $\hat{t}$ are shown in Figure 2.4. De Sitter horizons correspond to $r = H_\Lambda^{-1}$ and $\hat{t} = \pm \infty$. The static coordinates cover only half of de Sitter spacetime: regions I and III in Figure 2.4. They are singular on the horizons but can be continued beyond. For $r > H_\Lambda^{-1}$, the radial coordinate r plays the role of time and $\hat{t}$ becomes a space-like coordinate. Introduce the proper-time

$$d\tau = dr / \left[(H_\Lambda r)^2 - 1\right]^{1/2} \qquad (2.33)$$

and verify that in regions II and IV the "static" metric (2.32) describes contracting and expanding space respectively. We conclude that there exists no static coordinate system covering de Sitter spacetime on scales exceeding the curvature scale. Note that the trajectory $r = $ const is a geodesic only if $r = 0$.

2.3 Conformal diagrams

In a *flat de Sitter universe*, the scale factor grows as $a(\bar{t}) = H_\Lambda^{-1} \exp(H_\Lambda \bar{t})$, where the physical time $\bar{t}$ is related to the conformal time $\bar{\eta}$ via

$$\exp(H_\Lambda \bar{t}) = -1/\bar{\eta}. \tag{2.34}$$

Hence, in conformal coordinates the metric becomes

$$ds^2 = \frac{1}{H_\Lambda^2 \bar{\eta}^2}\left(d\bar{\eta}^2 - d\bar{\chi}^2 - \bar{\chi}^2 d\Omega^2\right), \tag{2.35}$$

where $0 > \bar{\eta} > -\infty$ and $+\infty > \bar{\chi} > 0$. Unlike the case of a closed de Sitter universe, here $\bar{\eta}, \bar{\chi}$ have infinite ranges and to draw the conformal diagram, we must first transform to coordinates which range over finite intervals. Fortunately, there is a natural choice for such coordinates: we simply use the η, χ coordinates of the closed de Sitter universe. The relation between $\bar{\eta}, \bar{\chi}$ and η, χ coordinates immediately follows from (1.99) if we express t and $\bar{t}$ in terms of η and $\bar{\eta}$ respectively. The result is

$$\bar{\eta} = \frac{\sin \eta}{\cos \eta + \cos \chi}, \qquad \bar{\chi} = \frac{\sin \chi}{\cos \eta + \cos \chi}. \tag{2.36}$$

Using these relations, one can draw the hypersurfaces of constant $\bar{\eta}$ and $\bar{\chi}$ (the coordinates in (2.35)) in the η–χ plane, as shown in Figure 2.5. We find that when $\bar{\eta}, \bar{\chi}$ coordinates run over their semi-infinite ranges, they cover only half of de Sitter

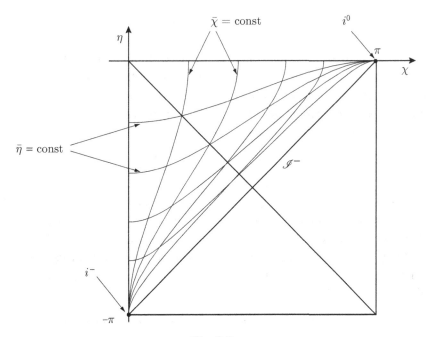

Fig. 2.5.

spacetime, a triangle whose lower boundary coincides with the particle horizon. On the particle horizon, $\bar{\eta} \to -\infty$, $\bar{\chi} \to +\infty$, and hence, the flat coordinates become singular.

Problem 2.8 To understand the shape of the constant $\bar{\eta}$ and constant $\bar{\chi}$ hypersurfaces near the corners of the triangular conformal diagram, $\chi = 0$, $\eta = -\pi$ and $\chi = \pi$, $\eta = 0$, calculate the derivatives $d\eta/d\chi$ along these hypersurfaces.

Viewing the flat de Sitter solution as describing an infinite space, we can categorize the types of infinities that arise. For instance, *space-like infinity*, where $\bar{\chi} \to +\infty$ along a hypersurface of constant $\bar{\eta}$, is represented on the conformal diagram by a point which is denoted as i^0. The *past time-like infinity*, from where all time-like lines emanate, occurs at $\bar{\eta} \to -\infty$ for finite $\bar{\chi}$ and is denoted by i^-. The lower diagonal boundary of the flat de Sitter diagram corresponds to the region from which incoming light-like geodesics originate. It is easy to verify that as we approach this boundary, $\bar{\chi} \to \infty$ and $\bar{\eta} \to -\infty$ but the sum $\bar{\chi} + \bar{\eta}$ remains finite. This infinity is called *past null infinity* and denoted by $\mathscr{I}^-$.

In an *open de Sitter universe*, the relation between physical and conformal times is

$$\sinh\left(H_\Lambda \tilde{t}\right) = -1/\sinh \tilde{\eta} \tag{2.37}$$

and the metric becomes

$$ds^2 = \frac{1}{H_\Lambda^2 \sinh^2 \tilde{\eta}} \left(d\tilde{\eta}^2 - d\tilde{\chi}^2 - \sinh^2 \tilde{\chi} d\Omega^2\right). \tag{2.38}$$

The coordinates run over the same range as in a flat de Sitter universe, $0 > \tilde{\eta} > -\infty$ and $+\infty > \tilde{\chi} > 0$, therefore the conformal diagrams of these two spaces will look similar. We can again use the closed coordinates to determine which part of the de Sitter spacetime is covered by the open coordinates. The corresponding relation between coordinate systems follows from (1.99):

$$\tanh \tilde{\eta} = \frac{\sin \eta}{\cos \chi}, \quad \tanh \tilde{\chi} = \frac{\sin \chi}{\cos \eta}. \tag{2.39}$$

In this case, the coordinates $\tilde{\eta}$, $\tilde{\chi}$ cover only one eighth of the whole de Sitter spacetime (Figure 2.6), and thus, cover an even smaller part of the de Sitter manifold than the flat coordinates. Of course, it only makes sense to compare the sizes of different diagrams when they describe the same spacetime, as in the case of the de Sitter manifold. Otherwise, as noted before, the size of the diagram has no invariant meaning.

Problem 2.9 Calculate the derivative $d\eta/d\chi$ along the hypersurfaces $\tilde{\eta} = const$ and $\tilde{\chi} = const$, near i^- and i^0 respectively.

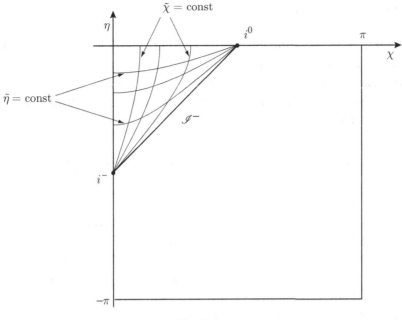

Fig. 2.6.

The conformal diagrams show explicitly that flat and open de Sitter universes are geodesically incomplete. For instance, following a geodesic for a photon, which arrives at $\chi = 0$, into its past, we find that this geodesic leaves first the open and then the flat de Sitter region.

Finally, we note that the hypersurfaces of constant time in all coordinate systems become increasingly flat and similar for $\chi \ll \pi/2$ as $\eta \to 0^-$. In this limit, the scale factor is inversely proportional to conformal time or, equivalently, increases exponentially with physical time.

The reader may naturally wonder why we need to study the same de Sitter spacetime in three different coordinate systems. As mentioned previously, the de Sitter spacetime is useful in a practical sense because it can be viewed as the leading order approximation to a universe undergoing inflationary expansion. In realistic inflationary models, time-translational invariance is broken and the energy density varies slightly with time. The hypersurface along which inflation ends is usually the hypersurface of constant energy density and the geometry of the future Friedmann universe depends on its shape. It can, in principle, be the surface of constant time in closed, flat or open de Sitter coordinates and, as a result of a graceful exit from inflation, one obtains a closed, flat or open Friedmann universe respectively.

The full cosmic history can be represented by gluing together the pieces of conformal diagrams describing different phases of the universe's evolution. When gluing these pieces, however, one should not forget that every point of the diagram corresponds to a 2-sphere and that the 3-geometries of hypersurfaces along which the diagrams are glued must match.

To complete the set of the diagrams needed in cosmology, we must also construct the conformal diagrams describing open and flat universes filled by matter and radiation. As a preliminary step, we first consider the conformal diagram for Minkowski spacetime, which turns out to be useful in drawing the diagrams of more complicated infinite spaces.

Minkowski spacetime In spherical coordinates the Minkowski metric takes the form

$$ds^2 = dt^2 - dr^2 - r^2 d\Omega^2. \tag{2.40}$$

It is trivially conformal but the time and radial coordinates range over infinite intervals, $+\infty > t > -\infty$ and $+\infty > r \geq 0$, and, therefore, have to be replaced by coordinates with finite ranges. There exist many such coordinate systems for Minkowski spacetime. One choice is to introduce η and χ coordinates which are related to the t and r coordinates in the same way as closed and open de Sitter coordinates are related (see (2.39)), namely,

$$\tanh t = \frac{\sin \eta}{\cos \chi}, \quad \tanh r = \frac{\sin \chi}{\cos \eta}. \tag{2.41}$$

The Minkowski metric in the new coordinates then becomes

$$ds^2 = \frac{1}{\cos^2 \chi - \sin^2 \eta} \left(d\eta^2 - d\chi^2 - \Psi^2(\eta, \chi) d\Omega^2\right), \tag{2.42}$$

where Ψ can be calculated but is not important for our purposes. Comparing the Minkowski time t to $\tilde{\eta}$ in an open de Sitter universe (2.39), we see that t runs from $-\infty$ to $+\infty$, while $\tilde{\eta}$ is restricted to negative values (because the scale factor in open de Sitter spacetime blows up as $\tilde{\eta} \to 0^-$). Therefore, in the $\eta-\chi$ plane, the hypersurfaces of constant t and r span a large triangle, which can be thought of as made from two smaller triangles describing the open de Sitter spacetime and its time-reversed copy (Figure 2.7). Minkowski spacetime possesses two additional types of infinities compared to an open de Sitter universe: a *future time-like infinity* i^+, where all time-like lines end ($t \to +\infty$, r is finite), and a *future null infinity* $\mathscr{I}^+$ ($t \to +\infty$, $r \to +\infty$ with $t - r$ finite), the region towards which outgoing radial light geodesics extend. Region I in Figure 2.7 corresponds to a future light cone which can also be covered by Milne coordinates. The Milne conformal diagram is geometrically similar to the Minkowski one, though it is four times smaller.

Problem 2.10 Draw the conformal diagram for the Milne universe and verify this last statement.

2.3 Conformal diagrams

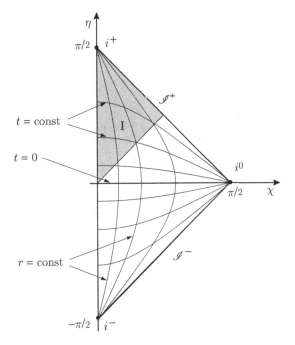

Fig. 2.7.

Open and flat universes Now we will use the Minkowski conformal diagram to construct the diagram for open and flat universes dominated by matter satisfying the strong energy dominance condition, $\varepsilon + 3p > 0$. The metric is

$$ds^2 = a^2(\tilde{\eta})\left(d\tilde{\eta}^2 - d\tilde{\chi}^2 - \Phi^2(\tilde{\chi})\,d\Omega^2\right), \qquad (2.43)$$

where the scale factor a vanishes at a singularity occurring at $\tilde{\eta} = 0$. (Here we have added tildes to the notation in (2.2) since η, χ are reserved for coordinates with finite ranges.) The conformal time $\tilde{\eta}$ is confined to the range $(0, +\infty)$. Since $\Phi(\tilde{\chi})$ is equal to $\tilde{\chi}$ for a flat universe and to $\sinh \tilde{\chi}$ for an open universe, in both cases $\tilde{\chi}$ changes from 0 to $+\infty$. For $\tilde{\eta} > 0$, the temporal–radial part of metric (2.43) is related to the Minkowski metric (2.40) by a nonsingular conformal transformation. The coordinates t and r considered in the upper half of Minkowski spacetime ($t > 0$) span the same range as the $\tilde{\eta}$, $\tilde{\chi}$ coordinates. Hence, the conformal diagrams of open and flat universes should have the same shape as the upper half of the Minkowski conformal diagram (Figure 2.8). The hypersurfaces of constant $\tilde{\eta}$ and constant $\tilde{\chi}$ can then be drawn in the η–χ plane, where η, χ are related to $\tilde{\eta}$, $\tilde{\chi}$ as in (2.41) with the substitutions $t \to \tilde{\eta}$ and $r \to \tilde{\chi}$. In open and flat universes, the lower boundary ($\tilde{\eta} = 0$) corresponds to a physical singularity.

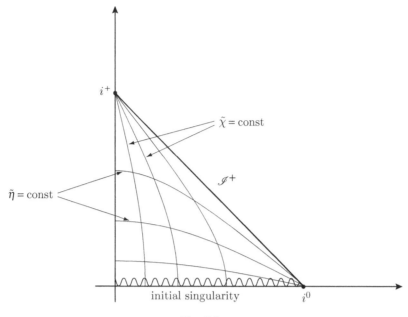

Fig. 2.8.

Problem 2.11 Draw the conformal diagram for open and flat universes where the scale factor changes as $a(t) \propto t^p$, $p > 1$. This is the situation for power-law inflation. Note that the strong energy condition is violated in this case. Indicate the particle and event horizons and the types of infinities. Draw the conformal diagram for a flat universe filled by matter with equation of state $p = -\varepsilon/3$. Compare this case with the Milne universe.

Problem 2.12 The metric of an eternal black hole in the Kruskal–Szekeres coordinate system takes the form

$$ds^2 = a^2(v, u)\left(dv^2 - du^2 - \Psi^2(v, u)\, d\Omega^2\right). \tag{2.44}$$

The only extra information we need to draw the conformal diagram is that the spacelike coordinate u ranges from $-\infty$ to $+\infty$ and that there is a physical singularity located at

$$v^2 - u^2 = 1.$$

The existence of a singularity means that, for every u, the spacetime cannot be extended outside the interval

$$-\sqrt{1+u^2} < v < +\sqrt{1+u^2}.$$

Draw the conformal diagram for the eternal black hole and identify the types of infinities. The Schwarzschild radius of the black hole is located at $v^2 = u^2$.

2.4 Redshift

The expansion of the universe leads to a redshift of the photon wavelength. To analyze this effect, let us consider a source of radiation with comoving coordinate χ_{em}, which at time η_{em} emits a signal of short conformal duration $\Delta \eta$ (Figure 2.9). According to (2.5), the trajectory of the signal is

$$\chi(\eta) = \chi_{em} - (\eta - \eta_{em})$$

and it reaches a detector located at $\chi_{obs} = 0$ at time $\eta_{obs} = \eta_{em} + \chi_{em}$. The conformal duration of the signal measured by the detector is the same as at the source, but the physical time intervals are different at the points of emission and detection. They are equal to

$$\Delta t_{em} = a(\eta_{em}) \Delta \eta \quad \text{and} \quad \Delta t_{obs} = a(\eta_{obs}) \Delta \eta$$

respectively. If Δt is the period of the light wave, the light is emitted with wavelength $\lambda_{em} = \Delta t_{em}$ but is observed with wavelength $\lambda_{obs} = \Delta t_{obs}$, so that

$$\frac{\lambda_{obs}}{\lambda_{em}} = \frac{a(\eta_{obs})}{a(\eta_{em})}. \tag{2.45}$$

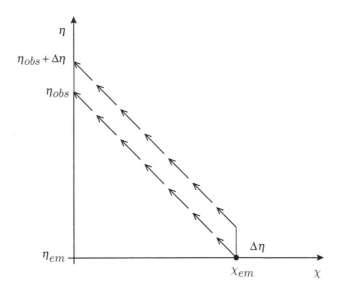

Fig. 2.9.

Thus, the wavelength of the photon changes in proportion to the scale factor, $\lambda(t) \propto a(t)$, and its frequency, $\omega \propto 1/\lambda$, decreases as $1/a$.

The Planck distribution, characterizing blackbody radiation, has the important property that it preserves its shape as the universe expands. However, because each photon is redshifted, $\omega \to \omega/a$, the temperature T scales as $1/a$. Therefore, the energy density of radiation, which is proportional to T^4, decreases as the fourth power of the scale factor, in complete agreement with what we obtained earlier for an ultra-relativistic gas with equation of state $p = \varepsilon/3$. The number density of the photons is proportional to T^3, and therefore decays as the third power of the scale factor so that the *total* number of photons is conserved.

Redshift as Doppler shift The cosmological redshift can be interpreted as a Doppler shift associated with the relative motion of galaxies due to Hubble expansion. If we begin with two neighboring galaxies separated by distance $\Delta l \ll H^{-1}$, then there exists a local inertial frame in which spacetime can be considered flat. According to the Hubble law, the relative recessional speed of the two galaxies is $v = H(t)\Delta l \ll 1$. Because of this, the frequency of a photon, $\omega(t_1)$, measured by an observer at galaxy "1" at the moment t_1, will be larger than the frequency of the same photon, $\omega(t_2)$, measured at $t_2 > t_1$ by an observer at galaxy "2", by a Doppler factor (Figure 2.10):

$$\Delta \omega \equiv \omega(t_1) - \omega(t_2) \approx \omega(t_1) v = \omega(t_1) H(t) \Delta l. \tag{2.46}$$

The time delay between measurements is $\Delta t = t_2 - t_1 = \Delta l$ and so we can rewrite (2.46) as a differential equation:

$$\dot{\omega} = -H(t)\omega. \tag{2.47}$$

This has the solution

$$\omega \propto 1/a. \tag{2.48}$$

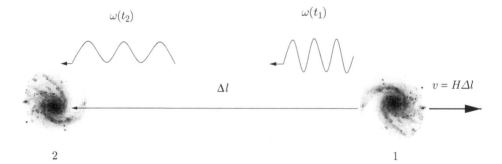

Fig. 2.10.

2.4 Redshift

Although the derivation above has been performed in a local inertial frame, it can be applied piecewise to a general geodesic photon trajectory. The result is therefore valid in curved spacetime as well. However, the interpretation of the redshift as a Doppler shift is not applicable for distances larger than the curvature scale. In this limit, as we have pointed out, distance and relative velocity do not have an invariant meaning, so the notion of Doppler shift becomes ill defined.

Redshift of peculiar velocities The peculiar velocities of massive particles (velocities with respect to the Hubble flow) are also redshifted as the universe expands. The peculiar velocity of a particle, $w(t_1)$, measured by observer "1" at time t_1, is different from the peculiar velocity of the same particle, $w(t_2)$, measured by observer "2", by the relative Hubble speed of the observers: $v = H(t)\Delta l$. Hence,

$$w(t_1) - w(t_2) \approx v = H(t)\,\Delta l. \tag{2.49}$$

Given that the particle needs time $\Delta t = t_2 - t_1 = \Delta l/w$ to make the journey between the two observers, we can rewrite this equation as

$$\dot{w} = -H(t)\,w. \tag{2.50}$$

Once again we have the solution

$$w \propto 1/a. \tag{2.51}$$

Thus, the expansion of the universe eventually brings particles to rest in the comoving frame.

The temperature of a nonrelativistic gas of particles is proportional to the peculiar velocity squared,

$$T_{gas} \propto w^2 \propto 1/a^2, \tag{2.52}$$

and therefore, if the gas and radiation are decoupled, gas will cool faster than radiation.

For the same reasons as in the case of radiation, the above derivation for peculiar velocities is rigorous and applicable in curved spacetime. This can also be verified directly by solving the geodesic equations for the particles.

Problem 2.13 Show that the geodesic equation

$$\frac{du^\alpha}{ds} + \Gamma^\alpha_{\beta\gamma} u^\beta u^\gamma = 0 \tag{2.53}$$

can be rewritten as

$$\frac{du_\alpha}{ds} - \frac{1}{2}\frac{\partial g_{\beta\gamma}}{\partial x^\alpha} u^\beta u^\gamma = 0. \tag{2.54}$$

One can always go to a coordinate system in which only the radial peculiar velocity of the particle, u^χ, is different from zero. Taking into account that in an isotropic, homogeneous universe the metric components $g_{\eta\eta}$ and $g_{\chi\chi}$ do not depend on χ, we infer from (2.54) that $u_\chi = $ const. Hence, the peculiar velocity,

$$w = au^\chi = ag^{\chi\chi}u_\chi \propto a^{-1}, \tag{2.55}$$

decays in inverse proportion to the scale factor.

2.4.1 Redshift as a measure of time and distance

The *redshift parameter* is defined as the fractional shift in wavelength of a photon emitted by a distant galaxy at time t_{em} and observed on Earth today:

$$z = \frac{\lambda_{obs} - \lambda_{em}}{\lambda_{em}}. \tag{2.56}$$

According to (2.45), the ratio $\lambda_{obs}/\lambda_{em}$ is equal to the ratio of the scale factors at the corresponding moments of time, and hence

$$1 + z = \frac{a_0}{a(t_{em})}, \tag{2.57}$$

where a_0 is the present value of the scale factor.

The light detected today was emitted at some earlier time t_{em} and, according to (2.57), there is a one-to-one correspondence between z and t_{em}. Therefore, the redshift z can be used instead of time t to parameterize the history of the universe. A given z corresponds to a time when our universe was $1 + z$ times smaller than now. We can express all time-dependent quantities as functions of z. For example, the formula for the energy density $\varepsilon(z)$ follows immediately from the energy conservation equation $d\varepsilon = -3(\varepsilon + p)d\ln a$:

$$\int_{\varepsilon_0}^{\varepsilon(z)} \frac{d\varepsilon}{\varepsilon + p(\varepsilon)} = 3\ln(1+z). \tag{2.58}$$

To obtain the expression for the Hubble parameter H in terms of z and the present values of H_0 and Ω_0, it is convenient to rewrite the Friedmann equation (1.67) in the form

$$H^2(z) + \frac{k}{a_0^2}(1+z)^2 = \Omega_0 H_0^2 \frac{\varepsilon(z)}{\varepsilon_0}, \tag{2.59}$$

2.4 Redshift

where the definitions in (1.21) and (2.57) have been used. At $z = 0$, this equation reduces to

$$\frac{k}{a_0^2} = (\Omega_0 - 1) H_0^2, \tag{2.60}$$

allowing us to express the current value of the scale factor a_0 in a spatially curved universe ($k \neq 0$) in terms of H_0 and Ω_0. Taking this into account, we obtain

$$H(z) = H_0 \left((1 - \Omega_0)(1 + z)^2 + \Omega_0 \frac{\varepsilon(z)}{\varepsilon_0} \right)^{1/2}. \tag{2.61}$$

Generically, the expressions for $a(t)$ are rather complicated and one cannot directly invert (2.57) to express the cosmic time $t \equiv t_{em}$ in terms of the redshift parameter z. It is useful, therefore, to derive a general integral expression for $t(z)$. Differentiating (2.57), we obtain

$$dz = -\frac{a_0}{a^2(t)} \dot{a}(t)\, dt = -(1 + z) H(t)\, dt, \tag{2.62}$$

from which it follows that

$$t = \int_z^\infty \frac{dz}{H(z)(1 + z)}. \tag{2.63}$$

A constant of integration has been chosen here so that $z \to \infty$ corresponds to the initial moment of time, $t = 0$. Thus, to determine $t(z)$, one should first find $\varepsilon(z)$ and, after substituting (2.61) into (2.63), perform the integration.

Knowing the redshift of light from a distant galaxy we can unambiguously determine its separation from us; that is, redshift can also be used as a measure of distance. The comoving distance to a galaxy that emitted a photon at time t_{em} which arrives today is

$$\chi = \eta_0 - \eta_{em} = \int_{t_{em}}^{t_0} \frac{dt}{a(t)}. \tag{2.64}$$

Substituting $a(t) = a_0/(1 + z)$ and the expression for dt in terms of dz from (2.62), we obtain

$$\chi(z) = \frac{1}{a_0} \int_0^z \frac{dz}{H(z)}. \tag{2.65}$$

In a universe with nonzero spatial curvature ($k \neq 0$), the current value of the scale factor a_0 can be expressed in terms of H_0 and Ω_0 via (2.60):

$$a_0^{-1} = \sqrt{|\Omega_0 - 1|} H_0.$$

Note that as $z \to \infty$, $\chi(z)$ approaches the particle horizon. Hence, the redshift parameter measures distance only within the particle horizon.

Finally, let us find the explicit expressions for $t(z)$ and $\chi(z)$ in a dust-dominated universe. In this case, $\varepsilon(z) = \varepsilon_0(1+z)^3$ and

$$H(z) = H_0(1+z)\sqrt{1+\Omega_0 z}.$$

For a flat universe ($\Omega_0 = 1$), the integrals in (2.63) and (2.65) are straightforward and we find

$$t(z) = \frac{2}{3H_0} \frac{1}{(1+z)^{3/2}}, \qquad \chi(z) = \frac{2}{a_0 H_0}\left(1 - \frac{1}{\sqrt{1+z}}\right). \tag{2.66}$$

Problem 2.14 Verify that in both open and closed dust-dominated universes

$$\Phi(\chi(z)) = \frac{2\sqrt{|\Omega_0 - 1|}}{\Omega_0^2(1+z)}\left[\Omega_0 z + (\Omega_0 - 2)\left(\sqrt{1+\Omega_0 z} - 1\right)\right], \tag{2.67}$$

where the function Φ is defined in (2.3). Note that if $\Omega_0 z \gg 1$, then $\Phi(\chi(z)) \to \Phi(\chi_p)$, given in (2.9). Derive the explicit expressions for $t(z)$.

2.5 Kinematic tests

For an object at a cosmological redshift, it is desirable to measure its angular size (the angle the object subtends on the sky) or its apparent luminosity. Given a class of objects of the same size (standard rulers), we find that the corresponding angular size changes with redshift in a specific way that depends on the values of the cosmological parameters. The same is also true for the apparent luminosities of objects with the same total brightness (standard candles). Therefore, if we know the appropriate dependencies for particular classes of standard rulers or standard candles, we can determine the cosmological parameters. Moreover, because the measurements refer to earlier times when the universe was $1 + z$ times smaller than now, we can study its recent expansion history and distinguish among models with different matter content.

2.5.1 Angular diameter–redshift relation

In a static, Euclidean space, the angle which an object with a given transverse size subtends on the sky is inversely proportional to the distance to this object. In an expanding universe, the relation between the distance and the angular size is not so trivial. Let us consider some extended object of given transverse size l at comoving distance χ_{em} from an observer (Figure 2.11). Without loss of generality, we can set $\varphi = $ const. Then, photons emitted by the endpoints of this object at time

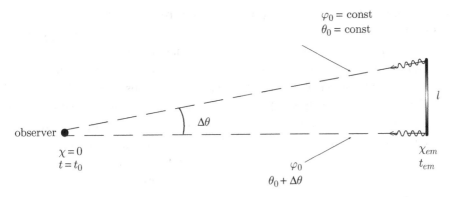

Fig. 2.11.

t_{em} propagate along radial geodesics and arrive today with an apparent angular separation $\Delta\theta$. The proper size of the object, l, is equal to the interval between the emission events at the endpoints:

$$l = \sqrt{-\Delta s^2} = a(t_{em})\,\Phi(\chi_{em})\,\Delta\theta, \qquad (2.68)$$

as obtained from metric (2.2). The angle subtended by the object is then

$$\Delta\theta = \frac{l}{a(t_{em})\,\Phi(\chi_{em})} = \frac{l}{a(\eta_0 - \chi_{em})\,\Phi(\chi_{em})}, \qquad (2.69)$$

where we have used the fact that the physical time t_{em} corresponds to the conformal time $\eta_{em} = \eta_0 - \chi_{em}$. If the object is close to us, that is, $\chi_{em} \ll \eta_0$, then

$$a(\eta_0 - \chi_{em}) \approx a(\eta_0), \quad \Phi(\chi_{em}) \approx \chi_{em},$$

and

$$\Delta\theta \approx \frac{l}{a(\eta_0)\,\chi_{em}} = \frac{l}{D}.$$

We see that in this case $\Delta\theta$ is inversely proportional to the distance, as expected. However, if the object is located far away, namely, close to the particle horizon, then $\eta_0 - \chi_{em} \ll \eta_0$, and

$$a(\eta_0 - \chi_{em}) \ll a(\eta_0), \quad \Phi(\chi_{em}) \to \Phi(\chi_p) = \text{const}.$$

The angular size of the object,

$$\Delta\theta \propto \frac{l}{a(\eta_0 - \chi_{em})},$$

increases with distance and as it approaches the horizon its image covers the whole sky. Of course, the apparent luminosity drops drastically with increasing distance, otherwise remote objects would completely outshine nearby ones.

To understand this unusual behavior of the angular diameter, it is again useful to turn to a low-dimensional analogy and consider how an observer on the north pole of the Earth would see an object of a given size at various distances. In this analogy, light propagates along meridians, which are geodesics on the Earth's surface, and we find that the angular size decreases with distance, but only if the object is north of the equator. If the object is south of the equator, the angular size increases with distance until, finally, an object at the south pole "covers the whole sky." This analogy, while illuminating, is not complete. The angular size of a very remote object also grows in a flat universe because of the time dependence of the scale factor; the 4-curvature of spacetime is responsible for the unusual behavior of the angular diameter.

The angular size $\Delta\theta$ can be expressed as a function of redshift z. Since $a_0/a(t_{em}) = 1 + z$, we can write (2.69) as

$$\Delta\theta(z) = (1+z)\frac{l}{a_0 \Phi(\chi_{em}(z))}, \qquad (2.70)$$

where $\chi_{em}(z)$ is given by (2.65). In a flat universe filled with dust, the function $\Phi(\chi_{em})$ equals χ_{em}, whose explicit dependence on z was given in (2.66). Hence, the angular diameter as a function of z is

$$\Delta\theta(z) = \frac{lH_0}{2}\frac{(1+z)^{3/2}}{(1+z)^{1/2} - 1}. \qquad (2.71)$$

At low redshifts ($z \ll 1$), the angular diameter decreases in inverse proportion to z, reaches a minimum at $z = 5/4$, and then scales as z for $z \gg 1$ (Figure 2.12).

The extension to more general cosmologies is straightforward. For example, substituting $\Phi(\chi_{em})$ from (2.67) into (2.70), we find that in a nonflat dust-dominated universe,

$$\Delta\theta(z) = \frac{lH_0}{2}\frac{\Omega_0^2(1+z)^2}{\Omega_0 z + (\Omega_0 - 2)((1+\Omega_0 z)^{1/2} - 1)}. \qquad (2.72)$$

In principle, having standard rulers distributed over a range of redshifts we could use the measurements of angular diameter versus redshift to test cosmological models. Unfortunately, the lack of reliable standard rulers has hampered progress in this technique for many years.

One spectacular success, though, has been a single standard ruler extracted from measurements of the cosmic microwave background. The temperature autocorrelation function measures how the microwave background temperature in two

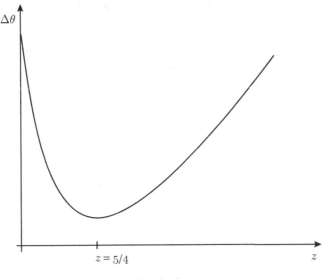

Fig. 2.12.

directions in the sky differs; this temperature difference depends on the angular separation. The power spectrum is observed to have a series of peaks as the angular separation is varied from large to small scales. The "first acoustic peak" is roughly determined by the sound horizon at recombination, the maximum distance that a sound wave in the baryon–radiation fluid can have propagated by recombination. This sound horizon serves as a standard ruler of length $l_s \sim H^{-1}(z_r)$. Recombination occurs at redshift $z_r \simeq 1100$. Since $\Omega_0 z_r \gg 1$, we can set $\chi_{em}(z_r) = \chi_p$ in (2.70) and in a *dust*-dominated universe, where $\Phi(\chi_p) = 2(a_0 H_0 \Omega_0)^{-1}$ (see (2.9)), we obtain

$$\Delta\theta_r \simeq \frac{z_r H_0 \Omega_0}{2 H(z_r)} \simeq \frac{1}{2} z_r^{-1/2} \Omega_0^{1/2} \simeq 0.87° \Omega_0^{1/2}. \tag{2.73}$$

We have substituted here $H_0/H(z_r) \simeq (\Omega_0 z_r^3)^{-1/2}$, as follows from (2.61). Note that in Euclidean space, the corresponding angular size would be $\Delta\theta_r \simeq t_r/t_0 \approx z_r^{-3/2}$, or about 1000 times smaller.

The remarkable aspect of this result is that the angular diameter depends directly only on Ω_0, which determines the spatial curvature, and is not very sensitive to other parameters. As we will see in Chapter 9, this is true not only for a dust-dominated universe, as considered here, but for a very wide range of cosmological models, containing multiple matter components. Hence, measuring the angular scale of the first acoustic peak has emerged as the leading and most direct method for determining the spatial curvature. Our best evidence that the universe is spatially flat ($\Omega_0 = 1$), as predicted by inflation, comes from this test.

2.5.2 Luminosity–redshift relation

A second method of recovering the expansion history is with the help of the luminosity–redshift relation. Let us consider a source of radiation with total luminosity (energy per unit time) L located at comoving distance χ_{em} from us. The total energy released by the source at time t_{em} within a conformal time interval $\Delta\eta$ is equal to

$$\Delta E_{em} = L\Delta t_{em}(\Delta\eta) = La(t_{em})\Delta\eta. \tag{2.74}$$

All of the emitted photons are located within a shell of constant conformal width $\Delta\chi = \Delta\eta$. The radius of this shell grows with time and the frequencies of the photons are redshifted. Therefore, when these photons reach the observer at time t_0, the total energy within the shell is

$$\Delta E_{obs} = \Delta E_{em}\frac{a(t_{em})}{a_0} = L\frac{a^2(t_{em})}{a_0}\Delta\eta. \tag{2.75}$$

At this moment, the shell has surface area

$$S_{sh}(t_0) = 4\pi a_0^2 \Phi^2(\chi_{em})$$

and physical width

$$\Delta l_{sh} = a_0\Delta\chi = a_0\Delta\eta.$$

The shell passes the observer's position over a time interval (measured by the observer) $\Delta t_{sh} = \Delta l_{sh} = a_0\Delta\eta$. Therefore, the measured bolometric flux (energy per unit area per unit time) is equal to

$$F \equiv \frac{\Delta E_{obs}}{S_{sh}(t_0)\Delta t_{sh}} = \frac{L}{4\pi \Phi^2(\chi_{em})}\frac{a^2(t_{em})}{a_0^4} \tag{2.76}$$

or, as a function of redshift,

$$F = \frac{L}{4\pi a_0^2 \Phi^2(\chi_{em}(z))(1+z)^2}. \tag{2.77}$$

Here $\chi_{em}(z)$ is given by (2.65). Instead of F, astronomers often use the apparent (bolometric) magnitude, m_{bol}, defined as

$$m_{bol}(z) \equiv -2,5\log_{10} F = 5\log_{10}(1+z) + 5\log_{10}(\Phi(\chi_{em}(z))) + \text{const}, \tag{2.78}$$

where const is z-independent.

For $z \ll 1$, we find that, irrespective of the spatial curvature and matter composition of the universe,

$$m_{bol}(z) = 5\log_{10} z + \frac{2.5}{\ln 10}(1-q_0)z + O(z^2) + \text{const}, \tag{2.79}$$

where $q_0 \equiv -(\ddot{a}/aH^2)_0$. In turn, the value of the deceleration parameter q_0 is determined by the equation of state. Using Friedmann equation (1.66), we obtain

$$q_0 = \frac{1}{2}\Omega_0\left(1 + 3\frac{p}{\varepsilon}\right)_0. \tag{2.80}$$

Thus, measuring the luminosity–redshift dependence for a set of standard candles, we can, in principle, determine the effective equation of state for the dominant matter components.

Measurements using Type IA supernovae as standard candles have produced a spectacular result. The expansion of the universe has been found to be accelerating, rather than decelerating. In other words, q_0 is negative. In a matter-dominated universe, the gravitational self-attraction of matter resists the expansion and slows it down. According to Friedmann equation (1.66), acceleration is possible only if a substantial fraction of the total energy density is a "dark *energy*" with negative pressure or, equivalently, negative equation of state $w \equiv p/\varepsilon$.

One possibility is that the dark energy component is a vacuum energy density or cosmological constant, which corresponds to $w = -1$. Alternatively, the dark energy can be dynamical, such as a slightly time-varying scalar field. The latter case is referred to as "quintessence." The discovery of cosmic acceleration raises a number of new problems in cosmology. At present, there is no convincing explanation as to why dark energy came to dominate so late in the history of the universe and exactly at the time to be observed. Additionally, because the nature of the dark energy is uncertain, the long-term future of the universe cannot be determined. If the dark energy is a cosmological constant, then the acceleration will continue forever and the universe will become empty. On the other hand, if the dark energy is a dynamical scalar field, then this field may decay, repopulating the universe with matter and energy. In summary, dark energy is one of the most enigmatic and challenging issues in cosmology today.

The supernovae that provide evidence for the dark energy component have redshifts of order unity and the expansion in (2.79), valid only for $z < 0.3$, is not applicable for them. Therefore, to describe the observations, we have to use the exact formula (2.78) and choose a particular class of cosmological models in order to compute $\Phi(\chi_{em}(z))$. For example, for a flat universe comprising only cold matter and a cosmological constant, so that $\Omega_0 = \Omega_\Lambda + \Omega_m = 1$, we have

$$\Phi(\chi_{em}(z)) = \chi_{em}(z) = \frac{1}{H_0 a_0}\int_0^z \frac{d\tilde{z}}{\sqrt{\Omega_m(1+\tilde{z})^3 + (1-\Omega_m)}}. \tag{2.81}$$

Calculating the integral numerically, we can find $m_{bol}(z)$ for different values of Ω_m (Figure 2.13). The best fit to the data is achieved for $\Omega_m \simeq 0.3$.

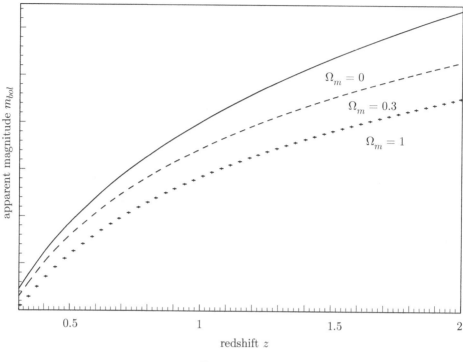

Fig. 2.13.

Problem 2.15 In Euclidean space, the observed flux F from an object of luminosity L at distance d is $F = L/4\pi d^2$ and the angular size of an object of known length l is $\Delta\theta = l/d$. Based on these relations, cosmologists sometimes *formally* define the luminosity distance d_L and the angular diameter distance d_A to an object in an expanding universe as

$$d_L \equiv \left(\frac{L}{4\pi F}\right)^{1/2}, \quad d_A \equiv \frac{l}{\Delta\theta}, \tag{2.82}$$

respectively. Calculate $d_L(z)$ and $d_A(z)$ in a dust-dominated universe. How are they related in general? Verify that the distances d_L and d_A coincide only to leading order in z and at small z revert to the Euclidean distance d. In contrast with d_A, the luminosity distance d_L increases with z at large redshift, as common sense would suggest. Both, however, are only formal for $z > 1$ where the notion of invariant physical distance does not exist.

2.5.3 Number counts

A further kinematic test is based on counting the number of cosmological objects with a given redshift. Suppose the number of galaxies or clusters per unit volume at a moment of time characterized by redshift z is spatially uniform and equal to $n(z)$. Then, the number of galaxies with redshifts between z and $z + \Delta z$, and within a solid angle $\Delta \Omega$, is

$$\Delta N = n(z) \, a^3(z) \, \Phi^2(\chi) \Delta \chi \Delta \Omega = n(z)(1+z)^{-3} a_0^2 H^{-1}(z) \, \Phi^2(\chi) \Delta z \Delta \Omega, \tag{2.83}$$

where we have used the relation between $\Delta \chi$ and Δz (see (2.65)). Substituting $\Phi^2(\chi)$ from (2.67), we find that in a dust-dominated universe

$$\frac{\Delta N}{\Delta z \, \Delta \Omega} = \frac{4n(z)}{H_0^3 \Omega_0^4} \frac{[\Omega_0 z + (\Omega_0 - 2)((1 + \Omega_0 z)^{1/2} - 1)]^2}{(1+z)^6 \, (1 + \Omega_0 z)^{1/2}}. \tag{2.84}$$

If we know $n(z)$, then measurement of $\Delta N / \Delta z \, \Delta \Omega$ can be used as a test of cosmological models. The difficulty in applying this method is that the number of galaxies varies with redshift not only because of the expansion but also as a result of dynamical evolution. For example, small galaxies merge to form large ones. Conceivably, this problem can be avoided if the number density of some subset of galaxies has predictable evolution.

2.5.4 Redshift evolution

The redshift of a given object drifts slowly with time due to the acceleration (or deceleration) of the universe. The effect is so small that it is not possible to measure it using today's technology. However, we introduce the concept as an example of a measurement that could be possible in the coming decades.

Light from a source located at comoving distance χ that we observe today at conformal time η_0 was emitted at conformal time $\eta_e = \eta_0 - \chi$. The appropriate redshift depends on η_0 and is equal to

$$z(\eta_0) = \frac{a_0}{a_e} = \frac{a(\eta_0)}{a(\eta_0 - \chi)}. \tag{2.85}$$

This redshift depends on the time of observation η_0 and since χ is constant, its time derivative is

$$\dot{z} \equiv \frac{dz}{dt} = \frac{1}{a(\eta_0)} \frac{\partial z}{\partial \eta_0} = \frac{\dot{a}_0}{a_e} - \frac{\dot{a}_e}{a_e} = (1+z) H_0 - H(z). \tag{2.86}$$

Taking into account that $\varepsilon(z) = \varepsilon_0^{cr}(\Omega_\Lambda + \Omega_m(1+z)^3)$ in a universe with a mixture of matter and vacuum density, and using this expression in (2.61) for $H(z)$, we obtain

$$\dot{z} = (1+z) H_0 \{ 1 - [1 - \Omega_0 + \Omega_m(1+z) + \Omega_\Lambda (1+z)^{-2}]^{1/2} \}. \tag{2.87}$$

In a flat universe, where $\Omega_0 = 1$ and $\Omega_\Lambda = 1 - \Omega_m$, the redshift drift, $\Delta v \equiv \Delta z/(1+z)$, is equal to

$$\Delta v \simeq \frac{\dot{z}\Delta t}{1+z} = -H_0 \Delta t \left\{ [\Omega_m(1+z) + (1-\Omega_m)(1+z)^{-2}]^{1/2} - 1 \right\}. \quad (2.88)$$

The drift is negative for a matter-dominated universe ($\Omega_m \to 1$) and positive if the cosmological constant dominates ($\Omega_m \to 0$). For $\Omega_m = 1$ and $\Omega_\Lambda = 0$, its magnitude is

$$\Delta v \approx -2(\sqrt{1+z} - 1) \text{ cm s}^{-1}$$

for observations made a period $\Delta t = 1$ year apart. Although the velocity shift is tiny and beyond current detection capabilities, redshift is one of the most precisely measured physical observables. Current technology would enable measurements of shifts of perhaps 10 m s^{-1} per year. The required improvement by a few orders of magnitude in the next few decades is conceivable. Such a measurement would represent a direct detection of acceleration which would complement the luminosity–redshift tests.

3
The hot universe

In the previous chapters we studied the geometrical properties of the universe. Now we turn to its thermal history. This history can be subdivided into several periods. Here we focus mainly on the period between neutrino decoupling and recombination. This period is characterized by a sequence of important departures from thermal and chemical equilibrium that shaped the present state of the universe.

We begin with an overview of the main thermal events and then turn to their detailed description. In particular, in this chapter we study the decoupling of neutrinos, primordial nucleosynthesis and recombination. Our considerations are based on well understood and tested laws of particle, nuclear and atomic physics below a few MeV and, as such, are not likely to be a rich source of future research. However, this is important background material which underlies the concept of the hot expanding universe.

3.1 The composition of the universe

According to the Friedmann equations, the expansion rate of the universe is determined by the energy density and equation of state of its constituents. The main components of the matter composition that played an important role at temperatures below a few MeV are primordial radiation, baryons, electrons, neutrinos, dark matter and dark energy.

Primordial radiation The cosmic microwave background (CMB) radiation has temperature $T_{\gamma 0} \simeq 2.73$ K. Its current energy density is about $\varepsilon_{\gamma 0} \simeq 10^{-34}$ g cm^{-3} and constitutes only 10^{-5} of the total energy density. The radiation has a perfect Planckian spectrum and appears to have been present in the very early universe at energies well above a GeV. Since the temperature of radiation scales in inverse proportion to the scale factor, it must have been very high in the past.

Baryonic matter This is the material out of which the planets, stars, clouds of gas and possibly "dark" stars of low mass are made; some of it could also form black holes. We will see later that the data on light element abundances and CMB fluctuations clearly indicate that the baryonic component contributes only a small percentage of the critical energy density ($\Omega_b \simeq 0.04$). The number of photons per baryon is of order 10^9.

Dark matter and dark energy The CMB fluctuations imply that at present the total energy density is equal to the critical density. This means that the largest fraction of the energy density of the universe is dark and nonbaryonic. It is not quite clear what constitutes this dark component. Combining the data on CMB, large scale structure, gravitational lensing and high-redshift supernovae it appears that the dark component is a mixture of two or more constituents. More precisely, it is composed of *cold dark matter* and *dark energy*. The cold dark matter has zero pressure and can cluster, contributing to gravitational instability. Various (supersymmetric) particle theories provide us with natural candidates for the cold dark matter, among which *weakly interacting massive particles* are most favored at present. The nonbaryonic cold dark matter contributes only about 25% of the critical density. The remaining 70% of the missing density comes in the form of nonclustered dark energy with negative pressure. It may be either a cosmological constant ($p_\Lambda = -\varepsilon_\Lambda$) or a scalar field (quintessence) with $p = w\varepsilon$, where w is less than $-1/3$ today.

Primordial neutrinos These are an inevitable remnant of a hot universe. If the three known neutrino species were massless, their temperature today would be $T_\nu \simeq 1.9$ K and they would contribute 0.68 times the radiation density (see Section 3.4.2). Atmospheric neutrino oscillation experiments suggest that the neutrinos have small masses. Even so, it appears that they cannot constitute more than 1% of the critical density.

The universe was hotter and denser in the past. The energy densities of radiation, cold matter and dark energy scale with redshift z as

$$\varepsilon_\gamma = \varepsilon_{\gamma 0}(1+z)^4, \quad \varepsilon_m = \varepsilon_0^{cr}\Omega_m(1+z)^3, \quad \varepsilon_Q = \varepsilon_0^{cr}\Omega_Q(1+z)^{3(1+w)}, \quad (3.1)$$

respectively. Here $\varepsilon_0^{cr} = 3H_0^2/8\pi G$ is the critical density today, Ω_m is the total contribution of baryons and cold dark matter to the current cosmological parameter and Ω_Q is the contribution of dark energy. When we go back in time the dark energy density grows the least quickly; its impact on the dynamics of the universe becomes less than that of cold matter at redshift (see Figure 3.1)

$$z_Q = (\Omega_Q/\Omega_m)^{-1/3w} - 1. \quad (3.2)$$

This occurs close to the present time, at $z_Q = 0.33$ to 1.33, for $-1 \leq w < -1/3$, $\Omega_m \approx 0.3$ and $\Omega_Q \approx 0.7$.

Problem 3.1 Find the value of z at which the accelerated expansion begins.

3.1 The composition of the universe

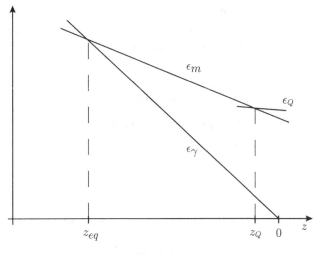

Fig. 3.1.

The radiation energy density grows faster than the density of cold matter and eventually becomes dominant at redshift

$$z_{eq} = \frac{\varepsilon_0^{cr} \Omega_m}{\varepsilon_{\gamma 0}} - 1 \simeq 2.26 \times 10^4 \Omega_m h_{75}^2, \quad (3.3)$$

where

$$h_{75} \equiv \frac{H_0}{75 \text{ km s}^{-1} \text{ Mpc}^{-1}}.$$

Thus, we can distinguish three dynamically different stages in the expansion history:

- *the radiation-dominated epoch* at $z > z_{eq} \sim 10^4$, where the universe is dominated by ultra-relativistic matter with $p = \varepsilon/3$ and scale factor increases as $a \propto t^{1/2}$;
- *the matter-dominated epoch* at $z_{eq} > z > z_Q$, where the pressureless components determine the expansion rate and $a \propto t^{2/3}$;
- *the dark-energy-dominated epoch* at $z < z_Q$, where the component with negative pressure, $p = w\varepsilon$, leads to an accelerated expansion and $a \propto t^{2/3(1+w)}$.

Note that the dark energy cannot begin to dominate too early because a substantial period of matter domination is needed for structure formation. In fact, it becomes relevant exactly at the present time. This astounding cosmic coincidence is one of the greatest mysteries of contemporary cosmology.

Problem 3.2 How do ultra-relativistic neutrinos influence an estimate for the redshift at which the ultra-relativistic matter begins to dominate?

Problem 3.3 Dark energy with equation of state $w = -1/3$ leads to a term $\propto 1/a^2$ in the Friedmann equation (1.67). How can we nevertheless distinguish it from the spatial curvature term, k/a^2, in an open universe?

3.2 Brief thermal history

The temperature of the cosmic radiation decreases as the universe expands. It is unambiguously related to the redshift,

$$T_\gamma(z) = T_{\gamma 0}(1 + z), \qquad (3.4)$$

and can be used as an alternative to time or redshift to parameterize the history of the universe. To obtain an estimate for the temperature expressed in MeV, at the time t measured in seconds, we can use the formula

$$T_{\text{MeV}} \simeq \frac{O(1)}{\sqrt{t_{\text{sec}}}},$$

which is valid during the radiation-dominated epoch (see Section 3.4.2).

Below we briefly summarize the sequence of main events constituting the history of our universe (in reverse chronological order):

- $\sim 10^{16}$–10^{17} s Galaxies and their clusters are formed from small initial inhomogeneities as a result of gravitational instability. Structure formation can be described using Newtonian gravity. However, it is still a very complicated nonlinear problem, which can only be solved numerically and it is likely to remain an active field of research for a long time. One of the main unresolved fundamental issues regarding this period is the nature of dark matter and dark energy.
- $\sim 10^{12}$–10^{13} s At this time nearly all free electrons and protons recombine and form neutral hydrogen. The universe becomes transparent to the background radiation. The CMB temperature fluctuations, induced by the slightly inhomogeneous matter distribution at recombination, survive to the present day and deliver direct information about the state of the universe at the last scattering surface. Helium, which constitutes about 25% of the baryonic matter, has recombined and become neutral before this time. After helium recombination there remain many free electrons and the universe is still opaque to radiation. Helium recombination, therefore, is not a very dramatic event, though we must take it properly into account when calculating the microwave background fluctuations because it influences the speed of sound.
- $\sim 10^{11}$ s ($T \sim$ eV) This time corresponds to matter–radiation equality which separates the radiation-dominated epoch from the matter-dominated epoch. The exact value of the cosmological time at equality depends on the constituents of the dark component and, therefore, is known at present only up to a numerical factor of order unity.
- ~ 200–300 s ($T \sim 0.05$ MeV) Nuclear reactions become efficient at this temperature. As a result, free protons and neutrons form helium and other light elements. The abundances

of the light elements resulting from *primordial nucleosynthesis* are in very good agreement with available observation data and this strongly supports our understanding of the universe's evolution back to the first second after the big bang.

- ~ 1 s ($T \sim 0.5$ MeV) The typical energy at this time is of order the electron mass. The numerous electron–positron pairs present in the very early universe begin to annihilate when the temperature drops below their rest mass and only a small excess of electrons over positrons, roughly one per billion photons, survives after annihilation. The photons produced are in thermal equilibrium and the radiation temperature increases compared to the temperature of neutrinos, which decoupled earlier.
- ~ 0.2 s ($T \sim 1$–2 MeV) Two important events take place during this period as certain weak interaction processes fall out of equilibrium. First, the primordial neutrinos decouple from the other particles and propagate without further scatterings. Second, the ratio of neutrons to protons "freezes out" because the interactions that keep neutrons and protons in chemical equilibrium become inefficient. Subsequently, the number of the surviving neutrons determines the abundances of the primordial elements.
- $\sim 10^{-5}$ s ($T \sim 200$ MeV) The quark–gluon transition takes place: free quarks and gluons become confined within baryons and mesons. The physics of the quark–gluon transition is not yet completely understood, though it is unlikely that this transition leaves any significant cosmological imprints.
- $\sim 10^{-10}$–10^{-14} s ($T \sim 100$ GeV–10 TeV) This range of energy scales can still be probed by accelerators. The Standard Model of electroweak and strong interactions appears to be applicable here. We expect that at temperatures above ~ 100 GeV the electroweak symmetry is restored and the gauge bosons are massless. Fermion and baryon numbers are strongly violated in topological transitions above the symmetry restoration scale.
- $\sim 10^{-14}$–10^{-43} s (10 TeV–10^{19} GeV) This energy range will probably not be reached by accelerators in the near future. Instead, the very early universe becomes, in Zel'dovich's words, "an accelerator for poor people" that can give us some rough information about fundamental physics. There is no reason to expect that nonperturbative quantum gravity plays any significant role below 10^{19} GeV. Therefore, we can still use General Relativity to describe the dynamics of the universe. The main uncertainty here is the matter composition of the universe. It might be that there are many more particle species than are evident today. For example, according to supersymmetry, the number of particles species must be doubled at least. Supersymmetry also provides us with good weakly interacting massive particle candidates for dark matter.

The origin of *baryon asymmetry* in the universe is also related to physics beyond the Standard Model. There are good reasons to expect that a *Grand Unification* of the electroweak and strong interactions takes place at energies about 10^{16} GeV. Topological defects, such as cosmic strings, monopoles, that occur naturally in unified theories might play some role in the early universe, though, according to the current microwave background anisotropy data, it is unlikely that they have any significance for large scale structure.

Perhaps the most interesting phenomenon in the above energy range is the accelerated expansion of the universe – *inflation* – which probably occurs somewhere near Grand

Unification scales. It is remarkable and fortunate that the most important robust predictions of inflation do not depend substantially on unknown particle physics. Therefore, the existence of such a stage may be observationally verified in the near future.

- **$\sim 10^{-43}$ s $(10^{19}$ GeV)** Near the Planckian scale, nonperturbative quantum gravity dominates and general relativity can no longer be trusted. However, at energies slightly below this scale, classical spacetime still makes sense and we expect that the universe is in a self-reproducing phase. Nevertheless, self-reproduction does not eliminate the fundamental issues of spacetime structure at the Planckian scale. In particular, the question of cosmic singularities still remains. It is expected that these problems will be properly addressed in an as yet unknown nonperturbative string/quantum gravity theory.

3.3 Rudiments of thermodynamics

To properly describe the physical processes in an expanding universe we need, strictly speaking, a full kinetic theory. Fortunately, the situation greatly simplifies in the very early universe, when the particles are in a state of *local* equilibrium with each other. We would like to stress that the universe cannot be treated as a usual thermodynamical system in equilibrium with an infinite thermal bath of given temperature: it is a nonequilibrium system. Therefore, by *local equilibrium* we simply mean that matter has maximal possible entropy. The entropy is well defined for any system even if this system is far from equilibrium and never decreases. Therefore, if within a typical cosmological time the particles scatter from each other many times, their entropy reaches the maximal possible value before the size of the universe changes significantly.

The reaction rate responsible for establishing equilibrium can be characterized by the *collision time*:

$$t_c \simeq \frac{1}{\sigma n v}, \tag{3.5}$$

where σ is the effective cross-section, n is the number density of the particles and v is their relative velocity. This time should be compared to the cosmic time, $t_H \sim 1/H$, and if

$$t_c \ll t_H, \tag{3.6}$$

local equilibrium is reached before expansion becomes relevant. Let us show that at temperatures above a few hundred GeV condition (3.6) is satisfied for both electroweak and strong interactions. At such high temperatures, all known particles are ultra-relativistic and the gauge bosons are all massless. Therefore, the cross-sections for strong and electroweak interactions have a similar energy dependence and they can be estimated (e.g. on dimensional grounds) as

$$\sigma \simeq O(1) \alpha^2 \lambda^2 \sim \frac{\alpha^2}{T^2}, \tag{3.7}$$

where $\lambda \sim 1/p$ is the de Broglie wavelength and $p = E \sim T$ is the typical momentum of the colliding ultra-relativistic particles. The corresponding *dimensionless* running coupling constants α vary only logarithmically with energy and are of order 10^{-1}–10^{-2}. Taking into account that the number density of the ultra-relativistic species is $n \sim T^3$, we find that

$$t_c \sim \frac{1}{\alpha^2 T}. \tag{3.8}$$

Comparing this time to the Hubble time,

$$t_H \sim \frac{1}{H} \sim \frac{1}{\sqrt{\varepsilon}} \sim \frac{1}{T^2}, \tag{3.9}$$

we find that at temperatures below $T \sim O(1)\alpha^2 \simeq 10^{15}$–$10^{17}$ GeV, but above a few hundred GeV (where (3.7) is applicable), (3.6) is satisfied and the electroweak as well as the strong interactions are efficient in establishing equilibrium between quarks, leptons and intermediate bosons.

The discerning reader might question whether one can apply the formulae for cross-sections derived in empty space to interactions which occur in extremely dense "plasma". To get an idea of the strength of the plasma effects, we have to compare the typical distance between the particles $1/n^{1/3} \sim 1/T$ to the "size" of the particles $\sqrt{\sigma} \sim \alpha/T$. If the coupling constant α is smaller than unity, the plasma effects are not very relevant.

Primordial gravitons and, possibly, other hypothetical particles that interact through the *dimensionful* gravitational constant already decouple from the rest of matter at Planckian times and propagate, subsequently, freely.

Below 100 GeV, the Z and $W^\pm$ bosons acquire mass ($M_W \simeq 80, 4$ GeV, $M_Z \simeq 91, 2$ GeV) and, thereafter, the cross-sections of the weak interactions begin to decrease as the temperature drops. As a result, the neutrinos decouple from the rest of matter. Finally the electromagnetic interactions also become inefficient and photons propagate freely. All these processes will be analyzed in detail later in the chapter, but first we would like to concentrate on the very early stages when known particles were in equilibrium with radiation and with each other. In this case, matter can be described in a very simple way: all particles are completely characterized by their temperature and corresponding chemical potential.

3.3.1 Maximal entropy state, thermal spectrum, conservation laws and chemical potentials

In this section, we outline an elegant derivation of the main formulae describing the maximal entropy state. This derivation is based entirely on the notion of entropy for a closed system and does not use any concepts from equilibrium thermodynamics. Therefore, it can also be applied to the expanding universe.

Let us assume that all possible states of some (complicated) closed system can be completely characterized and enumerated by a (composite) discrete variable α; different α correspond to microscopically different states. If we know that the system is in a certain state α, the information about this system is complete and its entropy should be zero. This follows from the general definition according to which the entropy characterizes the missing information. If, on the other hand, we know only the probability P_α of finding the system in state α, then the associated (nonequilibrium) entropy is

$$S = -\sum_\alpha P_\alpha \ln P_\alpha. \tag{3.10}$$

It takes its maximum value when all states are equally probable, that is, $P_\alpha = 1/\Gamma$, and is equal to

$$S = \ln \Gamma, \tag{3.11}$$

where Γ is the total number of possible microstates which the system can occupy. Note that the last expression gives a finite result only if the total energy is bounded, otherwise the number of possible states would be infinite.

Let us calculate the maximal possible entropy of an *ideal* gas of N bose particles with total energy E placed in a box of volume V. It is clear that maximal entropy occurs when there are no preferable directions and locations within the box. Therefore, given the total energy and number of particles, each state of the system is essentially given by the number of particles per mode of the one-particle energy spectrum. Let us denote by ΔN_ϵ the total number of particles, each having energy in the interval between ϵ and $\epsilon + \Delta\epsilon$, and by Δg_ϵ the total number of different possible microstates that a particle could occupy in the *one-particle* phase space. The total number of all possible configurations (microstates) for ΔN_ϵ bose particles is equal to the number of ways of redistributing ΔN_ϵ particles among Δg_ϵ cells (Figure 3.2):

$$\Delta G_\epsilon = \frac{(\Delta N_\epsilon + \Delta g_\epsilon - 1)!}{(\Delta N_\epsilon)!(\Delta g_\epsilon - 1)!}. \tag{3.12}$$

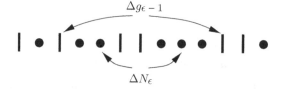

Fig. 3.2.

3.3 Rudiments of thermodynamics

The total number of states for the whole system, therefore, is

$$\Gamma(\{\Delta N_\epsilon\}) = \prod_\epsilon \Delta G_\epsilon. \tag{3.13}$$

Substituting (3.13) into (3.11), we find that the maximal possible entropy of the system with the *given energy spectrum* $\{\Delta N_\epsilon\}$ is

$$S(\{\Delta N_\epsilon\}) = \sum_\epsilon \ln \Delta G_\epsilon. \tag{3.14}$$

Let us assume that ΔN_ϵ and Δg_ϵ are much larger than unity. Using Stirling's formula,

$$\ln N! = \sum_{n=1}^N \ln n \approx \int_1^N \ln x\, dx + \frac{1}{2} \ln N = \left(N + \frac{1}{2}\right)\ln N - N, \tag{3.15}$$

we find from (3.12) and (3.14) that, to leading order,

$$S(\{\Delta N_\epsilon\}) \equiv S(\{n_\epsilon\}) = \sum_\epsilon [(n_\epsilon + 1)\ln(1 + n_\epsilon) - n_\epsilon \ln n_\epsilon]\, \Delta g_\epsilon, \tag{3.16}$$

where $n_\epsilon \equiv \Delta N_\epsilon / \Delta g_\epsilon$ are called occupation numbers. They characterize the average number of particles per microstate of a *single* particle. The entropy depends on the energy spectrum $\{n_\epsilon\}$ and we want to maximize it subject to the given total energy

$$E(\{n_\epsilon\}) = \sum_\epsilon \epsilon \Delta N_\epsilon = \sum_\epsilon \epsilon n_\epsilon \Delta g_\epsilon, \tag{3.17}$$

and total number of particles

$$N(\{n_\epsilon\}) = \sum_\epsilon \Delta N_\epsilon = \sum_\epsilon n_\epsilon \Delta g_\epsilon. \tag{3.18}$$

To extremize (3.16) with the two extra constraints (3.17) and (3.18), we apply the method of Lagrange multipliers. The variation of expression

$$S(\{n_\epsilon\}) + \lambda_1 E(\{n_\epsilon\}) + \lambda_2 N(\{n_\epsilon\})$$

with respect to n_ϵ vanishes for

$$n_\epsilon = \frac{1}{\exp(-\lambda_1 \epsilon - \lambda_2) - 1}. \tag{3.19}$$

Given spectrum (3.19), the Lagrange multipliers λ_1 and λ_2 are the parameters which allow us to satisfy the constraints. They can be expressed in terms of E and N, or, instead, in terms of temperature $T \equiv -1/\lambda_1$ and chemical potential

$\mu \equiv \lambda_2 T (k_B = 1)$. The distribution function (3.19) then takes the form

$$n_\epsilon = \frac{1}{\exp((\epsilon - \mu)/T) - 1}. \tag{3.20}$$

This spectrum describes bose particles in a state of maximal possible entropy and is known as the Bose–Einstein distribution. A similar derivation can be carried out for fermi particles, the only difference being that we have to take into account the Pauli exclusion principle, which forbids two fermions from simultaneously occupying the same microstate.

Problem 3.4 Derive the following expression for the entropy of fermi particles:

$$S(\{n_\epsilon\}) = \sum_\epsilon [(n_\epsilon - 1) \ln(1 - n_\epsilon) - n_\epsilon \ln n_\epsilon] \Delta g_\epsilon, \tag{3.21}$$

and show that it takes its maximal value for

$$n_\epsilon = \frac{1}{\exp((\epsilon - \mu)/T) + 1}. \tag{3.22}$$

Problem 3.5 According to (3.20) and (3.22), the energy of a single particle ϵ can, in principle, be larger than the total energy of the whole system E, which contradicts our assumptions. Where does the above derivation fail for ϵ comparable to or larger than E?

In quantum field theory particles can be created and annihilated, so their total number is generally not conserved. In this case the number of particles in equilibrium is determined solely by the requirement of maximal entropy for a given total energy. This removes the need to satisfy the second constraint (3.18). If there are no other constraints enforced by conservation laws, then the chemical potential μ is zero and there remains only one free parameter, λ_1, to fix the total energy. For example, the total number and the temperature of photons are entirely determined by their total energy E.

Because of the conservation of electric charge, electrons and positrons can be produced only in pairs. Therefore, the difference between the numbers of electrons and positrons $N_{e^-} - N_{e^+}$ does not change. With this extra constraint, the Lagrange variational principle takes the form

$$\delta \left[S\left(\{n_\epsilon^{e^-}\}\right) + S\left(\{n_\epsilon^{e^+}\}\right) + \lambda_1 (E_{e^-} + E_{e^+}) + \lambda_2 (N_{e^-} - N_{e^+}) \right] = 0, \tag{3.23}$$

where we vary separately with respect to $n_\epsilon^{e^-}$ and $n_\epsilon^{e^+}$. It is not hard to show that the variation vanishes only if the electrons and positrons both satisfy the Fermi distribution (3.22) with $T = -1/\lambda_1$ and $\mu_{e^-} = -\mu_{e^+} = T\lambda_2$. Thus, the chemical potentials of the electrons and positrons are equal in magnitude and have the opposite signs as a consequence of electric charge conservation. Only if the total electric charge of the electron–positron plasma is equal to zero do they vanish.

Problem 3.6 Assume particles of types A, B, C, D are in equilibrium with each other due to the reaction

$$A + B \rightleftarrows C + D.$$

It is easy to see that the following combinations are conserved: $N_A + N_C$, $N_A + N_D$, $N_B + N_C$, $N_B + N_D$. Using this fact, show that the chemical potentials satisfy the relation

$$\mu_A + \mu_B = \mu_C + \mu_D. \tag{3.24}$$

Note that, if electrons and positrons are in equilibrium with each other and with radiation due to the interaction $e^- + e^+ \rightleftarrows \gamma + \gamma$, then from (3.24) we recover the result, $\mu_{e^-} = -\mu_{e^+}$, since the chemical potential of radiation is equal to zero.

The above consideration can be directly applied to matter in a *homogeneous and isotropic* expanding universe. If the interaction rate is much larger than the rate of expansion, the entropy of matter reaches its maximal value very quickly. In a homogeneous universe there are no external sources of entropy, and therefore the total entropy of matter within a given comoving volume is conserved. If the interactions of some particles become inefficient, they decouple and evolve independently and their entropy is conserved separately. For example, after recombination photons propagate freely and they are not in thermal equilibrium. Nevertheless, they still have maximal possible entropy and hence satisfy the Bose–Einstein distribution as if they were in equilibrium. A similar situation occurs for neutrinos when they decouple from matter.

The simple arguments above are not valid when the universe becomes highly inhomogeneous as a result of gravitational instability. For this reason the initial state of the universe, which looks like a state of "thermal death" where nothing could happen, can evolve to a state where very complicated structures, such as biological systems, occur. Nonequilibrium processes and gravitational instability will be considered later in detail and here we concentrate on the local equilibrium state. It is rather remarkable that in this state general arguments involving only the entropy and conservation laws are sufficient to describe the system completely and we do not need to use a kinetic theory or go into the details of quantum field theory.

3.3.2 Energy density, pressure and the equation of state

To calculate the energy density and pressure for a given distribution function n_ϵ, we have to determine Δg_ϵ, the total number of possible microstates for a *single* particle having energy within the interval from ϵ to $\epsilon + \Delta\epsilon$. Let us first consider a particle with no internal degrees of freedom in one dimension. At any moment of time its state can be specified completely by the coordinate x and the momentum p. In classical mechanics two infinitesimally different coordinates or momenta correspond to microscopically different states. Therefore, the number of microstates is infinite and the entropy can be defined only up to an infinite additive factor. However, in quantum mechanics, two states within a cell of volume $2\pi\hbar$ in phase space are not distinguishable because of the uncertainty relation. Hence, there is only one possible microstate per corresponding phase volume. The generalization for the case of a particle with g internal degrees of freedom in three-dimensional space is straightforward:

$$\Delta g_\epsilon = g \int_\epsilon^{\epsilon+\Delta\epsilon} \frac{d^3x d^3\mathbf{p}}{(2\pi\hbar)^3} = \frac{gV}{(2\pi\hbar)^3} \int_\epsilon^{\epsilon+\Delta\epsilon} d^3\mathbf{p}, \quad (3.25)$$

where we have assumed homogeneity and integrated over the volume V. Henceforth, we use natural units where $c = \hbar = k_B = G = 1$. The energy ϵ depends on the momentum $|\mathbf{p}|$ and in the isotropic case we have

$$\Delta g_\epsilon = \frac{gV}{2\pi^2} \int_\epsilon^{\epsilon+\Delta\epsilon} |\mathbf{p}|^2 d|\mathbf{p}| \simeq \frac{gV}{2\pi^2} \sqrt{(\epsilon^2 - m^2)} \epsilon \Delta\epsilon, \quad (3.26)$$

where the relativistic relation,

$$\epsilon^2 = |\mathbf{p}|^2 + m^2,$$

has been used. Note that the state with the minimal possible energy, $\epsilon = m$, drops out when the approximate expression in (3.26) is used. This state becomes very important when the chemical potential of the bosons approaches the mass of the particles. In this case any new particles we add to the system occupy the minimal energy state and form a Bose condensate.

Taking the limit $\Delta\epsilon \to 0$ and considering a unit volume $(V = 1)$, we obtain the following expression for the *particle number density*:

$$n = \sum_\epsilon n_\epsilon \Delta g_\epsilon = \frac{g}{2\pi^2} \int_m^\infty \frac{\sqrt{(\epsilon^2 - m^2)}}{\exp((\epsilon - \mu)/T) \mp 1} \epsilon d\epsilon, \quad (3.27)$$

3.3 Rudiments of thermodynamics

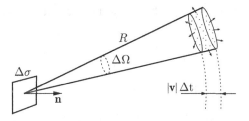

Fig. 3.3.

where the minus sign applies to bosons and the plus to fermions. The *energy density* is equal to

$$\varepsilon = \sum_\epsilon \epsilon n_\epsilon \Delta g_\epsilon = \frac{g}{2\pi^2} \int_m^\infty \frac{\sqrt{(\epsilon^2 - m^2)}}{\exp((\epsilon - \mu)/T) \mp 1} \epsilon^2 d\epsilon. \qquad (3.28)$$

Let us now calculate the pressure. To do so we consider a small area element $\Delta\sigma \mathbf{n}$, where $\mathbf{n}$ is the unit normal vector. All particles with velocity $|\mathbf{v}|$, striking this area element in the time interval between t and $t + \Delta t$, were located, at $t = 0$, in a spherical shell of radius $R = |\mathbf{v}| t$ with width $|\mathbf{v}| \Delta t$ (Figure 3.3). The total number of particles with energy $\epsilon(|\mathbf{v}|)$ within a solid angle $\Delta\Omega$ of this shell is equal to

$$\Delta N = n_\epsilon \Delta g_\epsilon R^2 |\mathbf{v}| \Delta t \Delta\Omega,$$

where Δg_ϵ is the number of states per unit spatial volume. Not all particles in the shell reach the target, only those with velocities directed to the area element. Taking into account the isotropy of the velocity distribution, we find the total number of particles striking the area element $\Delta\sigma \mathbf{n}$ with velocity $\mathbf{v}$ is

$$\Delta N_\sigma = \frac{(\mathbf{v} \cdot \mathbf{n})\Delta\sigma}{|\mathbf{v}| 4\pi R^2} \Delta N = \frac{(\mathbf{v} \cdot \mathbf{n})\Delta\sigma}{4\pi} n_\epsilon \Delta g_\epsilon \Delta t \Delta\Omega.$$

If these particles are reflected elastically, each transfers momentum $2(\mathbf{p} \cdot \mathbf{n})$ to the target. Therefore, the contribution of particles with velocity $|\mathbf{v}|$ to the pressure is

$$\Delta p = \int_\Omega \frac{2(\mathbf{p} \cdot \mathbf{n})\Delta N_\sigma}{\Delta\sigma \Delta t} = \frac{|\mathbf{p}|^2}{2\pi \epsilon} n_\epsilon \Delta g_\epsilon \int \cos^2\theta \sin\theta d\theta d\varphi = \frac{|\mathbf{p}|^2}{3\epsilon} n_\epsilon \Delta g_\epsilon,$$

where we have used the relation $|\mathbf{v}| = |\mathbf{p}|/\epsilon$ and integrated over the hemisphere. The total *pressure* is then

$$p = \sum_\epsilon \frac{|\mathbf{p}|^2}{3\epsilon} n_\epsilon \Delta g_\epsilon = \frac{\varepsilon}{3} - \frac{m^2 g}{6\pi^2} \int_m^\infty \frac{\sqrt{(\epsilon^2 - m^2)}}{\exp((\epsilon - \mu)/T) \mp 1} d\epsilon. \qquad (3.29)$$

Note that massless particles ($m = 0$) always have an ultra-relativistic equation of state,

$$p = \frac{\varepsilon}{3}, \qquad (3.30)$$

independent of their spin and chemical potential.

Problem 3.7 Substituting (3.20) into (3.16) and (3.22) into (3.21), verify that the *entropy density* is

$$s = \frac{\varepsilon + p - \mu n}{T}. \qquad (3.31)$$

(*Hint* Prove and then use the relation

$$\frac{p}{T} = \pm \sum \Delta g_\epsilon \ln(1 \pm n_\epsilon), \qquad (3.32)$$

where the plus and minus signs apply to bosons and fermions respectively. It follows that for $n_\epsilon \ll 1$ we have $p \simeq nT$.)

Verify the following useful relations

$$n = \frac{\partial p}{\partial \mu}, \qquad s = \frac{\partial p}{\partial T} \qquad (3.33)$$

The above integrals over energy cannot be calculated exactly when both the mass and chemical potential are different from zero. Therefore, we consider the limits of high and low temperature and expand the integral in terms of small parameters. At temperatures much larger than the mass the calculation of the leading terms can be performed by simply neglecting the mass. However, it is not so easy to derive the subleading corrections. The problem is that these corrections are nonanalytic in both the mass and the chemical potential. Because the corresponding results are not readily available in the literature, we provide below a derivation of the high-temperature expansion. The reader who is not interested in these mathematical details can skip the next subsection and go directly to the final formulae.

3.3.3 Calculating integrals

Changing the integration variable in (3.27), (3.28) and (3.29) from ϵ to $x = \epsilon/T$ and taking into account the fact that the chemical potentials of particles and antiparticles are equal in magnitudes and have opposite signs, the calculation of the basic thermodynamical quantities reduces to computing the integrals

$$J_\mp^{(\nu)}(\alpha, \beta) \equiv \int_\alpha^\infty \frac{(x^2 - \alpha^2)^{\nu/2}}{e^{x-\beta} \mp 1} dx + \int_\alpha^\infty \frac{(x^2 - \alpha^2)^{\nu/2}}{e^{x+\beta} \mp 1} dx, \qquad (3.34)$$

where
$$\alpha \equiv \frac{m}{T}, \qquad \beta \equiv \frac{\mu}{T}.$$

In particular, the total energy density of particles (p) and antiparticles ($\bar{p}$) is equal to
$$\varepsilon \equiv \varepsilon_p + \varepsilon_{\bar{p}} = \frac{gT^4}{2\pi^2}\left(J_\mp^{(3)} + \alpha^2 J_\mp^{(1)}\right), \tag{3.35}$$

and the total pressure is
$$p \equiv p_p + p_{\bar{p}} = \frac{gT^4}{6\pi^2} J_\mp^{(3)}. \tag{3.36}$$

Problem 3.8 Verify that the excess of particles over antiparticles is given by
$$n_p - n_{\bar{p}} = \frac{gT^3}{6\pi^2} \frac{\partial J_\mp^{(3)}}{\partial \beta}. \tag{3.37}$$

To find the expansions for the integrals $J_\mp^{(1)}$ and $J_\mp^{(3)}$ in the limits of high and low temperatures, we first calculate the auxiliary integral $J_\mp^{(-1)}$, which for $\beta < \alpha$ can be written as a convergent infinite series of the modified Bessel functions K_0:
$$J_\mp^{(-1)} = \sum_{n=1}^{\infty}(\pm 1)^{n+1} \int_\alpha^\infty \frac{(e^{n\beta} + e^{-n\beta})e^{-nx}}{\sqrt{x^2 - \alpha^2}} dx$$
$$= 2\sum_{n=1}^{\infty}(\pm 1)^{n+1} \cosh(n\beta) K_0(n\alpha). \tag{3.38}$$

Then, given the expansion for $J_\mp^{(-1)}(\alpha, \beta)$, the functions $J_\mp^{(\nu)}(\alpha, \beta)$ can be obtained by integrating the recurrence relation
$$\frac{\partial J_\mp^{(\nu)}}{\partial \alpha} = -\nu\alpha J_\mp^{(\nu-2)}, \tag{3.39}$$

which follows immediately from the definition of $J^{(\nu)}$ in (3.34). Note that this method works only for odd ν. The "initial conditions" for (3.39) can be determined by considering the limits $\alpha = 0$ or $\alpha \to \infty$, where the corresponding integrals can easily be calculated.

High temperature expansion At temperatures much larger than the mass of the particles, that is, for β and α much smaller than unity, every term in the series (3.38) contributes significantly. In this case we can use a known expansion for the sum of modified Bessel functions – formula (8.526) in I. Gradstein, I.Ryzhik, *Table* of *Integrals, Series, and Products* (San Diego: Academic Press, 1994). The result for purely imaginary β can be analytically continued to the real β and we obtain

$$J_{\mp}^{(-1)} = \begin{cases} \pi(\alpha^2 - \beta^2)^{-1/2} + (\ln(\alpha/4\pi) + \mathbf{C}) + O(\alpha^2, \beta^2), \\ -(\ln(\alpha/\pi) + \mathbf{C}) + O(\alpha^2, \beta^2), \end{cases} \quad (3.40)$$

for bosons and fermions respectively. Here $\mathbf{C} \approx 0.577$ is Euler's constant and $O(\alpha^2, \beta^2)$ denotes terms which are quadratic and higher order in α and β.

Problem 3.9 Verify that the next subleading correction to (3.40) is

$$\mp \frac{7\zeta(3)}{8\pi^2}(\alpha^2 + 2\beta^2),$$

where ζ is the Riemann zeta function.

To determine $J_{\mp}^{(1)}$ and $J_{\pm}^{(3)}$ from (3.39) and (3.40), we need the "initial conditions" $J_{\mp}^{(\nu)}(\alpha = 0, \beta)$. Setting $\alpha = 0$ and changing the integration variables, we can rewrite the expression in (3.34) as

$$J_{\mp}^{(\nu)}(0, \beta) = \int_0^\infty \frac{(y+\beta)^\nu + (y-\beta)^\nu}{e^y \mp 1} dy + \int_{-\beta}^0 \frac{(y+\beta)^\nu}{e^y \mp 1} dy - \int_0^\beta \frac{(y-\beta)^\nu}{e^y \mp 1} dy. \quad (3.41)$$

Replacing y by $-y$ in the last integral and noting that

$$\frac{1}{e^y \mp 1} + \frac{1}{e^{-y} \mp 1} = \mp 1,$$

we obtain for odd ν

$$J_{\mp}^{(\nu)}(0, \beta) = \int_0^\infty \frac{(y+\beta)^\nu + (y-\beta)^\nu}{e^y \mp 1} dy \mp \frac{\beta^{\nu+1}}{\nu+1}. \quad (3.42)$$

It follows that

$$J_{\mp}^{(1)}(0, \beta) = \begin{cases} \frac{1}{3}\pi^2 - \frac{1}{2}\beta^2, \\ \frac{1}{6}\pi^2 + \frac{1}{2}\beta^2. \end{cases} \quad (3.43)$$

Substituting (3.40) into (3.39) and taking into account (3.43), one finds

$$J_{\mp}^{(1)} = \begin{cases} \frac{1}{3}\pi^2 - \frac{1}{2}\beta^2 - \pi\sqrt{\alpha^2 - \beta^2} - \frac{1}{2}\alpha^2\left(\ln\left(\frac{\alpha}{4\pi}\right) + \mathbf{C} - \frac{1}{2}\right) + \alpha^2 O, \\ \frac{1}{6}\pi^2 + \frac{1}{2}\beta^2 + \frac{1}{2}\alpha^2\left(\ln\left(\frac{\alpha}{\pi}\right) + \mathbf{C} - \frac{1}{2}\right) + \alpha^2 O, \end{cases} \quad (3.44)$$

where $O \equiv O(\alpha^2, \beta^2)$. Similarly we obtain

$$
J_\mp^{(3)} = \begin{cases} \dfrac{2}{15}\pi^4 + \dfrac{1}{2}\pi^2(2\beta^2 - \alpha^2) + \pi(\alpha^2 - \beta^2)^{3/2} - \mathcal{A} + \alpha^4 O, \\ \dfrac{7}{60}\pi^4 + \dfrac{1}{4}\pi^2(2\beta^2 - \alpha^2) + \mathcal{A} - \dfrac{3}{4}(\ln 2)\alpha^4 + \alpha^4 O, \end{cases} \quad (3.45)
$$

where

$$
\mathcal{A} = \frac{1}{8}\left(2\beta^4 - 6\alpha^2\beta^2 - 3\alpha^4 \ln\left(\frac{e^C}{4\pi e^{3/4}}\alpha\right)\right).
$$

Low-temperature expansion In the limit of small temperatures we have $\alpha = m/T \gg 1$ and $K_0(n\alpha) \propto \exp(-n\alpha)$. Therefore, for $\alpha - \beta \gg 1$, all terms on the right hand side in (3.38) are negligible compared to the first term:

$$
J_\mp^{(-1)} \simeq 2K_0(\alpha)\cosh\beta. \quad (3.46)
$$

Integrating (3.39) and taking into account that $J_\mp^{(\nu)}$ must vanish as $\alpha \to \infty$, we obtain

$$
J_\mp^{(1)} \simeq 2\alpha K_1(\alpha)\cosh\beta = \sqrt{2\pi\alpha}\,e^{-\alpha}\cosh\beta\left[1 + \frac{3}{8\alpha} + O(\alpha^{-2})\right], \quad (3.47)
$$

and

$$
J_\mp^{(3)} \simeq 6\left(\alpha^2 K_0(\alpha) + 2\alpha K_1(\alpha)\right)\cosh\beta \simeq \sqrt{18\pi\alpha^3}\,e^{-\alpha}\cosh\beta\left(1 + \frac{15}{8\alpha}\right). \quad (3.48)
$$

These formulae allow us to calculate the basic thermodynamic properties of nonrelativistic particles when $\alpha - \beta \gg 1$. In such cases the exponential term dominates the denominator of the integrand in (3.34) and the difference between Fermi and Bose statistics becomes insignificant because the occupation numbers are much less than unity. We will see in the next section that this case is the situation most relevant for cosmological applications.

3.3.4 Ultra-relativistic particles

Bosons For bosons the maximal value of the chemical potential cannot exceed the mass, $\mu_b \leq m$. Assuming that both α and β are much smaller than unity and substituting (3.45) into (3.37), we find that at high temperatures the excess of particles over antiparticles to leading order is

$$
n_b - n_{\bar{b}} \simeq \frac{gT^3}{3}\frac{\mu_b}{T}. \quad (3.49)
$$

To estimate the number density of ultra-relativistic bosons we set $m = \mu_b = 0$ in (3.27) and then obtain

$$n_b \simeq \frac{\zeta(3)}{\pi^2} g T^3, \qquad (3.50)$$

where $\zeta(3) \approx 1.202$. From (3.49) and (3.50) one might be tempted to conclude that at high temperatures the excess of particles over antiparticles is always small compared to the number density of the particles themselves. This conclusion is wrong, however. The expression in (3.49) is applicable only if $\mu_b < m$. As $\mu_b \to m$, new particles added to the system fill the minimal energy state $\epsilon = m$, which is not taken into account in (3.49). These particles form a Bose condensate which can have an arbitrarily large particle excess.

Problem 3.10 Given a particle excess per unit volume Δn, find the temperature T_B below which the Bose condensate forms. Assume that $T_B \gg m$ and determine when this condition is actually satisfied. How much does a Bose condensate contribute to the total energy density, pressure and entropy?

If no Bose condensate is formed, the excess of bosons over antibosons is small compared to the number density. In this case the energy densities of particles and antiparticles are nearly equal and it follows from (3.35) and (3.45) that

$$\varepsilon_b \simeq \frac{\varepsilon_b + \varepsilon_{\bar{b}}}{2} \simeq \frac{\pi^2}{30} g T^4. \qquad (3.51)$$

The pressure and the entropy density are

$$p_b \simeq \frac{\varepsilon_b}{3}, \quad s_b \simeq \frac{4}{3}\frac{\varepsilon_b}{T} = \frac{2\pi^4}{45\zeta(3)} n_b, \qquad (3.52)$$

respectively. For massless bosons the chemical potential should be equal to zero. In this case (for example, for photons) all equations above are exact.

Fermions The chemical potential for fermions can be arbitrarily large and can exceed the mass. We first derive the *exact* formulae for an *arbitrary* μ_f in the limit of vanishing mass. Taking $\alpha \to 0$ in (3.44) and (3.45) and substituting the result into (3.35), we obtain

$$\varepsilon_f + \varepsilon_{\bar{f}} = \frac{7\pi^2}{120} g T^4 \left[1 + \frac{30\beta^2}{7\pi^2} + \frac{15\beta^4}{7\pi^4} \right], \qquad (3.53)$$

where $\beta = \mu_f/T$. The pressure is equal to one third of the energy density as expected for massless particles. It follows from (3.37) that the excess of fermions over antifermions is

$$n_f - n_{\bar{f}} = \frac{gT^3}{6}\beta\left[1 + \frac{\beta^2}{\pi^2}\right]. \tag{3.54}$$

Substituting the expressions above into (3.31) for the entropy density, we obtain

$$s_f + s_{\bar{f}} = \frac{7\pi^2}{90}gT^3\left[1 + \frac{15\beta^2}{7\pi^2}\right]. \tag{3.55}$$

If the chemical potential is much larger than the temperature, the main contribution to the total energy density comes from the degenerate fermions and is equal to $g\mu_f^4/8\pi^2$. These fermions fill the states with energies smaller than the Fermi energy $\varepsilon_F = \mu_f$, which determines the Fermi surface. The temperature correction to the energy, which to leading order is of order $gT^2\mu_f^2/4$, is due to the particles located in the shell of width T near this Fermi surface. One can see from (3.55) that the only states which contribute to the entropy are those near the Fermi surface. As the temperature approaches zero, the entropy vanishes. In this limit all fermions occupy definite states and information about the system is complete. Note that the antiparticles, for which $\mu_{\bar{f}} < 0$, disappear as the temperature vanishes.

If $\beta \ll 1$, then

$$n_f - n_{\bar{f}} \simeq \frac{gT^3}{6}\beta \tag{3.56}$$

and the excess of fermions over antifermions is small compared to the number density. In this case we can neglect the chemical potential in (3.27) and, to leading order, the number densities of fermions and antifermions are the same, namely,

$$n_f \simeq \frac{3\zeta(3)}{4\pi^2}gT^3. \tag{3.57}$$

The energy density, pressure and entropy density of the fermions are

$$\varepsilon_f \simeq \frac{7\pi^2}{240}gT^4, \quad p_f \simeq \frac{\varepsilon_f}{3}, \quad s_f \simeq \frac{4}{3}\frac{\varepsilon_f}{T}, \tag{3.58}$$

respectively. If the mass is small compared with the temperature but nonzero, there exist mass corrections, as can be inferred from the formulae derived in the previous subsection. They are nonanalytic in $\alpha \equiv m/T$ and if $\beta \equiv \mu/T \neq 0$, cross-terms simultaneously containing mass and chemical potential are also present.

Finally, we note the useful relation between the entropy of ultra-relativistic fermions with a small chemical potential and the entropy of ultra-relativistic bosons when the two types of particles have the same number of internal degrees of freedom:

$$s_f = \frac{7}{8}s_b. \tag{3.59}$$

3.3.5 Nonrelativistic particles

If the temperature is smaller than the rest mass and in addition

$$\frac{m-\mu}{T} \gg 1,$$

spin-statistics do not play an essential role and the formulae for bosons and fermions coincide to leading order. Substituting (3.48) into (3.37), we find that in this case

$$n - \bar{n} \simeq 2g \left(\frac{Tm}{2\pi}\right)^{3/2} \exp\left(-\frac{m}{T}\right) \sinh\left(\frac{\mu}{T}\right) \left[1 + \frac{15}{8}\frac{T}{m}\right]. \qquad (3.60)$$

It follows that the number density of particles is

$$n \simeq g \left(\frac{Tm}{2\pi}\right)^{3/2} \exp\left(-\frac{m-\mu}{T}\right) \left[1 + \frac{15}{8}\frac{T}{m}\right], \qquad (3.61)$$

and the number density of antiparticles, $\bar{n}$, is suppressed by a factor of $\exp(-2\mu/T)$ compared to n and if $\mu/T \gg 1$ the antiparticles can be neglected. In the early universe the number density of any type of nonrelativistic species never exceeds the number density of photons, that is, $n \ll n_\gamma \sim T^3$, and hence the inequality $(m-\mu)/T \gg 1$ is fulfilled. The energy density of particles is obtained by substituting (3.47) and (3.48) into (3.35), and can be expressed in terms of the particle number density as

$$\varepsilon \simeq mn + \frac{3}{2}nT. \qquad (3.62)$$

The pressure $p \simeq nT$ is much smaller than the energy density and can be neglected in the Einstein equations. The entropy density of the nonrelativistic particles can easily be calculated from (3.31) and is equal to

$$s \simeq \left(\frac{m-\mu}{T} + \frac{5}{2}\right) n. \qquad (3.63)$$

Problem 3.11 If $m/T \gg 1$ but $|m-\mu|/T \ll 1$, one cannot ignore spin-statistics. In this limit, however, the antiparticles are suppressed by a factor of $\exp(-2m/T)$ and hence can be neglected. Calculate the corresponding energy density, pressure and entropy for bosons and fermions in this case. Given a number density n, verify that at temperatures below $T_B = O(1) n^{2/3}/m$ a Bose condensate is formed.

The chemical potential of fermions can be arbitrarily large and may significantly exceed the mass. If $(\mu_f - m)/T \gg 1$, most fermions are degenerate. When $\mu_f \gg m$, fermions near the Fermi surface have momenta of order μ_f and are therefore relativistic, so we can use the results in (3.53)–(3.55). Otherwise, if $(\mu_f - m) \ll m$, the gas of degenerate fermions is nonrelativistic and the corresponding formulae are the standard ones found in any book on statistical physics. Having completed our brief review of relativistic statistical mechanics, we now apply the results derived to the early universe.

3.4 Lepton era

When the temperature in the universe drops below a hundred MeV (at $t > 10^{-4}$ s), the quarks and gluons are confined and form color-singlet bound states — baryons and mesons. We recall that baryons are made out of three quarks, each of which has baryon number 1/3, while the mesons are bound states of one quark and one antiquark, so that their resulting baryon number is zero.

The main ingredients of ordinary matter at temperatures below 100 MeV are primordial radiation (γ), neutrons (n), protons (p), electrons and positrons (e^-, e^+), and three neutrino species. Mesons, heavy baryons, μ- and τ-leptons are also present, but their number densities are very small and become increasingly negligible as the temperature decreases.

At energies of order a few MeV, the most important processes involve the weak interactions in which leptons, such as neutrinos, participate. Therefore, one calls this epoch the *lepton era*. At low energies, the baryon number and the lepton numbers are each conserved. The total electric charge is obviously also conserved. To enforce these conservation laws, a chemical potential is introduced for each particle species. The number of the independent potentials, however, is equal to the number of conserved quantities; any remaining potentials are expressed through these independent potentials using the chemical equilibrium conditions (3.24).

To demonstrate this, let us consider a medium containing the following ingredients: photons, leptons e, μ, τ, neutrinos ν_e, ν_μ, ν_τ, the lightest baryons p, n, Λ, and mesons π^0, $\pi^\pm$. The corresponding antiparticles are also present in the state of equilibrium. To enforce the conservation laws for electric charge, baryon number and the three different lepton numbers, we take as independent the following five chemical potentials: μ_{e^-}, μ_n, μ_{ν_e}, μ_{ν_μ}, μ_{ν_τ}. All other potentials will be written in terms of the members of this set. To start with,

$$\mu_{\pi^0} = 0 \qquad (3.64)$$

because, as a result of electromagnetic interaction, the π^0 meson quickly decays ($t_{\pi^0} \simeq 8.7 \times 10^{-17}$ s) into photons ($\pi^0 \to \gamma\gamma$) which have $\mu_\gamma = 0$. From $\Lambda \to n\pi^0$ ($t_\Lambda \simeq 2.6 \times 10^{-10}$ s), we find

$$\mu_\Lambda = \mu_n + \mu_{\pi^0} = \mu_n. \qquad (3.65)$$

The muon is unstable ($t_\mu \simeq 2.2 \times 10^{-6}$ s) and decays into an electron, an antineutrino and a neutrino, $\mu^- \to e^- \bar{\nu}_e \nu_\mu$, and hence

$$\mu_\mu = \mu_{e^-} - \mu_{\nu_e} + \mu_{\nu_\mu}. \qquad (3.66)$$

The τ-lepton also decays, for example into e^-, $\bar{\nu}_e$, ν_τ, therefore

$$\mu_\tau = \mu_{e^-} - \mu_{\nu_e} + \mu_{\nu_\tau}. \tag{3.67}$$

Finally, from the reactions $\pi^- \to \bar{\nu}_\mu \mu^-$ and $pe^- \rightleftarrows n\nu_e$, we deduce that

$$\mu_{\pi^-} = \mu_{e^-} - \mu_{\nu_e} \tag{3.68}$$

and

$$\mu_p = \mu_n + \mu_{\nu_e} - \mu_{e^-}. \tag{3.69}$$

All other possible reactions lead to relations that are consistent with the ones above. We recall that the chemical potentials for antiparticles are equal in magnitude to the chemical potentials for their corresponding particles but have opposite sign.

The five independent chemical potentials can be expressed through the five conserved quantities. The conservation of the total baryon number means that the number density of baryons minus antibaryons decreases in inverse proportion to the third power of the scale factor a. If matter is in equilibrium, the total entropy is conserved and, as the universe expands, the entropy density s also scales as a^{-3}. Therefore, the baryon-to-entropy ratio

$$B \equiv \frac{\Delta n_p + \Delta n_n + \Delta n_\Lambda}{s}, \tag{3.70}$$

remains constant. We denote here by $\Delta n \equiv n - \bar{n}$ the excess of the corresponding particles over their antiparticles. Similarly, the conservation law for total electric charge can be written as

$$Q \equiv \frac{\Delta n_p - \Delta n_e - \Delta n_\mu - \Delta n_\tau - \Delta n_{\pi^-}}{s} = \text{const}, \tag{3.71}$$

and for each type of lepton number we have

$$L_i \equiv \frac{\Delta n_i + \Delta n_{\nu_i}}{s} = \text{const}, \tag{3.72}$$

where $i \equiv e, \mu$ or τ. Because all Δn can be expressed through the temperature T and the corresponding chemical potentials, the system of equations (3.70), (3.71) and (3.72), together with the conservation law for the total entropy,

$$\frac{d(sa^3)}{dt} = 0, \tag{3.73}$$

allow us to determine the six unknown functions of time: $T(t), \mu_{e^-}(t), \mu_n(t), \mu_{\nu_e}(t), \mu_{\nu_\mu}(t), \mu_{\nu_\tau}(t)$.

What is known about the numerical values of B, Q and L_i? The universe appears to be electrically neutral and hence $Q = 0$. The baryon-to-entropy ratio is rather well established from observations and is of order $B \simeq 10^{-10}$–10^{-9}. This means that the entropy per one baryon or, equivalently, the number of photons $(n_\gamma \sim s \sim T^3)$ per baryon, is very large, $\sim 10^9$–10^{10}. The lepton-to-entropy ratios L_i are not so well established. The most severe limits on L_i are indirect. We will see in the next chapter that the total fermion number is not conserved at temperatures higher than 100 GeV and, as a result, the combination

$$B + a(L_e + L_\mu + L_\tau),$$

where $a \sim O(1)$, vanishes. Hence, if there are no special cancellations between the lepton numbers, their absolute values cannot significantly exceed the baryon number, that is, $|L_i| < 10^{-9}$. Limits from more direct observations are much weaker.

If the temperature is higher than the mass of a particular particle, the particle is relativistic and many particle–antiparticle pairs are created from the vacuum, so that the number density of pairs is of order the number density of photons, $n_\gamma \simeq T^3$. As the temperature drops below the mass, most of these pairs annihilate and finally only the particle excess survives. Let us determine when the numbers of particle–antiparticle pairs become negligible. The particle excess is characterized by a constant number

$$\beta = \frac{n - \bar{n}}{s}, \tag{3.74}$$

which can be either a baryon number, a lepton number or electric charge. Solving this together with the equation

$$\frac{n\bar{n}}{s^2} \sim \left(\frac{m}{T}\right)^3 \exp\left(-\frac{2m}{T}\right), \tag{3.75}$$

which follows from (3.61), we obtain

$$\frac{n}{s} \simeq \frac{\beta}{2} + \sqrt{\frac{\beta^2}{4} + \left(\frac{m}{T}\right)^3 \exp\left(-\frac{2m}{T}\right)};$$
$$\frac{\bar{n}}{s} \simeq -\frac{\beta}{2} + \sqrt{\frac{\beta^2}{4} + \left(\frac{m}{T}\right)^3 \exp\left(-\frac{2m}{T}\right)}. \tag{3.76}$$

It is clear that the number density of particle–antiparticle pairs becomes negligible compared to the particle excess when the second term under the square root becomes smaller than the first. For $\beta \ll 1$ this occurs at

$$\frac{m}{T} > \ln\left(\frac{2}{\beta}\right) + \frac{3}{2}\ln\left(\ln\left(\frac{2}{\beta}\right) + \cdots\right). \tag{3.77}$$

For example, if $\beta \simeq 10^{-9}$, the particle–antiparticle pairs can be neglected when the temperature drops by a factor of 25 below the mass. Thus, the number of baryon–antibaryon pairs becomes small compared to the baryon excess at temperatures below 40 MeV while positrons can be neglected at $T < 20$ keV.

At low temperatures, the conserved charge is mostly carried by the lightest particles possessing the given charge. For example, taking into account that $\mu_\Lambda = \mu_n$, from (3.60) we obtain

$$\frac{\Delta n_\Lambda}{\Delta n_n} \simeq \left(\frac{m_\Lambda}{m_n}\right)^{3/2} \exp\left(-\frac{m_\Lambda - m_n}{T}\right) \simeq \exp\left(-\frac{176 \text{ MeV}}{T}\right). \tag{3.78}$$

Thus, at $T < 176$ MeV, the contribution of Λ particles to the total baryon number can be discarded and the baryon asymmetry is due to the lightest baryons — protons and neutrons. Similarly, at temperatures below 100 MeV, the electric charge excess carried by leptons and mesons is mostly due to the overabundance of electrons, since μ- and τ-leptons and the lightest-charged mesons have relatively large masses, namely, $m_\mu \simeq 106$ MeV, $m_\tau \simeq 1.78$ GeV and $m_{\pi^\pm} \simeq 140$ MeV.

3.4.1 Chemical potentials

At temperatures higher than a few MeV, the weak and electromagnetic interactions are efficient and baryons, leptons and photons are in local *thermal* and *chemical equilibrium*. Note that in general, thermal and chemical equilibria are distinct. For example, while strong and electromagnetic interactions keep neutrons, protons and radiation at the same temperature, if the weak interaction rate is smaller than the expansion rate the chemical potentials of protons and neutrons do not need to satisfy a chemical equilibrium condition.

At temperatures below 100 MeV, we can neglect all heavy baryons and leptons. Let us estimate the chemical potentials of various matter components at these temperatures, beginning with neutrinos. Assuming that the lepton numbers L_i are much smaller than unity, we find from (3.72) and (3.54) that

$$\frac{\mu_{\nu_{\tau,\mu}}}{T} \sim L_{\tau,\mu}, \tag{3.79}$$

where the entropy density is estimated as $s \sim T^3$ and we have taken into account that the main contribution to $L_{\tau,\mu}$ comes from $\nu_{\tau,\mu}$ because the τ- and μ-leptons have large masses. The electrons are the lightest leptons which carry the electric charge needed to compensate the electric charge of the baryons. Therefore, their contribution to L_e is not negligible and the estimate, analogous to (3.79), applies to the *sum* $\mu_e + \mu_{\nu_e}$ rather than to the chemical potential of electron neutrinos alone. We see that the chemical potentials of relativistic particles decrease in proportion to the temperature as the universe expands.

3.4 Lepton era

As we found, at $T < 40$ MeV, antibaryons can be neglected and, therefore, $\Delta n_{p,n} \simeq n_{p,n}$. The conservation law of the total baryonic charge (3.70) implies that

$$B \simeq \frac{n_p}{s}\left(1 + \frac{n_n}{n_p}\right) \tag{3.80}$$

remains constant. The factor inside the parentheses is of order unity. Using formula (3.61) for n_p, we obtain

$$\frac{m_p - \mu_p(t)}{T(t)} \simeq \ln\left(\frac{1}{B}\left(\frac{m_p}{T}\right)^{3/2}\right). \tag{3.81}$$

For $B \simeq 10^{-10}$, the chemical potential μ_p changes from about -115 MeV to $+967$ MeV as the temperature drops from 40 MeV to 1 MeV. The number density of protons decays as $T^3 \propto a^{-3}$. Substituting (3.81) into (3.63), we obtain the following estimate for the contribution of protons to the total entropy,

$$\frac{s_p}{s} \simeq \left(\frac{m_p - \mu_p(t)}{T(t)} + \frac{5}{2}\right)\frac{n_p}{s} \simeq B \ln\left(\frac{1}{B}\left(\frac{m_p}{T}\right)^{3/2}\right). \tag{3.82}$$

Thus, *nonrelativistic* protons contribute only a small fraction of order 10^{-8} to the total entropy. It is interesting to note that the total entropy of protons themselves is not conserved. It follows from (3.82) that this entropy logarithmically increases as the temperature decreases. This has a simple physical explanation. If nonrelativistic protons were completely decoupled from the other components, their temperature would decrease faster than the temperature of the relativistic particles, namely, as $1/a^2$ instead of $1/a$. Therefore, to maintain thermal equilibrium with the dominant relativistic components, the protons borrow the energy and entropy needed from them.

Problem 3.12 Verify that the conservation of entropy and of the total number of nonrelativistic particles implies that their temperature decreases in inverse proportion to the second power of the scale factor. How does the chemical potential depend on the temperature in this case?

To estimate the chemical potential of electrons, μ_e, we use the conservation law for the electric charge (see (3.71)). Because the universe is electrically neutral, we have $Q = 0$. Taking into account that electrons are still relativistic at $T > 1$ MeV and skipping the negligible contributions from τ- and μ-leptons and π^- mesons in (3.71), we find

$$\frac{\mu_e}{T} \simeq \frac{\Delta n_e}{s} \simeq \frac{\Delta n_p}{s} \sim B \sim 10^{-10}. \tag{3.83}$$

94 *The hot universe*

This is not surprising because we need only a small excess of electrons to compensate the electric charge of the protons.

Finally, let us estimate the ratio of the number densities of neutrons and protons when they are still in chemical equilibrium with each other and with leptons. At the beginning of this section, we found that chemical equilibrium implies $\mu_p - \mu_n = \mu_{\nu_e} - \mu_e$. Using this relation together with (3.61), one immediately obtains

$$\frac{n_n}{n_p} = \exp\left(-\frac{m_n - m_p + \mu_{\nu_e} - \mu_e}{T}\right) \simeq \exp\left(-\frac{Q}{T}\right), \quad (3.84)$$

where $Q \equiv m_n - m_p \simeq 1.293$ MeV and we have neglected μ_{ν_e} and μ_e in the latter equality. The relation above will be used to set up the initial conditions for primordial nucleosynthesis.

3.4.2 Neutrino decoupling and electron–positron annihilation

At early times, the main contribution to the energy density comes from relativistic particles. Neglecting the chemical potentials, (3.51) and (3.53) imply that their total energy density is

$$\varepsilon_r = \kappa T^4, \quad (3.85)$$

where

$$\kappa = \frac{\pi^2}{30}\left(g_b + \frac{7}{8}g_f\right), \quad (3.86)$$

and g_b and g_f are the total numbers of internal degrees of freedom of all relativistic bosons and fermions respectively. Let us calculate κ in the universe when the only relativistic particles in equilibrium are photons, electrons, the three neutrino species and their corresponding antiparticles. Photons have two polarizations, and so $g_b = 2$. Electrons have two internal degrees of freedom, but each type of neutrino has only one because neutrinos are left-handed. The antiparticles double the total number of fermionic degrees of freedom and, therefore, $g_f = 10$. Thus, in this case, $\kappa \simeq 3.537$. Every extra bosonic or fermionic degree of freedom changes κ by $\Delta\kappa_b \simeq 0.329$ or $\Delta\kappa_f \simeq 0.288$ respectively.

Comparing (3.85) to (1.75), we find the relation between the temperature and the cosmological time in a flat, radiation-dominated universe:

$$t = \left(\frac{3}{32\pi\kappa}\right)^{1/2} T^{-2}. \quad (3.87)$$

3.4 Lepton era

Converting from Planckian units, we can rewrite this relation in the following useful form:

$$t_{\text{sec}} = t_{Pl} \left(\frac{3}{32\pi\kappa}\right)^{1/2} \left(\frac{T_{Pl}}{T}\right)^2 \simeq 1.39 \kappa^{-1/2} \frac{1}{T_{\text{MeV}}^2}, \qquad (3.88)$$

where the cosmological time and temperature are measured in seconds and MeV respectively.

When the temperature decreases to a few MeV, that is, about a second after the big bang, weak interactions become inefficient. These interactions are important in two respects. First, they keep neutrinos in thermal contact with each other and with the other particles, and second, they maintain the chemical equilibrium between protons and neutrons. The two events, namely, the thermal decoupling of neutrinos and the chemical decoupling of baryons, are somewhat separated in time. The first happens when the temperature is about 1.5 MeV, while the second occurs at $T \simeq 0.8$ MeV. The chemical decoupling of the baryons is essential for nucleosynthesis and it will be considered in detail in the next section. Here we concentrate on the thermal decoupling of neutrinos.

The main reactions responsible for the coupling of the electron neutrinos to the relativistic electron–positron plasma, and hence to radiation, are

$$e^+ + e^- \rightleftarrows \nu_e + \bar{\nu}_e, \quad e^\pm + \nu_e \to e^\pm + \nu_e, \quad e^\pm + \bar{\nu}_e \to e^\pm + \bar{\nu}_e. \qquad (3.89)$$

Some diagrams describing these interactions in electroweak theory (see next chapter) are shown in Figure 3.4. Both charged $W^\pm$-bosons and the neutral Z-boson contribute to these processes. At energies much smaller than the masses of the intermediate bosons the propagators of the Z- and W-bosons reduce to $1/M_{W,Z}^2$ and Fermi theory can be used to estimate the cross sections. For relativistic electrons

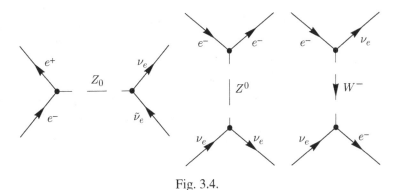

Fig. 3.4.

we have

$$\sigma_{e\nu} \simeq O(1) \frac{\alpha_w^2}{M_{W,Z}^4}(p_1 + p_2)^2, \tag{3.90}$$

where $\alpha_w \simeq 1/29$ is the weak fine structure constant and $p_{1,2}$ are the 4-momenta of the colliding particles. The neutrinos decouple from the electrons when the collision time,

$$t_\nu \simeq (\sigma_{e\nu} n_e)^{-1} \simeq O(1) \alpha_w^{-2} M_W^4 T^{-5}, \tag{3.91}$$

becomes of order the cosmological time t, which, in turn, is related to the temperature via (3.87). When deriving (3.91) we have assumed that the electrons are relativistic and hence $(p_1 + p_2)^2 \sim T^2$ and $n_e \sim T^3$. Comparing (3.91) to (3.87), one finds that the electron neutrinos ν_e decouple at temperature

$$T_{\nu_e} \simeq O(1) \alpha_w^{-2/3} M_W^{4/3}. \tag{3.92}$$

The exact calculation shows that the numerical coefficient in this formula is not much different from unity and hence $T_{\nu_e} \simeq 1.5$ MeV.

At temperatures of order MeV, the number densities of μ- and τ-leptons are negligibly small and the only reactions enforcing thermal contact between μ- and τ-neutrinos and the rest of matter are the elastic scatterings of $\nu_{\mu,\tau}$ on electrons ($e\nu_{\mu,\tau} \to e\nu_{\mu,\tau}$); these are entirely due to Z-boson exchange. As a consequence, the cross-sections for these reactions are smaller than the total cross-section of the $e\nu_e$ interactions and the μ- and τ-neutrinos decouple earlier than the electron neutrinos.

The most important conclusion from the above consideration is that all three neutrino species thermally decouple before the electron–positron pairs begin to annihilate at $T \sim m_e \simeq 0.5$ MeV. After decoupling, the neutrinos propagate without further scatterings, preserving the Planckian spectrum. Their temperature decreases in inverse proportion to the scale factor and is not influenced by the subsequent $e^\pm$ annihilation. The energy released in the electron–positron annihilation is thermalized and as a result the radiation is "heated." Therefore, the temperature of radiation must be larger than the neutrino temperature. Let us calculate the radiation-to-neutrino temperature ratio. After decoupling the neutrino entropy is conserved separately. The total entropy of the other components, which is dominated by radiation and the electron–positron plasma, is also conserved. Hence the ratio

$$\frac{s_\gamma + s_{e^\pm}}{s_\nu}$$

remains constant. Taking into account that $s_\gamma \propto T_\gamma^3$ and $s_\nu \propto T_\nu^3$, we have

$$\left(\frac{T_\gamma}{T_\nu}\right)^3 \left(1 + \frac{s_{e^\pm}}{s_\gamma}\right) = C, \tag{3.93}$$

where C is a constant. Just after neutrino decoupling, but before $e^\pm$ annihilation, $T_\gamma = T_\nu$ and $s_{e^\pm}/s_\gamma = 7/4$ (see (3.59)). Therefore, $C = 11/4$ and

$$\left(\frac{T_\gamma}{T_\nu}\right) = \left(\frac{11}{4}\right)^{1/3} \left(1 + \frac{s_{e^\pm}}{s_\gamma}\right)^{-1/3}. \tag{3.94}$$

When the electron–positron pairs begin to annihilate at $T \simeq 0.5$ MeV the ratio of entropies $s_{e^\pm}/s_\gamma$ decreases and finally becomes completely negligible (see (3.82), where one has to substitute m_e instead of m_p). Hence, after electron–positron annihilation we have

$$\left(\frac{T_\gamma}{T_\nu}\right) = \left(\frac{11}{4}\right)^{1/3} = 1.401. \tag{3.95}$$

Thus, the massless primordial neutrinos should have a temperature today of $T_\nu \simeq 2.73\,\text{K}/1.4 \simeq 1.95\,\text{K}$. Unfortunately it is not easy, if even possible, to detect the primordial neutrino background and verify this very robust prediction of the standard cosmological model.

Problem 3.13 Assuming that neutrinos have a small, but nonvanishing mass, estimate their temperature today.

Problem 3.14 Calculate the contribution of neutrinos to the energy density after $e^\pm$ annihilation and determine at which redshift z the total energy density of radiation and relativistic neutrinos is exactly equal to the energy density of cold (nonrelativistic) matter.

3.5 Nucleosynthesis

The most widespread chemical element in the universe is hydrogen, constituting nearly 75% of all baryonic matter. Helium-4 constitutes about 25%. The other light elements and metals have only very small abundances.

Simple arguments lead to the conclusion that the large amount of ^{4}He could not have been produced in stars. The binding energy of ^{4}He is 28.3 MeV, and therefore, when one nucleus of ^{4}He is formed, the energy released per one baryon is about 7.1 MeV $\simeq 1.1 \times 10^{-5}$ erg. Assuming that one quarter of all baryons has been fused into ^{4}He in stars during the last 10 billion years (3.2×10^{17}s), we obtain the following estimate for the luminosity-to-mass ratio:

$$\frac{L}{M_{bar}} \simeq \frac{1}{4} \frac{1.1 \times 10^{-5} \text{ erg}}{(1.7 \times 10^{-24} \text{ gm}) \times (3.2 \times 10^{17} \text{ s})} \simeq 5 \frac{\text{erg}}{\text{gm s}} \simeq 2.5 \frac{L_\odot}{M_\odot},$$

where $M_\odot$ and $L_\odot$ are the solar mass and luminosity respectively. However, the observed $L/M_{bar} \leq 0.05 L_\odot/M_\odot$, and therefore, if the luminosity of baryonic matter in the past was not much larger than at present, less than 0.5% of ^{4}He can be fused in stars.

The only plausible explanation of the helium abundance is that it was produced in the very early hot universe when the fusion energy constituted only a small fraction of the total energy. The energy released was then thermalized and redshifted long before the universe became transparent. It is obvious that a substantial amount of helium cannot be formed before the temperature drops below the binding energy ~28 MeV. Indeed, primordial nucleosynthesis took place at temperature roughly 0.1 MeV, that is, a few minutes after the big bang. The amount of helium produced depends on the availability of neutrons at this time, which, in turn, is determined by the weak interactions maintaining the chemical equilibrium between neutrons and protons. These weak interactions become inefficient when the temperature drops below a few MeV and, as a consequence, the neutron-to-proton ratio "freezes out." Thus, the processes responsible for the chemical abundances of primordial elements began seconds after the big bang and continued for the next several minutes.

In this section, we use analytical methods to calculate the abundances of the light primordial elements. Although more precise results are obtained with computer codes, the quasi-equilibrium approximation used here reproduces the numerical results with surprisingly good accuracy. In addition, the analytical methods allow us to understand why and how the primordial abundances depend on cosmological parameters.

3.5.1 Freeze-out of neutrons

We begin with the calculation of the neutron freeze-out concentration. The main processes responsible for the chemical equilibrium between protons and neutrons are the weak interaction reactions:

$$n + \nu \rightleftarrows p + e^-, \qquad n + e^+ \rightleftarrows p + \bar{\nu}. \qquad (3.96)$$

Here ν always refers to the electron neutrino. To calculate the reaction rates, we can use Fermi theory according to which the cross-sections can be expressed in terms of the matrix element for the four-fermion interaction represented in Figure 3.5:

$$|\mathcal{M}|^2 = 16(1 + 3g_A^2) G_F^2 (p_n \cdot p_\nu)(p_p \cdot p_e), \qquad (3.97)$$

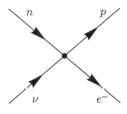

Fig. 3.5.

where
$$G_F = \frac{\pi \alpha_w}{\sqrt{2} M_W^2} \simeq 1.17 \times 10^{-5} \text{ GeV}^{-2}$$

is the Fermi coupling constant and $(p_i \cdot p_j)$ are the scalar products of the 4-momenta entering the vertex. The factor $g_A \simeq 1.26$ corrects the axial vector "weak charge" of the nucleon by accounting for the possibility that gluons inside the nucleon split into quark–antiquark pairs, thus contributing to the weak coupling. Note that the Fermi constant can be determined to very high accuracy by measuring the lifetime of the muon, while g_A can be measured only in interactions involving nucleons.

For the process $a + b \to c + d$, the differential cross-section is

$$\frac{d\sigma_{ab}}{d\Omega} = \frac{1}{(8\pi)^2} \frac{|\mathcal{M}|^2}{(p_a + p_b)^2} \left(\frac{(p_c \cdot p_d)^2 - m_c^2 m_d^2}{(p_a \cdot p_b)^2 - m_a^2 m_b^2} \right)^{1/2}. \tag{3.98}$$

This expression is manifestly Lorentz invariant and can be used in any coordinate system. The 4-momenta of the outgoing particles c and d are related to the 4-momenta of the colliding particles a and b by the conservation law: $p_c + p_d = p_a + p_b$.

Let us now consider the particular reaction

$$n + \nu \to p + e^-.$$

At temperatures of order a few MeV the nucleons are nonrelativistic and we have

$$(p_n + p_\nu)^2 \simeq m_n^2, \quad (p_n \cdot p_\nu) = m_n \epsilon_\nu,$$
$$\sqrt{(p_p \cdot p_e)^2 - m_p^2 m_e^2} \simeq m_p \epsilon_e \sqrt{1 - (m_e/\epsilon_e)^2} = m_p \epsilon_e v_e, \tag{3.99}$$

where ϵ_ν is the energy of the incoming neutrino and $\epsilon_e \simeq \epsilon_\nu + Q$ is the energy of the outgoing electron. The energy $Q \simeq 1.293$ MeV, introduced in (3.84), is released when the neutron is converted into the proton. Expression (3.98) is valid only in empty space. At temperatures above 0.5 MeV there are many electron–positron pairs and the allowed final states for the electron are partially occupied. As a result,

100   The hot universe

the cross-section is reduced by the factor

$$1 - n_{\epsilon_e} = \left[1 + \exp(-\epsilon_e/T)\right]^{-1}$$

to account for the Pauli exclusion principle. Given this factor, the substitution of (3.99) and (3.97) into (3.98) gives

$$\sigma_{n\nu} \simeq \frac{1 + 3g_A^2}{\pi} G_F^2 \epsilon_e^2 v_e \left[1 + \exp(-\epsilon_e/T)\right]^{-1}. \tag{3.100}$$

Because the number density of the nucleons is negligible compared to the number density of the light particles, the spectra of neutrinos and electrons are not significantly influenced by the above reactions and always remain thermal. Hence, the $n\nu$ interactions occurring within a time interval Δt in a given comoving volume containing N_n neutrons reduce the total number of neutrons by

$$\Delta N_n = -\left(\sum_{\epsilon_\nu} \sigma_{n\nu} n_{\epsilon_\nu} v_\nu \Delta g_{\epsilon_\nu}\right) N_n \Delta t, \tag{3.101}$$

where

$$n_{\epsilon_\nu} = \left[1 + \exp(\epsilon_\nu/T_\nu)\right]^{-1}$$

is the neutrino occupation number and Δg_{ϵ_ν} is the phase volume element (see (3.26), where $V = g = 1$). The velocity of neutrinos v_ν is equal to the speed of light: $v_\nu = 1$.

It is useful to introduce the relative concentration of neutrons

$$X_n = \frac{N_n}{N_n + N_p} = \frac{n_n}{n_n + n_p}. \tag{3.102}$$

Taking into account that the total number of baryons, $N_n + N_p$, is conserved, and substituting (3.100) into (3.101), we find that the rate of change of X_n due to the $n\nu$ reaction is

$$\left(\frac{dX_n}{dt}\right)_{n\nu} = -\lambda_{n\nu} X_n = -\frac{1 + 3g_A^2}{2\pi^3} G_F^2 Q^5 J(1;\infty) X_n, \tag{3.103}$$

where

$$J(a;b) \equiv \int_a^b \sqrt{1 - \frac{(m_e/Q)^2}{q^2}} \frac{q^2 (q-1)^2 \, dq}{\left(1 + e^{\frac{Q}{T_\nu}(q-1)}\right)\left(1 + e^{-\frac{Q}{T}q}\right)}, \tag{3.104}$$

and the integration variable is

$$q \equiv (\epsilon_\nu/Q) + 1 = \epsilon_e/Q.$$

Before electron–positron annihilation the temperatures of the electrons and neutrinos are equal, that is, $T = T_\nu$. To estimate the integral in (3.104) we note that $(m_e/Q)^2 \simeq 0.15$ and expand the square root in the integrand, keeping only the first two terms. Furthermore, ignoring the Pauli exclusion principle for the electrons or, equivalently, neglecting the second term in the denominator, we can calculate the resulting integrals and obtain

$$J(1;\infty) \simeq \frac{45\zeta(5)}{2}\left(\frac{T_\nu}{Q}\right)^5 + \frac{7\pi^4}{60}\left(\frac{T_\nu}{Q}\right)^4 + \frac{3\zeta(3)}{2}\left(1 - \frac{1}{2}\frac{m_e^2}{Q^2}\right)\left(\frac{T_\nu}{Q}\right)^3. \quad (3.105)$$

It is quite remarkable that this approximate expression reproduces the exact result with very good accuracy at all relevant temperatures. For example, for $T_\nu/Q > 1$, the error is about 2%, improving to 1% or better for $T_\nu/Q < 1$. Substituting (3.105) together with the values of G_F and Q into (3.103), and converting from Planckian to physical units, we find

$$\lambda_{n\nu} \simeq 1.63\left(\frac{T_\nu}{Q}\right)^3\left(\frac{T_\nu}{Q} + 0.25\right)^2 \text{ s}^{-1}. \quad (3.106)$$

Further simplifications made to obtain this last expression do not spoil the accuracy; at the temperatures relevant for freeze-out, $T_\nu \geq 0.5$ MeV, the error remains less than 2%.

Problem 3.15 Verify that the reaction rate for $ne^+ \to p\bar{\nu}$ is equal to

$$\lambda_{ne} = \frac{1 + 3g_A^2}{2\pi^3}G_F^2 Q^5 J\left(-\infty; -\frac{m_e}{Q}\right), \quad (3.107)$$

where J is the integral defined in (3.104). Check that if $T_\nu = T$ and $T > m_e$, then $\lambda_{ne} \simeq \lambda_{n\nu}$. Consider the inverse reactions $pe^- \to n\nu$ and $p\bar{\nu} \to ne^+$. Show that for $T_\nu = T$ their rates can be expressed through the rates of the direct reactions:

$$\lambda_{pe} = \exp(-Q/T)\lambda_{n\nu}, \quad \lambda_{p\bar{\nu}} = \exp(-Q/T)\lambda_{ne}. \quad (3.108)$$

Freeze-out The inverse reactions increase the neutron concentration at a rate $\lambda_{p \to n} X_p$. The balance equation for X_n is therefore

$$\frac{dX_n}{dt} = -\lambda_{n \to p} X_n + \lambda_{p \to n} X_p = -\lambda_{n \to p}\left(1 + e^{-\frac{Q}{T}}\right)(X_n - X_n^{\text{eq}}), \quad (3.109)$$

where $\lambda_{n \to p} \equiv \lambda_{ne} + \lambda_{n\nu}$ and $\lambda_{p \to n} \equiv \lambda_{pe} + \lambda_{p\bar{\nu}}$ are the total rates of the direct and inverse reactions respectively, and

$$X_n^{\text{eq}} = \frac{1}{1 + \exp(Q/T)} \quad (3.110)$$

is the equilibrium concentration of neutrons. To obtain the second equality in (3.109) we used the relations in (3.108), assuming $T_\nu = T$, as well as the fact that the proton concentration is $X_p = 1 - X_n$.

The exact solution of the linear differential equation (3.109), with the initial condition $X_n \to X_n^{\text{eq}}$ as $t \to 0$, is

$$X_n(t) = X_n^{\text{eq}}(t) - \int_0^t \exp\left(-\int_{\tilde{t}}^t \lambda_{n \to p}(\bar{t})\left(1 + e^{-\frac{Q}{\bar{t}}}\right) d\bar{t}\right) \dot{X}_n^{\text{eq}}(\tilde{t}) \, d\tilde{t}, \quad (3.111)$$

where the dot denotes the derivative with respect to time.

The second term on the right hand side in (3.111) characterizes the deviation from equilibrium and is negligible compared to the first term at small t. Integrating by parts, we can rewrite the solution (3.111) as an asymptotic series in increasing powers of the derivatives of X_n^{eq}:

$$X_n = X_n^{\text{eq}}\left(1 - \frac{1}{\lambda_{n \to p}(1 + \exp(-Q/T))} \frac{\dot{X}_n^{\text{eq}}}{X_n^{\text{eq}}} + \cdots\right). \quad (3.112)$$

If the reaction rate is much larger than the inverse cosmological time, that is, $\lambda_{n \to p} \gg t^{-1} \sim -\dot{X}_n^{\text{eq}}/X_n^{\text{eq}}$, then we have $X_n \approx X_n^{\text{eq}}$ in agreement with result (3.84). Subsequently, after the temperature has dropped significantly, $X_n^{\text{eq}} \to 0$, but the second term on the right hand side in (3.111) approaches a finite limit. Instead of vanishing, therefore, the neutron concentration freezes out at some finite value $X_n^* = X_n(t \to \infty)$. The freeze-out effectively occurs when the second term on the right hand side in (3.112) is of order the first one or, in other words, when the deviation from equilibrium becomes significant. This happens before $e^{\pm}$ annihilation and after the temperature drops below $Q \simeq 1.29$ MeV (as can be checked *a posteriori*). Consequently we can set $\lambda_{n \to p} \simeq 2\lambda_{n\nu}$ and neglect $\exp(-Q/T)$ in the equality $-\dot{X}_n^{\text{eq}}/X_n^{\text{eq}} \simeq \lambda_{n \to p}$, which determines the freeze-out temperature. Substituting into this equality expression (3.110) for X_n^{eq} and expression (3.106) for $\lambda_{n\nu}$ and using the temperature–time relation (3.88), the equation for the freeze-out temperature reduces to

$$\left(\frac{T_*}{Q}\right)^2 \left(\frac{T_*}{Q} + 0.25\right)^2 \simeq 0.18 \kappa^{1/2}. \quad (3.113)$$

In the case of three neutrino species we have $\kappa \simeq 3.54$ and the freeze-out temperature is $T_* \simeq 0.84$ MeV. The equilibrium neutron concentration at this time is $X_n^{\text{eq}}(T_*) \simeq 0.18$. Of course, this number gives only a rough estimate for the expected freeze-out concentration. One should not forget that at $T = T_*$ deviations from equilibrium are very significant and, in fact, $X_n(T_*)$ exceeds the equilibrium concentration by at least a factor of 2. Nevertheless, the above estimate enables us to see how the freeze-out concentration depends on the number of relativistic species present at the freeze-out time. Because $T_* \propto \kappa^{1/8}$, additional relativistic components increase T_* and, hence, more neutrons survive. Subsequently, nearly all neutrons fuse with protons to form ^{4}He and we anticipate, therefore, that additional relativistic species increase the primordial helium abundance. For example, in the extreme case of a very large number of unknown light particles, the temperature T_* would exceed Q and the neutron concentration at freeze-out would be almost 50%. This would lead to an unacceptably large abundance of ^{4}He. Thus, we see that primordial nucleosynthesis can help us to restrict the number of light species.

Problem 3.16 Find the freeze-out temperature using the simple criterion $t \simeq 1/\lambda$ and verify that in this approximation one obtains the result quoted in many books on cosmology, namely, $T_* \propto \kappa^{1/6}$. What accounts for the difference between this and the above result, $T_* \propto \kappa^{1/8}$?

Now we turn to a more accurate estimate for the freeze-out concentration. Since $X_n^{\text{eq}} \to 0$ as $T \to 0$, X_n^* is given by the integral term in (3.111) when we take the limit $t \to \infty$. The main contribution to the integral comes from $T > m_e$. Therefore we set $\lambda_{n \to p} \simeq 2\lambda_{nv}$, where λ_{nv} is given in (3.106). Using (3.88) to replace the integration variable t by $y = T/Q$, we obtain

$$X_n^* = \int_0^\infty \frac{\exp\left(-5.42\kappa^{-1/2} \int_0^y (x + 0.25)^2 (1 + e^{-1/x})\, dx\right)}{2y^2(1 + \cosh(1/y))} dy. \quad (3.114)$$

For the case of three neutrino species ($\kappa \simeq 3.54$) one finds $X_n^* \simeq 0.158$. This result is in very good agreement with more elaborate numerical calculations. The presence of an additional light neutrino, accompanied by the corresponding antineutrino, increases κ by amount $2 \cdot \Delta \kappa_f \simeq 0.58$, and the freeze-out concentration becomes $X_n^* \simeq 0.163$. Thus, two additional fermionic degrees of freedom increase X_n^* by about 0.5% and we conclude that

$$X_n^* \simeq 0.158 + 0.005(N_\nu - 3), \quad (3.115)$$

where N_ν is the number of light neutrino species.

Neutron decay Until now we have neglected neutron decay,

$$n \to p + e^- + \bar{\nu}. \quad (3.116)$$

This was justified because the lifetime of a free neutron $\tau_n \approx 886$ s is large compared to the freeze-out time, $t_* \sim O(1)$ s. However, after freeze-out the interactions (3.96) and the inverse three-body reaction (3.116) become inefficient and neutron decay is the sole remaining cause for a change in the number of neutrons. As a result, the neutron concentration decreases for $t > t_*$ as

$$X_n(t) = X_n^* \exp(-t/\tau_n). \tag{3.117}$$

Note that after freeze-out one can neglect the degeneracy of the leptons, which would increase the neutron lifetime, and use τ_n, as quoted above. We will see that nucleosynthesis, in which nearly all free neutrons are captured in the nuclei (where they become stable), occurs at $t \sim 250$ s. This is a rather substantial fraction of the neutron lifetime and hence the neutron decay significantly influences the final abundances of the light elements.

3.5.2 "Deuterium bottleneck"

Complex nuclei are formed as a result of nuclear interactions. Helium-4 could, in principle, be built directly in the four-body collision: $p + p + n + n \to {}^4\mathrm{He}$. However, the low number densities during the period in question strongly suppress these processes. Therefore, the light complex nuclei can be produced only through a sequence of two-body reactions. The first step is deuterium (D) production through the reaction

$$p + n \rightleftarrows \mathrm{D} + \gamma. \tag{3.118}$$

There is no problem with this step because for $t < 10^3$ s the corresponding reaction rate is much larger than the expansion rate.

Let us calculate the deuterium equilibrium abundance. We define the abundance by weight:

$$X_\mathrm{D} \equiv 2 n_\mathrm{D}/n_N,$$

where n_N is the total number of nucleons (baryons) including those in complex nuclei. The relation between X_D and the abundances of the free neutrons, $X_n \equiv n_n/n_N$, and protons, $X_p \equiv n_p/n_N$, can be found using (3.61) for each component. Because the deuterium nucleus with spin zero is metastable, its total statistical weight is $g_\mathrm{D} = 3$. Taking into account that $g_p = g_n = 2$ and the chemical potentials satisfy the condition $\mu_\mathrm{D} = \mu_p + \mu_n$, we find

$$X_\mathrm{D} = 5.67 \times 10^{-14} \eta_{10} T_{\mathrm{MeV}}^{3/2} \exp\left(\frac{B_\mathrm{D}}{T_{\mathrm{MeV}}}\right) X_p X_n, \tag{3.119}$$

where

$$B_\mathrm{D} \equiv m_p + m_n - m_\mathrm{D} \simeq 2.23 \text{ MeV}$$

is the binding energy of the deuterium. We have parameterized the baryon-to-photon ratio by

$$\eta_{10} \equiv 10^{10} \times \frac{n_N}{n_\gamma}. \qquad (3.120)$$

This parameter is related to Ω_b, the baryon contribution to the current critical density, via

$$\Omega_b h_{75}^2 \simeq 6.53 \times 10^{-3} \eta_{10}. \qquad (3.121)$$

At temperatures of order B_D, the abundance X_D is still extremely small and even at $T \sim 0.5$ MeV, for example, it is only about 2×10^{-13}. One of the reasons for this is the large number of energetic photons with $\epsilon > B_D$, which destroy the deuterium. The number of such photons per deuterium nucleus is

$$\frac{n_\gamma(\epsilon > B_D)}{n_D} \sim \frac{B_D^2 T e^{-B_D/T}}{n_N X_D} \sim 10^{10} \frac{1}{\eta_{10} X_D} \left(\frac{B_D}{T}\right)^2 e^{-B_D/T}, \qquad (3.122)$$

which becomes less than unity only at $T < 0.06$ MeV. Therefore, we expect that deuterium can constitute a significant fraction of baryonic matter only if the temperature is about 0.06 MeV. In fact, according to (3.119), for $\eta_{10} \sim O(1)$ the equilibrium deuterium abundance changes abruptly from 10^{-5} to of order unity as the temperature drops from 0.09 MeV to 0.06 MeV.

The rates of reactions converting deuterium into heavier elements are proportional to the deuterium concentration and these reactions are strongly suppressed until X_D has grown to a substantial value. This delays the formation of the other light elements, including ^{4}He. In fact, because of the large binding energy of ^{4}He (28.3 MeV), the *equilibrium* helium abundance would already be of order unity at temperature 0.3 MeV. However, this does not happen and the helium abundance is still negligible at $T \simeq 0.3$ MeV because the rate of the deuterium reactions, responsible for maintaining helium in chemical equilibrium with the nucleons, is much smaller than the expansion rate at this time. As a result, the heavier elements are chemically decoupled and present in completely negligible amounts despite their large binding energies. Only protons, neutrons and deuterium are in chemical equilibrium with each other. This situation is usually referred to as the "deuterium bottleneck."

Problem 3.17 Derive the formula for the equilibrium concentration of ^{4}He and verify that it is of order unity at $T \sim 0.3$ MeV.

Let us determine when the deuterium bottleneck opens up. This occurs when the main reactions converting deuterium into heavier elements,

$$(1)\ D + D \to {}^3He + n, \quad (2)\ D + D \to T + p, \qquad (3.123)$$

become efficient. Within the relevant temperature interval, 0.06 MeV to 0.09 MeV, the experimentally measured rates of these reactions are

$$\langle \sigma v \rangle_{DD1} = (1.3\text{--}2.2) \times 10^{-17} \text{ cm}^3 \text{ s}^{-1},$$
$$\langle \sigma v \rangle_{DD2} = (1.2\text{--}2) \times 10^{-17} \text{ cm}^3 \text{ s}^{-1}, \quad (3.124)$$

respectively. Due to reactions (3.123), the number of deuterium nuclei in a comoving volume containing N_D nuclei decreases during a time interval Δt by

$$\Delta N_D = -\langle \sigma v \rangle_{DD} n_D N_D \Delta t. \quad (3.125)$$

Rewriting this equation in terms of the concentration by weight, $X_D = 2N_D/N_N$, we obtain

$$\Delta X_D = -\tfrac{1}{2} \lambda_{DD} X_D^2 \Delta t, \quad (3.126)$$

where

$$\lambda_{DD} = (\langle \sigma v \rangle_{DD1} + \langle \sigma v \rangle_{DD2}) n_N \simeq 1.3 \times 10^5 K(T) T_{MeV}^3 \eta_{10} \text{ s}^{-1}. \quad (3.127)$$

The function $K(T)$ characterizes the temperature dependence of the reaction rate and it changes from 1 to 0.6 as the temperature drops from 0.09 MeV to 0.06 MeV. A substantial amount of the available deuterium is converted into helium-3 and tritium within a cosmological time t only if

$$|\Delta X_D| \simeq \left(\tfrac{1}{2}\right) \lambda_{DD} X_D^2 t \simeq X_D. \quad (3.128)$$

It follows that the deuterium bottleneck opens up when

$$X_D^{(bn)} \simeq \frac{2}{\lambda_{DD} t} \simeq \frac{1.2 \times 10^{-5}}{\eta_{10} T_{MeV}\left(X_D^{(bn)}\right)}, \quad (3.129)$$

where we have used the time–temperature relation (3.88) with $\kappa \simeq 1.11$. From (3.119), one can express the temperature as a function of X_D:

$$T_{MeV}(X_D) \simeq \frac{0.061}{\left(1 + 2.7 \times 10^{-2} \ln(X_D/\eta_{10})\right)}. \quad (3.130)$$

Substituting this expression into (3.129) and solving the resulting equation for $X_D^{(bn)}$ by the method of iteration for $10 > \eta_{10} > 10^{-1}$, we find

$$X_D^{(bn)} \simeq 1.5 \times 10^{-4} \eta_{10}^{-1} \left(1 - 7 \times 10^{-2} \ln \eta_{10}\right). \quad (3.131)$$

3.5 Nucleosynthesis

Problem 3.18 Verify that after electron–positron annihilation, the value of κ in (3.85) becomes

$$\kappa \simeq 1.11 + 0.15(N_\nu - 3), \qquad (3.132)$$

where N_ν is the number of neutrino species. (*Hint* Recall that the neutrino and radiation temperatures are different after $e^\pm$ annihilation.)

After the deuterium concentration reaches $X_D^{(bn)}$, everything proceeds very quickly. According to (3.130) the equilibrium concentration X_D increases from 10^{-4} to 10^{-2} as the temperature drops from 0.08 MeV to 0.07 MeV. As a result, the rate of deuterium conversion into heavier elements, proportional to X_D, becomes 100 times larger than the expansion rate. Such a system is far from equilibrium and nucleosynthesis is described by a complicated system of kinetic equations which are usually solved numerically. In Figures 3.7 and 3.8 below we present the results of highly precise numerical calculations for the time evolution of the light element concentrations and for their final abundances, respectively. We will now show how these results can be reproduced analytically with good accuracy. The system of kinetic equations will be solved using the quasi-equilibrium approximation. This will provide us with a solid physical understanding of primordial nucleosynthesis and will reveal the reasons for the dependence of final abundances on the cosmological parameters. To simplify our task we consider only the most abundant isotopes up to ^{7}Be, among which are ^{4}He, D, ^{3}He, T, lithium-7 (^{7}Li) and beryllium (^{7}Be) itself. Other elements such as ^{6}Li, ^{8}B etc. are produced in much smaller amounts and will be ignored.

The most important nuclear reactions are shown schematically in Figure 3.6. The reader is encouraged to keep a copy of this figure at hand throughout the rest of this section. Every element corresponds to a "reservoir." The reservoirs are connected by "one-way pipes", one for each nuclear reaction converting an element into another. To simplify the diagram, we include only the initial elements involved in the reaction; the outcome can easily be inferred from the diagram. The efficiency of the pipe is determined by the reaction rate. For example, for the rate of escape from the reservoir A due to the reaction AB $\to$ CD, we find

$$\dot{X}_A = -A_B^{-1} \lambda_{AB} X_A X_B, \qquad (3.133)$$

and the rate of increase of the element C is

$$\dot{X}_C = A_C A_A^{-1} A_B^{-1} \lambda_{AB} X_A X_B. \qquad (3.134)$$

Here $X_A \equiv A_A n_A / n_N$, etc. are the concentrations by weight of the corresponding elements, A are their mass numbers (for example, $A_D = 2$ and $A_T = 3$, etc.) and $\lambda_{AB} = \langle \sigma v \rangle_{AB} n_N$. The reaction is efficient only if $\dot{X}_A / X_A > t^{-1}$.

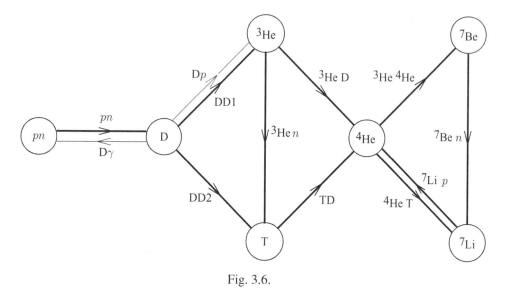

Fig. 3.6.

The general picture is as follows. Until the temperature drops to 0.08 MeV, the p, n and D reservoirs are in equilibrium with each other and decoupled from the rest (the deuterium bottleneck). However, as soon as the temperature drops to 0.08 MeV, the DD pipes become very efficient, rapidly converting the deuterium supply from the np reservoir into heavier elements. Finally, nearly all free neutrons have been bound in nuclei. Around this time the concentrations of the elements in the various "reservoirs" freeze out at their final abundances. Now we consider the build-up of each element in detail.

3.5.3 Helium-4

Once deuterium reaches the abundance $X_D^{(bn)}$ the bottleneck opens and nucleosynthesis begins. However, at the beginning, deuterium production in the reaction $pn \to D\gamma$ is still greater than its destruction in DD reactions. The ratio of the corresponding rates is

$$\frac{\lambda_{pn} X_p X_n}{\lambda_{DD} X_D^2} \simeq 10^4 \left(\frac{10^{-4}}{X_D}\right)^2, \qquad (3.135)$$

where the experimental value for $\lambda_{pn}/\lambda_{DD}$ is about 10^{-3} at $T_{\text{MeV}} \simeq 0.07$–$0.08$ and we have set $X_n \simeq 0.16$, $X_p \simeq 0.84$. Because of the very high supply rate, deuterium remains in chemical equilibrium with nucleons until its abundance rises to $X_D \simeq 10^{-2}$. After that, two-body DD reactions become dominant and X_D begins to decrease – see Figure 3.7, where the time dependence of abundances for $\eta_{10} \simeq 7$ is shown. (Note that the deuterium photodestruction can be ignored now because it

3.5 Nucleosynthesis

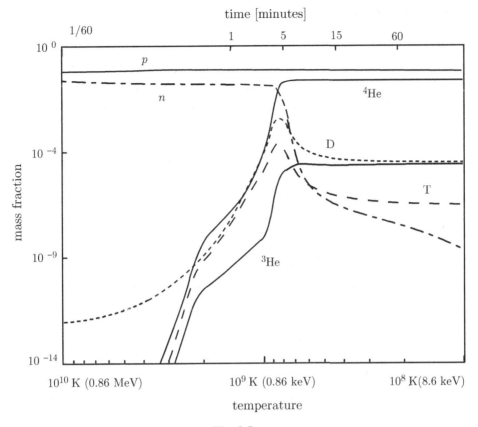

Fig. 3.7.

alone cannot prevent X_D from further growth.) Although the deuterium concentration ceases to grow, the concentration of free neutrons strongly decreases because they go first to the deuterium reservoir and then, without further delay, proceed down the pipes towards the heavier elements. For most of the neutrons, the final destination is the ^{4}He reservoir. In fact, the binding energy of ^{4}He (28.3 MeV) is four times larger than the binding energies of the intermediate elements, ^{3}He (7.72 MeV) and T(6.92 MeV) and, therefore, if ^{4}He were in equilibrium with these elements, it would dominate at low temperatures. A system always tends to equilibrium in the quickest possible way. Therefore, most of the free neutrons will form ^{4}He to fulfil its largest equilibrium demand.

Problem 3.19 Verify that at $T \sim 0.1$ MeV the equilibrium concentrations of D, ^{3}He and T are many orders of magnitude smaller than the ^{4}He concentration.

The reactions in which ^{4}He is formed proceed as follows. First, deuterium is converted into tritium and ^{3}He according to (3.123). Next, tritium combines with deuterium to produce ^{4}He:

$$\text{TD} \to {}^4\text{He}\,n. \tag{3.136}$$

In this sequence, two of the three neutrons end up in the newly formed ^{4}He nucleus and one neutron returns to the np reservoir. The ^{3}He nucleus can interact either with a free neutron and proceed to the T reservoir,

$$^3\text{He}\,n \to \text{T}\,p, \tag{3.137}$$

or with deuterium and go directly to the ^{4}He reservoir,

$$^3\text{He}\,\text{D} \to {}^4\text{He}\,p. \tag{3.138}$$

The ratio of the rates for these reactions is

$$\frac{\lambda_{^3\text{He}\,n} X_{^3\text{He}} X_n}{\lambda_{^3\text{He}\,\text{D}} X_{^3\text{He}} X_\text{D}} \sim 6 \frac{X_n}{X_\text{D}}. \tag{3.139}$$

Hence, until the concentration of free neutrons, X_n, drops below X_D (which never exceeds 10^{-2}), (3.137) is more efficient than (3.138). Therefore, most of the neutrons are fused into ^{4}He through the reaction chains $np \to \text{D} \to \text{T} \to {}^4\text{He}$ and $np \to \text{D} \to {}^3\text{He} \to \text{T} \to {}^4\text{He}$. Within a short interval around the time when the deuterium concentration reaches its maximal value $X_\text{D} \simeq 10^{-2}$, nearly all neutrons, except a very small fraction $\sim 10^{-4}$, end up in ^{4}He nuclei. Therefore, the final ^{4}He abundance is completely determined by the available free neutrons at this time. According to (3.130), X_D is of order 10^{-2} at temperature

$$T_{\text{MeV}}^{(N)} \simeq 0.07(1 + 0.03 \ln \eta_{10}), \tag{3.140}$$

or, equivalently, at time

$$t_{\text{sec}}^{(N)} \simeq 269(1 - 0.07(N_\nu - 3) - 0.06 \ln \eta_{10}), \tag{3.141}$$

where we have used (3.132) for κ in (3.88). Because half of the total weight of ^{4}He is due to protons, its final abundance by weight is

$$X_{^4\text{He}}^f = 2X_n(t^{(N)}) = 2X_n^* \exp\left(-\frac{t^{(N)}}{\tau_n}\right). \tag{3.142}$$

Substituting here X_n^* from (3.115) and $t^{(N)}$ from (3.141), one finally obtains

$$X_{^4\text{He}}^f \simeq 0.23 + 0.012(N_\nu - 3) + 0.005 \ln \eta_{10} \tag{3.143}$$

This result is in good agreement with the numerical calculations shown in Figure 3.8. The ^{4}He abundance depends on the number of ultra-relativistic species N_ν and the baryon density characterized by η_{10}. The presence of an additional massless neutrino increases the final abundance by about 1.2%. This increase comes from two

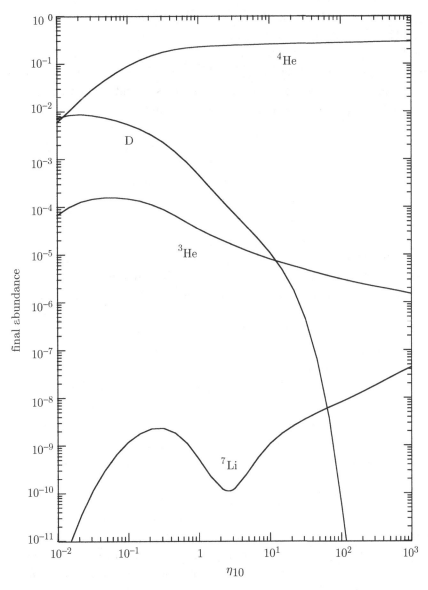

Fig. 3.8.

sources which give comparable contributions. First, the greater the number of ultra-relativistic species, the faster the universe expands for a given temperature. This means neutrons freeze out earlier, leading to larger X_n^*. Second, if there are more light species, the nucleosynthesis temperature is reached sooner and more neutrons avoid decay. Thus, given η_{10}, we can put rather strong bounds on the number of unknown light species using the observational data on helium-4 abundance. We will see later that η_{10} can be determined with high precision from data on deuterium abundance and CMB fluctuations.

It follows from (3.141) that nucleosynthesis begins earlier in a more dense universe and hence more neutrons are available. Therefore, the final helium-4 abundance depends logarithmically on the baryon density and, according to (3.143), increases by 1% or so if the baryon density is 10 times larger.

3.5.4 Deuterium

To calculate the time evolution and freeze-out abundance of deuterium, we make a series of assumptions which drastically simplify our task. The validity of these assumptions can be checked *a posteriori*.

First, we ignore ^{7}Be and ^{7}Li because their abundances turn out to be small compared to the abundances of ^{3}He and T. Second, we assume that the ^{3}He and T abundances take on their *quasi-equilibrium values*, that is, they are completely determined by the condition that "the total flux coming into each corresponding reservoir must be equal to the outgoing flux" (see Figure 3.6). Concretely, in the case of ^{3}He, the amount of ^{3}He produced within a given time interval via DD and Dp reactions should be equal to the amount of ^{3}He destroyed within the same time in ^{3}HeD and ^{3}Hen reactions.

Let us describe the primordial nucleosynthesis process once more, but this time in greater detail. When the deuterium concentration reaches $X_D \simeq 10^{-2}$ the DD reactions become efficient and the deuterium produced in the pn reaction is quickly converted into ^{3}He and T. Thus, further deuterium accumulation stops and, in fact, its concentration begins to decrease. As a result, neutrons are taken from the np reservoir and sent, without delay in the D reservoir, directly to the ^{3}He and T reservoirs along the DD and Dp pipes. From there they proceed through ^{3}HeD and TD pipes to their final destination − the ^{4}He reservoir.

Not all the neutrons reach the ^{4}He reservoir on their first attempt; some of them "leak out" on the way there. Concretely, neutrons are released in the reactions DD → ^{3}Hen and TD → ^{4}Hen and they return to the np reservoir. From there, they again try to reach the ^{4}He reservoir. Thus, after the beginning of nucleosynthesis, there is a steady flux of neutrons from the np reservoir to the ^{4}He reservoir through the intermediate D, ^{3}He and T reservoirs. The system of pipes is self-regulating and maintains the ^{3}He and T concentrations in accordance with the demands of quasi-equilibrium. To be precise, the rate of destruction of ^{3}He and T is proportional to their concentrations and if, for example, the abundance of ^{3}He becomes larger or smaller than the quasi-equilibrium concentration, then the size of the ^{3}Hen pipe grows or shrinks respectively, and the concentration quickly returns to its quasi-equilibrium value.

3.5 Nucleosynthesis

If the universe were not expanding, then nearly all free neutrons would end up in ^{4}He nuclei and there would be negligible abundances of the other light elements. However, in an expanding universe, expansion acts as a "shut-off valve" for the pipes. At the moment the expansion rate becomes larger than a particular reaction rate, the corresponding pipe closes. When all pipes entering a reservoir have closed, the abundance of that light element freezes out. The final abundances of ^{3}He and T are determined by the freeze-out concentration of deuterium, which we now calculate.

Let us derive the system of kinetic equations for the abundances by weight X_n, X_D, X_T and $X_{^3He}$. The concentration of free neutrons decreases due to the reactions $pn \to D\gamma$ and $^3Hen \to Tp$ but increases in the processes $DD \to {}^3Hen$ and $DT \to {}^4Hen$. Therefore, taking into account (3.133) and (3.134), we obtain

$$\frac{dX_n}{dt} = -\lambda_{pn} X_p X_n - \tfrac{1}{3}\lambda_{^3Hen} X_{^3He} X_n + \tfrac{1}{4}\lambda_{DD1} X_D^2 + \tfrac{1}{6}\lambda_{DT} X_D X_T. \qquad (3.144)$$

Deuterium is produced only in the reaction $pn \to D\gamma$ and destroyed in the reactions $DD \to {}^3Hen$, $DD \to Tp$, $Dp \to {}^3He\gamma$, $^3HeD \to {}^4Hep$, $DT \to {}^4Hen$. Hence,

$$\frac{dX_D}{dt} = 2\lambda_{pn} X_p X_n - \tfrac{1}{2}\lambda_{DD} X_D^2 - \lambda_{Dp} X_D X_p - \tfrac{1}{3}\lambda_{DT} X_D X_T - \tfrac{1}{3}\lambda_{^3HeD} X_{^3He} X_D, \qquad (3.145)$$

where $\lambda_{DD} = \lambda_{DD1} + \lambda_{DD2}$. The equation for tritium is obtained similarly:

$$\frac{dX_T}{dt} = \tfrac{3}{4}\lambda_{DD2} X_D^2 + \lambda_{^3Hen} X_{^3He} X_n - \tfrac{1}{2}\lambda_{DT} X_D X_T. \qquad (3.146)$$

We assume that the tritium concentration satisfies the quasi-equilibrium condition, that is, the rate of its overall change is much smaller than the rates of the individual reactions on the right hand side of (3.146). Therefore, we set $dX_T/dt \approx 0$ and (3.146) reduces to

$$\tfrac{3}{4}\lambda_{DD2} X_D^2 + \lambda_{^3Hen} X_{^3He} X_n \approx \tfrac{1}{2}\lambda_{DT} X_D X_T. \qquad (3.147)$$

The quasi-equilibrium condition for helium-3 takes the form

$$\tfrac{3}{4}\lambda_{DD1} X_D^2 + \tfrac{3}{2}\lambda_{Dp} X_D X_p \approx \tfrac{1}{2}\lambda_{^3HeD} X_{^3He} X_D + \lambda_{^3Hen} X_{^3He} X_n. \qquad (3.148)$$

Using (3.147) and (3.148) to express $X_{^3He}$ and X_T through the neutron and deuterium concentrations, (3.144) and (3.145) become

$$\frac{dX_n}{dt} = \tfrac{1}{4}\lambda_{DD} X_D^2 - \lambda_{pn} X_p X_n, \qquad (3.149)$$

$$\frac{dX_D}{dt} = 2\lambda_{pn} X_p X_n - \lambda_{DD} X_D^2 - 2\lambda_{Dp} X_D X_p. \qquad (3.150)$$

It is convenient to rewrite these equations using a temperature variable instead of a time variable (see (3.88)). Substituting the explicit value for λ_{DD} from (3.127) then gives

$$\frac{dX_n}{dT_{\text{MeV}}} = \alpha \eta_{10}\left(R_1 X_n - X_D^2\right), \tag{3.151}$$

$$\frac{dX_D}{dT_{\text{MeV}}} = 4\alpha \eta_{10}\left(X_D^2 + R_2 X_D - \tfrac{1}{2} R_1 X_n\right), \tag{3.152}$$

where

$$\alpha \equiv \alpha(T) = 0.86 \times 10^5 K(T)$$

and the coefficient $K(T)$ describes the temperature dependence of $\langle \sigma v \rangle_{DD}$. Its value changes from 1 to 0.5 when the temperature drops from 0.09 MeV to 0.04 MeV. Over the same temperature interval the coefficients R_1 and R_2 are

$$R_1 \equiv 4X_p \frac{\lambda_{pn}}{\lambda_{DD}} \simeq (3\text{–}8) \times 10^{-3}, \quad R_2 \equiv 2X_p \frac{\lambda_{pD}}{\lambda_{DD}} \simeq (2.5\text{–}2.3) \times 10^{-5}, \tag{3.153}$$

where the experimental value for the ratio of the corresponding reaction rates has been used. The system of equations (3.151) and (3.152) has attractor solutions.

First we consider the initial stage of nucleosynthesis when $X_D \ll X_n$. It turns out that in this case the deuterium concentration satisfies the quasi-equilibrium condition and we can set $dX_D/dT \approx 0$ in (3.152). Since $R_2 \ll R_1$, the term $R_2 X_D$ is small compared to $R_1 X_n$, and it follows from (3.152) and (3.151) that

$$X_D = \sqrt{\frac{R_1 X_n}{2}} \left[1 + O\left(\frac{X_D}{X_n}\right)\right]. \tag{3.154}$$

This solution is valid after the deuterium concentration reaches its maximal value of order 10^{-2} and begins to decrease (Figure 3.7). It fails as soon as X_n drops to X_D and, at this time, $X_n \sim X_D \sim R_1$. Note that, according to (3.154), the maximal concentration of deuterium is equal to $X_D \simeq 10^{-2}$ for $X_n \simeq 0.12$. This is in agreement with the naive estimate derived earlier by comparing the rates of pn and DD reactions. Substituting (3.154) into (3.151), we obtain

$$\frac{dX_n}{dT_{\text{MeV}}} \simeq \tfrac{1}{2} \alpha \eta_{10} R_1 X_n. \tag{3.155}$$

In this regime the neutrons determine their own fate and also dictate the quasi-equilibrium concentrations to the other elements, including deuterium. In other words, they regulate the shut-off valves between the reservoirs in Figure 3.6. At the beginning of nucleosynthesis, at $T = T_{\text{MeV}}^{(N)}$, most of the neutrons are still free, and hence $X_n \simeq 0.12$. Neglecting the temperature dependence of α, we then find the approximate solution of (3.155):

$$X_n(T) \simeq 0.12 \exp\left(\tfrac{1}{2}\alpha\eta_{10}R_1\left(T - T_{\text{MeV}}^{(N)}\right)\right), \tag{3.156}$$

where $T_{\text{MeV}}^{(N)}$ is given in (3.140). It follows that the neutron concentration becomes comparable to the deuterium concentration, $X_n \sim X_D \sim R_1$, at the temperature

$$T_{\text{MeV}}^* \sim 0.07 + 0.002 \ln \eta_{10} - 0.02 K^{-1}\eta_{10}^{-1}. \tag{3.157}$$

In a universe with very low baryon density, $K\eta_{10} < 0.3$, the abundance of free neutrons (neglecting their decay) does not decrease below X_D and freezes out at the value

$$X_n^f \simeq 0.12 \exp\left(-\tfrac{1}{2}\alpha\eta_{10}R_1 T_{\text{MeV}}^{(N)}\right) \sim 0.12 \exp(-10K\eta_{10}). \tag{3.158}$$

The remaining free neutrons then decay. This explains why, for instance, the ^{4}He abundance is less than 1% in a universe with $\eta_{10} \simeq 10^{-2}$ (Figure 3.8).

Problem 3.20 At which value does the deuterium concentration freeze out in low baryon density universe? How does it depend on η_{10}?

In the derivation of the ^{4}He abundance presented above, we tacitly assumed that the reactions converting the neutrons into ^{4}He are very efficient in transferring most of the available neutrons into heavier elements. This means that (3.143) is valid only for $K\eta_{10} > 0.3$. Observations suggest that $10 > \eta_{10} > 1$ and therefore we will assume below that $\eta_{10} > 1$.

When the neutron concentration becomes of order the deuterium concentration, (3.154) fails and the system quickly reaches another attractor. Afterwards, the neutron concentration satisfies the quasi-equilibrium condition, $dX_n/dT \approx 0$, and it follows from (3.151) that

$$X_n = \frac{1}{R_1}X_D^2\left[1 + O\left(\frac{X_n}{X_D}\right)\right]. \tag{3.159}$$

Equation (3.152) then becomes

$$\frac{dX_D}{dT_{\text{MeV}}} = 2\alpha\eta_{10}\left(X_D^2 + 2R_2 X_D\right). \tag{3.160}$$

Now the deuterium determines its own fate and regulates the quasi-equilibrium concentrations of the neutrons and the other light elements. Since R_2 changes insignificantly within the relevant temperature interval (see (3.153)), it can be taken to be a constant. Equation (3.160) is then readily integrated:

$$\left(1 + \frac{2R_2}{X_D(T)}\right) = \left(1 + \frac{2R_2}{X_D(T^*)}\right) \exp\left(4R_2\eta_{10} \int_T^{T^*} \alpha(T)\,dT\right), \quad (3.161)$$

where the temperature is expressed in MeV. As the temperature decreases, the deuterium concentration freezes out at $X_D^f \equiv X_D(T \to 0)$. Taking into account that $X_D(T^*) \sim R_1 \gg R_2$, we obtain

$$X_D^f \simeq \frac{2R_2}{\exp(A\eta_{10}) - 1}, \quad (3.162)$$

where

$$A \equiv 4R_2 \int_T^{T^*} \alpha(T)\,dT \sim 4R_2\alpha(T^*)\, T^*_{\text{MeV}}. \quad (3.163)$$

The coefficient A depends only weakly on η_{10}; it increases by a factor of 2 as η_{10} goes from 1 to 10^2. Taking as an estimate $A \simeq 0.1$, we find good agreement with the results of the numerical calculations shown in Figure 3.8.

For $\eta_{10} < 1/A \sim 10$, (3.162) simplifies to

$$X_D^f \simeq \frac{2R_2}{A\eta_{10}} \sim 4 \times 10^{-4} \eta_{10}^{-1}. \quad (3.164)$$

For this range of η_{10} the deuterium freeze-out abundance decreases in inverse proportion to η_{10}. This dependence on η_{10} can easily be understood. For $\eta_{10} < 10$, the freeze-out concentration X_D^f is larger than $R_2 \simeq 2 \times 10^{-5}$ and, according to (3.160), DD reactions dominate in destroying deuterium. The deuterium freeze-out is then determined by the condition $\dot{X}_D/X_D \sim \lambda_{DD} X_D^f \sim t^{-1}$. Since $\lambda_{DD} \propto n_N \propto \eta_{10}$, we find that, to leading order, $X_D^f \propto \eta_{10}^{-1}$.

For $\eta_{10} > 10$, (3.162) becomes

$$X_D^f \simeq 2R_2 \exp(-A\eta_{10}). \quad (3.165)$$

In this case the deuterium abundance decays exponentially with η_{10} and decreases by five orders of magnitude, from 10^{-5} to 10^{-10}, when η_{10} changes from 10 to 100 (Figure 3.8). In a universe with high baryon density, the reaction $Dp \to {}^3\text{He}\gamma$ dominates in the destruction of deuterium when $X_D < R_2 \simeq 2 \times 10^{-5}$. Hence, the freeze-out concentration is determined by the term linear in X_D in (3.160).

Thus, deuterium turns out to be an extremely sensitive indicator of the baryon density in the universe. The observational data certainly rule out the possibility of a flat universe composed only of baryonic matter.

3.5.5 The other light elements

Now we can calculate the final abundances of the other light elements by simply using the quasi-equilibrium conditions.

Helium-3 The expression for the quasi-equilibrium concentration of ^{3}He follows from (3.148):

$$X_{^3\text{He}} \approx \frac{3}{2}\left(\frac{\lambda_{\text{DD1}}}{\lambda_{^3\text{HeD}}}X_\text{D} + 2\frac{\lambda_{\text{D}p}}{\lambda_{^3\text{HeD}}}X_p\right)\left(1 + 2\frac{\lambda_{^3\text{He}n}}{\lambda_{^3\text{HeD}}}\frac{X_n}{X_\text{D}}\right)^{-1}. \tag{3.166}$$

If the baryon density is not too large, the rate of the dominant reaction in which ^{3}He is destroyed is larger than the rate of the deuterium destruction. Therefore, the freeze-out of ^{3}He occurs a little bit later than the freeze-out of deuterium. After deuterium freeze-out, a small leakage from the D reservoir to the ^{3}He reservoir still maintains a stationary flow through the ^{3}He reservoir and the quasi-equilibrium condition for ^{3}He is roughly satisfied at the time of its freeze-out. Substituting $X_n \simeq X_\text{D}^2/R_1$ into (3.166) and the experimental values for the ratios of the corresponding reaction rates, taken for definiteness at $T \simeq 0.06$ MeV, we obtain

$$X_{^3\text{He}}^f \simeq \frac{0.2 X_\text{D}^f + 10^{-5}}{1 + 4 \times 10^3 X_\text{D}^f}, \tag{3.167}$$

where X_D^f is given in (3.162). This result is in good agreement with the numerical calculations shown in Figure 3.8. For example, for $\eta_{10} = 1$, we have $X_\text{D}^f \simeq 4 \times 10^{-4}$ and $X_{^3\text{He}}^f \simeq 3 \times 10^{-5}$, that is, the final ^{3}He abundance is 10 times smaller than the deuterium abundance.

The difference between X_D^f and $X_{^3\text{He}}^f$ decreases for larger η_{10}. For $\eta_{10} \simeq 10$, the freeze-out concentrations of the deuterium and helium-3 are about the same and equal to 10^{-5}. In a universe with $\eta_{10} > 10$, the reaction D$p \to{}^3$Heγ dominates in producing ^{3}He around the freeze-out time and nearly all deuterium is destroyed in favor of ^{3}He, which thus becomes more abundant than deuterium. In this case, the freeze-out of ^{3}He is determined by two competing reactions, D$p \to{}^3$Heγ and ^{3}HeD $\to{}^4$Hen, and, irrespective of how large X_D is, they give rise to the final ^{3}He abundance, $X_{^3\text{He}}^f \simeq \lambda_{\text{D}p}/\lambda_{^3\text{HeD}} \simeq 10^{-5}$. The weak dependence of $X_{^3\text{He}}^f$ on the baryon density for $\eta_{10} > 10$ is due to the temperature dependence of the reaction rates, which we have ignored.

Tritium The quasi-equilibrium condition (3.147) gives

$$X_\text{T} = \left(\frac{3}{2}\frac{\lambda_{\text{DD2}}}{\lambda_{\text{DT}}} + 2\frac{\lambda_{^3\text{He}n}}{\lambda_{\text{DT}}}\frac{X_n}{X_\text{D}^2}X_{^3\text{He}}\right)X_\text{D}. \tag{3.168}$$

Assuming that tritium freeze-out occurs at about the same time as for deuterium, and substituting $X_n \simeq X_D^2/R_1$ into (3.168), we find

$$X_T^f \simeq \left(0.015 + 3 \times 10^2 X_{^3\text{He}}^f\right) X_D^f, \quad (3.169)$$

where the experimental values $\lambda_{\text{DD2}}/\lambda_{\text{DT}} \simeq 0.01$ and $\lambda_{^3\text{He}n}/\lambda_{\text{DT}} \simeq 1$ have been used. For $\eta_{10} \simeq 1$, we have $X_T^f \simeq 10^{-5}$. Note that, for any η_{10}, the tritium final abundance is several times smaller than the deuterium abundance.

Problem 3.21 When does tritium freeze-out take place? For which η_{10} can we use X_D^f in (3.168) to estimate X_T^f? Which value of X_D should be used otherwise?

Problem 3.22 Explain why the ^{3}He concentration increases monotonically in time (see Figure 3.7) but the tritium concentration first rises to a maximum and then decreases until it freezes out.

Lithium-7 and beryllium-7 The quasi-equilibrium conditions for ^{7}Li and ^{7}Be result from the dominant reactions in which ^{7}Li and ^{7}Be are produced and destroyed (see Figure 3.6):

$$\frac{7}{12}\lambda_{^4\text{HeT}} X_{^4\text{He}} X_T + \lambda_{^7\text{Be}n} X_{^7\text{Be}} X_n = \lambda_{^7\text{Li}p} X_{^7\text{Li}} X_p, \quad (3.170)$$

$$\frac{7}{12}\lambda_{^4\text{He}^3\text{He}} X_{^4\text{He}} X_{^3\text{He}} = \lambda_{^7\text{Be}n} X_{^7\text{Be}} X_n. \quad (3.171)$$

One can check that other reactions, such as ^{7}Li + D $\to$ 2^4He + n and ^{7}Be + D $\to$ 2^4He + p, can be ignored for $\eta_{10} > 1$. It follows from these equations that

$$X_{^7\text{Li}} = \frac{7}{12} \frac{X_{^4\text{He}}}{X_p} \left(\frac{\lambda_{^4\text{HeT}}}{\lambda_{^7\text{Li}p}}\right)\left(X_T + \frac{\lambda_{^4\text{He}^3\text{He}}}{\lambda_{^4\text{HeT}}} X_{^3\text{He}}\right). \quad (3.172)$$

The ratio $\lambda_{^4\text{HeT}}/\lambda_{^7\text{Li}p}$ is nearly constant over a broad temperature interval, increasing only from 2.2×10^{-3} to 3×10^{-3} as the temperature drops from 0.09 MeV to 0.03 MeV, while

$$r(T) \equiv \frac{\lambda_{^4\text{He}^3\text{He}}}{\lambda_{^4\text{HeT}}}$$

changes significantly over the same temperature interval, namely, $r \simeq 5 \times 10^{-2}$ for $T \simeq 0.09$ MeV and $r \simeq 6 \times 10^{-3}$ for $T \simeq 0.03$ MeV. With these values of the reaction rates we obtain

$$X_{^7\text{Li}} \sim 10^{-4}(X_T + r(T) X_{^3\text{He}}). \quad (3.173)$$

To estimate the freeze-out concentration for ^{7}Li we must know the values of X_T, $r(T)$ and $X_{^3\text{He}}$ at ^{7}Li freeze-out. For $5 > \eta_{10} > 1$, freeze-out occurs after the

3.5 Nucleosynthesis

deuterium reaches its final abundance, and we can substitute into (3.173) the values of $X^f_{^3\text{He}}$ and X^f_T obtained previously. For $\eta_{10} \simeq 1$, the first term on the right hand side in (3.173) dominates and, using the estimate $X^f_\text{T} \simeq 10^{-5}$, we obtain $X^f_{^7\text{Li}} \sim 10^{-9}$. For larger η_{10}, the tritium abundance X^f_T is smaller and, consequently, $X^f_{^7\text{Li}}$ decreases as η_{10} grows, but only until the second term on the right hand side in (3.173) starts to dominate. The minimum final ^{7}Li abundance, $X^f_{^7\text{Li}} \sim 10^{-10}$, is reached for η_{10} between 2 and 3 (see Figure 3.8). Then, further increase in η_{10} causes the ^{7}Li abundance to rise. This rise is mostly due to the temperature dependence of r; for $\eta_{10} > 3$, the freeze-out temperature is determined by the efficiency of the ^{7}Ben reaction, which in turn depends on the neutron concentration. In a universe with high baryon density, the deuterium and free neutrons burn more efficiently and disappear earlier (at a higher temperature) than in a universe with low baryon density. Therefore, the ^{7}Li concentration freezes out at a higher temperature at which r is larger. Note also that, for $\eta_{10} > 5$, the ^{7}Ben reaction becomes inefficient before ^{3}He reaches its freeze-out concentration, and hence, to estimate $X^f_{^7\text{Li}}$ properly we have to substitute in (3.173) the actual value of $X^3_{^3\text{He}}$ at ^{7}Li freeze-out, which is larger than $X^f_{^3\text{He}}$. Numerical calculations show that after passing through a relatively deep minimum with $X^f_{^7\text{Li}} \sim 10^{-10}$, the lithium concentration comes back to 10^{-9} at $\eta_{10} \simeq 10$.

In summary, the trough in the $X^f_{^7\text{Li}} - \eta_{10}$ curve is due to the competition of two reactions. In a universe with $\eta_{10} < 3$, most of the ^{7}Li is produced directly in the ^{4}HeT reaction. For $\eta_{10} > 3$, the reaction ^{7}Ben is more important and ^{7}Li is produced mainly through the intermediate ^{7}Be reservoir.

Beryllium-7 is not so important from the observational point of view, so, simply to gain a feeling for its abundance, we estimate it in the range $5 > \eta_{10} > 1$, in which ^{7}Be freeze-out occurs after that of deuterium. The quasi-equilibrium solution for free neutrons is valid at this time and, substituting $X_n \simeq X_\text{D}^2/R_1$ into (3.171), we find

$$X^f_{^7\text{Be}} = \frac{7}{12} R_1 X_{^4\text{He}} \left(\frac{\lambda_{^3\text{He}^4\text{He}}}{\lambda_{^7\text{Be}n}}\right) \frac{X^f_{^3\text{He}}}{\left(X^f_\text{D}\right)^2} \sim 10^{-12} \frac{X^f_{^3\text{He}}}{\left(X^f_\text{D}\right)^2}, \quad (3.174)$$

where the experimental values for the ratios of the relevant reactions have been used. In this case, the product of the corresponding ratios changes by a factor of 5 over the relevant temperature interval, so (3.174) is merely an estimate. For $\eta_{10} = 1$, we have $X^f_\text{D} \simeq 4 \times 10^{-4}$, $X^f_{^3\text{He}} \simeq 3 \times 10^{-5}$ and, hence, $X^f_{^7\text{Be}} \sim 2.5 \times 10^{-10}$.

The observed light element abundances are in very good agreement with theoretical predictions, thus lending strong support to the standard cosmological model. Observations suggest that $7 > \eta_{10} > 3$ at 95% confidence level.

3.6 Recombination

The most important matter ingredients in thermodynamical processes after nucleosynthesis are thermal radiation, electrons, protons p (hydrogen nuclei) and fully ionized helium nuclei, He^{2+}. The concentrations of the other light elements are very small and we neglect them here. As the temperature decreases, the ionized helium and hydrogen nuclei begin to capture the available free electrons and become electrically neutral. In a short period of time, nearly all free electrons and nuclei have combined to form neutral atoms and the universe becomes transparent to radiation. Since this process occurs so quickly, we refer to this epoch as the recombination *moment*.

We must, however, distinguish the helium and hydrogen recombinations, because they happen at different times. Helium has significantly larger ionization potentials than hydrogen and therefore becomes neutral earlier. However, after helium recombination, many free electrons remain and the universe is still opaque to radiation. Only after hydrogen recombination have most photons decoupled from matter; these are the photons that give us a "baby photo" of the universe. As a result, hydrogen recombination is a more interesting and dramatic event from an observational point of view.

Helium recombination, nevertheless, has some cosmological relevance. When helium becomes neutral it decouples from the plasma thus altering the speed of sound in the radiation–baryon fluid. We will see in Chapter 9 that this speed influences the CMB temperature fluctuations.

Recombination is not an equilibrium process. Hence, the formulae derived under the assumption of local equilibrium can only be used to estimate when recombination occurs. This is sufficient when we consider helium recombination. However, the subtleties of hydrogen recombination are very important for the calculation of the CMB temperature fluctuations. Therefore, after estimating the hydrogen recombination temperature based on the equilibrium equations, we will use kinetic theory to reveal the details of nonequilibrium recombination.

3.6.1 Helium recombination

The electric charge of the helium nucleus is 2, so it must capture two electrons to become neutral. This occurs in two steps. First, the helium captures one electron, becoming a singly charged, hydrogen-like ion He^+. The binding energy of this ion is four times larger than the binding energy for hydrogen:

$$B_+ = m_e + m_{2+} - m_+ = 54.4 \, \text{eV}, \quad (3.175)$$

where m_{2+} and m_+ are the masses of He^{2+} and He^+ respectively. This energy corresponds to a temperature of 632 000 K. To estimate the temperature at which most of the helium nuclei are converted into helium ions, we assume that the reaction

$$\text{He}^{2+} + e^- \leftrightharpoons \text{He}^+ + \gamma \tag{3.176}$$

is efficient in maintaining the chemical equilibrium between He^{2+} and He^+. Then, the chemical potentials satisfy

$$\mu_{2+} + \mu_e = \mu_+, \tag{3.177}$$

and considering the ratio $(n_{2+}n_e)/n_+$, where the number densities are given by (3.61), we obtain the Saha formula:

$$\frac{n_{2+}n_e}{n_+} = \frac{g_{2+}g_e}{g_+} \left(\frac{Tm_e}{2\pi}\right)^{3/2} \exp\left(-\frac{B_+}{T}\right). \tag{3.178}$$

The ratio of the statistical weights here is equal to unity. Even complete recombination of helium reduces the number of free electrons by 12% at most. Therefore, before hydrogen recombination, the number density of free electrons is

$$n_e \simeq (0.75 \text{ to } 0.88)n_N \simeq 2 \times 10^{-11} \eta_{10} T^3. \tag{3.179}$$

Substituting this into (3.178), we obtain

$$\frac{n_{2+}}{n_+} \simeq \exp\left(35.6 + \frac{3}{2}\ln\left(\frac{B_+}{T}\right) - \frac{B_+}{T} - \ln\eta_{10}\right). \tag{3.180}$$

If the expression in the exponent is positive, the concentration of He^+ ions is small compared to the concentration of completely ionized helium. Using the method of iteration, we find that at the temperature

$$T_+ \simeq \frac{B_+}{42 - \ln\eta_{10}} \simeq 15\,000 \times \left(1 + 2.3 \times 10^{-2} \ln\eta_{10}\right) \text{K}, \tag{3.181}$$

the ratio n_{2+}/n_+ is of order unity. At this time, the He^+ ions constitute about 50% of all helium and the rest is completely ionized. Very soon after this, nearly all helium nuclei capture an electron and are converted into He^+. Expanding the expression in the exponent in (3.180) about $T = T_+$, we find, to leading order in $\Delta T \equiv T_+ - T \ll T_+$,

$$\frac{n_{2+}}{n_+} \sim \exp\left(-\frac{B_+}{T_+}\frac{\Delta T}{T_+}\right) \sim \exp\left(-42\frac{\Delta T}{T_+}\right). \tag{3.182}$$

When the temperature falls only 20% below T_+ (going from 15 000 K to 12 000 K), the number density of He^{2+} reduces to $n_{2+} \sim 10^{-4} n_+$. We see in (3.181) that the temperature T_+ varies logarithmically with the baryon density η_{10}; the larger the baryon density, the earlier recombination occurs (i.e., at a higher temperature).

After most of the helium is converted into He$^+$ ions, the singly charged ions capture a second electron and become neutral. The second electron also ends up in the first orbit. The electron–electron interaction substantially reduces the binding energy: it is only 24.62 eV for the second electron. Therefore, the second stage of helium recombination occurs at a lower temperature than the first. For example, at $T \simeq 12\,000$ K, the density of neutral helium atoms is still negligible; only after the temperature decreases below $T \sim 5000$ K does helium become neutral and decouple from radiation. At this time hydrogen is still fully ionized and the universe remains opaque to radiation.

Problem 3.23 Assuming chemical equilibrium, derive the expression for the ratio of the number densities of He$^+$ and neutral He. Verify that for $\eta_{10} \simeq 5$ this ratio is equal to unity at $T \simeq 6800$ K and is about 10^{-4} at $T \simeq 5600$ K.

Problem 3.24 Explain why the recombination temperature is significantly smaller than the corresponding ionization potential energies.

3.6.2 Hydrogen recombination: equilibrium consideration

The main reaction responsible for maintaining hydrogen and radiation in equilibrium is

$$p + e^- \rightleftarrows H + \gamma, \qquad (3.183)$$

where H is a neutral hydrogen atom. For the ground($1S$) state, the binding energy of neutral hydrogen,

$$B_H = m_p + m_e - m_H = 13.6 \text{ eV}, \qquad (3.184)$$

corresponds to a temperature of 158 000 K. In this case, the Saha formula can be derived in the same manner as (3.178) and takes the form

$$\frac{n_p n_e}{n_H} = \left(\frac{T m_e}{2\pi}\right)^{3/2} \exp\left(-\frac{B_H}{T}\right), \qquad (3.185)$$

where n_H is the number density of the hydrogen atoms in the ground state and we have taken into account that the corresponding ratio of the statistical weights g_i is equal to unity. At equilibrium, neutral hydrogen atoms are also present in the excited states: $2S, 2P\ldots$. However, at $T < 5000$ K, their relative concentrations are negligible: for example,

$$\frac{n_{2P}}{n_H} = \frac{g_{2P}}{g_{1S}} \exp\left(-\frac{3}{4}\frac{B_H}{T}\right) < 10^{-10}. \qquad (3.186)$$

Therefore, for now, we neglect the excited hydrogen atoms and introduce the ionization fraction:

$$X_e \equiv \frac{n_e}{n_e + n_H}. \tag{3.187}$$

Since $n_p = n_e$ and

$$n_e + n_H \simeq 0.75 \times 10^{-10} \eta_{10} n_\gamma \simeq 3.1 \times 10^{-8} \eta_{10} (T/T_{\gamma 0})^3 \, \text{cm}^{-3}, \tag{3.188}$$

(3.185) becomes

$$\frac{X_e^2}{1 - X_e} = \exp\left(37.7 + \frac{3}{2}\ln\left(\frac{B_H}{T}\right) - \frac{B_H}{T} - \ln \eta_{10}\right). \tag{3.189}$$

The expression in the exponent vanishes at

$$T_{rec} \simeq \frac{B_H}{43.4 - \ln \eta_{10}} \simeq 3650\left(1 + 2.3 \times 10^{-2} \ln \eta_{10}\right) \text{K}. \tag{3.190}$$

At this time, the ionization fraction is $X_e \simeq 0.6$ and the temperature T_{rec} characterizes the moment when hydrogen recombination begins. At earlier times $X_e \to 1$; for example, $1 - X_e \simeq 10^{-5}$ at $T \simeq 5000$ K. As soon as the temperature decreases below T_{rec}, recombination proceeds very rapidly. According to (3.189), a 10% fall in temperature reduces the ionization fraction by a factor of 10; at $T \sim 2500$ K, we have $X_e \sim 10^{-4}$.

The equilibrium Saha formula tells us that the ionization fraction should continue to decrease exponentially as the temperature drops. However, this does not occur in an expanding universe; the ionization fraction freezes out instead. More importantly, the equilibrium description fails almost immediately after the beginning of recombination. The main reason for the failure of the Saha formula is the large number of energetic photons emitted when the nuclei and electrons combine. These nonthermal photons significantly distort the high-energy tail of the thermal radiation spectrum exactly at energies crucial to recombination. As a result, it becomes essential to take into account deviation from equilibrium and we must use kinetic theory.

3.6.3 Hydrogen recombination: the kinetic approach

Direct recombination to the ground state accompanied by the emission of a photon does not substantially increase the number of neutral atoms because the emitted photon has enough energy to immediately ionize the first neutral hydrogen atom it meets. These two competing processes occur at very high rates and result in no net change in n_H. More efficient is cascading recombination, in which neutral hydrogen is first produced in an excited state and then decays to the ground state in a sequence of steps. However, even in cascading recombination, at least one very energetic photon is emitted corresponding to the energy difference between the $2P$ and $1S$ states of the hydrogen atom. This so-called Lyman-α photon, L_α, has energy $3B_\mathrm{H}/4 = 117\,000$ K and a rather large resonance absorption cross-section, $\sigma_\alpha \simeq 10^{-17}$–$10^{-16}$ cm^2, at the recombination temperature. Therefore, the L_α photons are reabsorbed in a short time, $\tau_\alpha \simeq (\sigma_\alpha n_\mathrm{H})^{-1} \sim 10^3$–$10^4$ s, after emission. We have to compare this time to the cosmological time at recombination. During the matter-dominated epoch, the cosmological time can easily be expressed in terms of the radiation temperature by equating the energy density of cold particles (see (3.1)) with the critical density $\varepsilon^{cr} = 1/(6\pi t^2)$ and noting that $T = T_{\gamma 0}(1+z)$. We obtain

$$t_\mathrm{sec} \simeq 2.75 \times 10^{17} \left(\Omega_m h_{75}^2\right)^{-1/2} \left(\frac{T_{\gamma 0}}{T}\right)^{3/2}. \qquad (3.191)$$

At the moment of recombination, $\tau_\alpha \ll t \sim 10^{13}$ s and, consequently, the L_α photons are not significantly redshifted before reabsorption. To simplify our considerations we will neglect the redshift effect.

The presence of a large number of L_α photons and other energetic photons results in a greater abundance of electrons, protons and hydrogen atoms in excited $2S$ and $2P$ states than expected according to the equilibrium Saha formula. This delays recombination so that, for a given temperature, the actual ionization fraction exceeds its equilibrium value. The full system of kinetic equations describing nonequilibrium recombination is rather complicated and is usually solved numerically. To solve them analytically, we use the method of quasi-equilibrium concentrations, as was applied to the problem of nucleosynthesis. The results obtained by this method are in good agreement with the numerical calculations.

We will make a series of simplifying assumptions whose validity can be checked *a posteriori*. First of all, we neglect all highly excited hydrogen states and retain only the $1S$, $2S$ and $2P$ states of neutral hydrogen. The remaining ingredients are electrons, protons and thermal photons, as well as L_α and other nonthermal photons emitted during recombination. The main reactions in which these components participate are symbolically represented in Figure 3.9. The direct recombination to the ground state can be ignored because it results in no net change of neutral hydrogen, as explained above. Thermal radiation dominates in ionizing the $2S$ and $2P$ states. In fact, to ionize an excited atom, the energy of a photon must only be larger than $B_\mathrm{H}/4$. The number of such thermal photons is much greater than the number of energetic nonthermal photons, and hence, when considering the ionization of excited atoms, we can ignore the distortion of the thermal radiation spectrum.

In contrast, thermal photons play no significant role in transitions between $1S$ and $2P$ states after the beginning of recombination. These transitions are mostly due

3.6 Recombination

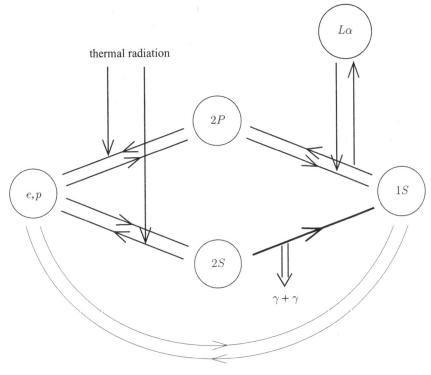

Fig. 3.9.

to the nonthermal L_α photons. When the deviation from equilibrium becomes large and the $2S$ level is overpopulated, we can ignore the transition $1S + \gamma + \gamma \to 2S$ compared to the two-photon decay $2S \to 1S + \gamma + \gamma$. (A transition with a single photon is forbidden by angular momentum conservation.) The rate for the two-photon decay, $W_{2S \to 1S} \simeq 8.23 \text{ s}^{-1}$, is very small (compare, for example, with $W_{2P \to 1S} \simeq 4 \times 10^8 \text{ s}^{-1}$); nevertheless, this decay plays the dominant role in non-equilibrium recombination. In terms of the pipe-and-reservoir picture, the two-photon transition is the main source of irreversible leakage from the e, p reservoir to the $1S$ reservoir. Because all other processes result in high-energy photons which reionize neutral hydrogen and return electrons to the e, p reservoir, the rate of net change in the ionization fraction is

$$\frac{dX_e}{dt} = -\frac{dX_{1S}}{dt} = -W_{2S} X_{2S}, \tag{3.192}$$

where $X_e \equiv n_e/n_T$, $X_{2S} \equiv n_{2S}/n_T$, and n_T is the total number density of neutral atoms plus electrons, given in (3.188). Once a substantial fraction ($\sim 50\%$) of neutral hydrogen has formed, (3.192) is a good approximation to use until nearly the end of recombination.

To express X_{2S} in terms of X_e, we use the quasi-equilibrium condition for the intermediate $2S$ reservoir; this is justified by the high rate of the reactions, shown in Figure 3.9. For the $2S$ reservoir, this condition takes the following form:

$$\langle \sigma v \rangle_{ep \to \gamma 2S} n_e n_p - \langle \sigma \rangle_{\gamma 2S \to ep} n_\gamma^{eq} n_{2S} - W_{2S \to 1S} n_{2S} = 0, \quad (3.193)$$

where n_γ^{eq} is the number density of thermal photons. The relation between the cross-sections for the direct and inverse reactions, $ep \leftrightarrows \gamma 2S$, can be found if one notes that in a state of equilibrium these reactions compensate each other. Then, we have

$$\frac{\langle \sigma \rangle_{\gamma 2S \to ep} n_\gamma^{eq}}{\langle \sigma v \rangle_{ep \to \gamma 2S}} = \frac{n_e^{eq} n_p^{eq}}{n_{2S}^{eq}} = \left(\frac{Tm_e}{2\pi}\right)^{3/2} \exp\left(-\frac{B_H}{4T}\right), \quad (3.194)$$

where the Saha formula has been used to obtain the latter equality (recall that the binding energy of $2S$ state is $B_H/4$). With the help of this relation, we can express X_{2S}, from (3.193), as

$$X_{2S} = \left[\frac{W_{2S}}{\langle \sigma v \rangle_{ep \to 2S}} + \left(\frac{Tm_e}{2\pi}\right)^{3/2} \exp\left(-\frac{B_H}{4T}\right)\right]^{-1} n_T X_e^2, \quad (3.195)$$

and (3.192) becomes

$$\frac{dX_e}{dt} = -W_{2S} \left[\frac{W_{2S}}{\langle \sigma v \rangle_{ep \to 2S}} + \left(\frac{Tm_e}{2\pi}\right)^{3/2} \exp\left(-\frac{B_H}{4T}\right)\right]^{-1} n_T X_e^2. \quad (3.196)$$

When the first term inside the square brackets is small compared to the second, the electrons and *excited* hydrogen atoms are in equilibrium with each other and with the thermal radiation. Therefore, the ratio of the e, p and $2S$ number densities satisfies a Saha-type relation (see the second equality in (3.194)). The ionization fraction, however, does not obey (3.189) because, as mentioned above, the ground state is not in equilibrium with the other levels after the beginning of recombination. The excited states are more abundant than one expects in full equilibrium and the ionization fraction significantly exceeds that given in (3.189).

Problem 3.25 The cross-section for recombination to the $2S$ level is well approximated by the formula

$$\langle \sigma v \rangle_{ep \to \gamma 2S} \simeq 6.3 \times 10^{-14} \left(\frac{B_H}{4T}\right)^{1/2} \text{ cm}^3 \text{ s}^{-1}. \quad (3.197)$$

Using this expression, verify that the two terms inside the square brackets in (3.196) become comparable at the temperature $T \simeq 2450$ K.

Hence, only at $T > 2450$ K is the reaction $\gamma 2S \rightleftarrows ep$ efficient in maintaining chemical equilibrium between the electrons, protons and hydrogen $2S$ states. After

3.6 Recombination

the temperature drops below $T \simeq 2450$ K, thermal radiation no longer plays an essential role, and the quasi-equilibrium concentration of the $2S$ states is determined by equating the rates for recombination to the $2S$ level and two-photon decay (see (3.193), where the second term can be neglected). At $T < 2450$ K, the second term in the square brackets in (3.196) can be neglected and (3.196) simplifies to

$$\frac{dX_e}{dt} \simeq -\langle \sigma v \rangle_{ep \to 2S} n_T X_e^2. \tag{3.198}$$

Thus, in this regime, the rate of recombination is entirely determined by the rate of recombination to the $2S$ level and does not depend on W_{2S}. At this time, the number density of L_α photons is almost completely depleted due to two-photon decays and the $2P$ states also drop out of equilibrium with electrons, protons and thermal radiation. Consequently, nearly every recombination event to the $2P$ state or any other excited state succeeds in producing a neutral hydrogen atom. This effect becomes relevant only at late stages and can be incorporated in (3.198) and (3.196) by replacing $\langle \sigma v \rangle_{ep \to 2S}$ with the cross-section for recombination to all excited states. The latter is well approximated by the fitting formula

$$\langle \sigma v \rangle_{rec} \simeq 8.7 \times 10^{-14} \left(\frac{B_H}{4T} \right)^{0.8} \text{ cm}^3 \text{ s}^{-1}. \tag{3.199}$$

It is convenient to rewrite the corrected (3.196) using the redshift parameter $z = T/T_{\gamma 0} - 1$, instead of cosmological time (see (3.191)). After some elementary algebra, we obtain

$$\frac{dX_e}{dz} \simeq 0.1 \frac{\eta_{10}}{\sqrt{\Omega_m h_{75}^2}} \left[0.72 \left(\frac{z}{14400} \right)^{0.3} + 10^4 z \exp\left(-\frac{14400}{z} \right) \right]^{-1} X_e^2. \tag{3.200}$$

This equation is readily integrated:

$$X_e(z) \simeq 6.9 \times 10^{-4} \frac{\sqrt{\Omega_m h_{75}^2}}{\eta_{10}} \left[\int_{\frac{z}{14400}} \frac{dy}{0.72 y^{0.3} + 1.44 \times 10^8 y \exp(-1/y)} \right]^{-1}. \tag{3.201}$$

The solution $X_e(z)$ is not very sensitive to the initial conditions when $X_e(z_{in}) \gg X_e(z)$ because the main contribution to the integral comes from $z < z_{in}$. For $z > 900$, or equivalently, at $T > 2450$ K, the first term in the denominator of the integrand can be neglected and expression (3.201) is well approximated by

$$X_e(z) \simeq 1.4 \times 10^9 \frac{\sqrt{\Omega_m h_{75}^2}}{\eta_{10}} z^{-1} \exp\left(-\frac{14400}{z} \right). \tag{3.202}$$

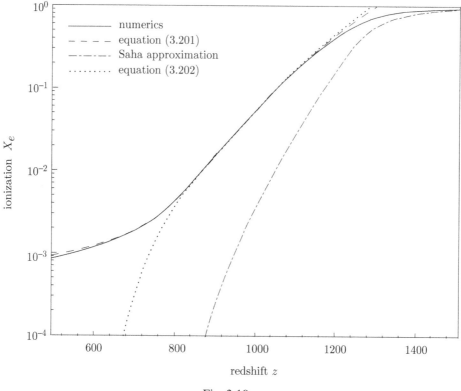

Fig. 3.10.

In this regime the rate of recombination is completely determined by the rate of two-photon decay. Obviously, (3.201) and (3.202) are valid only after the ionization fraction decreases significantly below unity and the deviation from the equilibrium becomes significant. Compared with numerical results, they become efficiently accurate after the concentration of neutral hydrogen has reached about 50% (Figure 3.10). According to (3.202), for realistic values of the cosmological parameters $\left(\Omega_m h_{75}^2 \simeq 0.3 \text{ and } \eta_{10} \simeq 5\right)$, this occurs at $z \simeq 1220$ or, equivalently, at $T \simeq 3400$ K. Hence, the range of applicability of (3.202) is not very wide, namely, $1200 > z > 900$. During this time, however, the temperature drops only from 3400 K to 2450 K but the ionization fraction decreases very substantially, to $X_e(900) \simeq 2 \times 10^{-2}$. It is interesting to compare this result with the prediction of the equilibrium Saha formula (3.189), according to which $X_e(2450 \text{ K}) \sim 10^{-5}$. Thus, at $z \simeq 900$, the actual ionization fraction is a thousand times larger than the equilibrium one. It is also noteworthy that the equilibrium ionization fraction is completely determined by the baryon density and the temperature but the nonequilibrium $X_e(z)$, given in (3.201), also depends on the total density of nonrelativistic matter. This is not surprising because nonrelativistic matter determines the

3.6 Recombination

cosmological expansion rate, which is an important factor in the kinetic description of nonequilibrium recombination.

Problem 3.26 Compare nonequilibrium recombination with the predictions of the Saha formula for various values of the cosmological parameters $\Omega_m h_{75}^2$ and η_{10}. In which cases is the deviation from the Saha result large immediately after the beginning of recombination?

At $z < 900$, when the temperature drops below 2450 K, the approximate formula (3.202) is no longer valid and we should use (3.201). The ionization fraction continues to drop at first and then freezes out. For example, (3.201) predicts $X_e(z = 800) \simeq 5 \times 10^{-3}$, $X_e(400) \simeq 7 \times 10^{-4}$ and $X_e(100) \simeq 4 \times 10^{-4}$, for $\Omega_m h_{75}^2 \simeq 0.3$ and $\eta_{10} \simeq 5$. To calculate the freeze-out concentration, we note that the integral in (3.201) converges to 0.27 as z goes to zero; hence,

$$X_e^f \simeq 2.5 \times 10^{-3} \frac{\sqrt{\Omega_m h_{75}^2}}{\eta_{10}} \simeq 1.6 \times 10^{-5} \frac{\sqrt{\Omega_m}}{\Omega_b h_{75}}. \tag{3.203}$$

After the ionization fraction drops below unity, the approximate results given in (3.201) and (3.202) are in excellent agreement with the numerical solutions of the kinetic equations, while the Saha approximation fails completely (see Figure 3.10).

Problem 3.27 Freeze-out of the electron concentration occurs roughly when the rate of the reaction $ep \to H\gamma$ becomes comparable to the cosmological expansion rate. Using this simple criterion, estimate the freeze-out concentration.

At the beginning of recombination, most of the neutral hydrogen atoms are formed as a result of cascading transitions, and the number of L_α photons is about the same as the number of hydrogen atoms. What happens to all these L_α photons afterwards? Do they survive and, if so, can we observe them today as a (redshifted) narrow line in the spectrum of the CMB? During recombination, the number density of L_α photons, n_α, satisfies the quasi-equilibrium condition for the L_α reservoir:

$$W_{2P \to 1S} n_{2P} = \langle \sigma_\alpha \rangle n_\alpha n_{1S}. \tag{3.204}$$

Since $n_{1S} \to n_T$ and $n_{2P} \propto X_e^2$, the number of L_α photons decreases with the ionization fraction and nearly all of them disappear by the end of recombination. Their number density is depleted due to two-photon decays of the $2S$ states. Hence, there will be no sharp line in the primordial radiation spectrum. Nevertheless, as a result of recombination, the CMB is warped in this part of the Wien region. This region is significantly obscured, however, by radiation from other astrophysical sources.

Finally, let us find out when exactly the universe becomes transparent to radiation. This occurs when the typical time for photon scattering begins to exceed the cosmological time. The Rayleigh cross-section for scattering on neutral hydrogen is negligibly small and, therefore, despite of the low concentration of electrons, the opacity is due to Thomson scattering from free electrons. Substituting $\sigma_T \simeq 6.65 \times 10^{-25}$ cm^2, cosmological time t and total number density n_t from (3.191) and (3.188) respectively, into

$$t \sim \frac{1}{\sigma_T n_t X_e}, \tag{3.205}$$

we find that photon decoupling occurs when

$$X_e^{dec} \sim 6 \times 10^3 \frac{\sqrt{\Omega_m h_{75}^2}}{\eta_{10}} \left(\frac{T_{\gamma 0}}{T_{dec}} \right)^{3/2}. \tag{3.206}$$

It follows that $T_{dec} \sim 2500$ K, or equivalently, $z_{dec} \sim 900$, independent of the cosmological parameters. For $\Omega_m h_{75}^2 \simeq 0.3$ and $\eta_{10} \simeq 5$, the ionization fraction at this time is about 2×10^{-2}. It is interesting to note that this time coincides with the moment when e, p and $2S$ levels fall out of equilibrium and the approximate (3.202) becomes inapplicable.

Radiation decoupling does not mean that matter and radiation lose all thermal contact. In fact, the interaction of a small number of photons with matter keeps the temperatures of matter and radiation equal down to redshifts $z \sim 100$. Only after that does the temperature of baryonic matter begin to decrease faster than that of radiation. There is no trace of this temperature in baryons seen today because most of them are bound to galaxies where they are heated during gravitational collapse.

4
The very early universe

The laws of particle interactions are well established only below the energy currently reached by accelerators, which is about a few hundred GeV. The next generation of accelerators will allow us to go a couple of orders of magnitude further, but even in the remote future it will be impossible to overcome the existing gap of about seventeen orders of magnitude to reach the Planckian scale. Therefore, the only "laboratories" for testing particle theories at very high energies are the very early universe and astrophysical sources of highly energetic particles. The quality of cosmological information is much worse than that gained from accelerators. However, given the lack of choice, we can still hope to learn essential features of high-energy physics based on cosmological and astrophysical observations.

The particle theory describing interactions below the TeV scale is called the Standard Model and it comprises the unified electroweak theory and quantum chromodynamics, both based on the idea of local gauge symmetry. Attempts to incorporate the electroweak and strong interactions in some larger symmetry group and thus unify them have not yet met with success. Unfortunately, there are too many ways to extend the theory beyond the Standard Model while remaining in agreement with available experimental data. Only further experiments can help us in selecting the "correct theory of nature."

This situation determines our selection of topics for this chapter. First, we consider the Standard Model, and explore the most interesting consequences of this theory for cosmology. In particular, the quark–gluon transition, restoration of electroweak symmetry and nonconservation of the fermion number will be discussed in great detail.

Two important cosmological issues beyond the Standard Model are the generation of baryon asymmetry in the universe and the nature of weakly interacting massive particles, a possible component of cold dark matter. In the following chapter we will see that any initial baryon asymmetry is washed out during inflation and its generation is a crucial element of inflationary cosmology. The general conditions under which this asymmetry occurs are rather simple and model-independent. However, the particular realization of these conditions depends on the particle theory involved. At present, there exists no preferable scenario for the origin of baryon asymmetry. There are many possibilities and the problem is, as always, to select the correct one. For these reasons, we will only demonstrate that the important *single* number, characterizing baryon excess, can be easily "explained." The situation for the origin of cold dark matter is very similar, and we likewise concentrate on general ideas here.

Almost all plausible extensions of the Standard Model have a number of features in common, which are rather insensitive to the details of any particular theory. Among these features is a nontrivial vacuum structure, potentially responsible for phase transitions in the very early universe. As a result, topological defects, such as domain walls, strings, or monopoles, could also have been formed. There is no doubt that such good physics belongs to a primary course on cosmology.

We begin with a brief overview of the *elements* of the Standard Model, which should by no means be considered a substitute for standard textbooks in particle physics. It serves as a reminder of the basic ideas we need in cosmological applications. To shorten the presentation, we follow an "antihistorical" approach: the theory is formulated in its "final" form, and then its consequences for cosmology are explored. However, the reader should not forget that the numerous building blocks of the Standard Model were discovered as a result of concerted − and rarely straightforward − efforts to understand and interpret an enormous amount of experimental data.

4.1 Basics

Elementary particles are the fundamental indivisible components of matter. They are completely characterized by their masses, spins and charges. Different charges are responsible for different interactions and the interaction strength is proportional to the corresponding charge. There are four known forces: *gravitational, electromagnetic, weak, and strong*. The first two are long-range forces whose strength decays following an inverse square law. The weak and strong interactions are short-range forces. They are effective only over short distances and then decay exponentially quickly outside this range. Gravity is described by Einstein's theory of General Relativity, and the other three interactions by the Standard Model, based on the idea of local gauge invariance.

4.1.1 Local gauge invariance

Particles are interpreted as elementary excitations of fields. The field describing free fermions of spin one half (for instance, electrons) obeys the Dirac equation

$$i\gamma^\mu \partial_\mu \psi - m\psi = 0, \tag{4.1}$$

where ψ is the four-component Dirac spinor and γ^μ are the 4×4 Dirac matrices. This equation can be derived from the Lorentz-invariant Lagrangian density

$$\mathcal{L} = i\bar\psi \gamma^\mu \partial_\mu \psi - m\bar\psi\psi, \tag{4.2}$$

where $\bar\psi \equiv \psi^\dagger \gamma^0$. This Lagrangian is also invariant under *global* gauge transformations: that is, it does not change when we multiply ψ by an arbitrary complex number with unit norm, for example $\exp(-i\theta)$, where θ is constant in space and time. What happens, however, if we allow θ to vary from point to point, taking $\theta = e\lambda(x^\alpha)$ to be an arbitrary function of space and time? Will the Lagrangian still remain invariant under such *local* gauge transformation? Obviously *not*. Acting on $\lambda(x^\alpha)$, the derivative ∂_μ generates an extra term,

$$\partial_\mu \psi \rightarrow \partial_\mu (e^{-ie\lambda}\psi) = e^{-ie\lambda}(\partial_\mu - ie(\partial_\mu \lambda))\psi, \tag{4.3}$$

and the invariance of the original Lagrangian (4.2) can only be preserved if we modify it by introducing an extra field. Under gauge transformations this field should change in such a way as to cancel the extra term in (4.3). Let us consider a *vector gauge field* A_μ and replace the derivative ∂_μ in (4.3) with the "covariant derivative"

$$\mathcal{D}_\mu \equiv \partial_\mu + ieA_\mu. \tag{4.4}$$

If we assume that under gauge transformations $A_\mu \rightarrow \tilde A_\mu$, then

$$\mathcal{D}_\mu \psi \rightarrow \tilde{\mathcal{D}}_\mu (e^{-ie\lambda}\psi) = e^{-ie\lambda}(\partial_\mu + ie\tilde A_\mu - ie(\partial_\mu \lambda))\psi.$$

Therefore we *postulate* the transformation law

$$A_\mu \rightarrow \tilde A_\mu = A_\mu + \partial_\mu \lambda, \tag{4.5}$$

and find

$$\tilde{\mathcal{D}}_\mu (e^{-ie\lambda}\psi) = e^{-ie\lambda}\mathcal{D}_\mu \psi.$$

Hence

$$\bar\psi \gamma^\mu \mathcal{D}_\mu \psi \rightarrow (\bar\psi e^{ie\lambda})\gamma^\mu \tilde{\mathcal{D}}_\mu (e^{-ie\lambda}\psi) = \bar\psi \gamma^\mu \mathcal{D}_\mu \psi.$$

and Lagrangian (4.2), when we substitute $\mathcal{D}_\mu$ for ∂_μ, is invariant under local gauge transformations.

The gauge field A_μ can be a dynamical field. To find the Lagrangian describing its dynamics, we have to build a gauge-invariant Lorentz scalar out of the field strength A_μ and its derivatives. As follows from (4.5),

$$F_{\mu\nu} \equiv \mathcal{D}_\mu A_\nu - \mathcal{D}_\nu A_\mu = \partial_\mu A_\nu - \partial_\nu A_\mu \qquad (4.6)$$

does not change under gauge transformations and therefore the Lorentz scalar $F_{\mu\nu}F^{\mu\nu}$ is the simplest Lagrangian we can construct. The scalar $A_\mu A^\mu$, which would give mass to the field, is not allowed because it would spoil gauge invariance. In the resulting full Lagrangian,

$$\mathcal{L} = i\bar{\psi}\gamma^\mu \partial_\mu \psi - m\bar{\psi}\psi - \tfrac{1}{4}F_{\mu\nu}F^{\mu\nu} - e(\bar{\psi}\gamma^\mu \psi)A_\mu, \qquad (4.7)$$

in which the reader will immediately recognize electrodynamics with the coupling constant proportional to the electric charge e. Because the fine structure constant $\alpha = e^2/4\pi \simeq 1/137$ is small, one can consider the interaction term as a small correction and hence develop perturbation theory.

It is convenient to represent this perturbation theory by Feynman diagrams, where the interaction term $e(\bar{\psi}\gamma^\mu \psi)A_\mu$ corresponds to a vertex where electron lines ψ, $\bar{\psi}$ meet photon line A. The incoming solid line corresponds to ψ and the outgoing to $\bar{\psi}$. Assuming that time runs "horizontally to the right," Figure 4.1(a) is read as follows: the electron enters the vertex, emits (or absorbs) the photon, and goes on. A rule, which is justified in quantum field theory, is the following: an electron "running backward in time" on the same diagram, but reoriented as in Figure 4.1(b), is interpreted as its antiparticle, a positron, running forward in time.

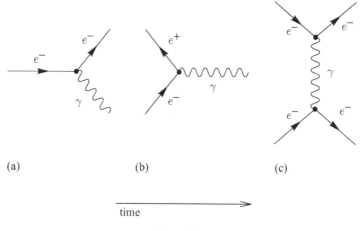

Fig. 4.1.

Therefore, this diagram describes electron–positron annihilation with the emission of a photon. Because the photon is its own antiparticle, we do not need an arrow on its line. More complicated processes can be described by simply combining primitive vertices. For instance, Figure 4.1(c) is responsible for the Coulomb repulsion of two electrons.

The replacement of all particles by antiparticles (charge conjugation C) corresponds to the reversal of all arrows on the diagrams. Lagrangian (4.7) is invariant with respect to charge conjugation.

Problem 4.1 Consider a complex scalar field φ with Lagrangian

$$\mathcal{L} = \tfrac{1}{2}(\partial^\mu \varphi^* \partial_\mu \varphi - m^2 \varphi^* \varphi). \tag{4.8}$$

How should this Lagrangian be generalized to become *locally* gauge-invariant? Write down the interaction terms and draw the corresponding vertices.

4.1.2 Non-Abelian gauge theories

The gauge transformations we have considered so far can be thought of as a multiplication of ψ by 1×1 unitary matrices $\mathbf{U} \equiv \exp(-i\theta)$, satisfying $\mathbf{U}^\dagger \mathbf{U} = \mathbf{1}$. The group of all such matrices is called $U(1)$. The local $U(1)$ gauge invariance of electrodynamics was realized a long time ago. However, the importance of such symmetry was not fully appreciated until 1954, when Yang and Mills extended it to $SU(2)$ local gauge transformations. This symmetry was later used to construct the electroweak theory.

The transformations generated by $N \times N$ unitary matrices $\mathbf{U}$ are called $U(N)$ gauge transformations. Generalizing from $U(1)$ gauge transformations is very straightforward. Let us consider N free Dirac fields with equal masses. Then the Lagrangian is

$$\mathcal{L} = \sum_{a=1}^{N} (i\bar{\psi}_a \gamma^\mu \partial_\mu \psi^a - m\bar{\psi}_a \psi^a) = i\bar{\psi}\gamma^\mu \partial_\mu \psi - m\bar{\psi}\psi, \tag{4.9}$$

where $a = 1, \ldots, N$ and in the second equality we have introduced the matrix notation

$$\psi = \begin{pmatrix} \psi^1 \\ \cdots \\ \psi^N \end{pmatrix}, \quad \bar{\psi} = (\bar{\psi}_1, \cdots, \bar{\psi}_N).$$

One should not forget that every element of these matrices is in its turn a four-component Dirac spinor. The fields ψ^a have the same spins and masses, and therefore differ only by charges (for instance, in quantum chromodynamics these charges are called "colors"). Lagrangian (4.9) is obviously invariant with respect to the *global* gauge transformation generated by the unitary, spacetime-independent matrix **U**:

$$\psi \to \mathbf{U}\psi,$$

because $\bar\psi \to \bar\psi \mathbf{U}^\dagger$ and $\mathbf{U}^\dagger \mathbf{U} = \mathbf{1}$. This is no longer true if we assume that matrix **U** is a function of x^α. As in (4.3), the derivative ∂_α induces an extra term:

$$\partial_\mu \psi \to \partial_\mu(\mathbf{U}\psi) = \mathbf{U}\big(\partial_\mu + \mathbf{U}^{-1}(\partial_\mu \mathbf{U})\big)\psi,$$

which needs to be compensated for if we want to preserve gauge invariance. With this purpose, let us introduce the gauge fields $\mathbf{A}_\mu$, which are Hermitian $N \times N$ matrixes, and replace ∂_α by the "covariant derivative"

$$\mathbf{D}_\mu \equiv \partial_\mu + ig\mathbf{A}_\mu, \qquad (4.10)$$

where g is the gauge coupling constant. If we assume that under gauge transformations $\mathbf{A}_\mu \to \tilde{\mathbf{A}}_\mu$, we obtain

$$\mathbf{D}_\mu \psi \to \tilde{\mathbf{D}}_\mu(\mathbf{U}\psi) = \mathbf{U}\big(\partial_\mu + ig\mathbf{U}^{-1}\tilde{\mathbf{A}}_\mu \mathbf{U} + \mathbf{U}^{-1}(\partial_\mu \mathbf{U})\big)\psi.$$

(Note that one must be careful with the order of multiplication because the matrices do not generally commute.) Therefore, we *postulate* the transformation law

$$\mathbf{A}_\mu \to \tilde{\mathbf{A}}_\mu = \mathbf{U}\mathbf{A}_\mu \mathbf{U}^{-1} + \frac{i}{g}(\partial_\mu \mathbf{U})\mathbf{U}^{-1}. \qquad (4.11)$$

Then

$$\mathbf{D}_\mu \psi \to \tilde{\mathbf{D}}_\mu(\mathbf{U}\psi) = \mathbf{U}\mathbf{D}_\mu \psi,$$

and the Lagrangian

$$\mathcal{L} = i\bar\psi \gamma^\mu \mathbf{D}_\mu \psi - m\bar\psi\psi \qquad (4.12)$$

is invariant under $U(N)$ local gauge transformations. To derive the Lagrangian for the gauge fields, we note that

$$\mathbf{F}_{\mu\nu} \equiv \mathbf{D}_\mu \mathbf{A}_\nu - \mathbf{D}_\nu \mathbf{A}_\mu = \partial_\mu \mathbf{A}_\nu - \partial_\nu \mathbf{A}_\mu + ig(\mathbf{A}_\mu \mathbf{A}_\nu - \mathbf{A}_\nu \mathbf{A}_\mu) \qquad (4.13)$$

transforms as $\mathbf{F}_{\mu\nu} \to \tilde{\mathbf{F}}_{\mu\nu} = \mathbf{U}\mathbf{F}_{\mu\nu}\mathbf{U}^{-1}$, and hence the simplest gauge-invariant Lorentz scalar is $\text{tr}\,(\mathbf{F}_{\mu\nu}\mathbf{F}^{\mu\nu})$. The full Lagrangian is then

$$\mathcal{L} = i\bar\psi \gamma^\mu \partial_\mu \psi - m\bar\psi\psi - g\bar\psi\gamma^\mu \mathbf{A}_\mu \psi - \frac{1}{2}\text{tr}\,(\mathbf{F}_{\mu\nu}\mathbf{F}^{\mu\nu}), \qquad (4.14)$$

where we have used the standard normalization for the last term in cases where $N \geq 2$.

Problem 4.2 Verify the transformation law for $\mathbf{F}_{\mu\nu}$. (*Hint* To simplify the calculation, justify and use the following commutation rule: $\tilde{\mathbf{D}}_\mu \mathbf{U} = \mathbf{U}\mathbf{D}_\mu$.)

Thus, starting from a simple idea, we have achieved a significant result. Namely, the interactions between fermions and gauge fields, as well as the simplest possible Lagrangian for the gauge fields, were completely determined by the requirement of gauge invariance. We would like to stress once more that the gauge fields are massless because a mass term would spoil the gauge invariance.

There is an important difference between $U(1)$ and $U(N)$ groups. All elements of the $U(1)$ group (complex numbers) commute with each other (Abelian group), while in the case of the $U(N)$ group the elements do not generally commute (non-Abelian group). This has an important consequence. The $U(1)$ gauge field has no self-coupling and interacts only with fermions (the last term in (4.13) vanishes when $N = 1$), or, in other words, this field does not carry the group charge (photons are electrically neutral). The non-Abelian $U(N)$ fields do carry group charges and the last term in (4.13) induces their self-interaction.

Problem 4.3 Consider N complex scalar fields instead of fermions and find the interaction terms in this case. Draw corresponding diagrams including those describing the self-interaction of the gauge fields.

To find the minimum number of compensating fields needed to ensure gauge invariance, we have to count the number of generators of the $U(N)$ group or, in other words, the number of independent elements of an $N \times N$ unitary matrix. Any unitary matrix can be written as

$$\mathbf{U} = \exp(i\mathbf{H}), \qquad (4.15)$$

where $\mathbf{H}$ is an *Hermitian* matrix ($\mathbf{H} = \mathbf{H}^\dagger$).

Problem 4.4 Verify that the number of independent *real* numbers characterizing $N \times N$ Hermitian matrix is equal to N^2.

In turn, an Hermitian $\mathbf{H}$ can always be decomposed into a linear superposition of N^2 independent *basis matrices*, one of which is the unit matrix

$$\mathbf{H} = \theta\mathbf{1} + \sum_{C=1}^{N^2-1} \theta^C \mathbf{T}_C = \theta\mathbf{1} + \theta^C \mathbf{T}_C, \qquad (4.16)$$

where $\mathbf{T}_C$ are traceless matrices and θ^C are real numbers; hence

$$\mathbf{U} = e^{i\theta} \exp\left(i\theta^C \mathbf{T}_C\right). \qquad (4.17)$$

The first multiplier corresponds to the $U(1)$ Abelian subgroup of $U(N)$ and the second term belongs to the $SU(N)$ subgroup consisting of all unitary matrixes with det $\mathbf{U} = 1$. Therefore we can write $U(N) = U(1) \times SU(N)$, and consider the local $SU(N)$ gauge groups separately. The $SU(N)$ group has $N^2 - 1$ independent generators and hence we need at least that number of independent compensating fields A_μ^C. The Hermitian matrix $\mathbf{A}_\mu$ can then be written as

$$\mathbf{A}_\mu = A_\mu^C \mathbf{T}_C. \qquad (4.18)$$

For $SU(2)$ and $SU(3)$ groups it is convenient to use as the basis matrices $\sigma_C/2$ and $\lambda_C/2$ respectively, where σ_C are three familiar Pauli matrices and λ_C are eight Gell-Mann matrices, the explicit form of which will not be needed here.

4.2 Quantum chromodynamics and quark–gluon plasma

The strong force is responsible for binding neutrons and protons within nuclei. The particles participating in strong interactions are called hadrons. They can be either fermions or bosons. The fermions have half-integer spin and they are called baryons, while the bosons have integer spin and are called mesons. The hadron family is extremely large. To date, several hundred hadrons have been discovered. It would be a nightmare if all these particles were elementary. Fortunately, they are composite and built out of fermions of spin $1/2$ called quarks. This is similar to the way all chemical elements are made of protons and neutrons. In contrast to the chemical elements, each of which has its own name, only the lightest and most important hadrons have names reflecting their "individuality." To classify hadrons (or, in other words, put them in their own "Periodic Table") we need five different kinds (*flavors*) of quarks, which are accompanied by appropriate antiquarks. The sixth quark, needed for cancellation of anomalies in the Standard Model, was also discovered experimentally. The quarks have different masses and electric charges. Three of them, namely u (up), c (charm), and t (top) quarks, have a positive electric charge, which is $+2/3$ of the elementary charge. The other three quarks, d (down), s (strange), and b (bottom), have a negative electric charge equal to $-1/3$.

Strong interactions of quarks are described by an $SU(3)$ gauge theory called quantum chromodynamics. *According to this theory*, every quark of a given flavor comes in three different colors: "red" (r), "blue" (b), and "green" (g). The colors are simply names for the charges of the $SU(3)$ gauge group, which acts on triplets of spinor fields of the same flavor but different colors. The gauge-invariant quantum chromodynamics Lagrangian is

$$\mathcal{L} = \sum_{f,C} \left(i\bar{\psi}_f \gamma^\mu \partial_\mu \psi_f - m_f \bar{\psi}_f \psi_f - \frac{1}{2} g_s \left(\bar{\psi}_f \gamma^\mu \lambda_C \psi_f \right) A_\mu^C \right)$$
$$- \frac{1}{2} \mathrm{tr}\left(\mathbf{F}_{\mu\nu} \mathbf{F}^{\mu\nu} \right), \qquad (4.19)$$

4.2 Quantum chromodynamics and quark–gluon plasma

where g_s is the strong coupling constant, λ_C (where $C = 1, \ldots, 8$) are the eight Gell-Mann 3×3 matrices and f runs over the quark flavors u, d, s, c, t, b. There are eight gauge fields A_μ^C, called *gluons*, which are responsible for strong interactions. The symbol ψ_f denotes the column of three quark spinor fields:

$$\psi_f \equiv \begin{pmatrix} r_f \\ b_f \\ g_f \end{pmatrix}, \qquad (4.20)$$

where r_f is the Dirac spinor describing the red quark with flavor f, etc. The bare quark masses m_f are determined from experiments. They are very different and not so well known. The lightest is the u quark with mass of 1.5–4.5 MeV. The d quark is a bit heavier: $m_d = 5$–8.5 MeV. The strange quark has a mass of 80–155 MeV, and the remaining three quarks are much heavier: $m_c = 1.3 \pm 0.3$ GeV, $m_b = 4.3 \pm 0.2$ GeV, and $m_t \sim 170$ GeV.

The antiparticles of the quarks are called antiquarks and they can have "antired" ($\bar{r}$), "antiblue" ($\bar{b}$), and "antigreen" ($\bar{g}$) colors. As distinct from photons, gluons are also charged. They carry one unit of color and one of anticolor. For instance, using the explicit form of Gell-Mann matrix λ_1, we find that the first interaction term in the Lagrangian (4.19) is

$$g_s(\bar{b}r + \bar{r}b)A^1,$$

where we have omitted flavor and spacetime indices together with the Dirac matrices. The appropriate quark–gluon vertices describing this interaction are shown in Figure 4.2. When a quark changes its color, the color difference is carried off by a gluon, which in this case is either $r\bar{b}$ or $b\bar{r}$ colored. The state $(r\bar{b} + b\bar{r})$ is the first state of the "color octet" of gluons. Using the explicit form of the Gell-Mann matrices, the reader can easily find the remaining seven states of the octet. In principle, however, from three colors and three anticolors, we can compose *nine* independent color–anticolor combinations: $r\bar{r}, r\bar{b}, r\bar{g}, b\bar{r}, b\bar{b}, b\bar{g}, g\bar{r}, g\bar{b}, g\bar{g}$. Therefore, one is led to ask which particular combination of colors does not occur in Lagrangian (4.19) and hence does not participate in strong interactions. The answer is the "color singlet" $(r\bar{r} + b\bar{b} + g\bar{g})$, which is invariant under $SU(3)$ gauge transformations.

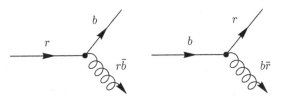

Fig. 4.2.

This color combination would only occur if the unit matrix were among the λ matrices. But the unit matrix was excluded when we decided to restrict ourselves to the $SU(3)$ group instead of the $U(3)$ group. The $U(3)$ group would have an extra $U(1)$ gauge boson decoupled from the other gluons. This boson would induce long range interactions between all hadrons regardless of electrical charge, in obvious contradiction with experiments.

In contrast to photons, gluons interact with each other. The Lagrangian for non-Abelian gauge fields contains third and fourth powers in the field strength A and the corresponding interaction vertices have three and four legs respectively.

Conservation laws are easily determined from the elementary vertices. First of all, we see that quark flavor does not change in strong interactions and this leads to numerous flavor conservation laws. The total number of quarks minus the number of antiquarks also remains unchanged, and hence the total baryon number is conserved (by convention a quark has baryon number $1/3$ and an antiquark $-1/3$). In addition, there is a color conservation law which is analogous to electric charge conservation in electrodynamics.

At first glance, the number of quarks (6 flavors $\times$ 3 colors = 18 quarks) seems too large to give an elegant explanation of the "Periodic Table of the hadrons": the Periodic Table of chemical elements is built out of only two elementary constituents — protons and neutrons. However, one should not forget that unlike the chemical elements, which can be composed from an arbitrary number of protons and neutrons, the few hundred hadrons consist only of quark–antiquark pairs or of three quarks. To be precise, all mesons are composed of quark–antiquark pairs and all baryons consist of three quarks. For instance, the lightest baryons, the proton and the neutron, are composed of uud and udd quarks respectively. The lightest meson, π^+, is made of a u quark and a $\bar{d}$ antiquark.

There is a deep reason why two or four quark bound systems do not exist as free "particles." *Every naturally occurring particle should be a color singlet.* This statement is known as the *confinement* hypothesis, according to which colored particles, irrespective of whether they are elementary or composite, cannot be observed below the confinement scale. In particular, quarks are always bound within mesons and baryons.

As we have seen, the colorless gluon state $(r\bar{r} + b\bar{b} + g\bar{g})$ does not enter the fundamental Lagrangian (4.19). Therefore, it is natural to assume that the appropriate colorless composite particles are "neutral" with respect to strong interaction and can exist at any energy scale. The above color singlet can be built only as a quark–antiquark pair and corresponds to mesons. Another possible color singlet is a three-quark combination: $(rbg - rgb + grb - gbr + bgr - brg)$, and it corresponds to baryons. All other colorless states can be interpreted as describing few mesons or baryons.

Problem 4.5 Verify that color–anticolor and three-quark combinations are really colorless, that is, they do not change under gauge transformation. (*Hint* The different anticolors can be thought of as three element rows, $\bar{r} = (1, 0, 0)$, $\bar{b} = (0, 1, 0)$, $\bar{g} = (0, 0, 1)$ and different colors as corresponding columns.)

Confinement should, in principle, be derivable from the fundamental Lagrangian (4.19) but until now this has not been achieved. Nevertheless, there are strong experimental and theoretical indications that this hypothesis is valid. In particular, the increase of the strong interaction strength at low energy strongly supports the idea of confinement. The energy dependence of the strong coupling constant has another important feature: the interaction strength vanishes in the limit of very high energies (or, correspondingly, at very small distances). This is called *asymptotic freedom*. As a consequence, this allows us to use perturbation theory to calculate strong interaction processes with highly energetic hadrons. We explain below why the coupling constant should be scale-dependent and then calculate how it "runs." To introduce the concept of a running coupling, we begin with familiar electromagnetic interactions, and then derive the results for a general renormalizable field theory and apply them to quantum chromodynamics.

4.2.1 Running coupling constant and asymptotic freedom

Let us consider two electrically charged particles. According to quantum electrodynamics, their interaction via photon exchange can be represented by a set of diagrams, some of which are shown in Figure 4.3. The contribution of a particular diagram to the total interaction strength is proportional to the number of primitive vertices in the diagram. Each vertex brings a factor of e. Because $e \ll 1$, the largest contribution to the interaction ($\propto e^2$) comes from the simplest (tree-level) diagram with only two vertices. The next order (one-loop) diagrams have four vertices and their contribution is proportional to e^4. Hence, the diagrams in Figure 4.3 are simply a graphical representation of the perturbative expansion in powers of the fine structure constant

$$\alpha \equiv \frac{e^2}{4\pi} \simeq \frac{1}{137}.$$

The interaction strength depends not only on the charge but also on the distance between the particles, characterized by the 4-momentum transfer

$$q^\mu = p_2^\mu - p_1^\mu. \tag{4.21}$$

Note that for virtual photons $q^2 \equiv |q^\mu q_\mu| \neq 0$, that is, they do not lie on their mass shell.

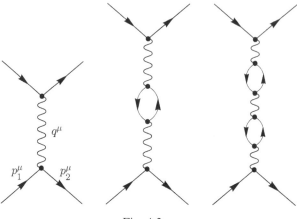

Fig. 4.3.

The diagrams containing closed loops are generically divergent. Fortunately, in so-called renormalizable theories, the divergences can be "isolated and combined" with bare coupling constants, bare masses, etc. What we measure in experiment is not the value of the bare parameter, but only the finite outcome of "its combination with infinities." For instance, given a distance characterized by the momentum transfer $q^2 = \mu^2$ (called the normalization point), we can measure the interaction force and thus determine the renormalized coupling constant $\alpha(\mu^2)$ which becomes the actual parameter of the perturbative expansion. After removing and absorbing the infinities, there remain finite q^2-dependent loop contributions to the interaction force (the vacuum polarization effect). They too can be absorbed by redefining the coupling constant which becomes q^2-dependent, or in other words, begins to run. In the limit of vanishing masses (or for $q^2 \gg m^2$), the expansion of the *dimensionless running coupling "constant"* $\alpha(q^2)$ in powers of the renormalized coupling constant $\alpha(\mu^2)$ can be expressed on dimensional grounds as

$$\alpha(q^2) = \alpha(\mu^2) + \alpha^2(\mu^2) f_1\left(\frac{q^2}{\mu^2}\right) + \cdots = \sum_{n=0}^{\infty} \alpha^{n+1}(\mu^2) f_n\left(\frac{q^2}{\mu^2}\right), \qquad (4.22)$$

where $f_0 = 1$ and the other functions f_n are determined by appropriate *n-loop* diagrams. Since $\alpha(q^2) = \alpha(\mu^2)$ at $q^2 = \mu^2$, we have

$$f_n(1) = 0$$

for $n \geq 1$.

If we consider a process with q-momentum transfer, we can use the running constant $\alpha(q^2)$ instead of $\alpha(\mu^2)$ as a small expansion parameter in the remaining finite diagrams. This corresponds to the resummation of finite contributions from divergent diagrams. However, to take advantage of this resummation, we have to

4.2 Quantum chromodynamics and quark–gluon plasma

figure out the structure of perturbative expansion (4.22) and find a way to resum this series, at least partially. This can be done using simple physical arguments. Let us note that the value of the coupling constant $\alpha(q^2)$ should not depend on the normalization point μ^2, which is arbitrary. Therefore, the derivative of the right hand side of (4.22) with respect to μ^2 should be equal to zero:

$$\frac{d}{d\mu^2}\left(\sum_{n=0}^{\infty} \alpha^{n+1}(\mu^2) f_n\left(\frac{q^2}{\mu^2}\right)\right) = 0. \tag{4.23}$$

Differentiating and rearranging the terms, we obtain the following differential equation for $\alpha(\mu^2)$:

$$\frac{d\alpha(\mu^2)}{d\ln\mu^2} = \alpha^2 \left(\frac{\sum_{l=0}^{\infty} x f'_{l+1}(x) \alpha^l}{\sum_{l=0}^{\infty} (l+1) f_l(x) \alpha^l}\right) = \alpha^2(\mu^2)\left(\sum_{l=0}^{\infty} f'_{l+1}(1) \alpha^l(\mu^2)\right), \tag{4.24}$$

where a prime denotes the derivative with respect to $x \equiv q^2/\mu^2$. The ratio of sums in (4.24) should not depend on x because the left hand side of the equation is x-independent. Therefore, to obtain the second equality in (4.24) we set $x = 1$. The requirement that the ratio does not depend on x imposes rather strong restrictions on the admissible functions $f_n(x)$. From the second equality, we derive the following recurrence relations:

$$\frac{df_{n+1}(x)}{d\ln x} = \sum_{k=0}^{n} (k+1) f'_{n+1-k}(1) f_k(x). \tag{4.25}$$

Problem 4.6 Verify that the general solution of these recurrence relations is given by

$$f_n(x) = \sum_{l=0}^{n} c_l (\ln x)^l, \tag{4.26}$$

where the numerical coefficient in front of the leading logarithm is equal to $c_n = (f'_1(1))^n$.

The running constant $\alpha(q^2)$ depends on q^2 in the same way that $\alpha(\mu^2)$ depends on μ^2. Hence $\alpha(q^2)$ satisfies the equation

$$\frac{d\alpha(q^2)}{d\ln q^2} = \alpha^2(q^2)\left(\sum_{l=0}^{\infty} f'_{l+1}(1) \alpha^l(q^2)\right), \tag{4.27}$$

which follows from (4.24) by the substitution $\mu^2 \to q^2$. Equations (4.24) and (4.27) are the well known Gell-Mann–Low *renormalization group* equations and the expression on the right hand side of these equations,

$$\beta(\alpha) = f_1'(1)\alpha^2 + f_2'(1)\alpha^3 + \cdots, \tag{4.28}$$

is called the β function.

The results obtained are generic and valid in any renormalizable quantum field theory. The only input we need from concrete theory is the numerical values of the coefficients $f_n'(1)$. For instance, to determine $f_1'(1)$, one has to calculate the appropriate one-loop diagrams. Other coefficients require the calculations of higher-order diagrams.

Let us assume that $\alpha(q^2) \ll 1$ for q^2 interest of (this assumption should be checked *a posteriori*). In this case we may retain in the β function only the leading one-loop term $f_1'(1)\alpha^2$, and neglect all higher order contributions. Equation (4.27) is then easily integrated, with the result

$$\alpha(q^2) = \frac{\alpha(\mu^2)}{1 - f_1'(1)\alpha(\mu^2)\ln(q^2/\mu^2)}, \tag{4.29}$$

where $\alpha(\mu^2)$ reappears as an integration constant. The expression obtained corresponds to the partial resummation of series (4.22). As is clear from (4.26), this resummation takes into account only the leading $(\ln x)^n$ contributions of all *n-loop* diagrams. Knowledge of the β-function to two loops, combined with the Gell-Mann–Low equations, would allow us to resum next-to-leading logarithms.

Problem 4.7 Find the behavior of the running coupling constant in the two-loop approximation. (The coefficient $f_1'(1)$ does not depend on the renormalization scheme, while $f_2'(1)$, $f_3'(1)$ etc. can be scheme-dependent).

In quantum electrodynamics the coefficient $f_1'(1)$ is positive and equals $1/3\pi$. In this case the coupling $\alpha(q^2)$ increases as the charges get closer together (q^2 increases). It is a straightforward consequence of vacuum polarization. In fact, the vacuum can be thought of as a kind of "dielectric media" where negative charge attracts positive charges and repel negative ones. As a result, the charge is surrounded by a polarized "halo", which screens it. Therefore, a negative charge, observed from far away (small q^2), will be reduced by the charge of the surrounding halo. At higher q^2 we approach the charge more closely, penetrating inside the halo, and see a diminished screening of the charge.

In quantum chromodynamics we have one-loop diagrams of the kind shown in Figure 4.3, where quarks and gluons should be substituted for the electrons and photons respectively. They also give a positive contribution to $f_1'(1)$, proportional to the number of possible diagrams of this kind and hence to the number of quark flavors. As previously noted, gluons unlike photons, are charged. Hence, in quantum chromodynamics, in addition to the diagrams in Figure 4.3 there are also one-loop

4.2 Quantum chromodynamics and quark–gluon plasma

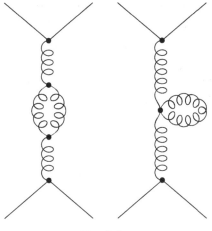

Fig. 4.4.

diagrams with virtual *gluon* bubbles (Figure 4.4). Their contribution to $f_1'(1)$ is *negative* and the number of possible diagrams of this kind is proportional to the number of colors. For non-Abelian gauge theory with f massless flavors and n colors,

$$f_1'(1) = \frac{1}{12\pi}(2f - 11n). \tag{4.30}$$

Problem 4.8 Why does the fermion contribution to $f_1'(1)$ not depend on the number of colors? Why is the gluon contribution proportional to the number of colors, but not to the number of different gluons? Why do one-loop gluon diagrams, which are due to the coupling A^4, not contribute to $f_1'(1)$?

The formula for the running coupling constant was derived in the limit when fermion masses are negligible compared to q. It turns out that the contribution of fermions with mass m to $f_1'(1)$ becomes significant only when q^2 becomes larger than m^2. Hence the β function coefficients change by discrete amounts as quark masses are crossed. In quantum chromodynamics, $f = 6$ for energies larger than the top quark mass ($\sim 170\,\text{GeV}$) and $f = 5$ in the range $5\,\text{GeV} \ll q \ll 170\,\text{GeV}$. On the other hand, since the number of colors is $n = 3$, $f_1'(1)$ is always less than zero. This has far-reaching consequences. As follows from (4.29), the running "strong fine structure constant", $\alpha_s(q^2) \equiv g_s^2/4\pi$, decreases as q^2 increases. This is opposite to the situation in electrodynamics. The strength of strong interactions decreases at very high energies (small distances), so that α_s becomes much smaller than unity and we can use perturbation theory to calculate highly energetic hadron processes. The approximation we used to derive $\alpha_s(q^2)$ becomes more and more reliable as q^2 grows and in the limit that $q^2 \to \infty$, interactions disappear. This property of quantum chromodynamics is known as *asymptotic freedom*. The decrease in the coupling constant is due to the gluon loops, which dominate over the fermion loop contribution and thus lead to antiscreening of the colors. Quark colors are mainly due to the polarized halo of gluons.

The normalization point μ^2 in (4.29) is arbitrary, and the value of $\alpha_s(q^2)$ does not depend on it. Introducing the physical scale Λ_{QCD}, defined by

$$\ln\frac{\Lambda_{QCD}^2}{\mu^2} = -\frac{12\pi}{(11n - 2f)\alpha_s(\mu^2)},$$

we can rewrite (4.29) for the running strong fine structure constant in terms of a single parameter:

$$\alpha_s(q^2) = \frac{12\pi}{(11n - 2f)\ln(q^2/\Lambda_{QCD}^2)}. \tag{4.31}$$

Experimental data suggest that Λ_{QCD} is about 220 MeV (to 10% accuracy). The strong coupling constant α_s is 0.13 at $q \simeq 100$ GeV and increases to 0.21 when the energy decreases to 10 GeV (in this energy range $f = 5$). According to (4.31), the strength of strong interactions should become infinite at $q^2 = \Lambda_{QCD}^2$. However, this is not more than an informed estimate consistent with the confinement hypothesis. We should not forget that (4.31) was derived in the one-loop approximation and is applicable only if $\alpha_s(q^2) \ll 1$, that is, at $q^2 \gg \Lambda_{QCD}^2$. At $q^2 \sim \Lambda_{QCD}^2$, all loops give comparable contributions to the β-function and when α_s becomes the order of unity, (4.31) fails. To go further we have to apply nonperturbative methods, for instance, numerical lattice calculations. These methods also strongly support the idea of confinement.

Quantum chromodynamics is a quantitative theory only when we consider highly energetic processes with $q \gg O(1)$ GeV. The strong force binding baryons in the nuclei is a low-energy process and cannot be calculated perturbatively. It can only be *qualitatively* explained as the result of collective multi-gluon and pion exchange.

4.2.2 Cosmological quark–gluon phase transition

At high temperature and/or baryon density we can expect a transition from hadronic matter to a quark–gluon plasma. In the very early universe at temperatures exceeding $\Lambda_{QCD} \simeq 220$ MeV, the strong coupling $\alpha_s(T^2)$ is small and most quarks and gluons only interact with each other weakly. They are no longer confined within particular hadrons and their degrees of freedom are liberated. In this limit the quark–gluon plasma consists of free noninteracting quarks and gluons, which can be described in the ideal gas approximation. Of course, there always exist soft modes with momenta $q^2 \leq \Lambda_{QCD}^2$, which can by no means be treated as noninteracting particles;

4.2 Quantum chromodynamics and quark–gluon plasma

but at $T \gg \Lambda_{QCD}$ they constitute only a small fraction of the total energy density. Baryon number is very small and therefore we can neglect the appropriate chemical potentials. The contribution of the quark–gluon plasma to the total pressure is then

$$p_{qg} = \frac{\kappa_{qg}}{3} T^4 - B(T), \qquad (4.32)$$

where function $B(T)$ represents the *correction* due to the soft, low-energy modes and

$$\kappa_{qg} = \frac{\pi^2}{30}\left(2 \times 8 + \frac{7}{8} \times 3 \times 2 \times 2 \times N_f\right). \qquad (4.33)$$

The first term on the right hand side here accounts for the contribution of eight gluons (with two polarizations each) and the second one is due to N_f light quark flavors with $m_q \ll T$ (every flavor has three colors, two polarization states and the extra factor 2 accounts for antiquarks).

Unfortunately the correction term $B(T)$ cannot be calculated analytically from first principles. To get an idea of how it may look, we can use a phenomenological description of confinement, for instance, the MIT bag model. According to this model, quarks and gluons are described by free fields inside bags (bounded regions of space), identified with hadrons, and these fields vanish outside the bags. To account for appropriate boundary conditions in a relativistically invariant way, one adds to the Lagrangian "a cosmological constant" B_0 (called the bag constant), which is assumed to vanish outside the bag. This "cosmological constant" induces negative pressure and prevents quarks escaping from the bag. In a quark–gluon plasma, where the bags "overlap", $B(T) = B_0 = \text{const}$ everywhere.

Given the pressure, the energy density and entropy can be derived using the thermodynamical relations (3.33) and (3.31):

$$s_{qg} = \frac{4}{3}\kappa_{qg} T^3 - \frac{\partial B}{\partial T}, \quad \varepsilon_{qg} = \kappa_{qg} T^4 + B - T\frac{\partial B}{\partial T}. \qquad (4.34)$$

As soon as the temperature drops below some critical value T_c, which on quite general grounds is expected to be about $\Lambda_{QCD} \simeq 200\,\text{MeV}$, most quarks and gluons will be trapped and confined within the lightest hadrons – pions ($\pi^0, \pi^\pm$). Their masses are about 130 MeV and at the time of the phase transition they can still be treated as ultra-relativistic particles. After quarks and gluons are captured, the total number of degrees of freedom drops drastically, from 16 (for gluons) + $12N_f$ (for quarks) to only 3 for pions. The pressure and entropy density of the ultra-relativistic pions are then

$$p_h = \frac{\kappa_h}{3} T^4, \quad s_h = \frac{4}{3}\kappa_h T^3, \qquad (4.35)$$

where $\kappa_h = \pi^2/10$.

Whether the transition from the quark–gluon plasma to hadronic matter is characterized by truly singular behavior of the basic thermodynamical quantities or their derivatives (first or second order phase transitions, respectively), or whether it is merely a cross-over with rapid continuous change of these quantities, crucially depends on the quark masses. A first order phase transition is usually related to a discontinuous change of the symmetries characterizing the different phases. In $SU(3)$ pure gauge theory, without dynamical quarks, the expected first order phase transition has been verified in numerical lattice calculations. In the case of two quark flavors one expects a second order phase transition (continuous change of symmetry). In the limit of three massless quarks, again for reasons of symmetry, we expect a first order transition. When the quark masses do not vanish, the appropriate symmetries can be broken explicitly and a cross-over is expected. This is the situation most likely realized in nature: of the three quarks relevant for dynamics, two (u, d) are very light and one (s) is relatively heavy. However, despite more than 20 years of efforts, the character of the cosmological quark–gluon transition has not yet been firmly established. This is due to the great difficulty of computing with light dynamical quarks on the lattice. The possibility of a true phase transition, therefore, has not been ruled out.

Irrespective of the nature of the transition, there is a very sharp change in the energy density and entropy in the narrow temperature interval around T_c. This result is confirmed in lattice calculations and clearly indicates the liberation of the quark degrees of freedom. A first order phase transition is the most interesting for cosmology and therefore we briefly discuss it, assuming $B(T) = B_0 = $ const, as in the bag model. This reproduces the bulk features of the equation of state obtained in numerical lattice calculations. The pressure and entropy density as functions of temperature are shown in Figure 4.5. At T_c, even if the phase transition is first order,

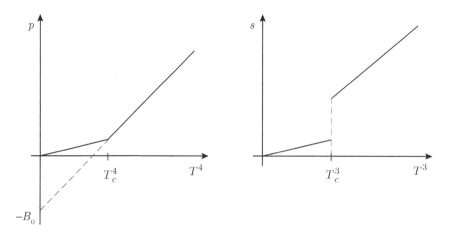

Fig. 4.5.

the pressure should be continuous, allowing both phases (hadrons and quark–gluon plasma) to coexist. Hence, equating (4.32) and (4.35) at $T = T_c$, we can express the critical temperature T_c through the bag constant B_0:

$$T_c = \left(\frac{3B_0}{\kappa_{qg} - \kappa_h}\right)^{1/4} = \left(\frac{180}{(26 + 21N_f)\pi^2}\right)^{1/4} B_0^{1/4}. \qquad (4.36)$$

For $B_0^{1/4} \simeq 220$ MeV and for $N_f = 3$ light quark flavors, $T_c \simeq 150$ MeV.

Problem 4.9 How should (4.32) be modified if the baryon number is different from zero? Using the condition $p_{qg}(T_c, \mu_B) \simeq 0$, where μ_B is the baryon's chemical potential, as an approximate criterion for the phase transition, draw in the T_c–μ_B plane the shape of the transition line separating the hadron and quark–gluon phases. Why does the above criterion give us a good estimate?

In the case of a first order phase transition, the entropy density is discontinuous at the transition and its jump, $\Delta s_{qg} = (4/3)(\kappa_{qg} - \kappa_h)T_c^3$, is directly proportional to the change in the number of active degrees of freedom. A first order phase transition occurs via the formation of bubbles of hadronic phase in the quark–gluon plasma. As the universe expands, these bubbles take up more and more space and when what is left is mainly the hadronic phase the transition is over. During a first order phase transition the temperature is strictly constant and is equal to T_c. The released latent heat, $\Delta\varepsilon = T_c \Delta s_{qg}$, keeps the temperature of the radiation and leptons unchanged in spite of expansion. To estimate the duration of the transition one can use the conservation law for the total entropy.

Problem 4.10 Taking into account that in the quantum chromodynamics epoch, in addition to quarks and gluons, there are photons, three flavors of neutrinos, electrons and muons, verify that the scale factor increases by a factor of about 1.5 during the phase transition.

If the transition is of second order or a cross-over, the entropy is a continuous function of temperature, which changes very sharply in the vicinity of T_c. As the universe expands the temperature always drops, but during the transition it remains nearly constant. For the case of a cross-over transition, the notion of phase is not defined during the transition.

As we have already mentioned, only a first order quantum chromodynamics phase transition has interesting cosmological consequences. This is due to its "violent nature." In particular, it could lead to inhomogeneities in the baryon distribution and hence influence nucleosynthesis. However, calculations show that this effect is too small to be relevant. There could be other, more speculative consequences, for instance, the formation of quark nuggets, the generation of magnetic fields and gravitational waves, black hole formation, etc. These are still the subjects of investigation. At present, however, it is unlikely that a quantum chromodynamics phase transition leaves the observationally important "imprint." It looks like an interesting but rather "silent" epoch in the evolution of the universe.

4.3 Electroweak theory

The most familiar weak interaction process is neutron decay: $n \to pe\bar{\nu}_e$. In Section 3.5 we described it using the Fermi four-fermion interaction theory, which is very successful at low energies. However, this theory is not self-consistent because it is not renormalizable and violates unitarity (conservation of probability) at high energies (above 300 GeV). The Fermi constant G_F, characterizing the strength of the weak interactions, has dimension of inverse mass squared. Therefore, it is natural to assume that the four-fermion vertex is simply the low-energy limit of a diagram made of three-legged vertices with dimensionless coupling g_w, which describes the exchange of a massive vector boson W (Figure 4.6).

At energies much smaller than the mass of the intermediate boson, one can replace the boson propagator with $1/M_W^2$ and the diagram shrinks to the four-legged diagram with effective coupling constant $G_F = O(1) g_w^2/M_W^2$. Noting the vectorial nature of the weak interactions, we are led to a description using gauge symmetries. However, an obstacle immediately arises. We have noted above that gauge bosons should be massless because the mass term spoils gauge invariance, renormalizability and unitarity. This problem was finally resolved by the renormalizable standard electroweak theory, where the masses of all particles (including intermediate gauge bosons) emerge as a result of their interaction with a classical scalar field. The electroweak theory is based on a unification (or, more precisely, on a "mixing") of electromagnetic and weak interactions within a model based on $SU(2) \times U(1)$

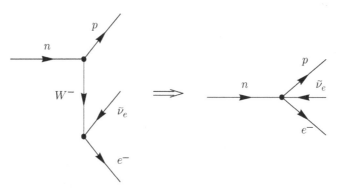

Fig. 4.6.

4.3 Electroweak theory

gauge symmetry. The gauge coupling constants of the $SU(2)$ and $U(1)$ groups should be taken as independent, and therefore the $SU(2) \times U(1)$ group cannot be "unified" in a single $U(2)$ group.

4.3.1 Fermion content

In contrast to quarks, leptons are not involved in strong interactions, but both *leptons* and *quarks* participate in weak interactions. The three electrically charged leptons, the electron e, the muon μ, and the τ-lepton, are partnered by neutrinos ν_e, ν_μ and ν_τ respectively.

The neutrino masses are very small and we will first consider them as if they were massless. The neutrino has spin $1/2$ and, in principle, the normalized component of spin in the direction of motion, called the *helicity*, can take the value $+1$ or -1. It has been found in experiments, however, that all neutrinos are *left-handed*: they have helicity -1, that is, their spins are always directed antiparallel to their velocity. All antineutrinos are *right-handed*. Hence, in weak interactions, the symmetry between right- and left-handedness (parity-P) is broken and the corresponding theory is *chiral*. Note that the notion of helicity is Lorentz-invariant only for massless particles which move with the speed of light, otherwise one can always go to a frame of reference moving faster than the particle and change its helicity.

The quarks and leptons are massive. However, because of the chiral nature of the theory, mass terms cannot be introduced directly without spoiling gauge invariance. In electroweak theory the masses arise as a result of interaction with a classical scalar field. They will be considered later; until then, we will treat *all fermions* as if they were massless particles.

In weak interactions, the charged leptons can be converted into their corresponding electrically neutral neutrinos. As a consequence the intermediate vector boson must carry electric charge. Its antiparticle has the opposite charge and hence there should be at least two gauge bosons responsible for weak interactions. The simplest gauge group which can incorporate them is the $SU(2)$ group.

Only the left-handed electron e_L can be converted into the left-handed neutrino ν_e. They form an $SU(2)$ doublet and transform as

$$\psi_L^e \equiv \begin{pmatrix} \nu_e \\ e \end{pmatrix}_L \to \mathbf{U}\psi_L^e, \qquad (4.37)$$

where $\mathbf{U}$ is a unitary 2×2 matrix with $\det \mathbf{U} = 1$, and ν_e and e_L are Dirac spinors describing the massless left-handed neutrino and electron. The right-handed electron is a singlet with respect to the $SU(2)$ group: $\psi_R^e \equiv e_R \to \psi_R^e$. The concrete form of Dirac spinors for chiral states depends on the Dirac matrix representation used. For instance, in the chiral representation the left-handed fermions are described by four component spinors with the first two components equal to zero. To make concrete calculations of processes, the reader should be familiar with the standard algebra of Dirac matrices, which can be found in any book on field theory. We will not need it here.

Other leptons also come in doublets and singlets:

$$\begin{pmatrix} \nu_\mu \\ \mu \end{pmatrix}_L, \mu_R; \quad \begin{pmatrix} \nu_\tau \\ \tau \end{pmatrix}_L, \tau_R. \tag{4.38}$$

The three different *generations* of leptons have very similar properties. Because weak interactions convert particles only within a particular generation, the lepton numbers are conserved separately.

The six quark flavors also form three generations under weak interaction:

$$\begin{pmatrix} u \\ d' \end{pmatrix}_L, u_R, d'_R; \quad \begin{pmatrix} c \\ s' \end{pmatrix}_L, c_R, s'_R; \quad \begin{pmatrix} t \\ b' \end{pmatrix}_L, t_R, b'_R. \tag{4.39}$$

We have skipped the color indices which are irrelevant for electroweak interactions. The flavors d', s', b' entering the doublets are linear superpositions of the flavors d, s, b, conserved in strong interactions. As a consequence, the weak interactions violate all flavor conservation laws.

Since the individual Lagrangians for each generation have the same form, the fermionic part of electroweak Lagrangian is obtained essentially by replication of the Lagrangian for one particular lepton generation. Therefore we consider, for example, only the electron and its corresponding neutrino. However, the importance of the quark generations should not be underestimated. The anomalies which would spoil renormalizability are canceled only if the number of quark generations is equal to the number of lepton generations.

The $SU(2)$ group has three gauge bosons. As we have already mentioned, two of them are responsible for the charged weak interaction. The third boson is electrically neutral since only then is it its own antiparticle. However, it cannot be identified with the photon, because the photon should be an Abelian $U(1)$ gauge boson. Because one of the partners in the doublet (4.37) is electrically charged, it makes sense to try to incorporate both the electromagnetic and weak interactions into the $SU(2) \times U(1)$ group. The corresponding Lagrangian

$$\mathcal{L}_f = i\bar{\psi}_L \gamma^\mu \left(\partial_\mu + ig\mathbf{A}_\mu + ig'Y_L B_\mu\right)\psi_L + i\bar{\psi}_R \gamma^\mu \left(\partial_\mu + ig'Y_R B_\mu\right)\psi_R \tag{4.40}$$

is invariant under both $SU(2)$ transformations,

$$\psi_L \to \mathbf{U}\psi_L, \quad \psi_R \to \psi_R,$$

and $U(1)$ transformations,

$$\psi_L \to e^{-ig'Y_L\lambda(x)}\psi_L, \quad \psi_R \to e^{-ig'Y_R\lambda(x)}\psi_R,$$

if the gauge fields $\mathbf{A}_\mu$ and B_μ transform according to (4.11) and (4.5) respectively. The $U(1)$ hypercharges Y_L and Y_R can be different for the right- and left-handed electrons. The only requirement is that they should be able to reproduce the correct values of the observed electric charges.

In electroweak theory, three out of four gauge bosons should acquire masses and one boson should remain massless. Additionally, the fermions should become massive. These masses can be generated in a soft way via interaction with the classical scalar field. In this case the theory remains gauge-invariant and renormalizable. To demonstrate how this mechanism works, let us first consider the simplest $U(1)$ Abelian gauge field which interacts with a complex scalar field.

4.3.2 "Spontaneous breaking" of $U(1)$ symmetry

The Lagrangian

$$\mathcal{L} = \tfrac{1}{2}((\partial^\mu + ieA^\mu)\varphi)^*((\partial_\mu + ieA_\mu)\varphi) - V(\varphi^*\varphi) - \tfrac{1}{4}F^2(A), \tag{4.41}$$

where $F^2 \equiv F_{\mu\nu}F^{\mu\nu}$, is invariant under the gauge transformations

$$\varphi \to e^{-ie\lambda}\varphi, \quad A_\mu \to A_\mu + \partial_\mu\lambda.$$

For $\varphi \neq 0$, we can parameterize the complex scalar field φ by two real scalar fields χ and ζ, defined via

$$\varphi = \chi \exp(ie\zeta). \tag{4.42}$$

The field χ is gauge-invariant and the field ζ transforms as $\zeta \to \zeta - \lambda$. We can combine the field A_μ and ζ to form the gauge-invariant variable

$$G_\mu \equiv A_\mu + \partial_\mu\zeta. \tag{4.43}$$

Lagrangian (4.41) can then be rewritten entirely in terms of the gauge-invariant fields χ and G_μ:

$$\mathcal{L} = \frac{1}{2}\partial^\mu\chi\partial_\mu\chi - V(\chi^2) - \frac{1}{4}F^2(G) + \frac{e^2}{2}\chi^2 G^\mu G_\mu. \tag{4.44}$$

If the potential V has a minimum at $\chi_0 = \text{const} \neq 0$, we can consider small perturbations around this minimum, $\chi = \chi_0 + \phi$, and expand the Lagrangian in powers of ϕ. It then describes the real scalar field ϕ of mass $m_H = \sqrt{V_{,\chi\chi}(\chi_0)}$, which interacts with the massive vector field G_μ. The mass of the vector field is $M_G = e\chi_0$. If Lagrangian (4.41) is renormalizable, one expects that after rewriting it in explicitly gauge-invariant form, it will remain renormalizable. This is what really happens in spite of the fact that the vector field acquires mass. If $\chi_0 \neq 0$, the physical fields corresponding to observable particles are the gauge-invariant real scalar field (Higgs field) and the massive vector field. Of course, after we have rewritten the Lagrangian in terms of the new variables, the total number of physical degrees of freedom does not change. In fact, the system described by (4.41) has four degrees of freedom per point in space: namely, two for the complex scalar field and two corresponding to the transverse components of the vector field. The Lagrangian expressed in terms of gauge-invariant variables describes a real scalar field with one degree of freedom and a massive vector field with three degrees of freedom.

This method of generating the mass term is known as the Higgs mechanism and its main advantage is that it does not spoil renormalizability. The vector field acquires mass and its longitudinal degree of freedom becomes physical at the expense of a classical scalar field. Of course, (4.44) is invariant with respect to the original gauge transformations which become trivial: $\chi \to \chi$ and $G_\mu \to G_\mu$. However, if one tries to interpret the gauge-invariant field G_μ as a gauge field like A_μ, then one erroneously concludes that gauge invariance is gone. This is why one often says that the symmetry is spontaneously broken. Such a statement is somewhat misleading but we will nevertheless use this wide accepted, standard terminology.

The gauge-invariant variables can be introduced and interpreted as physical degrees of freedom only if $\chi_0 \neq 0$ and only when the perturbations around χ_0 are small, so that $\chi = \chi_0 + \phi \neq 0$ everywhere in the space. Otherwise (4.42) becomes singular at $\chi = 0$ and the fields χ and ζ used to construct the gauge-invariant variables become ill-defined. In the case $\chi_0 = 0$, one has to work directly with the Lagrangian in its original form (4.41).

4.3.3 Gauge bosons

In electroweak theory, the masses of the gauge bosons can also be generated using the Higgs mechanism. Let us consider the $SU(2) \times U(1)$ gauge-invariant Lagrangian

$$\mathcal{L}_\varphi = \tfrac{1}{2}(\mathbf{D}^\mu \varphi)^\dagger (\mathbf{D}_\mu \varphi) - V(\varphi^\dagger \varphi), \tag{4.45}$$

where φ is an $SU(2)$ doublet of complex scalar fields, † denotes the Hermitian conjugation and

$$\mathbf{D}_\mu \equiv \partial_\mu + ig\mathbf{A}_\mu - \frac{i}{2}g' B_\mu. \tag{4.46}$$

We have assumed here that the hypercharge of the scalar doublet is $Y_\varphi = -1/2$, so that under the $U(1)$ group it transforms as $\varphi \to e^{\frac{i}{2}g'\lambda} \varphi$. The scalar field can be written as

$$\varphi = \chi \begin{pmatrix} \zeta_1 \\ \zeta_2 \end{pmatrix} = \chi \begin{pmatrix} \zeta_2^* & \zeta_1 \\ -\zeta_1^* & \zeta_2 \end{pmatrix} \begin{pmatrix} 0 \\ 1 \end{pmatrix} \equiv \chi \zeta \varphi_0, \qquad (4.47)$$

where χ is a real field and ζ_1, ζ_2 are two complex scalar fields satisfying the condition $|\zeta_1|^2 + |\zeta_2|^2 = 1$. The definition of the $SU(2)$ matrix ζ and the constant vector φ_0 can easily be read off the last equality. Substituting $\varphi = \chi \zeta \varphi_0$ in (4.45), we obtain

$$\mathcal{L}_\varphi = \frac{1}{2} \partial^\mu \chi \partial_\mu \chi - V(\chi^2) + \frac{\chi^2}{2} \varphi_0^\dagger \left(g\mathbf{G}_\mu - \frac{1}{2} g' B_\mu \right) \left(g\mathbf{G}^\mu - \frac{1}{2} g' B^\mu \right) \varphi_0, \qquad (4.48)$$

where

$$\mathbf{G}_\mu \equiv \zeta^{-1} \mathbf{A}_\mu \zeta - \frac{i}{g} \zeta^{-1} \partial_\mu \zeta \qquad (4.49)$$

are $SU(2)$ gauge-invariant variables.

Problem 4.11 Consider the $SU(2)$ transformation

$$\begin{pmatrix} \zeta_1 \\ \zeta_2 \end{pmatrix} \rightarrow \begin{pmatrix} \tilde{\zeta}_1 \\ \tilde{\zeta}_2 \end{pmatrix} = \mathbf{U} \begin{pmatrix} \zeta_1 \\ \zeta_2 \end{pmatrix} \qquad (4.50)$$

accompanied by the $U(1)$ transformation $\tilde{\zeta} \rightarrow e^{\frac{i}{2} g' \lambda} \tilde{\zeta}$ and verify that

$$\zeta \rightarrow \begin{pmatrix} e^{-\frac{i}{2} g' \lambda} \tilde{\zeta}_2^* & e^{\frac{i}{2} g' \lambda} \tilde{\zeta}_1 \\ -e^{-\frac{i}{2} g' \lambda} \tilde{\zeta}_1^* & e^{\frac{i}{2} g' \lambda} \tilde{\zeta}_2 \end{pmatrix} = \mathbf{U} \zeta \mathbf{E}, \qquad (4.51)$$

where

$$\zeta \equiv \begin{pmatrix} \zeta_2^* & \zeta_1 \\ -\zeta_1^* & \zeta_2 \end{pmatrix}, \quad \mathbf{E} \equiv \begin{pmatrix} e^{-\frac{i}{2} g' \lambda} & 0 \\ 0 & e^{\frac{i}{2} g' \lambda} \end{pmatrix} \qquad (4.52)$$

(*Hint* Note that an arbitrary $SU(2)$ matrix has the same form as matrix ζ with ζ_1, ζ_2 replaced by some complex numbers α, β.)

Using this result, it is easy to see that the field $\mathbf{G}_\mu$ is $SU(2)$ gauge-invariant, that is,

$$\mathbf{G}_\mu \rightarrow \mathbf{G}_\mu \qquad (4.53)$$

as

$$\zeta \rightarrow \mathbf{U}\zeta, \quad \mathbf{A}_\mu \rightarrow \mathbf{U}\mathbf{A}_\mu \mathbf{U}^{-1} + (i/g)(\partial_\mu \mathbf{U})\mathbf{U}^{-1}.$$

Thus, we have rewritten our original Lagrangian (4.45) in terms of $SU(2)$ gauge-invariant variables $\chi, \mathbf{G}_\mu, B_\mu$.

156   *The very early universe*

The fields B_μ and $\mathbf{G}_\mu$ change under $U(1)$ transformations. Field B_μ transforms as

$$B_\mu \to B_\mu + \partial_\mu \lambda. \tag{4.54}$$

Taking into account that $\mathbf{A}_\mu \to \mathbf{A}_\mu$ and $\zeta \to \zeta \mathbf{E}$, where matrix $\mathbf{E}$ is defined in (4.52), we find that

$$\mathbf{G}_\mu \to \tilde{\mathbf{G}}_\mu = \mathbf{E}^{-1} \mathbf{G}_\mu \mathbf{E} - \frac{i}{g} \mathbf{E}^{-1} \partial_\mu \mathbf{E} \tag{4.55}$$

under the $U(1)$ transformations.

Matrix $\mathbf{G}_\mu$ is the Hermitian traceless matrix

$$\mathbf{G}_\mu \equiv \begin{pmatrix} -G^3_\mu/2 & -W^+_\mu/\sqrt{2} \\ -W^-_\mu/\sqrt{2} & G^3_\mu/2 \end{pmatrix}, \tag{4.56}$$

where $W^\pm_\mu$ are a conjugate pair of complex vector fields and G^3_μ is a real vector field. In parameterizing matrix $\mathbf{G}_\mu$ we have used the standard sign convention and normalization adopted in the literature. Substituting this expression in (4.48) and replacing fields G^3_μ and B_μ with the "orthogonal" linear combinations Z_μ and A_μ,

$$\begin{pmatrix} A_\mu \\ Z_\mu \end{pmatrix} \equiv \begin{pmatrix} \cos\theta_w & \sin\theta_w \\ -\sin\theta_w & \cos\theta_w \end{pmatrix} \begin{pmatrix} B_\mu \\ G^3_\mu \end{pmatrix}, \tag{4.57}$$

where θ_w is the Weinberg angle and

$$\cos\theta_w = \frac{g}{\sqrt{g^2 + g'^2}}, \tag{4.58}$$

we can rewrite (4.48) in the following form:

$$\mathcal{L}_\varphi = \frac{1}{2} \partial^\mu \chi \partial_\mu \chi - V(\chi^2) + \frac{(g^2 + g'^2)\chi^2}{8} Z_\mu Z^\mu + \frac{g^2 \chi^2}{4} W^+_\mu W^{-\mu}. \tag{4.59}$$

Because

$$\operatorname{tr} \mathbf{F}^2(\mathbf{A}) = \operatorname{tr} \mathbf{F}^2(\mathbf{G}),$$

where $\mathbf{F}^2 \equiv \mathbf{F}_{\mu\nu} \mathbf{F}^{\mu\nu}$, the Lagrangian for the gauge fields is

$$\mathcal{L}_F = -\tfrac{1}{4} F^2(B) - \tfrac{1}{2} \operatorname{tr} \mathbf{F}^2(\mathbf{G}). \tag{4.60}$$

Problem 4.12 Substituting (4.56) in (4.60) and using the definitions (4.13) and (4.57), verify that (4.60) can be rewritten as

$$\mathcal{L}_F = -\tfrac{1}{4} F^2(A) - \tfrac{1}{4} F^2(Z) - \tfrac{1}{2} F_{\mu\nu}(W^+) F^{\mu\nu}(W^-), \tag{4.61}$$

where

$$F_{\mu\nu}(A) \equiv \partial_\mu A_\nu - \partial_\nu A_\mu + ig\sin\theta_w\left(W_\mu^- W_\nu^+ - W_\nu^- W_\mu^+\right),$$
$$F_{\mu\nu}(Z) \equiv \partial_\mu Z_\nu - \partial_\nu Z_\mu + ig\cos\theta_w\left(W_\mu^- W_\nu^+ - W_\nu^- W_\mu^+\right),$$
(4.62)

and

$$F_{\mu\nu}(W^\pm) \equiv \mathcal{D}_\mu^\pm W_\nu^\pm - \mathcal{D}_\nu^\pm W_\mu^\pm,$$
$$\mathcal{D}_\mu^\pm \equiv \partial_\mu \mp ig\sin\theta_w A_\mu \mp ig\cos\theta_w Z_\mu.$$
(4.63)

The terms which are third and fourth order in the field strength describe the interactions of the gauge fields. Draw the corresponding vertices.

We now turn to the renormalizable scalar field potential

$$V(\chi^2) = \frac{\lambda}{4}\left(\chi^2 - \chi_0^2\right)^2. \tag{4.64}$$

In this case, the field χ acquires a vacuum expectation value χ_0, corresponding to the minimum of this potential. Let us consider small perturbations around this minimum, so that $\chi = \chi_0 + \phi$. It is obvious that the Lagrangian $\mathcal{L}_\varphi + \mathcal{L}_F$ then describes the Higgs scalar field ϕ of mass

$$m_H = \sqrt{V_{,\chi\chi}(\chi_0)} = \sqrt{2\lambda}\chi_0, \tag{4.65}$$

massive vector fields Z_μ and $W_\nu^\pm$, with masses

$$M_Z = \sqrt{g^2 + g'^2}\frac{\chi_0}{2}, \quad M_W = \frac{g\chi_0}{2} = M_Z\cos\theta_w, \tag{4.66}$$

and the massless field A_μ. This massless field is responsible for long-range interactions and should be identified with the electromagnetic field.

Problem 4.13 Using (4.55), verify that under $U(1)$ transformations

$$W_\mu^\pm \to e^{\pm ig'\lambda} W_\mu^\pm, \quad A_\mu \to A_\mu + \frac{1}{\cos\theta_w}\partial_\mu\lambda, \quad Z_\mu \to Z_\mu. \tag{4.67}$$

Thus we see that $W_\mu^\pm$ transform as electrically charged fields. Comparing these transformation laws with those of electrodynamics, we can identify the electric charge of the $W_\mu^\pm$ bosons (see also (4.63)) as

$$e = g'\cos\theta_w = g\sin\theta_w. \tag{4.68}$$

The boson Z_μ is electrically neutral. As we will see, W and Z bosons are responsible for the charged and neutral weak interactions respectively, and the "weakness" of these interactions is due to the large masses of the intermediate bosons rather than the smallness of the weak coupling constant g. It follows from (4.68) that the "weak fine structure constant"

$$\alpha_w \equiv \frac{g^2}{4\pi} = \frac{e^2}{4\pi \sin^2 \theta_w} \tag{4.69}$$

is in fact larger than $\alpha \equiv e^2/4\pi \simeq 1/137$.

4.3.4 Fermion interactions

Combining the left-handed doublet ψ_L^e with the scalar field, we can easily build the corresponding fermionic $SU(2)$ gauge-invariant variables:

$$\Psi_L^e = \zeta^{-1} \psi_L^e. \tag{4.70}$$

The right-handed electron, $\psi_R^e \equiv e_R$, is a singlet with respect to the $SU(2)$ group. Because under $U(1)$ transformations $\psi_L^e \to e^{-ig'Y_L\lambda}\psi_L^e$ and $\zeta \to \zeta \mathbf{E}$, we obtain

$$\Psi_L^e \to e^{-ig'Y_L\lambda} \mathbf{E}^{-1} \Psi_L^e.$$

Defining the $SU(2)$ gauge-invariant left-handed electron and neutrino as

$$\Psi_L^e \equiv \begin{pmatrix} \nu_L \\ e_L \end{pmatrix}, \tag{4.71}$$

we have

$$\nu_L \to e^{ig'(\frac{1}{2}-Y_L)\lambda}\nu_L, \quad e_L \to e^{-ig'(\frac{1}{2}+Y_L)\lambda}e_L, \quad e_R \to e^{-ig'Y_R\lambda}e_L. \tag{4.72}$$

The neutrino has no electrical charge and therefore should not transform. Hence the hypercharge of the left-handed doublet should be taken to be $Y_L = 1/2$. In this case the left-handed electron transforms as $e_L \to e^{-ig'\lambda}e_L$. To ensure the same value for the electric charges of the right- and left-handed electrons we have to put $Y_R = 1$. Taking into account the transformation law for the vector potential A_μ (see (4.67)), we conclude that the electric charge of the electron is equal to e given in (4.68).

Substituting $\psi_L^e = \zeta \Psi_L$ in (4.40) and using definition (4.49), we can rewrite the Lagrangian for fermions in terms of the gauge-invariant variables:

$$\mathcal{L}_f = i\bar{\Psi}_L^e \gamma^\mu \left(\partial_\mu + ig\mathbf{G}_\mu + \frac{i}{2}g'B_\mu \right) \Psi_L^e + i\bar{\psi}_R \gamma^\mu \left(\partial_\mu + ig'B_\mu \right) \psi_R. \tag{4.73}$$

4.3 Electroweak theory

Alternatively, using definitions (4.56), (4.57) and (4.71), we obtain

$$\mathcal{L}_f = i\left(\bar{e}\gamma^\mu \partial_\mu e + \bar{\nu}_L \gamma^\mu \partial_\mu \nu_L\right)$$
$$-e(\bar{e}\gamma^\mu e) A_\mu + \frac{g}{\sqrt{2}}\left((\bar{\nu}_L \gamma^\mu e_L) W^+_\mu + (\bar{e}_L \gamma^\mu \nu_L) W^-_\mu\right)$$
$$+\left[\frac{\sin^2\theta_w}{\cos\theta_w} g(\bar{e}_R \gamma^\mu e_R) - \frac{\cos 2\theta_w}{2\cos\theta_w} g(\bar{e}_L \gamma^\mu e_L) + \frac{g}{2\cos\theta_w}(\bar{\nu}_L \gamma^\mu \nu_L)\right] Z_\mu,$$
(4.74)

where the well known properties of the Dirac matrices have been used to write

$$\bar{e}_L \gamma^\mu e_L + \bar{e}_R \gamma^\mu e_R = \bar{e}\gamma^\mu e.$$

The first cubic term is the familiar electromagnetic interaction and the next terms describe the charged and neutral weak interactions due to the exchange of $W^\pm$ and Z bosons respectively. Note that the right-handed electrons participate only in electromagnetic and neutral weak interactions, and not in the charged interactions.

Replacing e, ν_e in (4.74) by μ, ν_μ /τ, ν_τ, we obtain the Lagrangian for the second/third generation of leptons. Let us consider muon decay. The appropriate tree diagram is shown in Figure 4.7 (the reader must take care to correctly identify the notation used for wave functions and particles in diagrams; for example, the conjugated wave function $\bar{\nu}$ can describe a neutrino as well as an antineutrino depending on the orientation of the diagram). Because the muon mass is much smaller than the mass of the W boson, the boson propagator can be replaced by $ig_{\mu\nu}/M_W^2$ and the diagram in Figure 4.7 reduces to the four-fermion diagram corresponding to the coupling term

$$2\sqrt{2} G_F (\bar{\nu}_\mu \gamma^\alpha \mu_L)(\bar{e}_L \gamma_\alpha \nu_e),$$

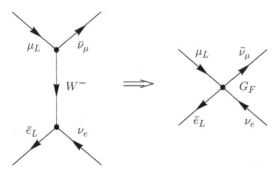

Fig. 4.7.

where

$$G_F \equiv \frac{1}{4\sqrt{2}} \frac{g^2}{M_W^2} = \frac{\pi}{\sqrt{2}} \frac{\alpha_w^2}{M_W^2} \simeq 1.166 \times 10^{-5} \,\text{GeV}^{-2} \qquad (4.75)$$

is Fermi coupling constant. The experimental value of G_F, the Weinberg angle $\theta_w \simeq 28.7°$ ($\sin^2 \theta_w \simeq 0.23$), determined in the neutral current interactions, and the measured masses

$$M_W \simeq 80.4 \,\text{GeV}, \quad M_Z \simeq 91.2 \,\text{GeV},$$

are in very good agreement with the theoretical predictions of the standard electroweak model. From (4.69), "the weak fine structure constant" $\alpha_w \simeq 1/29$ is seen to be 4.5 times larger than the fine structure constant. As follows from (4.66) and (4.68), the expectation value for the Higgs field is

$$\chi_0 = \frac{2 M_W \sin \theta_w}{e} \simeq 250 \,\text{GeV}. \qquad (4.76)$$

However, since the quartic coupling constant λ can be arbitrary, the Higgs mass, given in (4.65), is not predicted. Higgs particles have not yet been discovered. The experimental lower bound on their masses is $m_H > 114$ GeV. Requiring self-consistency of the theory, namely, the validity of the perturbative expansion in λ, one could expect that $\lambda < 1$ and hence m_H cannot greatly exceed 350 GeV.

Problem 4.14 Given that the electric charges of u and d quarks equal $+2/3$ and $-1/3$ respectively, determine their hypercharges. Derive the Lagrangian for the first quark generation. (Note that both u_R and d_R are present in the Lagrangian as $SU(2)$ singlets.) Draw the vertices describing the quark weak interactions.

The neutron decay can be interpreted as the underlying quark process: $d \to u + W$. As a result the neutron, which is a bound state of three quarks udd, is converted into the proton consisting of uud quarks. Because quarks always appear in bound systems, the calculation of the weak interactions of hadrons is more complicated and suffers from various uncertainties.

4.3.5 Fermion masses

Until now we have treated all fermions as massless. This is obviously in disagreement with experiment. However, fermion masses, introduced by hand, would spoil gauge invariance and renormalizability. Therefore, the only way to generate them is again to use the Higgs mechanism.

4.3 Electroweak theory

The gauge-invariant Yukawa coupling between scalars and fermions for the first lepton generation is

$$\mathcal{L}_Y^e = -f_e\left(\bar{\psi}_L^e \varphi e_R + \bar{e}_R \varphi^\dagger \psi_L^e\right), \qquad (4.77)$$

where f_e is the dimensionless Yukawa coupling constant. This term is obviously $SU(2)$-invariant and, if the hypercharges satisfy the condition

$$Y_L = Y_R + Y_\varphi, \qquad (4.78)$$

it is also invariant with respect to $U(1)$ transformations. Substituting $\varphi = \chi \zeta \varphi_0$ (see (4.47)) in (4.77), we can rewrite the Yukawa coupling in terms of $SU(2)$-invariant variables as

$$\mathcal{L}_Y^e = -f_e \chi \bar{\Psi}_L^e \varphi_0 e_R + \text{h.c.} = -f_e \chi (\bar{e}_L e_R + \bar{e}_R e_L) = -f_e \chi \bar{e} e, \qquad (4.79)$$

where h.c. denotes the Hermitian conjugated term. To write the last equality we have used a well known relation from the theory of Dirac spinors. If the scalar field takes a nonzero expectation value, so that $\chi = \chi_0 + \phi$, the electron acquires the mass

$$m_e = f_e \chi_0. \qquad (4.80)$$

With χ_0 given in (4.76) and $f_e \simeq 2 \times 10^{-6}$, we get the correct value for the electron mass. The appearance of such a small coupling constant has no natural explanation in the electroweak theory, where f_e is a free parameter. The term $f_e \phi \bar{e} e$ describes the interaction of Higgs particles with electrons. Note that the particular form of the Yukawa coupling (4.77) gives mass only to the lower component of the doublet. The neutrino remains massless.

For quarks some complications arise. First of all, both components of the doublets should acquire masses. Second, to explain flavor nonconservation in the weak interactions, we have to assume that the lower components of the $SU(2)$ doublets are superpositions of the lower quark flavors and hence are not quark mass eigenstates. This suggests we should simultaneously consider all three quark generations. Let us denote the upper and lower components of the $SU(2)$ *gauge-invariant* quark doublets by

$$u^i \equiv (u, c, t)$$

and

$$d'^i \equiv (d', s', b')$$

respectively, where $i = 1, 2, 3$ is the generation index. The lower components are linear superpositions of the appropriate flavors,

$$d'^i = V^i_j d^j, \tag{4.81}$$

where V^i_j is the unitary 3×3 *Kobayashi–Maskawa matrix*. The general quark Yukawa term can then be written as

$$\mathcal{L}^q_Y = -f^d_{ij} \chi \bar{\mathbf{Q}}^i_L \varphi_0 d'^j_R - f^u_{ij} \chi \bar{\mathbf{Q}}^i_L \varphi_1 u^j_R + \text{h.c.}, \tag{4.82}$$

where

$$\mathbf{Q}^i_L = \begin{pmatrix} u^i \\ d'^i \end{pmatrix}, \quad \varphi_1 = \begin{pmatrix} 1 \\ 0 \end{pmatrix}. \tag{4.83}$$

The second term on the right hand side in (4.82) is also gauge-invariant and generates the masses of the upper components of the doublets. Expression (4.82), rewritten in terms of the original flavors, gives the mass term

$$\mathcal{L}_{m_q} = -\left(V^{i*}_m f^d_{ij} V^j_k\right) \chi \bar{d}^m_L d^k_R - f^u_{ij} \chi \bar{u}^i_L u^j_R + \text{h.c.} \tag{4.84}$$

Taking the matrices f^d_{ij} and f^u_{ij} such that

$$\left(V^{i*}_m f^d_{ij} V^j_k\right) \chi_0 = m^d_k \delta_{mk} \tag{4.85}$$

and

$$f^u_{ij} \chi_0 = m^u_i \delta_{ij}, \tag{4.86}$$

we get the usual quark mass terms. The Yukawa coupling constant is largest for the top quark, $f^t \simeq 0.7$ ($m_t \simeq 170$ GeV). For the other quarks it is more than 10 times smaller.

Neutrino masses, which – according to measurements – are different from zero, can be generated in an almost identical manner. In this case, the neutrino flavors naturally mix in a similar way to the quark generations and this leads to neutrino oscillations.

4.3.6 CP violation

The parity operation P corresponds to reflection $(t, x, y, z) \to (t, -x, -y, -z)$ and converts left-handed particles into right-handed particles without changing their other properties. Charge conjugation C replaces particles by antiparticles without changing handedness. For instance, the C operation converts a left-handed electron to a left-handed positron. Any chiral gauge theory is not invariant with respect to

4.3 Electroweak theory

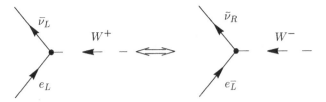

Fig. 4.8.

P and *C* operations applied separately and in the Standard Model these symmetries are violated in a maximal possible way.

It is obvious that (4.40) is not invariant with respect to the replacement $L \leftrightarrow R$ because parity operation converts left-handed neutrinos into nonexistent right-handed neutrinos. Similarly, charge conjugation converts left-handed neutrinos into nonexistent left-handed antineutrinos. However, the combined operation *CP*, which interchanges left-handed particles with right-handed antiparticles, seems to be a symmetry of the electroweak theory. In fact, *CP* is a symmetry of the Lagrangian without quarks. As an example, let us find out what happens to the charged weak interaction coupling term in (4.74),

$$\frac{g}{\sqrt{2}}\left((\bar{\nu}_L\gamma^\mu e_L)\,W_\mu^+ + (\bar{e}_L\gamma^\mu \nu_L)\,W_\mu^-\right), \qquad (4.87)$$

under *CP* transformations. The first term here corresponds to the vertex in Figure 4.8 and can be interpreted as describing left-handed electron and right-handed antineutrino "annihilation" with the emission of a W^- boson. Recall that an arrow entering a vertex corresponds to a wave function ψ while an outgoing line corresponds to the conjugated function $\bar{\psi}$. If the arrow coincides with the direction of time, then the corresponding line describes the particle; otherwise it describes the antiparticle. Hence the W^+ boson line entering the vertex in Figure 4.8 corresponds to its antiparticle, that is, the W^- boson. The wave function e_L describes the left-handed electron e_L^-, and $\bar{\nu}_L$ corresponds to the right-handed antineutrino $\tilde{\nu}_R$. Under charge conjugation *C* all arrows on the diagram are reversed (Figure 4.9). The right-handed antineutrino goes to the right-handed neutrino, $\tilde{\nu}_R \to \nu_R$, and the left-handed electron converts to the left-handed positron, $e_L^- \to e_L^+ \leftrightarrow \bar{e}_R$. Thus

$$\frac{g}{\sqrt{2}}(\bar{\nu}_L\gamma^\mu e_L)\,W_\mu^+ \to \frac{g}{\sqrt{2}}(\bar{e}_R\gamma^\mu \nu_R)\,W_\mu^-. \qquad (4.88)$$

After applying the *P* operation, this term coincides with the second term in Lagrangian (4.87). Likewise, the second term in (4.87) converts to the first one. Therefore expression (4.87) is *CP*-invariant. The reader can verify that the other terms in (4.74) are also *CP*-invariant.

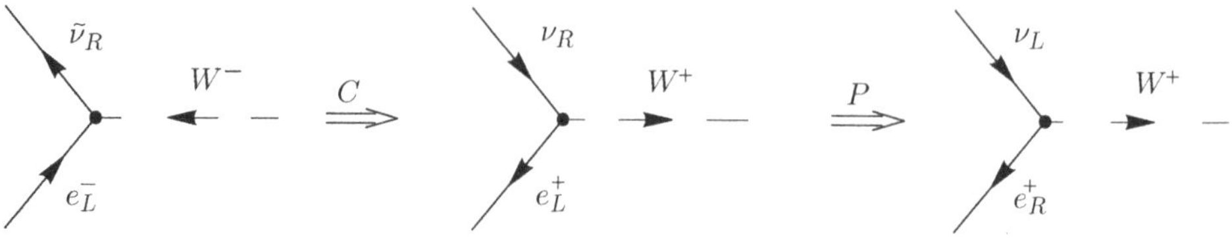

Fig. 4.9.

The term describing the charged weak interactions for quarks can be written as (4.87):

$$\frac{g}{\sqrt{2}} \sum_i \left(\left(\bar{u}_L^i \gamma^\mu d_L'^i \right) W_\mu^+ + \left(\bar{d}_L'^i \gamma^\mu u_L^i \right) W_\mu^- \right), \tag{4.89}$$

or, after rewriting it in terms of quark flavors, as

$$\frac{g}{\sqrt{2}} \left(V_j^i \left(\bar{u}_L^i \gamma^\mu d_L^j \right) W_\mu^+ + \left(V_j^i \right)^* \left(\bar{d}_L^j \gamma^\mu u_L^i \right) W_\mu^- \right). \tag{4.90}$$

Under CP transformation, the first term becomes

$$V_j^i \left(\bar{u}_L^i \gamma^\mu d_L^j \right) W_\mu^+ \to V_j^i \left(\bar{d}_L^j \gamma^\mu u_L^i \right) W_\mu^-, \tag{4.91}$$

and coincides with the second term only if $V_j^i = \left(V_j^i \right)^*$. In other words, (4.90) is CP-invariant only if the Kobayashi–Maskawa matrix is real-valued; otherwise CP is violated. An arbitrary 3×3 unitary matrix

$$V_j^i = r_j^i \exp(i\theta_{ij})$$

is characterized by three independent real numbers r and by six independent phases θ. The quark Lagrangian is invariant with respect to global quark rotations: $q_i \to \exp(i\alpha_i) q_i$. Using the six independent parameters for six quarks, we can eliminate five θ phases as having no physical meaning. One phase is left over, however, because bilinear quark combinations are insensitive to the overall phase of the quark rotation. Because of this one remaining phase factor the Kobayashi–Maskawa matrix will generally have complex elements and therefore one can expect CP violation. This CP violation is due to the complex-valued coupling constant in the charged weak interaction term.

Problem 4.15 Could we expect CP violation in a model with only two quark generations, where the quark mixing is entirely characterized by Cabibbo angle?

The violation of CP symmetry was first observed in 1964 in kaon $K^0(d\bar{s})$ decay and then, in 2001, in the $B^0(d\bar{b})$ meson system. There is strong evidence that the Kobayashi–Maskawa mechanism is responsible for this CP violation.

As we will see, CP violation plays a very important role in baryogenesis, ensuring the possibility of different decay rates for particles and their antiparticles in particular decay channels. If this were not the case, generation of baryon asymmetry would be impossible.

Note that if we accompany the CP transformation by time reversal $T(t \to -t)$, which reverses the direction of arrows on the diagrams and changes the handedness, (4.90) remains invariant. The Lagrangian of the Standard Model is CPT-invariant. This invariance guarantees that the *total* decay rates, which include all decay channels, are the same for particles and their antiparticles.

4.4 "Symmetry restoration" and phase transitions

A classical scalar field χ interacts with gauge fields which influence its behavior. In the early universe this influence can be described using an effective potential. At very high temperatures the effective potential has only one minimum, at $\chi = 0$, and the homogeneous component of χ disappears. As a result all fermions and intermediate bosons become massless and one says that the symmetry is restored. In fact, as we have pointed out, the gauge symmetry is never broken by the Higgs mechanism. Nevertheless, in deference to the commonly used terminology we use the term symmetry restoration to designate the disappearance of the homogeneous scalar field.

As the universe expands the temperature decreases. Below a critical temperature the effective potential acquires an energetically favorable local minimum, at $\chi(T) \neq 0$, and the transition to this state becomes possible. Depending on the parameters of the theory this can be either a phase transition or a simple cross-over.

In this section we investigate symmetry restoration and phase transitions in gauge theories.

4.4.1 Effective potential

To introduce an idea of the effective potential we first consider a simple model describing a self-interacting real scalar field, which satisfies the equation

$$\chi^{;\alpha}_{;\alpha} + V'(\chi) = 0, \tag{4.92}$$

where $V'(\chi) \equiv \partial V/\partial \chi$. The field χ can always be decomposed into homogeneous and inhomogeneous components:

$$\chi(t, \mathbf{x}) = \bar{\chi}(t) + \phi(t, \mathbf{x}), \tag{4.93}$$

so that the spatial average of $\phi(\mathbf{x}, t)$ is equal to zero. Substituting (4.93) into (4.92), expanding the potential V in powers of ϕ, and averaging over space, we obtain the following equation for $\bar\chi(t)$:

$$\bar\chi^{;\alpha}_{;\alpha} + V'(\bar\chi) + \frac{1}{2}V'''(\bar\chi)\langle\phi^2\rangle = 0, \qquad (4.94)$$

where the higher-order terms $\sim \langle\phi^3\rangle$, etc. have been neglected. In quantum field theory this corresponds to the so called one-loop approximation. We will now show that in a hot universe the last term in (4.94) can be combined with $V'(\bar\chi)$ and rewritten as the derivative of an effective potential $V_{\text{eff}}(\bar\chi, T)$. To this purpose, let us calculate the average $\langle\phi^2\rangle$.

Scalar field quantization In the lowest (linear) order, the inhomogeneous modes ϕ obey the equation

$$\phi^{;\alpha}_{;\alpha} + V''(\bar\chi)\phi = 0, \qquad (4.95)$$

obtained by linearizing (4.92). Assuming that the mass

$$m^2_\phi(\bar\chi) \equiv V''(\bar\chi) \geq 0$$

does not depend on time, and neglecting the expansion of the universe (this is a good approximation for our purposes), the solution of (4.95) is

$$\phi(\mathbf{x},t) = \int \frac{1}{\sqrt{2\omega_k}}\left(e^{-i\omega_k t + i\mathbf{k}\mathbf{x}}a^-_\mathbf{k} + e^{i\omega_k t - i\mathbf{k}\mathbf{x}}a^+_\mathbf{k}\right)\frac{d^3k}{(2\pi)^{3/2}}, \qquad (4.96)$$

where

$$\omega_k = \sqrt{k^2 + V''(\bar\chi)} = \sqrt{k^2 + m^2_\phi},$$

$k \equiv |\mathbf{k}|$, and $a^-_\mathbf{k}$, $a^+_\mathbf{k} = \left(a^-_\mathbf{k}\right)^*$ are the integration constants. Our task is to calculate both quantum and thermal contributions to $\langle\phi^2\rangle$.

In quantum theory, the field $\phi(\mathbf{x}, t) \equiv \phi_\mathbf{x}(t)$ becomes a "position" operator $\hat\phi_\mathbf{x}(t)$ and the spatial coordinates $\mathbf{x}$ can be considered simply as enumerating the degrees of freedom of the physical system. That is, at each point in space, we have one degree of freedom – a field strength – which plays the role of position in a configuration space. Hence, a quantum field is a quantum mechanical system with an infinite number of degrees of freedom. As in usual quantum mechanics, the position operators $\hat\phi_\mathbf{x}(t)$ and their conjugated momenta

$$\hat\pi_\mathbf{y} \equiv \partial\mathcal{L}/\partial\dot\phi = \partial\hat\phi_\mathbf{y}/\partial t$$

should satisfy the Heisenberg commutation relations:

$$[\hat\phi_\mathbf{x}(t), \hat\pi_\mathbf{y}(t)] = \left[\hat\phi_\mathbf{x}(t), \frac{\partial\hat\phi_\mathbf{y}(t)}{\partial t}\right] = i\delta(\mathbf{x} - \mathbf{y}), \qquad (4.97)$$

4.4 "Symmetry restoration" and phase transitions

where

$$[\hat{\phi}_{\mathbf{x}}(t), \hat{\pi}_{\mathbf{y}}(t)] \equiv \hat{\phi}_{\mathbf{x}}(t)\hat{\pi}_{\mathbf{y}}(t) - \hat{\pi}_{\mathbf{y}}(t)\hat{\phi}_{\mathbf{x}}(t)$$

and Planck's constant is set to unity. The field operator $\hat{\phi}_{\mathbf{x}}(t)$ obeys (4.95) and its solution is given in (4.96), but now the integration constants should be considered as time-independent operators $\hat{a}_{\mathbf{k}}^-$, $\hat{a}_{\mathbf{k}}^+$. Substituting (4.96) in (4.97), we find that the operators $\hat{a}_{\mathbf{k}}^+$, $\hat{a}_{\mathbf{k}}^-$ satisfy the commutation relations

$$[\hat{a}_{\mathbf{k}}^-, \hat{a}_{\mathbf{k}'}^+] = \delta(\mathbf{k} - \mathbf{k}'), \quad [\hat{a}_{\mathbf{k}}^-, \hat{a}_{\mathbf{k}'}^-] = [\hat{a}_{\mathbf{k}}^+, \hat{a}_{\mathbf{k}'}^+] = 0. \tag{4.98}$$

Except for the appearance of the δ function, these behave like the creation and annihilation operators of harmonic oscillators. The Hilbert space in which these operators act then resembles the Hilbert space of a set of harmonic oscillators. The vacuum state $|0\rangle$ is defined via

$$\hat{a}_{\mathbf{k}}^- |0\rangle = 0 \tag{4.99}$$

for all $\mathbf{k}$, and corresponds to the minimal energy state. The vectors

$$|n_{\mathbf{k}}\rangle = \frac{(\hat{a}_{\mathbf{k}}^+)^n}{\sqrt{n!}} |0\rangle \tag{4.100}$$

are interpreted as describing $n_{\mathbf{k}}$ particles *per single quantum state* characterized by the wave vector $\mathbf{k}$.

Problem 4.16 The operator $\hat{N}_{\mathbf{k}} \equiv \hat{a}_{\mathbf{k}}^+ \hat{a}_{\mathbf{k}}^-$ corresponds to the total number of particles with wave vector $\mathbf{k}$. Using commutation relations (4.98), verify that $\hat{N}_{\mathbf{k}} |n_{\mathbf{k}}\rangle = \delta(0) n_{\mathbf{k}} |n_{\mathbf{k}}\rangle$. The appearance of $\delta(0)$ can easily be understood if we take into account that the total number of quantum states with a given momentum is proportional to the volume. Because we consider an infinite volume, the factor $\delta(0)$ simply reflects this infinity. The number of particles *per unit volume* is finite and equal to the occupation number $n_{\mathbf{k}}$.

Verify that

$$\langle \hat{a}_{\mathbf{k}}^+ \hat{a}_{\mathbf{k}'}^- \rangle_Q \equiv \frac{\langle n_{\mathbf{k}}| \hat{a}_{\mathbf{k}}^+ \hat{a}_{\mathbf{k}'}^- |n_{\mathbf{k}}\rangle}{\langle n_{\mathbf{k}} | n_{\mathbf{k}}\rangle} = n_{\mathbf{k}} \delta(\mathbf{k} - \mathbf{k}'),$$
$$\langle \hat{a}_{\mathbf{k}}^+ \hat{a}_{\mathbf{k}'}^+ \rangle_Q = \langle \hat{a}_{\mathbf{k}}^- \hat{a}_{\mathbf{k}'}^- \rangle_Q = 0. \tag{4.101}$$

Now we can proceed with the calculation of $\langle \phi^2 \rangle$. Let us take a quantum state with occupation numbers $n_{\mathbf{k}}$ in every mode $\mathbf{k}$. In a homogeneous isotropic universe the spatial average can be replaced by the quantum average. Squaring (4.96) and using the results (4.101), we find that

$$\langle \phi^2(\mathbf{x}) \rangle = \frac{1}{2\pi^2} \int \frac{k^2}{\sqrt{k^2 + m_\phi^2(\bar{\chi})}} \left(\frac{1}{2} + n_\mathbf{k} \right) dk. \tag{4.102}$$

Vacuum contribution to V_{eff} First we consider only vacuum fluctuations and set $n_\mathbf{k} = 0$. The integral in (4.102) is divergent as $k \to \infty$. To reveal the nature of the divergence we regularize this integral by introducing the cut-off scale $k_c = M$. Taking into account that $m_\phi^2(\bar{\chi}) = V''$, we can rewrite the third term in (4.94) as

$$\frac{1}{2} V''' \langle \phi^2 \rangle^{reg}_{vac} = \frac{1}{8\pi^2} \frac{\partial m_\phi^2(\bar{\chi})}{\partial \bar{\chi}} \int_0^M \frac{k^2 dk}{\sqrt{k^2 + m_\phi^2(\bar{\chi})}} = \frac{\partial V_\phi}{\partial \bar{\chi}}, \tag{4.103}$$

where

$$V_\phi = \frac{1}{4\pi^2} \int_0^M \sqrt{k^2 + m_\phi^2(\bar{\chi})} k^2 dk \equiv \frac{I(m_\phi(\bar{\chi}))}{4\pi^2} \tag{4.104}$$

is simply equal to the energy density of the vacuum fluctuations.

Using (4.103), (4.94) becomes

$$\bar{\chi}^{;\alpha}_{;\alpha} + V'_{\text{eff}}(\bar{\chi}) = 0, \tag{4.105}$$

where $V_{\text{eff}}(\bar{\chi}) = V + V_\phi$ is the one-loop effective potential. The integral I which enters (4.104) can be calculated exactly:

$$I(m) = \frac{1}{8} \left[M(2M^2 + m^2) \sqrt{M^2 + m^2} + m^4 \ln \frac{m}{M + \sqrt{M^2 + m^2}} \right].$$

Taking the limit $M \to \infty$, we obtain the following expression for the effective potential:

$$V_{\text{eff}} = V + V_\infty + \frac{m_\phi^4(\bar{\chi})}{64\pi^2} \ln \frac{m_\phi^2(\bar{\chi})}{\mu^2}, \tag{4.106}$$

where the divergent term

$$V_\infty = \frac{M^4}{4\pi^2} + \frac{m_\phi^2}{16\pi^2} M^2 - \frac{m_\phi^4}{32\pi^2} \ln \frac{2M}{e^{3/4} \mu} + O\left(\frac{1}{M^2} \right)$$

can be absorbed by a redefinition of constants in the original potential. For instance, in the case of the renormalizable quartic potential

$$V(\bar{\chi}) = \frac{\lambda_0}{4} \bar{\chi}^4 + \frac{m_0^2}{2} \bar{\chi}^2 + \Lambda_0, \tag{4.107}$$

4.4 "Symmetry restoration" and phase transitions

the mass is

$$m_\phi^2 = V'' = 3\lambda_0 \bar\chi^2 + m_0^2.$$

The divergent terms come to be multiplied with $\bar\chi^4$, $\bar\chi^2$, $\bar\chi^0$, and can be combined with appropriate terms in $V(\bar\chi)$. As a result, the bare constants λ_0, m_0^2 and Λ_0 are replaced by the finite renormalized constants λ_R, m_R^2 and Λ_R, measured in experiments. The term V_∞, therefore, can be omitted in (4.106).

Problem 4.17 Find the explicit relations between the bare and renormalized parameters.

The potential (4.106) looks peculiar because it seems to depend explicitly on an arbitrary scale μ. However, it is easy to see that the change in μ induces a term proportional to $m^4(\bar\chi)$, which, for a renormalizable theory, has the same structure as the original potential V and hence leads to finite renormalization of constants. These constants, therefore, become scale-dependent (running), reflecting the renormalization group properties of the quantum field theory. The physics remains the same; it is only our way of interpreting the constants that changes. For pure χ^4 theory with $m_\phi^2(\bar\chi) = 3\lambda\bar\chi^2$, we have

$$V_{\text{eff}} = \frac{1}{4}\lambda\bar\chi^4 + \frac{9\lambda^2\bar\chi^4}{32\pi^2}\ln\frac{\bar\chi}{\chi_0}, \qquad (4.108)$$

where $\lambda = \lambda(\chi_0)$ and χ_0 is some normalization scale. The logarithmic corrections to the potential are proportional to λ^2 and become comparable to the leading $\lambda\bar\chi^4$ term only when $\lambda \ln(\bar\chi/\chi_0) \sim O(1)$. At these large values of $\bar\chi$, however, the higher-loop contributions we have neglected thus far become crucial.

Problem 4.18 Requiring that potential (4.108) should not depend on χ_0, derive the renormalization group equation for $\lambda(\bar\chi)$. Solve this equation, keeping only the leading term in the β function, and verify that $\lambda(\bar\chi)$ blows up when $\lambda(\chi_0)\ln(\bar\chi/\chi_0) \sim O(1)$.

Thermal contribution to V_{eff} In a hot universe the field ϕ is no longer in its vacuum state. The occupation numbers n_k are given by the Bose–Einstein formula (3.20), where $\epsilon = \omega_k = \sqrt{k^2 + m_\phi^2}$ and the chemical potential can be neglected. Substituting the Bose–Einstein distribution in (4.102), we obtain the following expression for the thermal contribution to $\langle\phi^2\rangle$:

$$\langle\phi^2\rangle_T = \frac{1}{2\pi^2}\int_0^\infty \frac{k^2 dk}{\omega_k(e^{\omega_k/T} - 1)} = \frac{T^2}{4\pi^2}J^{(1)}\left(\frac{m_\phi(\bar\chi)}{T}, 0\right). \qquad (4.109)$$

170   *The very early universe*

In deriving this formula, we have changed the integration variable $k \to \omega_k/T$ to express the result through the integral $J_-^{(1)}$ defined in (3.34). For thermal fluctuations the third term on the left hand side in (4.94) can be rewritten as

$$\frac{1}{2} V''' \langle \phi^2 \rangle_T = \frac{\partial m_\phi}{\partial \bar{\chi}} m_\phi \frac{T^2}{4\pi^2} J_-^{(1)} = \frac{\partial V_\phi^T}{\partial \bar{\chi}}, \tag{4.110}$$

where

$$V_\phi^T = \frac{T^4}{4\pi^2} \int_0^{m_\phi/T} \alpha J_-^{(1)}(\alpha, 0) \, d\alpha \equiv \frac{T^4}{4\pi^2} F_- \left(\frac{m_\phi}{T} \right) \tag{4.111}$$

is the temperature-dependent contribution of scalar particles to V_{eff}.

The final result, which includes both quantum and thermal contributions, is

$$V_{\text{eff}} = V + \frac{m_\phi^4(\bar{\chi})}{64\pi^2} \ln \frac{m_\phi^2(\bar{\chi})}{\mu^2} + \frac{T^4}{4\pi^2} F_- \left(\frac{m_\phi(\bar{\chi})}{T} \right), \tag{4.112}$$

where $m_\phi^2(\bar{\chi}) = V''(\bar{\chi})$.

4.4.2 U(1) model

Now we calculate the effective potential in a $U(1)$ gauge model. The equation for the scalar field immediately follows from (4.44),

$$\chi^{;\alpha}_{;\alpha} + V'(\chi) - e^2 \chi G^\mu G_\mu = 0. \tag{4.113}$$

In this case the calculation of the contribution of the scalar particles to V_{eff} is a bit more complicated because the field χ is unambiguously defined only for $\chi > 0$. To avoid the complications we consider, therefore, only the most interesting case when the contribution of the vector particles dominates that of the scalar particles. For the quartic potential in (4.107) this means $e^2 \gg \lambda$, that is, the mass of the gauge boson $m_G(\bar{\chi}) = e\bar{\chi}$ is much bigger than the mass of the Higgs particle. Note that the calculation of the one-loop contribution of the vector particles to V_{eff} can still be trusted, even when it becomes comparable to the $\lambda \bar{\chi}^4$ term. Neglecting the contribution of field ϕ, we find that the homogeneous component of the scalar field satisfies the equation

$$\bar{\chi}^{;\alpha}_{;\alpha} + V'(\bar{\chi}) - e^2 \bar{\chi} \langle G^\mu G_\mu \rangle = 0. \tag{4.114}$$

The term $\langle G^\mu G_\mu \rangle_{\text{vac}}^{\text{reg}}$ can be calculated similarly to $\langle \phi^2 \rangle_{\text{vac}}^{\text{reg}}$ and it is easy to show that

$$-e^2 \bar{\chi} \langle G^\mu G_\mu \rangle_{\text{vac}}^{\text{reg}} = \frac{\partial}{\partial \bar{\chi}} \left(\frac{3I(m_G(\bar{\chi}))}{4\pi^2} \right) \equiv \frac{\partial V_G}{\partial \bar{\chi}}, \tag{4.115}$$

4.4 "Symmetry restoration" and phase transitions

where the integral I is defined in (4.104) and V_G is the energy density of the vacuum fluctuations of the vector field with mass

$$m_G(\bar{\chi}) = e\bar{\chi}.$$

The factor 3 in (4.115) is due to the fact that the massive vector field has three degrees of freedom at every point in space. The calculation of the temperature-dependent contribution of the vector field essentially repeats the calculation for the scalar field and the final result, which includes both quantum and thermal contributions, is

$$V_{\text{eff}}(\bar{\chi}, T) = V + \frac{3m_G^4(\bar{\chi})}{64\pi^2} \ln \frac{m_G^2(\bar{\chi})}{\mu^2} + \frac{3T^4}{4\pi^2} F_-\left(\frac{m_G(\bar{\chi})}{T}\right), \quad (4.116)$$

where $m_G(\bar{\chi}) = e\bar{\chi}$ and F_- is defined in (4.111).

At zero temperature the last term vanishes and for the quartic potential in (4.107) we obtain the following result:

$$V_{\text{eff}} = \frac{\lambda_R}{4} \bar{\chi}^4 + \frac{m_R^2}{2} \bar{\chi}^2 + \Lambda_R + \frac{3e^4}{32\pi^2} \bar{\chi}^4 \ln \frac{\bar{\chi}}{\chi_0}, \quad (4.117)$$

where the renormalized constants $\lambda_R, m_R, \Lambda_R$ can be expressed through experimentally measurable parameters. The concrete set of these parameters depends on the normalization conditions used.

Problem 4.19 Assume that the potential V_{eff} has its minimum at some $\chi_0 \neq 0$ and is equal to zero at this minimum, that is, there is no cosmological constant in the broken symmetry phase. Solving the equations $V_{\text{eff}}(\chi_0) = 0$, $V'_{\text{eff}}(\chi_0) = 0$ and $V''_{\text{eff}}(\chi_0) = m_H^2$, verify that

$$\lambda_R = \frac{m_H^2}{2\chi_0^2} - \frac{9e^4}{32\pi^2}, \quad m_R^2 = -\frac{m_H^2}{2} + \frac{3e^4 \chi_0^2}{16\pi^2},$$

$$\Lambda_R = \frac{\chi_0^2}{4} \left[\frac{m_H^2}{2} - \frac{3e^4 \chi_0^2}{32\pi^2}\right]. \quad (4.118)$$

Thus we have expressed the renormalized constants in (4.117) in terms of χ_0, the gauge coupling constant e and the Higgs mass m_H.

Given that in the broken symmetry phase

$$M_G \equiv m_G(\chi_0) = e\chi_0, \quad (4.119)$$

we note that if $m_H^2 < 3e^2 M_G^2/8\pi^2$, then $m_R^2 > 0$, and the potential (4.117) acquires a second local minimum at $\bar{\chi} = 0$. Moreover, for $m_H^2 < 3e^2 M_G^2/16\pi^2$, this minimum is even deeper than the minimum at χ_0, because $V_{\text{eff}}(\bar{\chi} = 0) = \Lambda_R < 0$. Therefore symmetry breaking becomes energetically unfavorable. We will see later that symmetry is restored in the very early universe. Hence, if the mass of the Higgs particle did not satisfy the inequality

$$m_H^2 > 3e^2 M_G^2 / 16\pi^2, \tag{4.120}$$

known as the Linde–Weinberg bound, the symmetry would remain unbroken and the gauge bosons would be massless.

Let us consider a special case: $m_R^2 = 0$, or equivalently, $V''_{\text{eff}}(\bar\chi = 0) = 0$. Then it follows from (4.118) that potential (4.117) reduces to

$$V_{\text{eff}} = \frac{3e^4}{32\pi^2}\left(\bar\chi^4 \ln\frac{\bar\chi}{\chi_0} - \frac{1}{4}\bar\chi^4 + \frac{1}{4}\chi_0^4\right), \tag{4.121}$$

which is the Coleman–Weinberg potential. Such potentials may arise in unified particle theories. They are especially interesting for cosmological applications and can be used to construct the so called new inflationary scenario.

Now we derive the asymptotic behavior of the potential in the limit of very high temperatures. To calculate F_- for large temperatures, $T \gg m_G(\bar\chi)$, one can use the high-temperature expansion (3.44) for $J_-^{(1)}$. Then, taking into account (4.118), potential (4.116) reduces to

$$V_{\text{eff}}(\bar\chi, T) \simeq \frac{\lambda_T}{4}\bar\chi^4 - \frac{e^3}{4\pi}T\bar\chi^3 + \frac{m_T^2}{2}\bar\chi^2 + \Lambda_R, \tag{4.122}$$

where

$$\lambda_T = \frac{m_H^2}{2\chi_0^2} + \frac{3e^4}{16\pi^2}\ln\frac{bT^2}{(e\chi_0)^2}, \quad \ln b = 2\ln 4\pi - 2\mathbf{C} \simeq 3.5 \tag{4.123}$$

is the effective coupling constant and

$$m_T^2 = \frac{e^2}{4}(T^2 - T_0^2), \quad T_0^2 = \frac{2m_H^2}{e^2} - \frac{3e^2\chi_0^2}{4\pi^2} \tag{4.124}$$

is the temperature-dependent mass. Formula (4.122) is applicable only if $m_G = e\bar\chi \ll T$ or, in other words, for $\bar\chi \ll T/e$.

Note that our investigation so far was based on the one-loop approximation. Higher order corrections can modify a detailed structure of the effective potential at small $\bar\chi$. In particular, it can be shown that when account is taken of these corrections, the cubic term in (4.122) should be multiplied by the coefficient 2/3. This effect, however, goes beyond the scope of our consideration and will be ignored in what follows.

Problem 4.20 Using (3.47) for $J_-^{(1)}(\alpha, 0)$, find the first few terms in the low-temperature expansion of effective potential (4.116). (*Hint* In this case it is more convenient to use m/T and ∞ as the limits of integration in (4.111). Why?)

4.4 "Symmetry restoration" and phase transitions

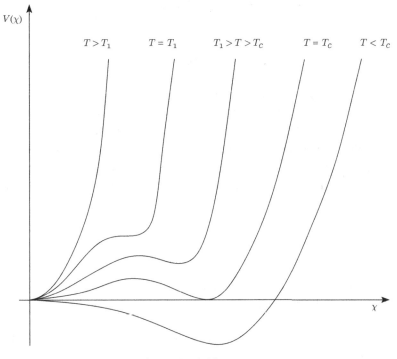

Fig. 4.10.

4.4.3 Symmetry restoration at high temperature

Potential (4.122) is shown in Figure 4.10 for a few different temperatures. For very large T it has only one minimum at $\bar{\chi} = 0$; the symmetry is restored and the gauge bosons and fermions are massless. When the temperature drops below

$$T_1 = \frac{T_0}{\sqrt{1 - (9e^4/16\pi^2 \lambda_{T_1})}}, \qquad (4.125)$$

the second minimum appears; first it is located at $\bar{\chi}_1 = (3e^3 T_1/8\pi \lambda_T)$, and then it moves to the right as the temperature drops. The values of V_{eff} at the two minima become equal at the critical temperature

$$T_c = \frac{T_0}{\sqrt{1 - (e^4/2\pi^2 \lambda_{T_c})}}. \qquad (4.126)$$

At this time the second minimum, located at

$$\bar{\chi}_c = \frac{e^3 T_c}{2\pi \lambda_T}, \qquad (4.127)$$

is separated from $\bar{\chi} = 0$ by a potential barrier of height,

$$\Delta V_c = \frac{e^{12} T_c^4}{4(4\pi)^4 \lambda_{T_c}^3}, \tag{4.128}$$

with the maximum at $\bar{\chi} = \bar{\chi}_c/2$. Note that the coupling constant λ_T cannot be taken as arbitrarily small, because for large T the temperature corrections to it exceed e^4. Hence, at the critical temperature the second minimum, located at $\bar{\chi}_c < T/e$, is always within the region of applicability of the high-temperature expansion. As the temperature drops below T_0, m_T^2 becomes negative and the minimum at $\bar{\chi} = 0$ disappears. Finally, at very low temperatures (see Problem 4.20) the potential converges to (4.117).

4.4.4 Phase transitions

As the temperature drops below the critical value T_c, the minimum at $\bar{\chi}_m \neq 0$ becomes energetically favorable. Therefore, the field $\bar{\chi}$ can change its value and evolve to this second minimum. If at this time the two minima are separated by a potential barrier, the transition occurs with bubble nucleation. Inside the bubbles the scalar field acquires a nonvanishing expectation value. If the bubble nucleation rate exceeds the universe's expansion rate, the bubbles collide and eventually fill all space. As a result, gauge bosons and fermions become massive. Such a transition is called a first order phase transition. It is very violent and one can expect large deviations from thermal equilibrium.

The other possible scenario takes place if $\bar{\chi} = 0$ and $\bar{\chi}_m \neq 0$ are never separated by a potential barrier. In this case, the field $\bar{\chi}$ gradually changes its value and the transition is smooth. It can be either a second order phase transition or simply a cross-over. As we have pointed out a second order phase transition is usually characterized by a continuous change of some symmetry. Because the gauge symmetry is never broken by the Higgs mechanism, we expect that in gauge theories a smooth transition is a cross-over. From the point of view of cosmological scenarios, a cross-over is not very different from a second order phase transition and for our purposes we simply have to distinguish a violent from a smooth transition.

Let us now discuss what kind of transition one could expect in $U(1)$ theory. To answer this question we consider the high-temperature expansion of V_{eff}, given in (4.122). First of all we note that the barrier is due entirely to the $\bar{\chi}^3$ term, which in turn appears because of the nonanalytic term $\propto \sqrt{(m_G/T)^2}$ in (3.44) for $J_-^{(1)}$. If the mass m_G were zero, this term would be absent. Therefore, to establish the character of the transition, we need to know when we can trust our calculation of the $\bar{\chi}^3$ contribution to V_{eff}. As follows from (4.109), the temperature fluctuations

4.4 "Symmetry restoration" and phase transitions

of the scalar field are about

$$\delta\phi = \sqrt{\langle\phi^2\rangle_T} \simeq T/\sqrt{24}.$$

For $\bar{\chi} < T/\sqrt{24}$, the vector bosons can no longer be treated as massive particles. Therefore, for this range, *the perturbative consideration of the mass corrections in $J_-^{(1)}$ fails* and one expects the $\bar{\chi}^3$ term to be absent.

The following simple criteria provide a sense of when the barrier will be present: if, at critical temperature T_c, the value of the scalar field at the expected location of the barrier maximum,

$$\bar{\chi}_c/2 = e^3 T_c / 4\pi \lambda_T,$$

exceeds $T/\sqrt{24}$, then the barrier really exists. Indeed, in this case the calculation of $\bar{\chi}^3$ term is reliable. Thus, we conclude that if the coupling constant λ is small enough, namely

$$\frac{\sqrt{6}}{2\pi} e^3 > \lambda > \frac{9}{16\pi^2} e^4, \tag{4.129}$$

the maximum of the barrier is located at

$$T_c/e > \bar{\chi}_c/2 > T_c/\sqrt{24}$$

and the first order phase transition with bubble nucleation should take place. Because the Higgs mass is proportional to λ, this situation can be realized only if the Higgs particle is not too heavy.

On the other hand, if the coupling constant is large,

$$e^2 > \lambda > \frac{\sqrt{6}}{2\pi} e^3, \tag{4.130}$$

the barrier should be located at

$$\bar{\chi}_c/2 < T_c/\sqrt{24}.$$

However, for this range of $\bar{\chi}$, the bosons should be treated as massless particles and the $\bar{\chi}^3$ term should be left out of the potential. Thus, we expect that the barrier does not arise at all and the effective potential changes as shown in Figure 4.11. In this case the symmetry breaking occurs smoothly via a gradual increase of the mean value of the scalar field. Therefore, the transition has no dramatic cosmological consequences. We remind the reader that the contribution of the scalar particles can be neglected only if the first inequality in (4.130) is fulfilled.

The criteria derived above are no more than rough estimates. However, more sophisticated analysis shows that these estimates reproduce the more rigorous results rather well.

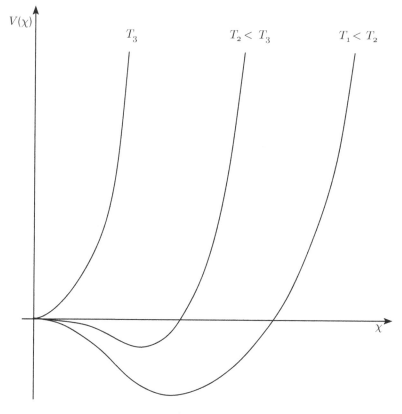

Fig. 4.11.

4.4.5 Electroweak phase transition

The above considerations can easily be generalized to study the electroweak phase transition in the early universe. In electroweak theory, the equation for the scalar field is obtained by variation of χ-dependent terms in the electroweak Lagrangian given in (4.59), (4.79) and (4.84). If we assume that the Higgs mass is small and *neglect the scalar particles*, then the equation for the homogeneous field $\bar{\chi}$ is

$$\bar{\chi}^{;\alpha}_{;\alpha} + V'(\bar{\chi}) - \frac{g^2 + g'^2}{4}\bar{\chi}\langle Z_\mu Z^\mu\rangle - \frac{g^2}{2}\bar{\chi}\langle W^+_\mu W^{-\mu}\rangle + f_t\langle t\bar{t}\rangle = 0. \quad (4.131)$$

We have retained here only the top quarks, which dominate over the contributions from the other fermions because of their large Yukawa coupling constant f_t. The contributions of Z and W bosons to V_{eff} can immediately be written down using the formulae derived in the previous section. We simply note that the charged W bosons have twice as many degrees of freedom as the neutral vector field and hence give twice the contribution to V_{eff}.

4.4 "Symmetry restoration" and phase transitions

Problem 4.21 Verify that $f_t \langle t\bar{t} \rangle = \partial V_t / \partial \bar{\chi}$, where

$$V_t \equiv 3 \times 4 \left(-\frac{I(m_t)}{4\pi^2} + \frac{T^4}{4\pi^2} F_+\left(\frac{m_t}{T}\right) \right). \tag{4.132}$$

Here $m_t = f_t \bar{\chi}$ and

$$F_+\left(\frac{m_t}{T}\right) \equiv \int_0^{m_t/T} \alpha J_+^{(1)}(\alpha, 0) \, d\alpha. \tag{4.133}$$

The integrals $I(m_t)$ and $J_+^{(1)}(\alpha, 0)$ are defined in (4.104) and (3.34) respectively. The factor 3 in (4.132) accounts for the three different colors of t quarks, the factor 4 for four degrees of freedom of the fermions of each color, and the negative sign in front of I indicates that the vacuum energy density of fermions is negative.

Using this result we obtain

$$V_{\text{eff}} = V + \frac{3}{64\pi^2} \left(m_Z^4 \ln \frac{m_Z^2}{\mu^2} + 2 m_W^4 \ln \frac{m_W^2}{\mu^2} - 4 m_t^4 \ln \frac{m_t^2}{\mu^2} \right)$$
$$+ \frac{3T^4}{4\pi^2} \left[F_-\left(\frac{m_Z}{T}\right) + 2 F_-\left(\frac{m_W}{T}\right) + 4 F_+\left(\frac{m_t}{T}\right) \right], \tag{4.134}$$

where

$$m_Z = \frac{\sqrt{g^2 + g'^2}}{2} \bar{\chi}, \quad m_W = \frac{g}{2} \bar{\chi}, \quad m_t = f_t \bar{\chi}.$$

This formula resembles (4.116) and the numerical coefficients in front of the different terms can easily be understood by counting the number of degrees of freedom of the corresponding fields.

Problem 4.22 Using the same normalization conditions as in Problem 4.19, verify that at $T = 0$ K the effective potential is given by (4.117) and (4.118), where we have to substitute

$$e^4 \to \frac{M_Z^4 + 2 M_W^4 - 4 M_t^4}{\chi_0^4}.$$

Since in the broken symmetry phase the masses of the gauge bosons and t quark are $M_Z \equiv m_Z(\chi_0) \simeq 91.2$ GeV, $M_W \simeq 80.4$ GeV and $M_t \simeq 170$ GeV, this combination of masses is negative and the Linde–Weinberg arguments do not lead to a lower bound on the Higgs mass. However, in this case the top quark contribution causes the coefficient in front of the logarithmic term in (4.117) to be negative. At very large $\bar{\chi}$ this term dominates and the potential becomes negative and unbounded from below.

Taking different values for the Higgs mass (10 GeV, 30 GeV, 100 GeV), find out when the potential V_{eff} becomes negative. In this way, assuming that the standard electroweak theory is valid up to the scale $\bar{\chi}_m$ and requiring the absence of the dangerous second minimum in V_{eff} at very large $\bar{\chi} > \chi_0$, one can obtain a lower bound on the Higgs mass. However, this bound is not as robust as the Linde–Weinberg bound.

The high-temperature expansion of the potential (4.134) is derived using the same methods as for (4.122). The result is

$$V_{\text{eff}}(\bar{\chi}, T) \simeq \frac{\lambda_T}{4}\bar{\chi}^4 - \frac{\Theta}{3}T\bar{\chi}^3 + \frac{\Upsilon(T^2 - T_0^2)}{2}\bar{\chi}^2 + \Lambda_R, \quad (4.135)$$

where the temperature-dependent coupling constant

$$\lambda_T = \frac{m_H^2}{2\chi_0^2} + \frac{3}{16\pi^2 \chi_0^4}\left(M_Z^4 \ln\frac{bT^2}{M_Z^2} + 2M_W^4 \ln\frac{bT^2}{M_W^2} - 4M_t^4 \ln\frac{b_F T^2}{M_t^2}\right) \quad (4.136)$$

is expressed through the masses of the gauge bosons and t quark in the broken symmetry phase, e.g. $M_Z \equiv m_Z(\chi_0)$. The constant b is defined in (4.123) and $\ln b_F = 2\ln\pi - 2\mathbf{C} \simeq 1.14$.

The dimensionless constants Θ and Υ are

$$\Theta = \frac{3(M_Z^3 + 2M_W^3)}{4\pi \chi_0^3} \simeq 2.7 \times 10^{-2}, \quad \Upsilon = \frac{M_Z^2 + 2M_W^2 + 2M_t^2}{4\chi_0^2} \simeq 0.3,$$

and the temperature

$$T_0^2 = \frac{1}{2\Upsilon}\left(m_H^2 - \frac{3(M_Z^4 + 2M_W^4 - 4M_t^4)}{8\pi^2 \chi_0^2}\right) \simeq 1.7\left(m_H^2 + (44\,\text{GeV})^2\right) \quad (4.137)$$

depends explicitly on the unknown Higgs mass.

The terms due to the vector bosons are of the same type as in (4.122). This becomes clear upon rewriting (4.122)–(4.124) using $M_G = e\chi_0$ instead of the coupling constant e. To find the contribution of the fermions we have used the high-temperature expansion (3.44) for $J_+^{(1)}$. Note that this expansion does not contain a nonanalytic term which would contribute to the numerical coefficient Θ in front of the $\bar{\chi}^3$ term.

Now we turn to the temperature behavior of potential (4.134) and study the symmetry breaking in the early universe. To get an idea of the expected character of the transition it is enough to consider the high-temperature expansion (4.135).

For a given temperature T, the contribution from different fields to (4.135) can be trusted only for those $\bar{\chi}$ for which the induced masses of the corresponding fields are smaller than the temperature. For instance, the t quark terms should be

4.4 "Symmetry restoration" and phase transitions

retained in (4.135) only when $m_t(\bar{\chi}) = M_t\bar{\chi}/\chi_0 \ll T$, that is, for $\bar{\chi} \ll (\chi_0/M_t)T$. Within the range

$$\frac{T}{M_{Z,W}}\chi_0 > \bar{\chi} > \frac{T}{M_t}\chi_0, \quad (4.138)$$

the Z and W bosons are relativistic, while the t quarks are nonrelativistic and hence their contribution should be omitted. As we have mentioned above, for very small $\bar{\chi}$ our derivation fails and we have to use more refined methods.

The analysis of the potential behavior follows almost exactly the consideration in the previous section. At very high temperatures, namely

$$T \gg T_1 = \frac{T_0}{\sqrt{1-(\Theta^2/4\Upsilon\lambda_{T_c})}}, \quad (4.139)$$

the potential has one minimum only, at $\bar{\chi} = 0$, the symmetry is restored and gauge bosons and fermions are massless. As the temperature drops below T_1, the second minimum appears and at

$$T_c = \frac{T_0}{\sqrt{1-(2\Theta^2/9\Upsilon\lambda_{T_c})}} \quad (4.140)$$

the depth of this second minimum, located at

$$\bar{\chi}_c = \frac{2\Theta T}{3\lambda_{T_c}}, \quad (4.141)$$

becomes the same as that of the minimum at $\bar{\chi} = 0$. Subsequently, the transition to the broken symmetry phase becomes possible. As noted above, the minimum at $\bar{\chi}_c$ is separated from $\bar{\chi} = 0$ by the barrier only if $\bar{\chi}_c/2 > T/\sqrt{24}$. Hence, we expect a strong first order phase transition only if $\lambda_{T_c} < \sqrt{8/3}\Theta$. Using (4.136), this condition can be rewritten in terms of the upper bound on the Higgs mass: $m_H < 75$ GeV. Thus, only if the Higgs particles were light enough would the electroweak phase transition be first order. For $m_H = 50$ GeV, this would occur at temperature $T_c \approx 88$ GeV.

The experimental bound on the Higgs mass is $m_H > 114$ GeV. Therefore in reality we expect that the breaking of electroweak symmetry happens smoothly. For large Higgs masses one can simply neglect the $\bar{\chi}^3$ term in (4.135). In this case, when the temperature drops below T_0 the only minimum of the potential is located at

$$\bar{\chi}_c = \left(\frac{\Upsilon}{\lambda_T}(T_0^2 - T^2)\right)^{1/2} \quad (4.142)$$

and the symmetry is broken. The transition is a cross-over with no dramatic cosmological consequences; in particular, no large deviation from the thermal equilibrium. The temperature at transition depends on the Higgs mass and it follows from (4.137) that, for instance, $T_0 \simeq 166$ GeV for $m_H \simeq 120$ GeV and $T_0 \simeq 240$ GeV for $m_H \simeq 180$ GeV. The above *estimates* are in good agreement with the results of more rigorous and elaborate calculations.

4.5 Instantons, sphalerons and the early universe

In gauge theories which incorporate the Higgs mechanism the vacuum has a nontrivial structure in several respects. First of all, as we have seen in the preceding section, in the early universe the effective potential of the Higgs field can have two local minima. If the transition from the symmetrical phase to the broken symmetry phase happens when the false and true vacua are still separated by a barrier, the transition is first order and accompanied by bubble nucleation. Although this situation seems not to occur in either quantum chromodynamics or the electroweak model, it is rather typical for unified field theories beyond the Standard Model.

Another interesting aspect of non-Abelian gauge theories is the existence of topologically different vacua. They are also separated by a barrier and topological transitions can take place in the early universe. These transitions are very important because they lead to anomalous nonconservation of fermion number in the Standard Model.

If the temperature at the time of transition is small compared to the height of the potential barrier, the transitions between different vacua occur as a result of subbarrier quantum tunneling. In this case the Euclidean solution of the field equations, called *an instanton*, gives the dominant contribution to the tunneling probability. On the other hand, if the temperature is high enough, thermal fluctuations can take the field over the barrier to the other vacuum without tunneling. In this situation the transitions are classical and their rate is determined by the static field configuration corresponding to the maximum of the potential, called *a sphaleron*.

In this section we will calculate the rates for the false vacuum decay and for topological vacuum transitions and determine under which conditions the sphalerons dominate the instantons and vice versa.

4.5.1 Particle escape from a potential well

To start with, let us consider as a "warm-up" the one-dimensional problem of a particle of mass M escaping from a potential well at $q_0 = 0$ (Figure 4.12). In this simple case we meet the main concepts we need for analyzing transitions in field theory.

4.5 Instantons, sphalerons and the early universe

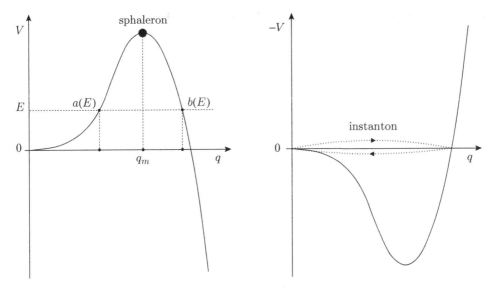

Fig. 4.12.

First, we neglect the thermal fluctuations and assume that the particle, with *fixed* energy $E < V(q_m)$, is initially localized in the potential well. The only way to escape from this well is via subbarrier tunneling. If the tunneling probability is small, the energy E is an *approximate* eigenvalue of the Hamiltonian, and we can use the stationary Schrödinger equation

$$\left(-\frac{1}{2M}\frac{\partial^2}{\partial q^2} + V(q)\right)\Psi \simeq E\Psi \qquad (4.143)$$

to *estimate* the tunneling amplitude. The approximate semiclassical solution of this equation is

$$\Psi \propto \exp\left(i\int \sqrt{2M(E-V)}\,dq\right). \qquad (4.144)$$

In the classically allowed regions, where $E > V$, the wave function simply oscillates, while under the barrier it decays exponentially. Hence, in the region $q > b(E)$, the wave function is suppressed by the factor

$$\exp\left(-\int_{a(E)}^{b(E)} \sqrt{2M(V-E)}\,dq\right) \qquad (4.145)$$

compared to its value inside the potential well. Expression (4.145) accounts for the dominant exponential contribution to the tunneling amplitude.

Instantons In the special case of $E = 0$ (the particle is at rest at the local minimum of the potential) we have

$$\int_0^{b(0)} \sqrt{2MV}\, dq = \int_{-\infty}^{\tau_b} \left(\frac{1}{2}M\dot{q}^2 + V\right) d\tau \equiv S_{b(0)}, \qquad (4.146)$$

where $q(\tau)$ satisfies the equation

$$M\ddot{q} + \frac{\partial(-V)}{\partial q} = 0. \qquad (4.147)$$

Integrating (4.147) once gives

$$M\dot{q}^2/2 - V = 0.$$

Taking into account time-translational invariance, we can set $\tau_b = 0$ in (4.146) without loss of generality.

Equation (4.147) describes the motion of the particle in the inverted potential $(-V)$ (Figure 4.12), and can be obtained from the original equation of motion if we make the Wick rotation $t \to \tau = it$. Note that the formal substitution $t = -i\tau$ converts the Minkowski metric

$$ds^2 = dt^2 - d\mathbf{x}^2$$

to the Euclidean metric

$$-ds_E^2 = d\tau^2 + d\mathbf{x}^2.$$

Therefore, τ is called "Euclidean time." The right hand side of (4.146) is the Euclidean action calculated for the trajectory satisfying (4.147), with boundary conditions $q(\tau \to -\infty) = 0$ and $q(\tau = 0) = b(0)$. We can "close" this trajectory by considering the "motion" back to $q = 0$ as $\tau \to +\infty$. The corresponding solution of (4.147), with boundary conditions $q(\tau \to \pm\infty) = 0$, is a baby version of the Euclidean field theory solutions called instantons. It is clear from symmetry that the instanton action S_I is just twice the action $S_{b(0)}$. Hence, for the ground state $(E = 0)$, the dominant contribution to the tunneling probability, which is the square of the amplitude (4.145), is

$$P_I \propto \exp(-S_I). \qquad (4.148)$$

Thermal fluctuations and sphalerons Now we consider a particle in equilibrium with a thermal bath of temperature T. The particle can acquire energy from the thermal bath and if this energy exceeds the height of the barrier the particle escapes from the potential well classically, without needing to "go under the barrier." The

4.5 Instantons, sphalerons and the early universe

probability that the particle gets energy E is given by the usual Boltzmann factor $\propto \exp(-E/T)$. Taking into account that for $E < V(q_m)$ the tunneling amplitude is given by (4.145), we obtain the following result for the *total* probability of escape:

$$P \propto \sum_E \exp\left(-\frac{E}{T} - 2\vartheta \int_{a(E)}^{b(E)} \sqrt{2M(V-E)} dq\right), \quad (4.149)$$

where $\vartheta = 1$ for $E < V(q_m)$ and $\vartheta = 0$ otherwise. The sum in (4.149) can be *estimated* using the saddle point approximation. Taking the derivative of the expression in the exponent, we find that this expression has its maximum value when E satisfies the equation

$$\frac{1}{T} = 2 \int_{a(E)}^{b(E)} \sqrt{\frac{M}{2(V-E)}} dq = 2 \int_{a(E)}^{b(E)} \frac{dq}{\dot{q}}. \quad (4.150)$$

Here $\dot{q} = dq/d\tau$ is the "Euclidean velocity" along the trajectory with total Euclidean energy $-E$. The term on the right hand side of (4.150) is equal to the period of oscillation in the inverted potential $(-V)$. Hence, for a given temperature T, the dominant contribution to the escape probability gives the periodic Euclidean trajectory describing oscillations with period $1/T$ in the potential $(-V)$.

Let us consider two limiting cases. From (4.149) and (4.150), it follows that for $T \ll V(q_m)/S_I$ the main contribution to the escape probability comes from the subbarrier trajectory with $E \ll V(q_m)$ and as a result the rate of escape is determined by the instanton. In the opposite case of very high temperatures,

$$T \gg \frac{V(q_m)}{S_I},$$

the "period of oscillation" tends to zero; hence the dominant trajectory comes very close to the top of the potential V and has energy $E \approx V(q_m)$. The unstable *static* solution $q = q_m$, which corresponds to the maximum of V, is a prototype of the field theory solutions called sphalerons (Greek name for "ready to fall"). For the sphaleron, the second term in the exponent in (4.149) can be neglected and the escape probability is given by

$$P \propto \exp\left(-\frac{E_{sph}}{T}\right), \quad (4.151)$$

where $E_{sph} = V(q_m)$ is the energy (or mass) of the sphaleron. We would like to stress that at very high temperatures the main contribution to the escape probability comes from the states which surmount the barrier classically. For $T > E_{sph}$ there is no exponential suppression and the particle very quickly leaves the potential well.

So far we have considered a particle with only one degree of freedom. For a system with N degrees of freedom the potential can depend on all coordinates $\mathbf{q} \equiv (q_1, q_2, \ldots, q_N)$. The generalization to this case, however, is rather straightforward. To calculate the tunneling probability at low temperatures we have to find an instanton with the minimal Euclidean action S_I. If the energy is normalized so that $V = 0$ at the bottom of the potential well, the dominant contribution to the tunneling probability is proportional to $\exp(-S_I)$. At very high temperatures we need to find the local extrema of the potential through which the particle can escape. The extremum with the minimal value of the potential determines the dominant contribution to the escape probability. This probability is given by (4.151), where E_{sph} is the value of the potential at the corresponding extremum.

4.5.2 Decay of the metastable vacuum

We consider a real scalar field in this section using the standard notation φ instead of χ and for simplicity neglect the expansion of the universe. Let us assume that at the time relevant for the transition the potential $V(\varphi)$ has the shape shown in Figure 4.13. For convenience we normalize the energy such that $V(0) = 0$ and $V(\varphi_0) = -\epsilon < 0$. Obviously the state $\varphi = 0$ is metastable and decays. If the transition takes place efficiently when the temperature is small, we can neglect thermal fluctuations and

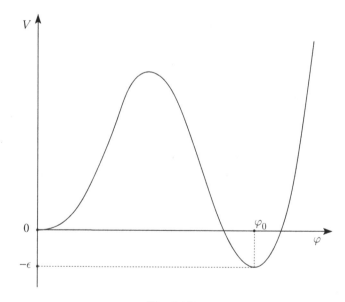

Fig. 4.13.

the metastable vacuum decays via quantum tunneling. If thermal fluctuations are not negligible on the other hand, they can push the field to the top of the potential and the transition occurs classically without tunneling. In both cases critical bubbles are formed, filled with the new phase $\varphi \neq 0$. If the bubble nucleation rate exceeds the expansion rate of the universe, the bubbles collide and finally fill all space with the new phase. Let us now calculate the decay rate of the metastable (false) vacuum.

The scalar field $\varphi(\mathbf{x}, t)$ is a system with an infinite number of degrees of freedom. We can treat the spatial coordinates $\mathbf{x}$ as a continuous index enumerating the degrees of freedom. In this case $\varphi(\mathbf{x}, t) \equiv \varphi_{\mathbf{x}}(t)$ plays the same role as $q_n(t)$ in the preceding discussion (the correspondence is obvious: $\varphi \Longleftrightarrow q$ and $\mathbf{x} \Longleftrightarrow n$). The action for the scalar field can be written as

$$S = \int (\mathcal{K} - \mathcal{V}) \, dt, \qquad (4.152)$$

where

$$\mathcal{K} \equiv \sum_{\mathbf{x}} \frac{1}{2} \left(\frac{\partial \varphi_{\mathbf{x}}}{\partial t} \right)^2 = \frac{1}{2} \int \left(\frac{\partial \varphi_{\mathbf{x}}}{\partial t} \right)^2 d^3x \qquad (4.153)$$

is the kinetic energy and

$$\mathcal{V}(\varphi_{\mathbf{x}}) \equiv \int \left(\frac{1}{2} (\partial_i \varphi_{\mathbf{x}})^2 + V(\varphi_{\mathbf{x}}) \right) d^3x \qquad (4.154)$$

is the potential energy or *potential* of the scalar field configuration $\varphi_{\mathbf{x}}$. As usual, ∂_i denotes the partial derivative with respect to the spatial coordinate x^i which, in the language adopted in this section, is the derivative with respect to the continuous index. We must stress that in the study of tunneling in field theory the potential $\mathcal{V}$, and *not* the scalar field potential $V(\varphi)$, plays the role of $V(\mathbf{q})$ from the previous discussion. To avoid confusion, the reader must distinguish between them carefully. The potential $\mathcal{V}(\varphi_{\mathbf{x}})$ is a *functional*. It depends on the infinite number of variables $\varphi_{\mathbf{x}}$ and takes a definite numerical value only when the field configuration $\varphi(\mathbf{x})$ is completely specified.

Decay via instantons For the scalar field potential $V(\varphi)$, shown in Figure 4.13, the state $\varphi(\mathbf{x}) = 0$ corresponds to a local minimum of the potential $\mathcal{V}$ with $\mathcal{V}(0) = 0$. The other static configuration $\varphi(\mathbf{x}) = \varphi_0$ has negative energy $\mathcal{V}(\varphi_0) = -\epsilon \times$ volume. Therefore the state $\varphi = 0$ is metastable and should decay. This decay can be described, in complete analogy with the preceding discussion, as an "escape" via tunneling of the infinitely many degrees of freedom from the local potential well in $\mathcal{V}$. The dominant contribution to the semiclassical tunneling probability, proportional to $\exp(-S_I)$, comes from the instanton with the action S_I. This instanton "connects" the metastable vacuum $\varphi(\mathbf{x}) = 0$ to some (classically allowed) configuration $\varphi_\mathbf{x}$ with $\mathcal{V}(\varphi_\mathbf{x}) \leq 0$, and satisfies the equation

$$\ddot{\varphi}_\mathbf{x}(\tau) + \frac{\delta(-\mathcal{V})}{\delta \varphi_\mathbf{x}} = 0, \quad (4.155)$$

where $\ddot{\varphi}_\mathbf{x} \equiv \partial^2 \varphi / \partial \tau^2$ and

$$\frac{\delta(-\mathcal{V})}{\delta \varphi_\mathbf{x}} = \Delta \varphi - V_{,\varphi}$$

is the functional derivative of the inverted potential. Equation (4.155) is an analog of (4.147) and it is obtained from the usual scalar field equation in Minkowski space under Wick rotation $t \to \tau = it$.

The Euclidean action is finite for those solutions describing tunneling in which the field φ changes its value from $\varphi = 0$ to $\varphi \neq 0$ only within a bounded region in space. On symmetry grounds one expects the most favorable emerging configuration of the scalar field to be a bubble with $\varphi_c \neq 0$ at its center and $\varphi \to 0$ far away from the center (Figure 4.14). To find the corresponding instanton relating the original metastable vacuum configuration $\varphi(\mathbf{x}) = 0$ to a bubble filled with a new phase we can again rely on symmetry. That is, we adopt the most symmetrical $O(4)$-invariant solution of the Euclidean equation (4.155), which describes the *four-dimensional* spherical "bubble" *in Euclidean "spacetime."* The scalar field then depends only on the radial coordinate

$$\tilde{r} = \sqrt{\mathbf{x}^2 + \tau^2}.$$

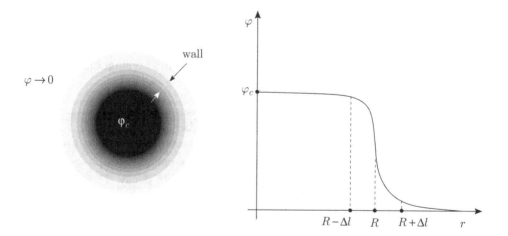

Fig. 4.14.

4.5 Instantons, sphalerons and the early universe

For $\varphi_{\mathbf{x}}(\tau) = \varphi(\tilde{r})$ (4.155) simplifies to the ordinary differential equation

$$\frac{d^2\varphi}{d\tilde{r}^2} + \frac{3}{\tilde{r}}\frac{d\varphi}{d\tilde{r}} - \frac{\partial V}{\partial \varphi} = 0. \tag{4.156}$$

The solution of this equation, with boundary conditions $\varphi \to 0$ as $\tilde{r} \to \infty$ and $d\varphi/d\tilde{r} = 0$ at $\tilde{r} = 0$, gives the desired instanton. The second boundary condition is needed to avoid a singularity at the center of the bubble.

The vacuum decay rate per unit time and unit volume is

$$\Gamma \simeq A \exp(-S_I), \tag{4.157}$$

where A is a pre-exponential factor that is very difficult to calculate. We can obtain a rough estimate for A using dimensional arguments. In units where $c = \hbar = 1$ and $G \neq 1$, the decay rate has the dimension cm^{-4} and hence

$$A \sim O\left(R_I^{-4}, V_{,\varphi\varphi}^2, \cdots\right), \tag{4.158}$$

where R_I is the size of the instanton and one uses typical instanton values for the derivatives of V. Usually all these quantities have the same order of magnitude.

Problem 4.23 Verify that the Euclidean action for the instanton can be reduced to

$$S_I = 2\pi^2 \int \left(\frac{1}{2}\left(\frac{d\varphi}{d\tilde{r}}\right)^2 + V\right) \tilde{r}^3 d\tilde{r}. \tag{4.159}$$

(*Hint* The Euclidean action S_E is related to the Lorentzian action S_L, given in (4.152), by $S_L \to iS_E$ as $t \to \tau = it$.)

Decay via sphalerons The above consideration is applicable only at low or zero temperature. If the temperature is very high (we will specify later what this means), the dominant contribution to the vacuum decay is given by over-barrier classical transitions. To estimate their rate we have to find the extremum of the potential V with the minimal possible energy. This extremum is reached for the *static* scalar field configuration (sphaleron), satisfying the equation

$$\delta V/\delta \varphi = -\Delta \varphi + V_{,\varphi} = 0. \tag{4.160}$$

For a spherical bubble this becomes

$$\frac{d^2\varphi}{dr^2} + \frac{2}{r}\frac{d\varphi}{dr} - \frac{\partial V}{\partial \varphi} = 0, \tag{4.161}$$

where $\varphi = \varphi(r)$ and $r = |\mathbf{x}|$ is the radial coordinate in *three-dimensional space*. The boundary conditions we need to impose are similar to those for (4.156), namely, $\varphi \to 0$ as $r \to \infty$ and $d\varphi/dr = 0$ at $r = 0$. Note that the sphaleron, in contrast with the instanton, is the unstable *static* solution which depends only on *the spatial coordinates* **x**.

For transitions dominated by sphalerons, the decay rate per unit time and unit volume is

$$\Gamma \simeq B \exp\left(-\frac{V_{sph}}{T}\right), \qquad (4.162)$$

where

$$V_{sph} = 4\pi \int \left(\frac{1}{2}\left(\frac{d\varphi}{dr}\right)^2 + V\right) r^2 dr \qquad (4.163)$$

is the energy (or *mass*) of the sphaleron and $B \sim O(T^4, \ldots)$ is a further pre-exponential factor.

Using the same reasoning as for particle escape, we find that for $T \ll V_{sph}/S_I$ tunneling is more important and to estimate the vacuum decay rate one has to use (4.157); otherwise, for $T \gg V_{sph}/S_I$ the decay rate is given by (4.162).

Thus, to calculate the vacuum decay rate we have to find the solution of either (4.156) or (4.161), depending on the temperature. One must usually resort to numerical calculation; however, for a wide class of scalar field potentials it is possible to find the explicit expression for Γ without specifying the shape of the potential V.

Thin wall approximation The first integral of (4.156) is

$$\frac{1}{2}\left(\frac{d\varphi}{d\tilde{r}}\right)^2 - V = \int_{\tilde{r}}^{\infty} \frac{3}{\tilde{r}'}\left(\frac{d\varphi}{d\tilde{r}'}\right)^2 d\tilde{r}', \qquad (4.164)$$

where the boundary condition $\varphi \to 0$ as $\tilde{r} \to \infty$ has been used. Taking into account the other boundary condition $(d\varphi/d\tilde{r})_{\tilde{r}=0} = 0$, we obtain the useful relation

$$-V(\varphi(\tilde{r}=0)) = \int_0^{\infty} \frac{3}{\tilde{r}'}\left(\frac{d\varphi}{d\tilde{r}'}\right)^2 d\tilde{r}'. \qquad (4.165)$$

Now let us assume that the instanton, which is a bubble in four-dimensional Euclidean "spacetime," has a thin wall. This means that inside the bubble of radius R_I the scalar field is nearly constant and equal to $\varphi(\tilde{r} = 0)$. The field φ changes very fast within the wall – a thin layer of width $2\Delta l \ll R_I$ – and tends to zero outside the bubble (Figure 4.14). Hence the integrand in (4.165) differs significantly from zero only inside the wall, and this gives the dominant contribution to the integral. Returning to (4.164) we see that the integral on the right hand side is suppressed inside the wall by a factor $\Delta l/R_I \ll 1$ compared to $(d\varphi/d\tilde{r})^2$ and V. Therefore, in the leading approximation we have

$$(d\varphi/d\tilde{r})^2 \approx 2V$$

for $R_I + \Delta l > \tilde{r} > R_I - \Delta l$. Using this result, (4.165) simplifies to

$$-V(\varphi_{\tilde{r}=0}) \approx \frac{3\sigma}{R_I}, \qquad (4.166)$$

where

$$\sigma \equiv \int_0^\infty \left(\frac{d\varphi}{d\tilde{r}'}\right)^2 d\tilde{r}' \approx \int_0^{\varphi_{\tilde{r}=0}} \sqrt{2V}\, d\varphi \qquad (4.167)$$

is the surface tension of the bubble. The instanton action (4.159) then becomes

$$S_I \approx \frac{\pi^2}{2} V(\varphi_{\tilde{r}=0}) R_I^4 + 2\pi^2 \sigma R_I^3 \approx \frac{27\pi^2 \sigma^4}{2|V(\varphi_{\tilde{r}=0})|^3}, \qquad (4.168)$$

where the first and second terms are the contributions of the internal region of the bubble and its wall respectively. For the potential V shown in Figure 4.13 the action takes the minimal value for $|V(\varphi_{\tilde{r}=0})| = \epsilon$. In this case the field inside the bubble is equal to φ_0 and, as follows from (4.166), the instanton has size $R_I \approx 3\sigma/\epsilon$. Hence the vacuum decay rate is equal to

$$\Gamma \simeq A \exp\left(-\frac{27\pi^2 \sigma^4}{2\epsilon^3}\right). \qquad (4.169)$$

At a "given moment of Euclidean time" τ the solution $\varphi\left(\sqrt{r^2 + \tau^2}\right)$ describes a three-dimensional bubble. The half of the instanton connecting the metastable vacuum with the classically allowed region "evolves" in Euclidean time as follows. At $\tau \to -\infty$ we have $\varphi = 0$ everywhere in space. "Later on" at $\tau \sim -R_I$ the bubble "appears" and its radius "grows" as $R(\tau) \approx \sqrt{R_I^2 - \tau^2}$ reaching the maximal value R_I at $\tau = 0$.

Problem 4.24 Calculate the potential energy of this bubble and verify that

$$\mathcal{V}(R(\tau)) \approx \frac{2\pi\epsilon}{3} R(\tau)\left(R_I^2 - R^2(\tau)\right). \qquad (4.170)$$

The total energy corresponding to the instanton solution is equal to zero. Therefore, a bubble with radius $0 < R(\tau) < R_I$ is in the classically forbidden region (under the barrier), where $\mathcal{V}(R) > 0$. A three-dimensional bubble of size $R_I \approx 3\sigma/\epsilon$ is on the border of the classically allowed region ($\mathcal{V}(R_I) = 0$) and it "materializes" in Minkowski spacetime. The picture described is a very close analog of subbarrier particle tunneling.

To understand when the thin wall approximation is valid we must determine when the condition $\Delta l / R_I \ll 1$ is satisfied. If V_m is the height of the *scalar field potential*, the positive energy inside the wall of the "emerging bubble" is of order $V_m R_I^2 \Delta l$. This is exactly compensated by the negative energy inside the bubble $\sim \epsilon R_I^3$, where $-\epsilon$ is the global minimum of the potential. Therefore, $\Delta l / R \sim \epsilon / V_m$ and hence the thin wall approximation is applicable only if $\epsilon / V_m \ll 1$.

The thin wall sphaleron which determines the decay rate at very high temperatures can be found in a similar way from (4.161).

Problem 4.25 Verify that the thin wall sphaleron has size $R_{sph} \approx 2\sigma/\epsilon$ and mass equal to

$$\mathcal{V}_{sph} \approx \frac{16\pi\sigma^3}{3\epsilon^2}. \qquad (4.171)$$

Thus, we find that if $\epsilon / V_m \ll 1$, the vacuum decay rate is about

$$\Gamma \simeq B \exp\left(-\frac{16\pi\sigma^3}{3\epsilon^2 T}\right) \qquad (4.172)$$

for $T \gg \mathcal{V}_{sph}/S_I \sim R_{sph}^{-1}$. On the other hand, for $T \ll R_{sph}^{-1} \sim R_I^{-1}$, the vacuum decay rate is given by (4.169).

After a bubble has emerged in Minkowski spacetime its behavior can easily be found if we analytically continue an appropriate solution of (4.156) back to Minkowski spacetime. The corresponding function $\varphi(\sqrt{r^2 - t^2})$ describes the expanding bubble. The field φ is constant along the hypersurfaces $r^2 - t^2 = \text{const}$. Figure 4.15 shows these hypersurfaces and makes it clear that from the point of view of an observer at rest, the thickness of the wall decreases with time and the speed of the wall approaches the speed of light.

Problem 4.26 Using the results of this section, derive and analyze the formulae describing first order phase transitions in the $U(1)$ model considered in Section 4.4.4.

4.5.3 The vacuum structure of gauge theories

In non-Abelian gauge theories the vacuum of gauge fields has a nontrivial structure. To understand how this comes about, we first consider pure $SU(N)$ theory without fermions and scalar fields. The gauge field $\mathbf{F}_{\mu\nu}$ should vanish in the vacuum. This does not mean, however, that the vector potential $\mathbf{A}_\mu$ also vanishes; the vanishing of $\mathbf{F}_{\mu\nu}$ only means that the vector potential is a gauge transform of zero (see (4.11)).

4.5 Instantons, sphalerons and the early universe

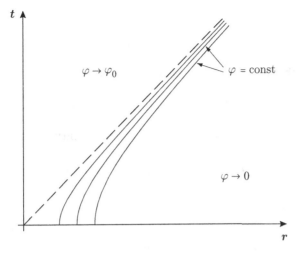

Fig. 4.15.

In particular, taking an arbitrary time-independent unitary matrix $\mathbf{U}(\mathbf{x})$, we find that

$$\mathbf{A}_0 = 0, \quad \mathbf{A}_i = \frac{i}{g}(\partial_i \mathbf{U})\,\mathbf{U}^{-1} \tag{4.173}$$

also describes the vacuum. If an arbitrary $\mathbf{U}(\mathbf{x})$ could be continuously transformed to the unit matrix *everywhere in space* all vacua would be equivalent. However this is not the case. Instead the set of all functions $\mathbf{U}(\mathbf{x})$ can be decomposed into homotopy classes. We say that two functions belong to the same homotopy class if there exists a nonsingular continuous transformation relating them; otherwise the functions belong to different homotopy classes.

Winding number To find and characterize the homotopy classes let us introduce the *winding number*, defined as

$$\nu \equiv -\frac{1}{24\pi^2} \int \mathrm{tr}\!\left(\varepsilon^{ijk}(\partial_i\mathbf{U})\,\mathbf{U}^{-1}(\partial_j\mathbf{U})\,\mathbf{U}^{-1}(\partial_k\mathbf{U})\,\mathbf{U}^{-1}\right) d^3x, \tag{4.174}$$

where ε^{ijk} is a totally antisymmetric Levi–Civita symbol: $\varepsilon^{123} = 1$ and it changes sign upon permutation of any two indices. We will show first that this number is a topological invariant and second that it takes integer values characterizing different homotopy classes. To prove the first statement let us consider a small nonsingular variation $\mathbf{U} \to \mathbf{U} + \delta\mathbf{U}$, and show that $\delta\nu = 0$. Taking into account that

$$(\delta\mathbf{U})\,\mathbf{U}^{-1} = -\mathbf{U}(\delta\mathbf{U}^{-1})$$

and

$$(\partial_i \mathbf{U})\mathbf{U}^{-1} = -\mathbf{U}(\partial_i \mathbf{U}^{-1}), \qquad (4.175)$$

the variation of the first term in the integrand can be written as

$$\delta((\partial_i \mathbf{U})\mathbf{U}^{-1}) = \mathbf{U}(\partial_i(\mathbf{U}^{-1}\delta\mathbf{U}))\mathbf{U}^{-1}. \qquad (4.176)$$

The variation of the other two terms gives a similar contribution, and therefore

$$\delta\nu \propto \int \mathrm{tr}(\varepsilon^{ijk}\partial_i(\mathbf{U}^{-1}\delta\mathbf{U})(\partial_j\mathbf{U}^{-1})(\partial_k\mathbf{U}))\,d^3x. \qquad (4.177)$$

Upon integration by parts there arise terms of the form $\varepsilon^{ijk}\partial_i\partial_j\ldots$, which vanish because of the antisymmetry of the ε symbol. Hence the winding number does not change under continuous nonsingular transformations.

To prove the existence of homotopy classes with different winding numbers, we construct them explicitly for the case of the $SU(2)$ group. Any $SU(2)$ matrix can be written as

$$\mathbf{U}(\chi, \mathbf{e}) = \cos m\chi\,\mathbf{1} - i(\mathbf{e}\cdot\boldsymbol{\sigma})\sin m\chi, \qquad (4.178)$$

where $\mathbf{e} = (e_1, e_2, e_3)$ is the unit vector and $\boldsymbol{\sigma} = (\sigma_1, \sigma_2, \sigma_3)$ are the three Pauli matrices combined as a vector.

Problem 4.27 Verify this last statement. (*Hint* An arbitrary $SU(2)$ matrix has the same form as the matrix ζ in (4.47).)

Thus, the elements of the $SU(2)$ group can be parameterized by the unit vector in four-dimensional Euclidean space,

$$l_\alpha = (\cos m\chi, \mathbf{e}\sin m\chi),$$

and they can be thought of topologically as the elements of the three-dimensional sphere. Let us take χ and $\mathbf{e}$ to be functions of the spatial coordinates $\mathbf{x}$ and identify the points at spatial infinity ($|\mathbf{x}| \to \infty$). Then $\mathbf{U}(\mathbf{x})$ is an unambiguous function of the spatial coordinates if $\chi(|\mathbf{x}| \to \infty) = \pi$ and m is an integer or zero. We can interpret the function $\mathbf{U}(\mathbf{x})$ as describing the mapping from the 3-sphere S^3 (Euclidean space $\mathbf{x}$ with infinity mapped to a point) to the 3-sphere of the elements of the $SU(2)$ group. For the mapping (4.178), the spatial coordinates $\mathbf{x}$ wrap around this sphere m times. The mappings with $+m$ and $-m$ correspond to different orientations and therefore should be distinguished. Using the identity (4.175), (4.174) simplifies to

$$\nu = -\frac{1}{8\pi^2}\int \mathrm{tr}\left\{\mathbf{U}^{-1}(\partial_r\mathbf{U})\left[(\partial_\varphi\mathbf{U}^{-1})(\partial_\theta\mathbf{U}) - (\partial_\theta\mathbf{U}^{-1})(\partial_\varphi\mathbf{U})\right]\right\}dr\,d\theta\,d\varphi, \qquad (4.179)$$

where r, θ, φ are the spherical coordinates in $\mathbf{x}$ space. Let us assume for simplicity that $\chi(\mathbf{x}) = \chi(r)$ and

$$\mathbf{e}(\mathbf{x}) = (\sin\theta\cos\varphi, \sin\theta\sin\varphi, \cos\theta).$$

Then, taking into account that $\mathbf{U}^{-1}(\chi) = \mathbf{U}(-\chi)$ and using the Pauli matrix property

$$\sigma_i \sigma_j = \delta_{ij} + i\varepsilon_{ijk}\sigma_k,$$

from which follows $(\mathbf{e}\boldsymbol{\sigma})^2 = 1$, we obtain

$$\mathbf{U}^{-1}(\partial_r \mathbf{U}) = -im(\mathbf{e}\boldsymbol{\sigma})\frac{d\chi}{dr}, \tag{4.180}$$

$$\left(\partial_\varphi \mathbf{U}^{-1}\right)\left(\partial_\theta \mathbf{U}\right) - \left(\partial_\theta \mathbf{U}^{-1}\right)\left(\partial_\varphi \mathbf{U}\right) = -2i\sin^2(m\chi)\sin\theta(\mathbf{e}\boldsymbol{\sigma}). \tag{4.181}$$

Substituting these expressions into (4.179) and integrating over the angles, we derive the desired result: $\nu = m$. Thus the homotopy class characterized by the winding number m corresponds to the mapping which wraps m times around the $SU(2)$ sphere.

To what extent are these results particular to the $SU(2)$ group? First of all note that every mapping of S^3 into the $U(1)$ Abelian group is continuously deformable to the trivial mapping. Hence the vacuum has a trivial structure in this case and there is no analog of winding number. As for a non-Abelian $SU(N)$ group, one can show by methods beyond the scope of this book that any continuous mapping of S^3 into an arbitrary $SU(N)$ group can be continuously deformed to a mapping into an $SU(2)$ subgroup of $SU(N)$. Therefore all results derived for the $SU(2)$ group are valid for an arbitrary $SU(N)$ group; in particular, (4.174) for the winding number requires no alteration.

Barrier height Two vacua, with different winding numbers, are separated by a barrier. To demonstrate this we will use the identity

$$\mathrm{tr}\big(\mathbf{F}\tilde{\mathbf{F}}\big) = \partial_\alpha\left[\mathrm{tr}\,\varepsilon^{\alpha\beta\gamma\delta}\left(\mathbf{F}_{\beta\gamma}\mathbf{A}_\delta - \frac{2}{3}ig\mathbf{A}_\beta\mathbf{A}_\gamma\mathbf{A}_\delta\right)\right], \tag{4.182}$$

where $\tilde{\mathbf{F}}^{\alpha\beta} \equiv \frac{1}{2}\varepsilon^{\alpha\beta\gamma\delta}\mathbf{F}_{\gamma\delta}$ is the tensor dual to $\mathbf{F}$.

Problem 4.28 Verify (4.182).

Let us consider two vacuum configurations (4.173) with winding numbers ν_0 and ν_1 specified on two different space-like hypersurfaces. Then, integrating (4.182) and using the Gauss theorem, we obtain

$$\int \mathrm{tr}\big(\mathbf{F}\tilde{\mathbf{F}}\big)\, d^4x = \frac{16\pi^2}{g^2}(\nu_1 - \nu_0). \tag{4.183}$$

Thus, the field configuration interpolating between two topologically different vacua has a nonvanishing field strength and hence, "in between," nonzero positive potential energy. Because the energy of both the initial and final states is equal to zero, the transition between the vacua can occur only as subbarrier tunneling via an instanton. To find the corresponding instanton we make the Wick rotation to Euclidean time: $t \to \tau = it$ and substitute $\mathbf{A}_0 \to i\mathbf{A}_0$; then $\mathbf{F}_{0i} \to i\mathbf{F}_{0i}$ and the form $\mathrm{tr}\mathbf{F}^2$ becomes nonnegative definite. By the Schwartz inequality,

$$\left(\int \mathrm{tr}(\mathbf{F}^2)\, d^4x\right)\left(\int \mathrm{tr}(\tilde{\mathbf{F}}^2)\, d^4x\right) \geq \left|\left(\int \mathrm{tr}(\mathbf{F}\tilde{\mathbf{F}})\, d^4x\right)\right|^2. \qquad (4.184)$$

Taking into account that $\mathrm{tr}(\tilde{\mathbf{F}}^2) = \mathrm{tr}(\mathbf{F}^2)$, this inequality together with (4.183) implies the lower bound for the Euclidean action of any field configuration connecting the two vacua:

$$S_E = \frac{1}{2}\int \mathrm{tr}(\mathbf{F}^2)\, d^4x \geq \frac{8\pi^2}{g^2}|\nu_1 - \nu_0|. \qquad (4.185)$$

The equality in (4.184) is attained if and only if $\mathbf{F} = \pm\tilde{\mathbf{F}}$. The corresponding interpolating solution with $\Delta\nu = 1$ is called the instanton. We do not need the explicit form of this solution here, but merely point out that it is characterized by a single parameter (a constant of integration), the instanton size ρ. The instanton action does not depend on the size and for any ρ is equal to

$$S_I = \frac{8\pi^2}{g^2} = \frac{2\pi}{\alpha} \qquad (4.186)$$

where $\alpha \equiv g^2/4\pi$ is the corresponding "fine structure constant."

Topological transitions Thus, the vacuum of a gauge theory generally has a complicated structure with many minima separated by potential barriers as shown in Figure 4.16. (This picture is, of course, no more than a *symbolic* representation of the vacuum structure and should not be taken too literally.) The instanton connects two adjacent minima. The *probability* of tunneling is proportional to $\exp(-2S_I)$ because as opposed to the particle tunneling, the instanton interpolates between the initial and final states only once. Hence the transition rate between topologically different vacua is

$$\Gamma \propto \exp\left(-\frac{4\pi}{\alpha}\right), \qquad (4.187)$$

In electroweak theory, $\alpha_w \simeq 1/29$ and we have $\Gamma \propto 10^{-160}$. Therefore, instanton transitions are strongly suppressed in electroweak theory.

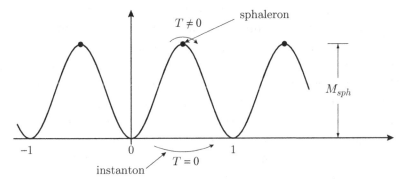

Fig. 4.16.

The existence of instantons of arbitrary size is not surprising since pure Yang–Mills theory is scale-invariant. Naively integrating over all instantons leads to a divergent transition probability. Moreover, the height of the barrier between two minima, which can be estimated on dimensional grounds as

$$\mathcal{V}_m \sim \frac{S_I}{\rho}, \qquad (4.188)$$

tends to zero as the instanton size ρ grows. Hence, one might expect the transition rate to be already very large at very small temperatures. In reality this does not happen because our consideration fails for instantons of large size. In fact, in pure non-Abelian Yang–Mills theory the gauge fields are confined and, hence, the instanton size cannot exceed the confinement scale.

Let us now turn to theories with the Higgs mechanism, where scale invariance is broken. For instance, in electroweak theory one has a natural infrared cut-off scale determined by the typical mass of the gauge bosons M_W, and a maximal instanton size equal to

$$\rho_m \sim M_W^{-1}. \qquad (4.189)$$

Vacuum transitions are strongly suppressed at zero temperature because the weak coupling constant is small. At high temperatures the transition rate is determined by sphalerons corresponding to the maximum of the potential (Figure 4.16). We can make a rough estimate of the sphaleron mass by considering the height of the potential in the "direction" where tunneling occurs via an instanton of the largest possible size $\rho_m \sim M_W^{-1}$. Then

$$M_{sph} \simeq \mathcal{V}_m \sim S_I/\rho_m \simeq 2\pi \frac{M_W}{\alpha_w} \sim 15 \text{ TeV}. \qquad (4.190)$$

This estimate is in good agreement with more elaborate calculations according to which $M_{sph} \simeq 7\text{--}13$ TeV. As in the case of metastable vacuum decay, the sphaleron size R_{sph} is comparable to the instanton size $\rho_m \sim M_W^{-1}$.

At high temperatures the rate of topological transitions is proportional to $\exp(-M_{sph}/T)$ and one expects that at $T > 10$ TeV they are no longer suppressed. In fact, the transitions become very efficient at much smaller temperatures. We have found that the expectation value of the Higgs field, and hence the masses of the gauge bosons, decrease as the temperature increases. As a consequence the height of the barrier, proportional to $M_W(T)$, also decreases. At the moment when $M_W(T) \sim \alpha_w T$, the exponential suppression,

$$\exp\left(-\frac{M_{sph}(T)}{T}\right) \sim \exp\left(-2\pi \frac{M_W(T)}{\alpha_w T}\right),$$

disappears and the rate of transitions per unit volume per unit time is

$$\Gamma \sim R_{sph}^{-4} \sim (\alpha_w T)^4. \tag{4.191}$$

This estimate, based on dimensional grounds, is also roughly valid at higher temperatures where symmetry is restored, the gauge bosons are massless and the barrier disappears. In the absence of the barrier there are no sphalerons and transitions occur via field configurations of typical size $\sim (\alpha_w T)^{-1}$. In electroweak theory the symmetry is restored when the temperature exceeds ~ 100 GeV (see (4.142)) and the topological transitions are very efficient. This leads to nonconservation of the total fermion number.

4.5.4 Chiral anomaly and nonconservation of the fermion number

Chiral anomaly The gauge interactions of the massless fermions preserve both left- and right-handed currents, $J_L^\mu \equiv \bar{\psi}_L \gamma^\mu \psi_L$ and $J_R^\mu \equiv \bar{\psi}_R \gamma^\mu \psi_R$, at the classical level. This means that the fermion numbers (equal to the difference between the numbers of fermions and antifermions) are conserved for each helicity separately. As a consequence, both the total current $J^\mu = J_L^\mu + J_R^\mu$ and the chiral current $J_5^\mu = J_R^\mu - J_L^\mu$ are conserved. In electrodynamics the conservation of the total current is equivalent to charge conservation and its violation would be a disaster. On the other hand, the violation of the chiral current would have no dramatic consequences. In fact, for massive fermions the chiral current is not conserved even classically. Quantum fluctuations lead to a violation of chiral current conservation for massless fermions also. In quantum electrodynamics the triangle diagram, shown in Figure 4.17, induces the *chiral anomaly*:

$$\partial_\mu \left(J_R^\mu - J_L^\mu \right) = \frac{e^2}{8\pi^2} F \tilde{F}. \tag{4.192}$$

4.5 Instantons, sphalerons and the early universe

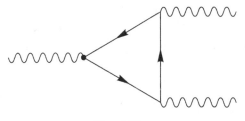

Fig. 4.17.

The situation is very similar in non-Abelian gauge theories, where the divergences of the left- and right-handed currents are given by

$$\partial_\mu J_L^\mu = -c_L \frac{g^2}{16\pi^2} \mathrm{tr}(\mathbf{F}\tilde{\mathbf{F}}), \quad \partial_\mu J_R^\mu = c_R \frac{g^2}{16\pi^2} \mathrm{tr}(\mathbf{F}\tilde{\mathbf{F}}), \quad (4.193)$$

respectively. In a theory where the right- and left-handed fermions are coupled to the gauge field with the same strength, as for instance in quantum chromodynamics, $c_L = c_R = 1$ for every flavor, and the total current and as a result the fermion number are conserved. On the other hand, the difference between the numbers of right- and left-handed fermions,

$$Q_5 \equiv \int \left(J_R^0 - J_L^0\right) d^3x = N_R - N_L, \quad (4.194)$$

changes in instanton transitions. Using Gauss's theorem and taking into account (4.183), we obtain from (4.193)

$$\Delta Q_5 = Q_5^f - Q_5^{in} = \frac{g^2}{8\pi^2} \int \mathrm{tr}(\mathbf{F}\tilde{\mathbf{F}}) \, d^4x = 2, \quad (4.195)$$

that is, the corresponding fermion flips its helicity in the instanton.

Violation of fermion number The situation is more interesting in *chiral* theories, where the right- and left-handed particles are coupled to the gauge fields differently. Let us consider electroweak theory at temperatures $T > 100\,\mathrm{GeV}$, where the symmetry is restored and the rate of topological transitions is very high. The $SU(2)$ gauge fields interact only with left-handed fermions and have the same strength for each doublet; therefore $c_L = 1$, $c_R = 0$ and

$$\partial_\mu^{(f)} J_L^\mu = -\frac{g^2}{16\pi^2} \mathrm{tr}(\mathbf{F}\tilde{\mathbf{F}}), \quad (4.196)$$

where f indicates the corresponding fermion doublet and runs from 1 to 12. For instance, $f = 1$ for the first lepton family,

$$^{(1)}J_L^\mu = \bar{e}_L \gamma^\mu e_L + \bar{\nu}_e \gamma^\mu \nu_e, \quad (4.197)$$

and f goes from 4 to 12 for three quark families, each of which is represented in three different colors. From (4.196) and (4.183), it follows that during a topological transition in which the winding number increases by $\Delta \nu$ units, the fermion number in *each* doublet decreases by $\Delta \nu$ units and hence the total fermion number is not conserved. Taking into account that there are nine quark doublets and that the baryon number of every quark is equal to 1/3, we obtain the following selection rule:

$$\Delta L_e = \Delta L_\mu = \Delta L_\tau = \tfrac{1}{3}\Delta B, \qquad (4.198)$$

where ΔL_i is the change of the lepton number in the corresponding family and ΔB is an overall change of the baryon number. Of course, the conservation laws for energy, total electric charge and color should be fulfilled. In an instanton/sphaleron transition the total number of fermions decreases by twelve units; correspondingly the total lepton and baryon numbers decrease by three units each: $\Delta L = \Delta B = -3$. The energy of the disappearing fermions is transferred to the remaining and newly created fermions and antifermions. One of the possible processes of this kind is shown in Figure 4.18.

Thus in chiral theories topological transitions lead to nonconservation of the total number of left-handed fermions. In electroweak theory there exist interactions which convert left-handed particles to right-handed ones. Hence the total fermion number is not conserved and some linear combination of baryon and lepton number, $B + aL$, should vanish at thermal equilibrium. The numerical coefficient a is of order unity and its exact value can be found taking into account the conservation laws and analyzing the conditions for the chemical equilibrium of all particles involved. In the Standard Model with three generations of fermions and one Higgs

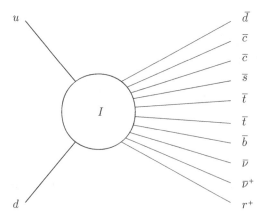

Fig. 4.18.

doublet, $a = 28/51$. On the other hand, it follows from (4.198) that the charges $L_i - (1/3)B$ and, as a result $B - L$, are conserved.

Topological transitions in the early universe can ensure equilibrium only if their rate per fermion, equal to $\Gamma/n_f \sim \alpha_w^4 T$ (see (4.191)), exceeds the expansion rate $H \sim T^2$, that is, they are efficient if $T < \alpha_w^4 \sim 10^{12}$ GeV. Thus, even if $B + aL$ were generated in the very early universe it would be washed out by topological transitions at 10^{12} Gev $> T > 10^2$ Gev. Hence if $B - L = 0$ then any pre-existent baryon number does not survive. To explain baryon asymmetry, therefore, one has to find a way to generate $B - L \neq 0$ in the very early universe. The other possibility is to generate $B + aL$ during a violent electroweak phase transition. However, this does not look very realistic because the electroweak transition seems to be rather smooth.

4.6 Beyond the Standard Model

Particle theories have been probed experimentally only to an energy scale of order a few hundred GeV. If we want to learn anything about the early universe at energies above 100 GeV, we inevitably have to rely on theories which are somewhere speculative. Fortunately they all have common features which may allow us to foresee possible solution of important cosmological problems. Among these problems are the origin of baryon asymmetry in the universe, the nature of dark matter and the mechanism for inflation. We devote a separate chapter to inflation; here we concentrate on the first two questions. Since these questions cannot be answered within the Standard Model, we are forced to go beyond it. In this section we begin by *outlining* the relevant general ideas behind the extensions of the Standard Model and then discuss the ways in which these ideas can be implemented in cosmology.

Grand unification The $SU(3) \times SU(2) \times U(1)$ Standard Model is characterized by the three coupling constants g_s, g and g'. They depend on the energy scale, and the corresponding "fine structure constants" $\alpha \equiv g^2/4\pi$ run according to (4.29). The strong interaction constant α_s is given by (4.31).

In the case of the $SU(2)$ group the coefficient $f_1'(1)$ in (4.29) can be inferred from (4.30). For $q > 100$ GeV all particles, including intermediate bosons, can be treated as massless and the number of "colors" n in (4.30) is equal to 2. The number of "flavors" f should be taken to be equal to *half* the number of left-handed fermion doublets (not forgetting that there is a quark doublet for each color), that is, $f = 12/2 = 6$. Hence, for the $SU(2)$ group,

$$f_1'(1) = \frac{1}{12\pi}(2 \times 6 - 11 \times 2) \approx -0.265$$

and

$$\alpha_w(q^2) \equiv \frac{g^2}{4\pi} \simeq \frac{\alpha_w^0}{1 + 0.265\alpha_w^0 \ln(q^2/q_0^2)}, \qquad (4.199)$$

where $q_0 \sim 100$ GeV and $\alpha_w^0 \equiv \alpha_w(q_0) \simeq 1/29$. Comparing (4.31) with (4.199), we find that the strong and weak coupling constants meet at $q \sim 10^{17}$ GeV. This suggests that above 10^{17} GeV, strong and weak interactions may be unified in a large gauge group and characterized by a *single* coupling constant g_U. The difference in their strength at low energies is then entirely explained by the different running of the coupling constants within the $SU(3)$ and $SU(2)$ subgroups, after the symmetry of this larger gauge group is broken at $\sim 10^{17}$ GeV. The simplest single group which incorporates all known fermions is the $SU(5)$ group. It also contains a $U(1)$ subgroup and could therefore include all gauge interactions of the Standard Model. However, to identify this subgroup with the $U(1)$ of electroweak theory, one has to verify that the properly normalized $U(1)$ fine structure constant, $\alpha_1 = (5/3)\, g'^2/4\pi$, meets the other two constants at the correct energy scale. This is a highly nontrivial task.

Problem 4.29 Find $\alpha_1(q^2)$. (*Hint* Do not forget hypercharges entering the vertices.)

With more accurate measurements of α_s and θ_w it has become clear that the three constants do not quite meet at a point (Figure 4.19). This fact, along with

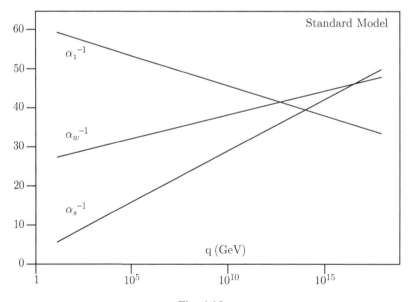

Fig. 4.19.

measurements of the proton lifetime and the discovery of neutrino masses, rules out the minimal $SU(5)$ model as a realistic theory. Nevertheless, we will take this model a little further in order to explain some important features common to more realistic theories.

The $SU(5)$ group has $5^2 - 1 = 24$ generators and correspondingly there are 24 gauge bosons. Eight of them should be identified with the gluons responsible for color transitions within the $SU(3)$ subgroup. Three bosons correspond to the $SU(2)$ subgroup and together with one $U(1)$ boson are responsible for the electroweak interactions. The remaining 12 bosons form two charged colored triplets,

$$X_i^{\pm 4/3}, \quad Y_i^{\pm 1/3},$$

where $i = r, b, g$ is the color index and the upper indices correspond to the electric charge. The X, Y bosons form a "bridge" between the $SU(2)$ and $SU(3)$ subgroups of $SU(5)$. After symmetry breaking (for instance, via the Higgs mechanism), they acquire masses of order $\sim 10^{15}$–10^{17} GeV and the transitions between $SU(2)$ and $SU(3)$ subgroups are strongly suppressed. At high energies, however, the X, Y bosons can "convert" quarks to leptons and vice versa very efficiently.

The X boson can decay either into a pair of quarks $(X \to qq)$ or into an antiquark–antilepton pair $(X \to \bar{q}\bar{l})$. The baryon numbers of the final products are $B = 2/3$ and $B = -1/3$ respectively; hence *baryon number is not conserved* in the $SU(5)$ model. On the other hand, the difference between baryon and lepton number is equal to $2/3$ in both cases and $B - L$ is not violated. Hence $B - L$ cannot be generated and any baryon asymmetry will be washed out in subsequent topological transitions. Thus the baryon asymmetry problem cannot be solved within the $SU(5)$ model.

The larger symmetry groups look more attractive for several different reasons. First of all, they have a richer fermion content than the minimal $SU(5)$ model. In particular, one can incorporate a right-handed neutrino, which is needed to explain the neutrino masses. The other attractive feature of these theories is that they not only violate baryon number, but also do not conserve $B - L$. This opens the door to an understanding of the origin of the baryon asymmetry of the universe. Finally, changing the fermion content influences the running of coupling constants, and thus one can hope that they will yet meet at one point. There is extensive literature dealing with large gauge groups, for example, $SO(10), SO(14), SO(22), \ldots, E_6, E_7, E_8, \ldots$. The corresponding theories contain many particles *not yet discovered* and hence many candidates for dark matter. However, in the absence of solid data, we see no point in going into further detail of unified theories here.

Supersymmetry The symmetries we have considered so far relate bosons to bosons and fermions to fermions. However, there may be a beautiful symmetry, known as *supersymmetry*, which relates bosons and fermions. If supersymmetry is a true symmetry of nature, then *every boson* should have *at least* one fermionic superpartner, with which it is paired in the *supermultiplet*. *Every fermion* should also be partnered with at least one boson. Under a supersymmetry transformation, bosons and fermions in the same supermultiplet are "mixed" with each other. It is clear that the supersymmetry generator which converts a boson to a fermion should be a spinor **Q**, and in the simplest case it is a *chiral* spinor of spin 1/2. Because bosons and fermions transform differently under the Poincaré group, supersymmetry transformations, unlike gauge transformations, cannot be completely decoupled from spacetime transformations. In fact, the algebra of the supersymmetry generators **Q** closes only when the generators of the space and time translations are included. Hence, if we try to make supersymmetry local, we are forced to deal with curved spacetime. *Local supersymmetry*, called *supergravity*, thus offers a possible way to unify gravity with the other forces.

Unfortunately, as in the case of Grand Unification, there are too many potential supersymmetric extensions of the Standard Model. First, the supersymmetry can be global or local. Second, we could include more than one boson–fermion pair in the same supermultiplet, an idea known as *extended* supersymmetry. In principle, all particles could be the members of a *single* multiplet. Extended supersymmetry is characterized by the number of supersymmetry generators $\mathbf{Q}^1, \mathbf{Q}^2, \ldots, \mathbf{Q}^N$ which determine the particle content of the supermultiplets. For instance, for $N=8$ the supermultiplet contains both *left- and right-handed* particles of spins 0, 1/2, 1, 3/2 and 2; thus $N=8$ *supergravity* would be an ideal candidate for unification. Unfortunately (or fortunately, depending on one's attitude) Nature does not act upon our wishes and in the absence of experimental data we must consider a diverse range of theoretical possibilities.

All supersymmetric theories have features in common and we concentrate on those which are relevant for cosmological applications. As we have mentioned, in these theories bosons and fermions are paired. Disappointingly, all combine known fermions with unknown bosons and vice versa. Hence, supersymmetric theories *predict* that the number of particles should be *at least* twice as big as the number of experimentally discovered particles. To understand why the supersymmetric partners of the known particles have not yet been discovered, we are forced to assume that supersymmetry is broken above the scale currently reached by accelerators. It is only when supersymmetry is broken that supersymmetric partners can have different masses; otherwise they are obliged to have the same mass.

In the minimal supersymmetric extension of the Standard Model, usually called MSSM, every quark and lepton has a supersymmetric scalar partner, called a *squark* and a *s*lepton respectively. Similarly, for every gauge particle we have a fermionic superpartner with spin 1/2, called a *gaugino*. Among these, *gluinos* are superpartners of gluons, and *winos* and the *bino* are the superpartners of the gauge bosons of the electroweak group. The gauginos mediate the interaction of the scalar particles and their fermionic partners, with a strength determined by the gauge coupling constant. The Higgs particle is accompanied by a *higgsino*. The lightest neutral combination of *-inos* (mass eigenstate), called the *neutralino*, must be stable; if supersymmetry were broken at the electroweak scale, it would interact weakly with ordinary matter. Therefore, the neutralino is an ideal candidate for cold dark matter. To conclude our brief excursion to the "*s*- and *-ino* zoo," we should mention the *gravitino* – the spin 3/2 superpartner of the graviton which could also serve as a dark matter particle. Thus we see that supersymmetric theories provide us with the weakly interacting massive particles necessary to explain dark matter in the universe.

Some remarkable properties of supersymmetric theories arise from the fact that the numbers of fermionic and bosonic degrees of freedom in these theories are equal. For instance, the superpartner of the left-handed fermion is a complex scalar field and they both have two degrees of freedom. The energy of the vacuum fluctuations per degree of freedom is the same in magnitude but opposite in sign for fermions and bosons of the same mass. Therefore, in supersymmetric theories the fermion and boson contributions cancel each other and the total vacuum energy density vanishes. In other words, the cosmological term is exactly zero. This is true, however, only if supersymmetry remains unbroken. But supersymmetry is broken and as a result the expected mismatch in vacuum energy densities, arising from the mass difference of the superpartners, is of order Λ_{SUSY}^4, where Λ_{SUSY} is the supersymmetry breaking scale. If $\Lambda_{SUSY} \sim 1$ TeV, the cosmological constant is about $\sim 10^{-64}$ in Planck units. This number is still about 60 orders of magnitude larger than the observational limit. Thus we see that supersymmetry, while a step in the right direction, does not quite solve the cosmological constant problem.

The last remark we wish to make here concerns the behavior of the running coupling constants in the minimal supersymmetric extension of the Standard Model. The additional supersymmetric particles influence the rate of running of the strong and electroweak coupling constants. As a result, the three constants meet at a single point with impressive accuracy (Figure 4.20). This revives our hopes of Grand Unification and gives an indication that we may be on the right track.

4.6.1 Dark matter candidates

Nucleosynthesis and CMB data clearly indicate that most of the matter in the universe is dark and nonbaryonic. There is no shortage of particle physics candidates

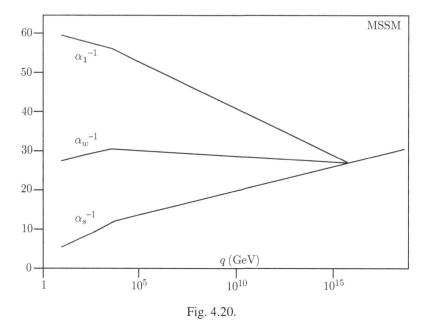

Fig. 4.20.

for this dark matter. These candidates can be classified according to whether the dark matter particles originated via decoupling from a thermal bath or were created in some nonthermal process.

In turn, thermal relics can be categorized further as to whether they were relativistic or nonrelativistic at the moment of decoupling. The relics which were relativistic at this time constitute hot dark matter, while those which were nonrelativistic constitute cold dark matter. The neutrino and the neutralino are examples of hot and cold relics respectively.

The simplest example of a nonthermal relic is the condensate of a weakly interacting massive scalar field. A well motivated particle physics candidate of this kind is the *axion*. Because the momentum of axions in the condensate is equal to zero, axions can also serve as a cold dark matter component even although their masses are very small. Below we describe some generic features characterizing different dark matter candidates.

Hot relics The freeze-out of hot relics occurs when they are still relativistic. After relics decouple from matter their number density n_ψ decreases in inverse proportion to the volume, that is, as a^{-3}. The total entropy density of the remaining matter, s, scales in the same way and hence the ratio n_ψ/s has remained constant until the present time.

4.6 Beyond the Standard Model

Problem 4.30 Using this fact and assuming that the hot relics are fermions of mass m_ψ with negligible chemical potential, verify that their contribution to the cosmological parameter is

$$\Omega_\psi h_{75}^2 \simeq \frac{g_\psi}{g_*} \left(\frac{m_\psi}{19 \text{ eV}} \right). \tag{4.200}$$

Here g_ψ is the total number of degrees of freedom of the hot relics and $g_* = g_b + (7/8) g_f$ is the effective number of bosonic and fermionic degrees of freedom at freeze-out for all particles which later convert their energy into photons.

Taking three left-handed light neutrino species of the same mass m_ψ, we have $g_\psi = 6$ (3 for neutrinos + 3 for antineutrinos). The neutrinos decouple at temperature $\sim O(1)$ MeV. At this time the only particles which contribute to g_* are the electrons, positrons and photons, and $g_* = (2 + (7/8) \times 4) = 5.5$. Therefore, if the mass of *every* neutrino were about 17 eV, then neutrinos would close the universe. According to observations, the contribution of dark matter to the total density does not exceed 30% (see, in particular, Chapter 9). Hence the *sum* of the neutrino masses should be smaller than 15 eV or, in the case of equal masses, $m_\nu < 5$ eV. The mass bound on hot relics species changes if we assume that they freeze out at a higher temperature when more relativistic particles are present. For instance, if decoupling happened before the quark–gluon phase transition, at $T \sim 300$ MeV, then $g_* = 53$ (for the number of degrees of freedom at this time, see (4.33) and do not forget to include photons and electron–positron pairs). In this case, it follows from (4.200) that $m_\psi < 151$ eV for $g_\psi = 2$. In reality the bound on the masses is even stronger because hot relics cannot explain all dark matter in the universe; they can constitute only a subdominant fraction of it. In fact, in models where hot dark matter dominates, inhomogeneities are washed out by free streaming on all scales up to the horizon scale at the moment when the relics became nonrelativistic (see Section 9.2). As a result the large scale structure of the universe cannot be explained. Therefore, more promising and successful models are those in which cold relics constitute the dominant part of the dark matter.

Thermal cold relics Cold relics χ decouple at temperatures T_* much less than their mass m_χ. Therefore, the number density n_χ is exponentially suppressed in comparison with the number density of photons. To deduce n_χ^* at the moment of decoupling, we simply equate the annihilation rate of relics χ to the Hubble expansion rate:

$$n_\chi^* \langle \sigma v \rangle_* \simeq H_* = (8\pi^3/90)^{1/2} \tilde{g}_*^{1/2} T_*^2, \tag{4.201}$$

where $\langle\sigma v\rangle_*$ is the thermally averaged product of the annihilation cross-section σ and the relative velocity v. The effective number of degrees of freedom $\tilde{g}_*$ accounts for all particles which are relativistic at T_*. Alternatively, we know from (3.61), setting $\mu = 0$, that

$$n_\chi^* \simeq g_\chi \frac{T_*^3}{(2\pi)^{3/2}} x_*^{3/2} e^{-x_*}, \qquad (4.202)$$

where $x_* \equiv m_\chi/T_*$. Then, using the estimate $\langle\sigma v\rangle_* \sim \sigma_* \sqrt{T_*/m_\chi}$, where σ_* is the effective cross-section at $T = T_*$, we find that at freeze-out

$$\begin{aligned}x_* &\simeq \ln\left(0.038 g_\chi \tilde{g}_*^{-1/2} \sigma_* m_\chi\right) \\ &\simeq 16.3 + \ln\left[g_\chi \tilde{g}_*^{-1/2} \left(\frac{\sigma_*}{10^{-38}\,\text{cm}^2}\right)\left(\frac{m_\chi}{\text{GeV}}\right)\right].\end{aligned} \qquad (4.203)$$

The relics are cold only if $x_* > O(1)$; to be definite we take $x_* > 3$. It follows from (4.203) that $\sigma_* m_\chi > 10^3 g_\chi^{-1} \tilde{g}_*^{1/2}$ (in Planck units) for cold relics. Their energy density today is

$$\varepsilon_\chi^0 \simeq m_\chi n_\chi^* \frac{s_0}{s_*} = \frac{2}{g_*} m_\chi n_\chi^* \left(\frac{T_{\gamma 0}}{T_*}\right)^3, \qquad (4.204)$$

where $T_{\gamma 0} \simeq 2.73$ K and s_0/s_* is the ratio of the present entropy density of radiation to the total entropy density at freeze-out of those components of matter (with g_* effective degrees of freedom) which later transfer their entropy to radiation. Substituting n_χ^* from (4.201) into (4.204) we finally obtain

$$\Omega_\chi h_{75}^2 \simeq \frac{\tilde{g}_*^{1/2}}{g_*} x_*^{3/2} \left(\frac{3 \times 10^{-38}\,\text{cm}^2}{\sigma_*}\right). \qquad (4.205)$$

Remarkably, the contribution of cold relics to the cosmological parameter depends only logarithmically on their mass (through x_*) and is mainly determined by the effective cross-section σ_* at decoupling.

Weakly interacting massive particles, which have masses between 10 GeV and a few TeV and cross-sections of approximately electroweak strength $\sigma_{EW} \sim 10^{-38}$ cm^2, are ideal candidates for cold dark matter. Their number density freezes out when $x_* \sim 20$ and, as is clear from (4.205), they may easily contribute the necessary 30% to the total density of the universe and thus constitute the dominant component of dark matter. Currently, the leading weakly interacting massive particle candidate is the lightest supersymmetric particle. Most supersymmetric theories have a discrete symmetry called *R-parity*, under which particles have eigenvalue $+1$ and superparticles -1. R-parity conservation guarantees the stability of the lightest supersymmetric particle. The lightest supersymmetric particle is most likely a *neutralino*, which could be (mostly) a bino or a photino.

Problem 4.31 Consider the annihilation reaction and its inverse: $\chi\bar{\chi} \leftrightarrows f\bar{f}$. Assuming that the annihilation products f and $\bar{f}$ always have thermal distribution with zero chemical potential (because of some "stronger" interactions with other particles), derive the following equation:

$$\frac{dX}{ds} = \frac{\langle \sigma v \rangle}{3H}(X^2 - X_{eq}^2), \qquad (4.206)$$

where $X \equiv n_\chi/s$ is the *actual* number of χ particles per comoving volume and X_{eq} is the equilibrium value. Total entropy is conserved and therefore the entropy density s scales as a^{-3}. Find the approximate solutions of (4.206) for hot and cold relics and compare them with the numerical solutions. Determine the corresponding freeze-out number densities and compare the results with the estimates made in this section (*Hint* In equilibrium the direct and inverse reactions balance each other exactly, so that $\langle \sigma v \rangle_{\chi\bar{\chi}} n_\chi^{eq} n_{\bar{\chi}}^{eq} = \langle \sigma v \rangle_{f\bar{f}} n_f^{eq} n_{\bar{f}}^{eq}$.)

Nonthermal relics In general, for nonthermal relics, interactions are so weak that they are never in equilibrium. There is no general formula which describes the contribution of these relics to the total energy density because this contribution depends on the concrete dynamics. As an example of nonthemal relics we consider the homogeneous condensate of massive scalar particles and neglect their interactions with other fields. The homogeneous scalar field satisfies the equation

$$\ddot{\varphi} + 3H\dot{\varphi} + m^2\varphi = 0. \qquad (4.207)$$

To solve it for generic $H(t)$, it is convenient to use the the conformal time $\eta \equiv \int dt/a$ as a temporal variable. Introducing the rescaled field $\varphi \equiv u/a$ we find that in terms of u, (4.207) becomes

$$u'' + \left(m^2 a^2 - \frac{a''}{a}\right) u = 0, \qquad (4.208)$$

where a prime denotes the derivative with respect to η. If $|a''/a^3| \sim H^2 \gg m^2$, the first term in the brackets can be neglected and the corresponding approximate solution of this equation is

$$u \simeq a\left(C_1 + C_2 \int \frac{d\eta}{a^2}\right), \qquad (4.209)$$

where C_1 and C_2 are integration constants. The dominant mode given by the first term yields $\varphi \simeq C_1 \sim$ const. Therefore, while the mass m is much smaller than the Hubble constant H, the scalar field is frozen and its energy density

$$\varepsilon_\varphi = \tfrac{1}{2}(\dot{\varphi}^2 + m^2\varphi^2) \qquad (4.210)$$

remains constant if $m =$ const, and resembles the cosmological constant.

As the Hubble constant becomes smaller than the mass, and eventually $H^2 \ll m^2$, we can neglect the second term inside the brackets in (4.208). The WKB solution of the simplified equation is then

$$u \propto (ma)^{-1/2} \sin\left(\int ma\, d\eta\right), \qquad (4.211)$$

and correspondingly

$$\varphi \propto m^{-1/2} a^{-3/2} \sin\left(\int m\, dt\right). \qquad (4.212)$$

Note that this solution is also valid for a slowly varying mass m. Substituting this into (4.210) we find, in the leading order, that the energy density of the scalar field decreases as ma^{-3}; hence it behaves as dust-like matter ($p \simeq 0$). This is easy to understand: after the value of the Hubble constant drops below that of the mass, the scalar field, which was frozen before, starts to oscillate and can be interpreted as a Bose condensate of many cold particles of mass m with zero momentum. For a slowly varying mass the particle number density, which is proportional to ε_φ/m, decays as a^{-3} and the total particle number is conserved.

Using these results, we can easily calculate the current energy density of the scalar field (in Planck units):

$$\varepsilon_\varphi^0 \simeq m_0 \left(\frac{\varepsilon_\varphi}{m}\right)_* \frac{s_0}{s_*} \simeq O(1) \frac{\tilde{g}_*^{3/4}}{g_*} \frac{m_0}{m_*^{1/2}} \varphi_{in}^2 T_{\gamma 0}^3, \qquad (4.213)$$

where m_0 is its mass at present, m_* is the mass at the moment when $H_* \simeq m_*$ and φ_{in} is the initial value of the scalar field when it was still frozen. For the case of constant mass ($m_0 = m_*$), the contribution of this field to the total energy density is

$$\Omega_\varphi h_{75}^2 \sim \frac{\tilde{g}_*^{3/4}}{g_*} \left(\frac{m_0}{100\,\text{GeV}}\right)^{1/2} \left(\frac{\varphi_{in}}{3 \times 10^9\,\text{GeV}}\right)^2. \qquad (4.214)$$

Thus, tuning two parameters, the mass and the initial value of the scalar field, we have a straightforward "explanation" for the observed cold dark matter in the universe.

Axions As previously mentioned, axions are an attractive nonthermal relic candidate. The axion field is introduced to solve the strong CP problem. Because the strong coupling constant is large at low energies, topological transitions are not suppressed in quantum chromodynamics. Therefore, one expects that the true quantum chromodynamics vacuum is the θ vacuum, which is a superposition of vacua with

4.6 Beyond the Standard Model

different winding numbers n:

$$|\theta\rangle = \sum e^{-in\theta} |n\rangle, \qquad (4.215)$$

where θ is an arbitrary parameter which must be determined experimentally. As a result, the effective Lagrangian possesses an additional *nonperturbative* term

$$\theta \frac{\alpha_s}{8\pi} \mathrm{tr}(\mathbf{F}\tilde{\mathbf{F}}), \qquad (4.216)$$

where $\mathbf{F}$ and $\tilde{\mathbf{F}}$ are the gluon field strength and its dual respectively. This term, being a total derivative, does not affect the equations of motion and conserves C. It violates CP, P and T, however, and produces a very large neutron dipole moment which contradicts experimental bounds unless $\theta < 10^{-10}$. One either has to accept the smallness of θ as fact or try to find a natural explanation for why this parameter is so small by introducing a new symmetry. The elegant known solution of the strong CP problem, suggested by Peccei and Quinn, involves an additional *global chiral* $U(1)_{PQ}$ symmetry imposed on the Standard Model. This symmetry, broken at a scale f, essentially serves to replace the θ parameter by a dynamical field – the axion field. In many axion models, a *new* complex scalar field $\varphi = \chi \exp(i\bar{\theta})$ is used to generate a $U(1)_{PQ}$-invariant mass term for some colored fermions, via Yukawa coupling. After symmetry breaking, the field χ acquires the expectation value f. In the case of local symmetry the field $\bar{\theta}$ would be "eaten" by the gauge field, but when the symmetry is global it becomes a massless degree of freedom and this is called the axion a. To be precise, $a = f\bar{\theta}$. At the quantum level the *chiral* $U(1)_{PQ}$ symmetry suffers from the chiral anomaly and as a result there is an effective interaction of the axion field with gluons:

$$\frac{a}{f} \frac{\alpha_s}{8\pi} \mathrm{tr}(\mathbf{F}\tilde{\mathbf{F}}), \qquad (4.217)$$

which shares the same structure as (4.216). The terms of this type generate an effective potential for $\theta + a/f$ with the minimum at $a = -f\theta$ and the overall CP violating term vanishes in the minimum of the potential. What is most important for us is that in the vicinity of the potential minimum the axion acquires a small mass of order

$$m_a = \frac{(m_u m_d)^{1/2}}{m_u + m_d} \frac{m_\pi f_\pi}{f} \sim \frac{6 \times 10^6}{f_{\mathrm{GeV}}} \mathrm{eV}, \qquad (4.218)$$

where m_u and m_d are the masses of the light quarks, $m_\pi \simeq 130$ MeV is the pion mass and $f_\pi \simeq 93$ MeV is the pion decay constant. The axion mass arises from quantum chromodynamics instanton effects and these are altered at finite temperatures. In particular, at $T \gg \Lambda_{QCD} \simeq 200$ MeV, we have

$$m_a(T) \sim m_a(\Lambda_{QCD}/T)^4. \tag{4.219}$$

For realistic values of f, the axion mass becomes of order the Hubble constant at these high temperatures. Using H_* in (4.201), we find $m_a(T_*) = H_*$ when

$$T_* \sim \tilde{g}_*^{-1/12} m_a^{1/6} \Lambda_{QCD}^{2/3}$$

and

$$m_* \sim \tilde{g}_*^{1/3} m_a^{1/3} \Lambda_{QCD}^{4/3}. \tag{4.220}$$

Substituting this value for m_* into (4.213) and taking into account that $a_{in} = f\theta_{in}$, we finally obtain the following expression for the axion contribution to the total energy density:

$$\Omega_a h_{75}^2 \sim O(1) \left(\frac{6 \times 10^{-6} \text{ eV}}{m_a} \right)^{7/6} \bar{\theta}_{in}^2 \sim O(1) \left(\frac{f}{10^{12} \text{ GeV}} \right)^{7/6} \bar{\theta}_{in}^2. \tag{4.221}$$

The axion is periodic in $\bar{\theta} = a/f$ and therefore the natural value for $\bar{\theta}_{in}$ is $\sim O(1)$. Thus, if $m_a \sim 10^{-5}$ eV, the axions can constitute the dominant component of dark matter. Because all couplings of the axion scale as $1/f$, axions interact only very weakly with ordinary matter and, in spite of their small mass, are cold particles. At first glance, only axions with masses within a very narrow window near 10^{-5} eV seem to be cosmologically interesting. However, one can argue that $\bar{\theta}_{in} \ll 1$ is not so unnatural in inflationary cosmology. This allows the universe to close even with axions of much smaller mass.

Particle theories beyond the Standard Model also provide us with other candidates for dark matter, among which are the gravitino, the axino (the superpartner of the axion), and the remnant of the inflaton. Their contributions can be determined using approaches similar to those outlined above.

4.6.2 Baryogenesis

The universe is asymmetric: there are more baryons than antibaryons. While antibaryons are produced in accelerators or in cosmic rays, "antigalaxies" are not observed. The relative excess of the baryons $B \equiv (n_b - n_{\bar{b}})/s \sim 10^{-10}$ is exactly what we need to explain the abundance of light elements and the observations of CMB fluctuations (see Chapter 9). In past the baryon asymmetry could be interpreted as simply due to the initial conditions in the universe. However, in the light inflationary cosmology, now widely accepted, this "solution" completely fails. We will see in the following chapter that an inflationary stage erases any pre-existing

4.6 Beyond the Standard Model

asymmetry and the possibility of its dynamical generation becomes an inevitable element of inflationary cosmology. Theories beyond the Standard Model provide us with many — perhaps too many — potential solutions to this problem. Fortunately, any particular model for baryogenesis should possess three basic ingredients which are independent of the details of the actual theory. These ingredients, formulated by A. Sakharov in 1967, are

(i) *baryon number violation*,
(ii) *C and CP violation*,
(iii) *departure from equilibrium*.

The first condition is obvious and does not require a long explanation. If baryon number is conserved and is equal to zero at the beginning, it will remain zero forever. If baryon number does not satisfy any conservation law, it vanishes in the state of thermal equilibrium. Therefore we need the third condition. The second condition is less trivial: it is a prerequisite for ensuring a different reaction (decay) rate for particles and antiparticles. If this condition is not met, the numbers of baryons and antibaryons produced are equal and no net baryon charge is generated even if the other two conditions are fulfilled.

Problem 4.32 Why must we require both C and CP violation? Why is CP violation alone not enough?

The Standard Model possesses all the ingredients necessary for the generation of baryon asymmetry. In fact, we have seen that baryon number is not conserved in topological transitions, CP is violated in weak interactions and the departure from thermal equilibrium naturally occurs in the expanding universe. It would be remarkable if the baryon number could be explained within the Standard Model itself. Unfortunately this seems not to work. The main obstacle is the third condition. For realistic values of the Higgs mass, the electroweak transition is a cross-over and cannot supply us with the necessary strong deviations from thermal equilibrium. Therefore, to explain baryon asymmetry we have to go beyond the Standard Model. There is a wide range of possibilities; below we outline the baryogenesis scenarios most commonly considered.

Baryogenesis in Grand Unified Theories Baryon number is generically not conserved in Grand Unified Theories. In the $SU(5)$ theory, for example, the heavy gauge boson X, responsible for "communication" between the quark and lepton sectors, can decay into either a qq pair or a $\bar{q}\bar{l}$ pair with baryon numbers 2/3 or $-1/3$ respectively. The antiboson $\bar{X}$ decays into a $\bar{q}\bar{q}$ or ql pair. The CPT invariance requires the equality of the *total* decay rates for X and $\bar{X}$. However, this does not mean that the decay rates are equal for every particular channel. If C and CP are

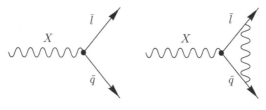

Fig. 4.21.

violated, the rate $\Gamma(X \to qq) \equiv r$ is generally not equal to $\Gamma(\bar{X} \to \bar{q}\bar{q}) \equiv \bar{r}$. In fact, let us assume that the only source for CP violation is the Kobayashi–Maskawa mechanism. Then the coupling constants can be complex: they enter the diagrams describing the decay of particles, but their conjugated values enter the diagrams for antiparticle decay. The difference in the decay rates can be seen in the interference of tree-level diagrams with higher-order diagrams such as in Figure 4.21. This difference, characterizing the degree of CP violation, is proportional to the imaginary part of the corresponding product of coupling constants and vanishes if all constants are real-valued (CP is not violated).

One expects that at temperatures much higher than m_X, the X and $\bar{X}$ bosons are in equilibrium and their abundances are the same, $n_X = n_{\bar{X}} \sim n_\gamma$. As the temperature drops to about m_X, the processes responsible for maintaining equilibrium become inefficient and the number density per comoving volume freezes out at some value $\gamma_* = n_X/s$ (see Section 4.6.1). Subsequently, only the *out of equilibrium decay* of the X and $\bar{X}$ bosons is important and net baryon charge can be produced. Let us estimate its size. We normalize the *overall* decay rate, which is the same for particles and antiparticles, to unity. Then $\Gamma(X \to \bar{q}\bar{l}) = 1 - r$ and $\Gamma(\bar{X} \to ql) = 1 - \bar{r}$. The mean net baryon number produced in the decay of the X boson is

$$B_X = (2/3)\,r + (-1/3)(1-r);$$

likewise

$$B_{\bar{X}} = (-2/3)\,\bar{r} + (1/3)(1-\bar{r}).$$

Hence the resulting baryon asymmetry is

$$B = \gamma_*(B_X + B_{\bar{X}}) = \gamma_*(r - \bar{r}). \qquad (4.222)$$

We see that B depends on the freeze-out concentration γ_* and on a parameter $\varepsilon \equiv (r - \bar{r})$ which characterizes the amount of CP violation. The term γ_* is mainly determined by the rates of reactions responsible for equilibrium (see Section 4.6.1) and does not exceed unity. The parameter ε comes from higher-order perturbation theory and therefore $B \ll 1$. For example, in the minimal $SU(5)$ model the parameter ε receives its first nontrivial contribution at the tenth order of perturbation theory and the resulting baryon asymmetry is many orders of magnitude smaller than the required 10^{-10}. Another unhappy feature of the $SU(5)$ model is that $B - L$ is conserved in this theory. In this case $B - L = 0$ even if $B \neq 0$ and any baryon asymmetry generated will be washed out in subsequent topological transitions. For successful baryogenesis we actually need to generate nonvanishing $B - L$. In more complex models both these obstacles can, in principle, be overcome. For example, in the $SO(10)$ model, neither B nor $B - L$ is conserved, and the necessary ε can be arranged.

In reality the situation with Grand Unified Theory baryogenesis is more complicated than described above. We will see later that inflation most likely ends at energy scales below the Grand Unified Theory scale and hence that relativistic X bosons were never in thermal equilibrium. However, one can produce them (out of equilibrium) during the preheating phase (see Section 5.5).

Baryogenesis via leptogenesis Baryon asymmetry can also be generated via leptogenesis. What is required is a nonvanishing initial value of $(B - L)_i$. Even if $B_i = 0$ and $L_i \neq 0$, then lepton number will be partially converted to baryon number in subsequent topological transitions. Since $B + aL$ vanishes in these transitions while $B - L$ is conserved, the final baryon number is

$$B_f = -\frac{a}{1+a} L_i, \tag{4.223}$$

where $a = 28/51$ in the Standard Model with one Higgs doublet. In turn, the initial nonvanishing lepton number L_i can be generated in out-of-equilibrium decay of heavy neutrinos.

Let us briefly discuss the motivation for the existence of such heavy neutrinos. The neutrino oscillations measured can be explained only if the neutrinos have nonvanishing masses. To generate the neutrino masses we need right-handed neutrinos ν_R. Then the Yukawa coupling term generating the Dirac masses can be written as in (4.82):

$$\mathcal{L}_Y^{(\nu)} = -f_{ij}^{(\nu)} \chi \bar{\mathbf{L}}_L^i \varphi_1 \nu_R^j + \text{h.c.} = -f_{ij}^{(\nu)} \chi \bar{\nu}_L^i \nu_R^j + \text{h.c.}, \tag{4.224}$$

where $i = 1, 2, 3$ is the lepton generation index and ν_L are the $SU(2)$ gauge-invariant left-handed neutrinos defined in Section 4.3.4. Under the $U(1)$ group, ν_L transforms according to (4.72) and, because $Y_L^{(\nu)} = 1/2$, it remains invariant; hence the Yukawa term (4.224) is gauge-invariant only if the hypercharges of the right-handed neutrinos are equal to zero. The right-handed neutrinos are $SU(2)$ singlets, have no color and do not carry hypercharge. Therefore a Majorana mass term,

$$\mathcal{L}_M^{(\nu)} = -\tfrac{1}{2} M_{ij} (\bar{\nu}_R^c)^i \nu_R^j, \tag{4.225}$$

is consistent with the gauge symmetries of the theory; here the ν_R^c are the charge-conjugated wave functions of the right-handed neutrinos. After symmetry breaking the field χ acquires the expectation value χ_0 and the Yukawa term induces the Dirac masses described by the mass matrix $(m_D)_{ij} = f_{ij}^{(\nu)} \chi_0$. Considering for simplicity the case of one generation, we can write the total mass term as

$$\mathcal{L}^{(\nu)} = -\frac{1}{2} \begin{pmatrix} \bar{\nu}_L & \bar{\nu}_R^c \end{pmatrix} \begin{pmatrix} 0 & m_D \\ m_D & M \end{pmatrix} \begin{pmatrix} \nu_L^c \\ \nu_R \end{pmatrix} + \text{h.c.}, \qquad (4.226)$$

taking into account that $\bar{\nu}_L \nu_R = \bar{\nu}_R^c \nu_L^c$. The mass matrix in (4.226) is not diagonal. When $m_D \ll M$, the mass eigenvalues

$$m_\nu \simeq -\frac{m_D^2}{M}, \qquad m_N \simeq M \qquad (4.227)$$

correspond to the eigenstates

$$\nu \simeq \nu_L + \nu_L^c, \qquad N \simeq \nu_R + \nu_R^c, \qquad (4.228)$$

which describe the Majorana fermions ($\nu = \nu^c$, $N = N^c$) – light and heavy neutrinos respectively. Choosing for m_D the largest known fermion mass of order the top quark mass, $m_D \sim m_t \sim 170$ GeV, and taking $M \simeq 3 \times 10^{15}$ GeV, we find from (4.227) that $m_\nu \simeq 10^{-2}$ eV, which is favored by neutrino oscillation measurements. If $m_D \sim m_e \sim 0.5$ MeV, then the mass of the heavy neutrino should be $\simeq 2 \times 10^6$ GeV. It is important to note that the Majorana mass terms are not generated via the Higgs mechanism, and therefore can be much larger than the masses of ordinary quarks and leptons. This leads to light neutrino masses that are very small, according to (4.227). Such a method of obtaining very small masses is known as the *seesaw mechanism*. If one were restricted to having only Dirac mass terms, the Yukawa couplings would have to be unnaturally small.

The Majorana mass terms violate lepton number by two units. The heavy Majorana neutrinos $N = N^c$ are coupled to the Higgs particles via (4.224) and they can decay into a lepton–Higgs pair, $N \to l\phi$, or into the CP conjugated state, $N \to \bar{l}\bar{\phi}$, thus violating lepton number (Figure 4.22). Returning to the case of three generations, we see that the neutrino mass eigenstates do not necessarily

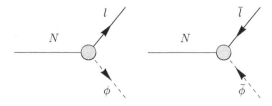

Fig. 4.22.

coincide with the flavor (weak) eigenstates. These states are related by the corresponding Kobayashi–Maskawa mixing matrix. This naturally explains neutrino oscillations and generically leads to *complex* Yukawa couplings which are a source of CP violation. As a result the decay rates

$$\Gamma(N \to l\phi) = \tfrac{1}{2}(1+\varepsilon)\,\Gamma_{tot}, \quad \Gamma\left(N \to \bar{l}\bar{\phi}\right) = \tfrac{1}{2}(1-\varepsilon)\,\Gamma_{tot} \qquad (4.229)$$

are different by the parameter $\varepsilon \ll 1$, which measures the amount of CP violation. The final products have different lepton numbers and the mean net lepton number from the decay of the N neutrino is equal to ε. Heavy neutrinos can be produced after inflation, either in the preheating phase or after thermalization. Subsequently, their concentration freezes out and the lepton asymmetry is produced in out-of-equilibrium decays of heavy neutrinos. In the topological transitions which follow, this asymmetry is partially transferred to baryon asymmetry in an amount given by (4.223). Detailed calculations show that for the range of parameters suggested by the measured neutrino oscillations, one can naturally obtain the observed baryon asymmetry via leptogenesis. At present this theory is considered the leading baryogenesis scenario.

Other scenarios In addition to Grand Unified Theory baryogenesis and leptogenesis there exist other mechanisms for explaining baryon asymmetry. Supersymmetry in particular opens a number of options. Since supersymmetry extends the particle content of the theory near the electroweak scale, the possibility of a strong electroweak phase transition cannot yet be completely excluded. This revives the hope of explaining baryon asymmetry entirely within the MSSM.

Another interesting consequence of supersymmetry is found in the Affleck–Dine scenario. This scenario is based on the observation that in supersymmetric theories ordinary quarks and leptons are accompanied by supersymmetric partners — squarks and sleptons — which are scalar particles. The corresponding scalar fields carry baryon and lepton number, which can in principle be very large in the case of a scalar condensate (classical scalar field). An important feature of supersymmetry theories is the existence of "flat directions" in the superpotential, along which the relevant components of the *complex* scalar fields φ can be considered as massless. Inflation displaces a massless field from its zero position (see Chapters 5 and 8) and establishes the initial conditions for subsequent evolution of the field. The condensate is frozen until supersymmetry breaking takes place. Supersymmetry breaking lifts the flat directions and the scalar field acquires mass. When the Hubble constant becomes of order this mass, the scalar field starts to oscillate and decays. At this time B, L and CP violating terms (for example, quartic couplings

$$\lambda_1 \varphi^3 \varphi^* + \text{c.c.} \quad \text{and} \quad \lambda_2 \varphi^4 + \text{c.c.}$$

with complex λ_1, λ_2) become important, and a substantial baryon asymmetry can be produced. The scalar particles decay into ordinary quarks and leptons, transferring to them the generated baryon number. The Affleck–Dine mechanism can be implemented at nearly any energy scale, even below 200 GeV. By suitable choice of the parameters, one can explain almost any amount of baryon asymmetry. This makes the Affleck–Dine scenario practically unfalsifiable and it is a very unattractive feature of this scenario.

More exotic possibilities have also been considered. Among them are baryogenesis via black hole evaporation and leptogenesis with very weakly coupled right-handed Dirac neutrinos. Although at present the accepted wisdom favors leptogenesis, it is not clear which scenario was actually realized in nature. Therefore the main lesson of this section is that there exist many ways to "solve" the baryogenesis problem.

4.6.3 Topological defects

Topological defects do not occur in the Standard Model. However, they are a rather generic prediction of theories beyond the Standard Model. Below we briefly discuss why unified theories lead to topological defects and what kind of defects can be produced in the early universe.

The Higgs mechanism has become an integral part of modern particle physics. The main feature of this mechanism is the existence of scalar fields used to break the original symmetry of the theory. Depending on the model, their Lagrangian can be written as

$$\mathcal{L}_\varphi = \frac{1}{2}(\partial_\alpha \phi)(\partial^\alpha \phi) - \frac{\lambda}{4}(\phi^2 - \sigma^2)^2, \quad (4.230)$$

where $\phi \equiv (\phi^1, \phi^2, \ldots, \phi^n)$ is an n-plet of real scalar fields. Complex scalar fields can be also rewritten in the form (4.230) if we use their real and imaginary parts. For example, in $U(1)$ gauge theory, $\varphi = \phi^1 + i\phi^2$ and $n = 2$. The doublet of the complex fields of electroweak theory corresponds to $n = 4$.

At very high temperatures symmetry is restored, that is, $\phi = 0$. As the universe cools, phase transitions take place. As a result the scalar fields acquire vacuum expectation values corresponding to the minimum of the potential in (4.230),

$$\phi^2 = (\phi^1)^2 + (\phi^2)^2 + \cdots + (\phi^n)^2 = \sigma^2.$$

This vacuum manifold $\mathcal{M}$ has a nontrivial structure. For example, for $n = 1$, both $\varphi = \sigma$ and $\varphi = -\sigma$ are states of minimal energy and so the vacuum manifold has the topology of a zero-dimensional sphere, $S^0 = \{-1, +1\}$. In $U(1)$ theory, the vacuum is isomorphic to a circle S^1 (the "bottom of the bottle"). In the case $n = 3$, the vacuum states form a two dimensional sphere S^2.

Topological defects are solitonic solutions of the classical equations for the scalar (and gauge) fields. They can be formed during a phase transition and since they interpolate between vacuum states they reflect the structure of the vacuum manifold. One real scalar field with two degenerate vacua ($n = 1$ and $\mathcal{M} = S^0$) leads to *domain walls*. In the case of a complex scalar field ($n = 2$, $\mathcal{M} = S^1$) *cosmic strings* can be formed. If the symmetry is broken with a triplet of real scalar fields ($n = 3$, $\mathcal{M} = S^2$), the topological defects are *monopoles*. Finally, in the case $n = 4$ (four scalar fields or, equivalently, a doublet of complex scalar fields) the vacuum manifold is a 3-sphere S^3 and the corresponding defects are *textures*. Depending on whether we consider a theory with or without local gauge invariance, the topological defects are called *local* or *global* respectively.

Domain walls Let us first consider one real scalar field which has the double-well potential in (4.230). The states $\phi = \sigma$ and $\phi = -\sigma$ correspond to two degenerate minima of the potential and during symmetry breaking the field ϕ acquires the values σ and $-\sigma$ with equal probability. The important thing is that the phase transition sets the maximum distance over which the scalar field is correlated. It is obvious that in the early universe the correlation length cannot exceed the size of the causally connected region. Let us consider two causally disconnected regions A and B, and assume that the field ϕ in region A went to the minimum at σ. The field in region B does not "know" what happened in region A and, with probability $1/2$, goes to the minimum at $-\sigma$. Since the scalar field changes continuously from $-\sigma$ to σ, it must vanish on some two-dimensional surface separating regions A and B. This surface, determined by the equation $\phi(x^1, x^2, x^3) = 0$, is called the domain wall (see Figure 4.23). The domain wall (Figure 4.24(a)) has a finite thickness l, which can be estimated with the following simple arguments. Let us assume, for simplicity, that the domain wall is static and not curved. The energy density of the scalar field is

$$\varepsilon = \tfrac{1}{2}(\partial_i \phi)^2 + V; \qquad (4.231)$$

it is distributed as shown in Figure 4.24(b). The total energy per unit surface area can be estimated as

$$E \sim \varepsilon l \sim \left(\frac{\sigma}{l}\right)^2 l + \lambda \sigma^4 l, \qquad (4.232)$$

where the first contribution comes from the gradient term. This energy is minimized for $l \sim \lambda^{-1/2} \sigma^{-1}$ and is equal to $E_w \sim \lambda^{1/2} \sigma^3$.

218   *The very early universe*

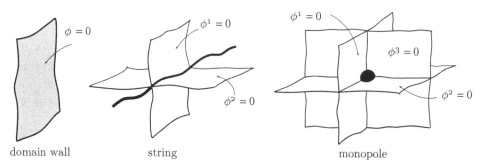

Fig. 4.23.

Problem 4.33 Consider an infinite wall in the x–y plane and verify that $\phi(z) = \sigma \tanh(z/l)$, where $l = (\lambda/2)^{-1/2}\sigma^{-1}$, is the solution of the scalar field equation with the potential in (4.230).

Domain walls are nonperturbative solutions of the field equations and they are stable with respect to small perturbations. To remove the wall described by the solution in Problem 4.33, one has to "lift" the scalar field over the potential barrier from $\phi = \sigma$ to $\phi = -\sigma$ in infinite space. This costs an infinite amount of energy.

It is clear from the previous discussion that on average at least one domain wall per horizon volume is formed during the cosmological phase transition. The subsequent evolution of the domain wall network is rather complicated and has been investigated numerically. The result is that one expects at least one domain

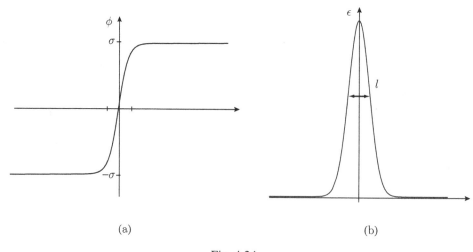

Fig. 4.24.

wall per present horizon scale $\sim t_0$. Its mass can be estimated as

$$M_{\text{wall}} \sim E_w t_0^2 \sim 10^{65} \lambda^{1/2} (\sigma/100 \text{ GeV})^3 \text{ g}.$$

For realistic values of λ and σ, the mass of the domain wall exceeds the mass of matter within the present horizon by many orders of magnitude. Such a wall would lead to unacceptably large CMB fluctuations. Therefore, domain walls are cosmologically admissible only if the coupling constant λ and the symmetry breaking scale σ are unjustifiably small.

Homotopy groups give us a useful unifying description of topological defects. Maps of the n-dimensional sphere S^n into a vacuum manifold $\mathcal{M}$ are classified by the homotopy group $\pi_n(\mathcal{M})$. This group counts the number of topologically inequivalent maps from S^n into $\mathcal{M}$ that cannot be continuously deformed into each other. In the language of homotopy groups, domain walls correspond to the group $\pi_0(\mathcal{M} = S^0)$, which describes the maps of a zero-dimensional sphere $S^0 = \{-1, +1\}$ to itself. This group is nontrivial and is isomorphic to the group of integers under addition modulo 2, that is, $\pi_0(S^0) = Z_2$.

Cosmic strings If the symmetry breaking occurs via a $U(1)$ complex scalar field $\varphi = \phi^1 + i\phi^2$ or, equivalently, with two real scalar fields ϕ^1, ϕ^2, cosmic strings are formed. In this case the vacuum manifold described by

$$(\phi^1)^2 + (\phi^2)^2 = \sigma^2 \tag{4.233}$$

is obviously a circle S^1.

Let us again consider two causally disconnected regions A and B and assume that inside region A the scalar fields acquired the expectation values $\phi_A^1 > 0$ and $\phi_A^2 > 0$ satisfying (4.233). Because of the absence of communication, the expectation values of the fields in region B are not correlated with those in region A and both can take negative values: $\phi_B^1 < 0$ and $\phi_B^2 < 0$ (the only restriction is that they have to satisfy (4.233)). The probability of this happening is $1/4$. The fields are continuous and, in changing from negative to positive values, must vanish somewhere between regions A and B. Namely, field ϕ^1 vanishes on the two-dimensional surface determined by the equation $\phi^1(x^1, x^2, x^3) = 0$, while field ϕ^2 is equal to zero on the surface described by $\phi^2(x^1, x^2, x^3) = 0$. These two surfaces generically cross each other along a curve which is either infinite or closed. On this curve, $\phi^1 = \phi^2 = 0$, and we have a false vacuum. Thus, as a result of symmetry breaking, one-dimensional topological objects – cosmic strings – are formed (see Figure 4.23). It is clear that they are produced with an abundance *at least* of order one per horizon volume.

As with domain walls, strings have a finite thickness. The field φ smoothly changes from zero in the core of the string and approaches the true vacuum, $|\varphi| = \sigma$, far away from it. Strings are topologically stable, classical solutions of the field equations.

In the language of homotopy groups, strings correspond to the mappings of a circle S^1 to the vacuum manifold $\mathcal{M} = S^1$. The corresponding group is nontrivial: $\pi_1(S^1) = Z$, where Z is the group of integers under addition. Taking a circle $\gamma(\tau)$ in **x** space, let us consider its map to $\mathcal{M}$: $\varphi(\tau) = \sigma \exp(i\theta(\tau))$. Because the complex field φ is an unambiguous function of the spatial coordinates **x**, the phase θ changes by $2\pi m$ around the circle γ, where m is an integer. If $m = 0$, the mapping is trivial and the contour γ can be continuously deformed to a point without passing through the region of false vacuum. Hence it does not contain any topological defects. The map with $m \neq 0$ wraps the circle γ around the vacuum manifold m times. For $m = 1$ there is a string inside the contour γ. In fact, considering two points with $\theta = 0$ and $\theta = \pi$, where φ is equal to σ and $-\sigma$ respectively, and shrinking the contour, one necessarily arrives at a place where the field φ vanishes because otherwise it would have infinite derivative. This is where the string lives.

Let us take for simplicity a straight global string (no gauge fields are present) and consider its energy per unit length. At a large distance r from the core of the string, the derivative $\partial_i \varphi$ can be estimated on dimensional grounds as σ/r. Hence the gradient term gives a logarithmically divergent contribution to the energy per unit length:

$$\mu_s \sim \sigma^2 \int \left(1/r^2\right) d^2x. \qquad (4.234)$$

In this case the natural regularization factor is the distance to the nearest string. As we have seen, axions assume a global $U_{PQ}(1)$ symmetry and therefore global axionic strings can exist.

In a theory with local gauge invariance the derivative $\partial_i \varphi$ is replaced by $\mathcal{D}_i \varphi = \partial_i \varphi + ieA_i \varphi$; the properties of the *local* strings are very different from those of global strings. Solving the coupled system of equations for φ and A_i, one can show that the gauge field compensates the leading term in $\partial_i \varphi$ and that the covariant derivative $\mathcal{D}_i \varphi$ decays faster than $1/r$ as $r \to \infty$. As a result the energy per unit length converges. Writing $\varphi = \chi \exp(i\theta)$, and assuming that $\partial_i \chi$ decays faster than $1/r$, we find that the compensation takes place only if

$$A_i \to (1/e) \partial_i \theta$$

as $r \to \infty$. Taking a contour γ, far away from the core of the string, and calculating the flux of the magnetic field ($\mathbf{B} = \nabla \times \mathbf{A}$) we immediately find

$$\int \mathbf{B} d^2 \sigma = \oint A_i dx^i = \frac{2\pi}{e} m. \qquad (4.235)$$

Here we take into account that θ changes around the contour by the integer m multiplied by 2π. Thus, in the gauged $U(1)$ theory, strings carry a magnetic flux inversely proportional to the electric charge e, and this flux is quantized.

After symmetry breaking the vector and scalar fields acquire masses $m_A = e\sigma$ and $m_\chi = \sqrt{2\lambda}\sigma$ respectively. These masses determine the "thickness" of the string. Outside the string core both fields tend to their true vacuum values exponentially quickly. This is not surprising because in the broken $U(1)$ gauge theory no massless fields remain after symmetry breaking (compare this to the global string, where one *massless* scalar field "survives"). The thickness of the string core is determined by the Compton wavelengths $\delta_\chi \sim m_\chi^{-1}$ and $\delta_A \sim m_A^{-1}$. For $m_\chi > m_A$, the size of the magnetic core ($\sim \delta_A$) exceeds the size of the false vacuum tube ($\sim \delta_\chi$). In this case, the scalar and magnetic fields give about the same contribution to the energy per unit length:

$$\mu(\chi) \sim \lambda\sigma^4\delta_\chi^2 \sim \sigma^2 \quad \text{and} \quad \mu(A) \sim B^2\delta_A^2 \sim \left(e\delta_A^2\right)^{-2}\delta_A^2 \sim \sigma^2.$$

The total mass of a string with length of order the present horizon is about

$$\sigma^2 t_0 \sim 10^{48}\left(\sigma/10^{15}\text{ GeV}\right)^2 \text{ g}.$$

Hence, even if symmetry breaking occurred at Grand Unified Theory scales, any strings produced would not be in immediate conflict with observations. Moreover, such Grand Unified Theory strings could, in principle, serve as the seeds for galaxy formation. CMB measurements, however, rule out this possibility; cosmic strings cannot play a dominant role in structure formation. Nevertheless this does not mean that they were not produced in the early universe. If they were detected, cosmic strings would reveal important features of the theory beyond the Standard Model.

Monopoles Monopoles arise if the vacuum manifold has the topology of a two-dimensional sphere; for example, in theories where the symmetry is broken with three real scalar fields. In this case the vacuum manifold described by

$$\left(\phi^1\right)^2 + \left(\phi^2\right)^2 + \left(\phi^3\right)^2 = \sigma^2 \tag{4.236}$$

is obviously S^2. Again considering two causally disconnected regions A and B, we find that with probability of order unity one can obtain after symmetry breaking $\phi_A^i > 0$ and $\phi_B^i < 0$, where $i = 1, 2, 3$ and the fields ϕ^i satisfy (4.236). Three two-dimensional hypersurfaces, determined by the equations $\phi^i(x^1, x^2, x^3) = 0$, generically cross each other at a point. This is the point of false vacuum because there all three fields ϕ^i vanish. Thus a zero-dimensional topological defect – a monopole – is formed (Figure 4.23). Solving the equations for the scalar field, one can find the classical spherically symmetric scalar field configuration which corresponds to the false vacuum in the center and approaches the true vacuum as $r \to \infty$. Without going into the detailed structure of the exact solutions, we can analyze the properties of monopoles using dimensional arguments. In the theory without gauge fields, two massless bosons survive after symmetry breaking. Therefore the fields ϕ^i, smoothly changing from zero in the core of the monopole, approach their true vacuum configuration only as some power of distance r. On dimensional grounds, $\partial \phi \sim \sigma/r$ (here and in the following formulae we skip all indices) and the mass of the global monopole

$$M \sim \sigma^2 \int \frac{1}{r^2} d^3 x \tag{4.237}$$

diverges linearly. The cut-off scale should be taken to be of order the correlation length; this can never exceed the horizon scale.

Local monopoles have quite different properties. As an example, let us consider $SO(3) \simeq SU(2)$ local gauge theory with the triplet of real scalar fields. After symmetry breaking, two of the three gauge fields acquire mass $m_W = e\sigma$. One gauge field, A, remains massless and $U(1)$ gauge invariance survives. The massive vector fields W decay exponentially quickly beyond their interaction distance, determined by the Compton wavelength $\delta_W \sim m_W^{-1}$. In the monopole solution the massless $U(1)$ gauge field A compensates the gradient term in the covariant derivative and $\mathcal{D}\phi = \partial \phi + eA\phi$ decays exponentially as $r \to \infty$. Therefore at large distances r we have

$$A \sim \frac{1}{e} \frac{\partial \phi}{\phi} \sim \frac{(1/e)}{r} \tag{4.238}$$

and the corresponding magnetic field is of order

$$B = \nabla \times A \sim \frac{(1/e)}{r^2}. \tag{4.239}$$

Thus the local monopole has a magnetic charge $g \sim 1/e$. The exact calculation gives $g = 2\pi/e$ in agreement with the result for the Dirac monopole. We have to stress, however, that in contrast with the Dirac monopole, which is a fundamental *point-like* magnetic charge, monopoles in gauge theories are *extended* objects. They are classical solutions of the field equations. As with the local string, the monopole has two cores — scalar and magnetic — with radii

$$\delta_s \sim m_s^{-1} = (2\lambda)^{-1/2} \sigma^{-1} \quad \text{and} \quad \delta_W \sim m_W^{-1} = e^{-1} \sigma^{-1}$$

respectively. Let us assume that the scalar core is smaller than the magnetic one, that is, $\delta_W > \delta_s$. It is easy to verify that in this case the dominant contribution to the mass comes from the gauge fields:

$$M \sim B^2 \delta_W^3 \sim \left(\frac{g}{\delta_W^2}\right)^2 \delta_W^3 \sim \frac{m_W}{e^2}. \tag{4.240}$$

One can show that this estimate, obtained for $\lambda > e^2$, still applies in the more complicated case $\lambda < e^2$.

The existence of magnetic monopoles is an inevitable consequence of Grand Unified Theories with (semi)simple gauge groups G. This general result follows immediately from the topological interpretation of monopole solutions. Monopoles exist if the vacuum manifold $\mathcal{M}$ contains noncontractible 2-surfaces, or equivalently, the group $\pi_2(\mathcal{M})$ is nontrivial. In the case above, $\mathcal{M} = S^2$ and $\pi_2(S^2) = Z$. In Grand Unified Theories, a semisimple group G is broken to $H = SU(3)_{QCD} \times U(1)_{em}$; the vacuum manifold is the quotient group G modulo H, that is, $\mathcal{M} = G/H$. Using a well known result from the theory of homotopy groups,

$$\pi_2(G/H) = \pi_1(H) = \pi_1\left(SU(3)_{QCD} \times U(1)_{em}\right) = Z, \tag{4.241}$$

we conclude that monopoles are unavoidable if the unification group incorporates electromagnetism. Furthermore, because the $SU(3)_{QCD}$ gauge fields are confined, the monopoles carry only the magnetic charge of the $U(1)_{em}$ group.

Problem 4.34 Why are the arguments presented above not applicable to electroweak symmetry breaking?

Let us estimate the abundance of Grand Unified Theory monopoles produced at $T_{GUT} \sim 10^{15}$ GeV. As we have already said, at least one defect per horizon volume is created during symmetry breaking. Taking into account that the horizon scale at this time is about $t_H \sim 1/T_{GUT}^2$, we immediately obtain the following estimate for the average number of monopoles per photon:

$$\frac{n_M}{n_\gamma} \gtrsim \frac{1}{T_{GUT}^3 t_H^3} \sim T_{GUT}^3. \tag{4.242}$$

This ratio does not change significantly during the expansion of the universe and so the present energy density of the GUT monopoles should be

$$\varepsilon_M^0 \sim M n_M(t_0) \sim \frac{m_W}{e^2} T_{GUT}^3 T_0^3$$

$$\sim 10^{-16} \left(\frac{m_W}{10^{15} \text{ GeV}}\right)\left(\frac{T_{GUT}}{10^{15} \text{ GeV}}\right)^3 \text{ g cm}^{-3}. \tag{4.243}$$

But this is at least 10^{13} times the critical density — obviously a cosmological disaster. Either one has to abandon Grand Unified Theories or find a solution to this *monopole problem*. Inflationary cosmology provides us with such a solution. If the monopoles were produced in the very early universe, then a subsequent inflationary stage would drastically dilute their number density, leaving less than one monopole per present horizon scale. Of course, this solution works only if the reheating temperature after inflation does not exceed the Grand Unified Theory scale; otherwise monopoles are produced in unacceptable amounts after the end of inflation. We will see in the following sections that this assumption about the energy scale of inflation is in agreement with contemporary ideas. Moreover, according to inflationary scenarios, it is likely that the temperature in the universe was never larger than the Grand Unified Theory scale and hence that monopoles were never produced according to the mechanism described above. This, however, does not mean that primordial Grand Unified Theory monopoles do not exist. In principle they could be produced during a preheating phase after inflation (see Section 5.5) in amounts allowed by present cosmological bounds. The search for primordial monopoles remains important.

Textures The other possible defect — a texture — arises when the symmetry is broken with four real scalar fields ϕ^i, $i = 1, \ldots, 4$. In this case the vacuum manifold is a 3-sphere S^3 and the textures are classified by the homotopy group $\pi_3(\mathcal{M})$. Because four equations

$$\phi^i(x^1, x^2, x^3) = 0$$

for three variables x^1, x^2, x^3 generically have no solutions, regions of false vacuum are not formed during the phase transition. However, the fields ϕ^i are uncorrelated on superhorizon scales and therefore $(\partial \phi)^2$ is generally different from zero even if $\phi^2 = \sigma^2$ and $V(\phi) = 0$. The resulting stable structure has positive energy and is called a *global* texture. Some time ago global textures were considered a compelling mechanism for explaining the structure of the universe. However, the texture scenario is in contradiction with measurements of CMB fluctuations and hence textures cannot play any significant role in structure formation.

Static textures do not "survive" in local gauge theories. In this case the gauge fields *exactly* compensate the spatial gradients of the scalar fields; as distinct from strings and monopoles, textures correspond to the true vacuum everywhere. As a result $(\mathcal{D}\phi)^2$ vanishes and the total energy of a local texture is equal to zero.

In the Standard Model, electroweak symmetry is broken with a doublet of complex scalar fields, or equivalently, with four real scalar fields. Hence, the only topological defects which could occur are textures. However, because this theory possesses local gauge invariance, the corresponding static textures have zero energy and they are not very interesting.

In concluding this section, we would like to warn the reader that the above considerations are simplified. To analyze defects in realistic theories with complicated symmetry breaking schemes, we have to use more powerful methods which go far beyond the scope of this book. In these theories hybrid topological defects, for example, strings with monopoles at their ends, or string-bounded domain walls, can exist.

5
Inflation I: homogeneous limit

Matter is distributed very homogeneously and isotropically on scales larger than a few hundred megaparsecs. The CMB gives us a "photograph" of the early universe, which shows that at recombination the universe was extremely homogeneous and isotropic (with accuracy $\sim 10^{-4}$) on all scales up to the present horizon. Given that the universe evolves according to the Hubble law, it is natural to ask which initial conditions could lead to such homogeneity and isotropy.

To obtain an exhaustive answer to this question we have to know the exact physical laws which govern the evolution of the very early universe. However, as long as we are interested only in the general features of the initial conditions it suffices to know a few simple properties of these laws. We will assume that inhomogeneity cannot be dissolved by expansion. This natural surmise is supported by General Relativity (see Part II of this book for details). We will also assume that nonperturbative quantum gravity does not play an essential role at sub-Planckian curvatures. On the other hand, we are nearly certain that nonperturbative quantum gravity effects become very important when the curvature reaches Planckian values and the notion of classical spacetime breaks down. Therefore we address the initial conditions at the Planckian time $t_i = t_{Pl} \sim 10^{-43}$ s.

In this chapter we discuss the initial conditions problem we face in a decelerating universe and show how this problem can be solved if the universe undergoes a stage of the accelerated expansion known as inflation.

5.1 Problem of initial conditions

There are two *independent* sets of initial conditions characterizing matter: (a) its spatial distribution, described by the energy density $\varepsilon(x)$ and (b) the initial field of velocities. Let us determine them given the current state of the universe.

Homogeneity, isotropy (horizon) problem The present homogeneous, isotropic domain of the universe is at least as large as the present horizon scale, $ct_0 \sim 10^{28}$ cm.

5.1 Problem of initial conditions

Initially the size of this domain was smaller by the ratio of the corresponding scale factors, a_i/a_0. Assuming that inhomogeneity cannot be dissolved by expansion, we may safely conclude that the size of the homogeneous, isotropic region from which our universe originated at $t = t_i$ was larger than

$$l_i \sim ct_0 \frac{a_i}{a_0}. \tag{5.1}$$

It is natural to compare this scale to the size of a causal region $l_c \sim ct_i$:

$$\frac{l_i}{l_c} \sim \frac{t_0}{t_i} \frac{a_i}{a_0}. \tag{5.2}$$

To obtain a rough estimate of this ratio we note that if the primordial radiation dominates at $t_i \sim t_{Pl}$, then its temperature is $T_{Pl} \sim 10^{32}$ K. Hence

$$(a_i/a_0) \sim (T_0/T_{Pl}) \sim 10^{-32}$$

and we obtain

$$\frac{l_i}{l_c} \sim \frac{10^{17}}{10^{-43}} 10^{-32} \sim 10^{28}. \tag{5.3}$$

Thus, at the initial Planckian time, the size of our universe exceeded the causality scale by 28 orders of magnitude. This means that in 10^{84} *causally disconnected* regions the energy density was smoothly distributed with a fractional variation not exceeding $\delta\varepsilon/\varepsilon \sim 10^{-4}$. Because no signals can propagate faster than light, no causal physical processes can be responsible for such an unnaturally fine-tuned matter distribution.

Assuming that the scale factor grows as some power of time, we can use an estimate $a/t \sim \dot{a}$ and rewrite (5.2) as

$$\frac{l_i}{l_c} \sim \frac{\dot{a}_i}{\dot{a}_0}. \tag{5.4}$$

Thus, the size of our universe was initially larger than that of a causal patch by the ratio of the corresponding expansion rates. Assuming that gravity was always attractive and hence was decelerating the expansion, we conclude from (5.4) that the homogeneity scale was always larger than the scale of causality. Therefore, the homogeneity problem is also sometimes called the *horizon* problem.

Initial velocities (flatness) problem Let us suppose for a minute that someone has managed to distribute matter in the required way. The next question concerns initial velocities. Only after they are specified is the Cauchy problem completely posed and can the equations of motion be used to predict the future of the universe unambiguously. The initial velocities must obey the Hubble law because otherwise the initial homogeneity is very quickly spoiled. That this has to occur in so many causally disconnected regions further complicates the horizon problem. Assuming that it has, nevertheless, been achieved, we can ask how accurately the initial Hubble velocities have to be chosen for a given matter distribution.

Let us consider a large spherically symmetric cloud of matter and compare its total energy with the kinetic energy due to Hubble expansion, E^k. The total energy is the sum of the positive kinetic energy and the negative potential energy of the gravitational self-interaction, E^p. It is conserved:

$$E^{tot} = E_i^k + E_i^p = E_0^k + E_0^p.$$

Because the kinetic energy is proportional to the velocity squared,

$$E_i^k = E_0^k (\dot a_i/\dot a_0)^2$$

and we have

$$\frac{E_i^{tot}}{E_i^k} = \frac{E_i^k + E_i^p}{E_i^k} = \frac{E_0^k + E_0^p}{E_0^k}\left(\frac{\dot a_0}{\dot a_i}\right)^2. \tag{5.5}$$

Since $E_0^k \sim |E_0^p|$ and $\dot a_0/\dot a_i \leq 10^{-28}$, we find

$$\frac{E_i^{tot}}{E_i^k} \leq 10^{-56}. \tag{5.6}$$

This means that for a given energy density distribution the initial Hubble velocities must be adjusted so that the huge negative gravitational energy of the matter is compensated by a huge positive kinetic energy to an unprecedented accuracy of $10^{-54}\%$. An error in the initial velocities exceeding $10^{-54}\%$ has a dramatic consequence: the universe either recollapses or becomes "empty" too early. To stress the unnaturalness of this requirement one speaks of the initial velocities problem.

Problem 5.1 How can the above consideration be made rigorous using the Birkhoff theorem?

In General Relativity the problem described can be reformulated in terms of the cosmological parameter $\Omega(t)$ introduced in (1.21). Using the definition of $\Omega(t)$ we can rewrite Friedmann equation (1.67) as

$$\Omega(t) - 1 = \frac{k}{(Ha)^2}, \tag{5.7}$$

and hence

$$\Omega_i - 1 = (\Omega_0 - 1)\frac{(Ha)_0^2}{(Ha)_i^2} = (\Omega_0 - 1)\left(\frac{\dot a_0}{\dot a_i}\right)^2 \leq 10^{-56}. \tag{5.8}$$

Note that this relation immediately follows from (5.5) if we take into account that $\Omega = |E^p|/E^k$ (see Problem 1.4). We infer from (5.8) that the cosmological parameter must initially be extremely close to unity, corresponding to a *flat* universe. Therefore the problem of initial velocities is also called the *flatness* problem.

Initial perturbation problem One further problem we mention here for completeness is the origin of the primordial inhomogeneities needed to explain the large-scale structure of the universe. They must be initially of order $\delta\varepsilon/\varepsilon \sim 10^{-5}$ on galactic scales. This further aggravates the very difficult problem of homogeneity and isotropy, making it completely intractable. We will see later that the problem of initial perturbations has the same roots as the horizon and flatness problems and that it can also be successfully solved in inflationary cosmology. However, for the moment we put it aside and proceed with the "more easy" problems.

The above considerations clearly show that the initial conditions which led to the observed universe are very unnatural and nongeneric. Of course, one can make the objection that naturalness is a question of taste and even claim that the most simple and symmetric initial conditions are "more physical." In the absence of a quantitative measure of "naturalness" for a set of initial conditions it is very difficult to argue with this attitude. On the other hand it is hard to imagine any measure which selects the special and degenerate conditions in preference to the generic ones. In the particular case under consideration the generic conditions would mean that the initial distribution of the matter is strongly inhomogeneous with $\delta\varepsilon/\varepsilon \gtrsim 1$ everywhere or, at least, in the causally disconnected regions.

The universe is unique and we do not have the opportunity to repeat the "experiment of creation". Therefore cosmological theory can claim to be a successful physical theory only if it can explain the state of the observed universe using simple physical ideas and starting with the most generic initial conditions. Otherwise it would simply amount to "cosmic archaeology," where "cosmic history" is written on the basis of a limited number of hot big bang remnants. If we are pretentious enough to answer the question raised by Einstein, "What really interests me is whether God had any choice when he created the World," we must be able to explain how a particular universe can be created starting with generic initial conditions. The inflationary paradigm seems to be a step in the right direction and it strongly restricts "God's choice." Moreover, it makes important predictions which can be verified experimentally (observationally), thus giving cosmology the status of a physical theory.

5.2 Inflation: main idea

We have seen so far that the same ratio, $\dot{a}_i/\dot{a}_0$, enters both sets of *independent* initial conditions. The large value of this ratio determines the number of causally disconnected regions and defines the necessary accuracy of the initial velocities. If gravity was always attractive, then $\dot{a}_i/\dot{a}_0$ is necessarily larger than unity because gravity decelerates an expansion. Therefore, the conclusion $\dot{a}_i/\dot{a}_0 \gg 1$ can be avoided *only* if we assume that during some period of expansion gravity acted as a "repulsive" force, thus accelerating the expansion. In this case we can have $\dot{a}_i/\dot{a}_0 < 1$ and the creation of our type of universe from a single causally connected domain may become possible. A period of accelerated expansion is a *necessary* condition, but whether is it also sufficient depends on the particular model in which this condition is realized. With these remarks in mind we arrive at the following general definition of inflation:

Inflation is a stage of accelerated expansion of the universe when gravity acts as a repulsive force.

Figure 5.1 shows how the old picture of a decelerated Friedmann universe is modified by inserting a stage of cosmic acceleration. It is obvious that if we do not want to spoil the successful predictions of the standard Friedmann model, such as nucleosynthesis, inflation should begin and end sufficiently early. We will see later that the requirement of the generation of primordial fluctuations further restricts the energy scale of inflation; namely, in the simple models inflation should be over at $t_f \sim 10^{-34}$–10^{-36} s. Successful inflation must also possess a smooth graceful exit into the decelerated Friedmann stage because otherwise the homogeneity of the universe would be destroyed.

Inflation explains the origin of the big bang; since it accelerates the expansion, small initial velocities within a causally connected patch become very large. Furthermore, inflation can produce the whole observable universe from a small homogeneous domain even if the universe was strongly inhomogeneous outside of

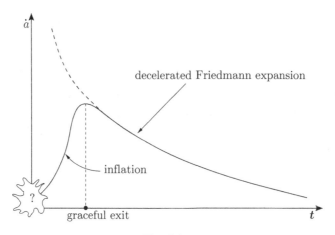

Fig. 5.1.

5.2 Inflation: main idea

this domain. The reason is that in an accelerating universe there *always* exists an event horizon. According to (2.13) it has size

$$r_e(t) = a(t) \int_t^{t_{\max}} \frac{dt}{a} = a(t) \int_{a(t)}^{a_{\max}} \frac{da}{\dot{a}a}. \tag{5.9}$$

The integral converges even if $a_{\max} \to \infty$ because the expansion rate $\dot{a}$ grows with a. The existence of an event horizon means that anything at time t a distance larger than $r_e(t)$ from an observer cannot influence that observer's future. Hence the future evolution of the region inside a ball of radius $r_e(t)$ is completely independent of the conditions outside a ball of radius $2r_e(t)$ centered at the same place. Let us assume that at $t = t_i$ matter was distributed homogeneously and isotropically only inside a ball of radius $2r_e(t_i)$ (Figure 5.2). Then an inhomogeneity propagating from outside this ball can spoil the homogeneity only in the region which was initially between the spheres of radii $r_e(t_i)$ and $2r_e(t_i)$. The region originating from the sphere of radius $r_e(t_i)$ remains homogeneous. This internal domain can be influenced only by events which happened at t_i between the two spheres, where the matter was initially distributed homogeneously and isotropically.

The physical size of the homogeneous internal region increases and is equal to

$$r_h(t_f) = r_e(t_i) \frac{a_f}{a_i} \tag{5.10}$$

at the end of inflation. It is natural to compare this scale with the particle horizon size, which in an accelerated universe can be estimated as

$$r_p(t) = a(t) \int_{t_i}^{t} \frac{dt}{a} = a(t) \int_{a_i}^{a} \frac{da}{\dot{a}a} \sim \frac{a(t)}{a_i} r_e(t_i), \tag{5.11}$$

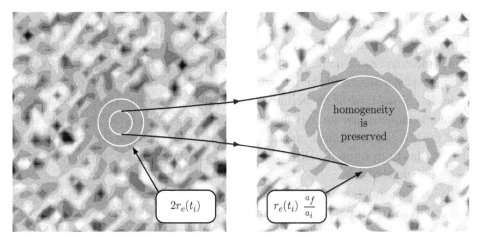

Fig. 5.2.

since the main contribution to the integral comes from $a \sim a_i$. At the end of inflation $r_p(t_f) \sim r_h(t_f)$, that is, the size of the homogeneous region, originating from a causal domain, is of order the particle horizon scale.

Thus, instead of considering a homogeneous universe in many causally disconnected regions, we can begin with a small homogeneous causal domain which inflation blows up to a very large size, preserving the homogeneity irrespective of the conditions outside this domain.

Problem 5.2 Why does the above consideration fail in a decelerating universe?

The next question is whether we can relax the restriction of homogeneity on the initial conditions. Namely, if we begin with a strongly inhomogeneous causal domain, can inflation still produce a large homogeneous universe?

The answer to this question is positive. Let us assume that the initial energy density inhomogeneity is of order unity on scales $\sim H_i^{-1}$, that is,

$$\left(\frac{\delta\varepsilon}{\varepsilon}\right)_{t_i} \sim \frac{1}{\varepsilon}\frac{|\nabla\varepsilon|}{a_i}H_i^{-1} = \frac{|\nabla\varepsilon|}{\varepsilon}\frac{1}{\dot{a}_i} \sim O(1), \tag{5.12}$$

where ∇ is the spatial derivative with respect to the comoving coordinates. At $t \gg t_i$, the contribution of this inhomogeneity to the variation of the energy density within the Hubble scale $H(t)^{-1}$ can be estimated as

$$\left(\frac{\delta\varepsilon}{\varepsilon}\right)_t \sim \frac{1}{\varepsilon}\frac{|\nabla\varepsilon|}{a(t)}H(t)^{-1} \sim O(1)\frac{\dot{a}_i}{\dot{a}(t)}, \tag{5.13}$$

where we have assumed that $|\nabla\varepsilon|/\varepsilon$ does not change substantially during expansion. This assumption is supported by the analysis of the behavior of linear perturbations on scales larger than the curvature scale H^{-1} (see Chapters 7 and 8). It follows from (5.13) that if the universe undergoes a stage of acceleration, that is, $\dot{a}(t) > \dot{a}_i$ for $t > t_i$, then the contribution of a large initial inhomogeneity to the energy variation on the curvature scale disappears. A patch of size H^{-1} becomes more and more homogeneous because the initial inhomogeneity is "kicked out": the physical size of the perturbation, $\propto a$, grows faster than the curvature scale, $H^{-1} = a/\dot{a}$, while the perturbation amplitude does not change substantially. Since inhomogeneities are "devalued" within the curvature scale, the name "inflation" fairly captures the physical effect of accelerated expansion. The consideration above is far from rigorous. However, it gives the flavor of the "no-hair" theorem for an inflationary stage.

To sum up, inflation demolishes large initial inhomogeneities and produces a homogeneous, isotropic domain. It follows from (5.13) that if we want to avoid the situation of a large initial perturbation re-entering the present horizon, $\sim H_0^{-1}$, and inducing a large inhomogeneity, we have to assume that the initial expansion rate was *much smaller* than the rate of expansion today, that is, $\dot{a}_i/\dot{a}_0 \ll 1$. More precisely, the CMB observations require that the variation of the energy density on the present horizon scale does not exceed 10^{-5}. The traces of an initial large inhomogeneity will be sufficiently diluted only if $\dot{a}_i/\dot{a}_0 < 10^{-5}$. Rewriting (5.8) as

$$\Omega_0 = 1 + (\Omega_i - 1)\left(\frac{\dot{a}_i}{\dot{a}_0}\right)^2, \tag{5.14}$$

we see that if $|\Omega_i - 1| \sim O(1)$ then

$$\Omega_0 = 1 \tag{5.15}$$

to very high accuracy. This important *robust prediction* of inflation has a kinematical origin and it states that the *total* energy density of all components of matter, *irrespective of their origin*, must be equal to the critical energy density today. We will see later that amplified quantum fluctuations lead to tiny corrections to $\Omega_0 = 1$, which are of order 10^{-5}. It is worth noting that, in contrast to a decelerating universe where $\Omega(t) \to 1$ as $t \to 0$, in an accelerating universe $\Omega(t) \to 1$ as $t \to \infty$, that is, $\Omega = 1$ is its *future* attractor.

Problem 5.3 Why does the consideration above fail for $\Omega_i = 0$?

5.3 How can gravity become "repulsive"?

To answer this question we recall the Friedmann equation (1.66):

$$\ddot{a} = -\frac{4\pi}{3}G(\varepsilon + 3p)a. \tag{5.16}$$

Obviously, if the strong energy dominance condition, $\varepsilon + 3p > 0$, is satisfied, then $\ddot{a} < 0$ and gravity decelerates the expansion. The universe can undergo a stage of accelerated expansion with $\ddot{a} > 0$ only if this condition is violated, that is, if $\varepsilon + 3p < 0$. One particular example of "matter" with a broken energy dominance condition is a positive cosmological constant, for which $p_V = -\varepsilon_V$ and $\varepsilon + 3p = -2\varepsilon_V < 0$. In this case the solution of Einstein's equations is a de Sitter universe – discussed in detail in Sections 1.3.6 and 2.3. For $t \gg H_\Lambda^{-1}$, the de Sitter universe expands exponentially quickly, $a \propto \exp(H_\Lambda t)$, and the rate of expansion grows as the scale factor. The *exact* de Sitter solution fails to satisfy all necessary conditions for successful inflation: namely, it does not possess a smooth graceful exit into the Friedmann stage. Therefore, in realistic inflationary models, it can be utilized only as a zero order approximation. To have a graceful exit from inflation we must allow the Hubble parameter to vary in time.

Let us now determine the general conditions which must be satisfied in a successful inflationary model. Because

$$\frac{\ddot{a}}{a} = H^2 + \dot{H}, \tag{5.17}$$

and $\ddot{a}$ should become negative during a graceful exit, the derivative of the Hubble constant, $\dot{H}$, must obviously be negative. The ratio $|\dot{H}|/H^2$ grows toward the end of inflation and the graceful exit takes place when $|\dot{H}|$ becomes of order H^2. Assuming that H^2 changes faster than $\dot{H}$, that is, $|\ddot{H}| < 2H\dot{H}$, we obtain the following generic estimate for the duration of inflation:

$$t_f \sim H_i/|\dot{H}_i|, \tag{5.18}$$

where H_i and $\dot{H}_i$ refer to the beginning of inflation. At $t \sim t_f$ the expression on the right hand side in (5.17) changes sign and the universe begins to decelerate.

Inflation should last long enough to stretch a small domain to the scale of the observable universe. Rewriting the condition $\dot{a}_i/\dot{a}_0 < 10^{-5}$ as

$$\frac{\dot{a}_i}{\dot{a}_f}\frac{\dot{a}_f}{\dot{a}_0} = \frac{a_i}{a_f}\frac{H_i}{H_f}\frac{\dot{a}_f}{\dot{a}_0} < 10^{-5},$$

and taking into account that $\dot{a}_f/\dot{a}_0$ should be larger than 10^{28}, we conclude that inflation is successful only if

$$\frac{a_f}{a_i} > 10^{33}\frac{H_i}{H_f}.$$

Let us assume that $|\dot{H}_i| \ll H_i^2$ and neglect the change of the Hubble parameter. Then the ratio of the scale factors can be roughly estimated as

$$a_f/a_i \sim \exp(H_i t_f) \sim \exp(H_i^2/|\dot{H}_i|) > 10^{33}. \tag{5.19}$$

Hence inflation can solve the initial conditions problem only if $t_f > 75 H_i^{-1}$, that is, it lasts *longer* than 75 Hubble times (e-folds). Rewritten in terms of the initial values of the Hubble parameter and its derivative, this condition takes the form

$$\frac{|\dot{H}_i|}{H_i^2} < \frac{1}{75}. \tag{5.20}$$

Using the Friedmann equations (1.67) and (1.68) with $k = 0$, we can reformulate it in terms of the bounds on the initial equation of state

$$\frac{(\varepsilon + p)_i}{\varepsilon_i} < 10^{-2}. \tag{5.21}$$

Thus, at the beginning of inflation the deviation from the vacuum equation of state must not exceed 1%. Therefore an exact de Sitter solution is a very good approximation for the initial stage of inflation. Inflation ends when $\varepsilon + p \sim \varepsilon$.

Problem 5.4 Consider an exceptional case where $|\dot{H}|$ decays at the same rate as H^2, that is, $\dot{H} = -pH^2$, where $p = $ const. Show that for $p < 1$ we have power-law inflation. This inflation has no natural graceful exit and in this sense is similar to a pure de Sitter universe.

5.4 How to realize the equation of state $p \approx -\varepsilon$

Thus far we have used the language of ideal hydrodynamics, which is an adequate phenomenological description of matter on large scales. Now we discuss a simple field-theoretic model where the required equation of state can be realized. The natural candidate to drive inflation is a scalar field. The name given to such a field is the "inflaton." We saw that the energy–momentum tensor for a scalar field can be rewritten in a form which mimics an ideal fluid (see (1.58)). The homogeneous classical field (scalar condensate) is then characterized by energy density

$$\varepsilon = \tfrac{1}{2}\dot{\varphi}^2 + V(\varphi), \tag{5.22}$$

and pressure

$$p = \tfrac{1}{2}\dot{\varphi}^2 - V(\varphi). \tag{5.23}$$

We have neglected spatial derivatives here because they become negligible soon after the beginning of inflation due to the "no-hair" theorem.

Problem 5.5 Consider a massive scalar field with potential $V = \tfrac{1}{2}m^2\varphi^2$, where $m \ll m_{Pl}$, and determine the bound on the allowed inhomogeneity imposed by the requirement that the energy density must not exceed the Planckian value. Why does the contribution of the spatial gradients to the energy–momentum tensor decay more quickly than the contribution of the mass term?

It follows from (5.22) and (5.23) that the scalar field has the desired equation of state only if $\dot{\varphi}^2 \ll V(\varphi)$. Because $p = -\varepsilon + \dot{\varphi}^2$, the deviation of the equation of state from that for the vacuum is entirely characterized by the kinetic energy, $\dot{\varphi}^2$, which must be much smaller than the potential energy $V(\varphi)$. Successful realization of inflation thus requires keeping $\dot{\varphi}^2$ small compared to $V(\varphi)$ during a sufficiently long time interval, or more precisely, for at least 75 e-folds. In turn this depends on the shape of the potential $V(\varphi)$. To determine which potentials can provide us with inflation, we have to study the behavior of a homogeneous classical scalar field in an expanding universe. The equation for this field can be derived either directly from the Klein–Gordon equation (1.57) or by substituting (5.22) and (5.23) into the conservation law (1.65). The result is

$$\ddot{\varphi} + 3H\dot{\varphi} + V_{,\varphi} = 0, \qquad (5.24)$$

where $V_{,\varphi} \equiv \partial V/\partial \varphi$. This equation has to be supplemented by the Friedmann equation:

$$H^2 = \frac{8\pi}{3}\left(\frac{1}{2}\dot{\varphi}^2 + V(\varphi)\right), \qquad (5.25)$$

where we have set $G = 1$ and $k = 0$. We first find the solutions of (5.24) and (5.25) for a free massive scalar field and then study the behavior of the scalar field in the case of a general potential $V(\varphi)$.

5.4.1 Simple example: $V = \frac{1}{2}m^2\varphi^2$.

Substituting H from (5.25) into (5.24), we obtain the closed form equation for φ,

$$\ddot{\varphi} + \sqrt{12\pi}\left(\dot{\varphi}^2 + m^2\varphi^2\right)^{1/2}\dot{\varphi} + m^2\varphi = 0. \qquad (5.26)$$

This is a nonlinear second order differential equation with no explicit time dependence. Therefore it can be reduced to a first order differential equation for $\dot{\varphi}(\varphi)$. Taking into account that

$$\ddot{\varphi} = \dot{\varphi}\frac{d\dot{\varphi}}{d\varphi},$$

(5.26) becomes

$$\frac{d\dot{\varphi}}{d\varphi} = -\frac{\sqrt{12\pi}\left(\dot{\varphi}^2 + m^2\varphi^2\right)^{1/2}\dot{\varphi} + m^2\varphi}{\dot{\varphi}}, \qquad (5.27)$$

which can be studied using the phase diagram method. The behavior of the solutions in the φ–$\dot{\varphi}$ plane is shown in Figure 5.3. The important feature of this diagram is the existence of an attractor solution to which all other solutions converge in time. One can distinguish different regions corresponding to different effective equations of state. Let us consider them in more detail. We restrict ourselves to the lower right quadrant ($\varphi > 0$, $\dot{\varphi} < 0$); solutions in the other quadrants can easily be derived simply by taking into account the symmetry of the diagram.

Ultra-hard equation of state First we study the region where $|\dot{\varphi}| \gg m\varphi$. It describes the situation when the potential energy is small compared to the kinetic energy, so that $\dot{\varphi}^2 \gg V$. It follows from (5.22) and (5.23) that in this case the equation of state

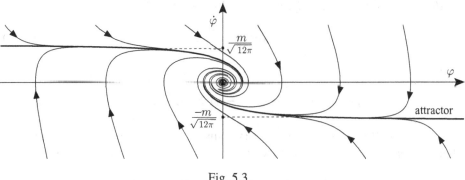

Fig. 5.3.

is ultra-hard, $p \approx +\varepsilon$. Neglecting $m\varphi$ compared to $\dot\varphi$ in (5.27), we obtain

$$\frac{d\dot\varphi}{d\varphi} \simeq \sqrt{12\pi}\,\dot\varphi. \tag{5.28}$$

The solution of this equation is

$$\dot\varphi = C \exp\left(\sqrt{12\pi}\,\varphi\right), \tag{5.29}$$

where $C < 0$ is a constant of integration. In turn, solving (5.29) for $\varphi(t)$ gives

$$\varphi = \text{const} - \frac{1}{\sqrt{12\pi}} \ln t. \tag{5.30}$$

Substituting this result into (5.25) and neglecting the potential term, we obtain

$$H^2 \equiv \left(\frac{\dot a}{a}\right)^2 \simeq \frac{1}{9t^2}. \tag{5.31}$$

It immediately follows that $a \propto t^{1/3}$ and $\varepsilon \propto a^{-6}$ in agreement with the ultra-hard equation of state. Note that the solution obtained is exact for a massless scalar field. According to (5.29) the derivative of the scalar field decays exponentially more quickly than the value of the scalar field itself. Therefore, the large initial value of $|\dot\varphi|$ is damped within a short time interval before the field φ itself has changed significantly. The trajectory which begins at large $|\dot\varphi|$ goes up very sharply and meets the attractor. This substantially enlarges the set of initial conditions which lead to an inflationary stage.

Inflationary solution If a trajectory joins the attractor where it is flat, at $|\varphi| \gg 1$, then afterwards the solution describes a stage of accelerated expansion (recall that we work in Planckian units). To determine the attractor solution we assume that

$d\dot{\varphi}/d\varphi \approx 0$ along its trajectory. It follows from (5.27) that

$$\dot{\varphi}_{\text{atr}} \approx -\frac{m}{\sqrt{12\pi}}, \tag{5.32}$$

and therefore

$$\varphi_{\text{atr}}(t) \simeq \varphi_i - \frac{m}{\sqrt{12\pi}}(t - t_i) \simeq \frac{m}{\sqrt{12\pi}}(t_f - t), \tag{5.33}$$

where t_i is the time when the trajectory joins the attractor and t_f is the moment when φ *formally* vanishes. In reality, (5.33) fails well before the field φ vanishes.

Problem 5.6 Calculate the corrections to the approximate attractor solution (5.32) and show that

$$\dot{\varphi}_{\text{atr}} = -\frac{m}{\sqrt{12\pi}}\left(1 - \frac{1}{2}\left(\sqrt{12\pi}\varphi\right)^{-2} + O\left(\left(\sqrt{12\pi}\varphi\right)^{-3}\right)\right). \tag{5.34}$$

The corrections to (5.32) become of order the leading term when $\varphi \sim O(1)$, that is, when the scalar field value drops to the Planckian value or, more precisely, to $\varphi \simeq 1/\sqrt{12\pi} \simeq 1/6$. Hence (5.33) is a good approximation only when the scalar field exceeds the Planckian value. This does not mean, however, that we require a theory of nonperturbative quantum gravity. Nonperturbative quantum gravity effects become relevant only if the curvature or the energy density reaches the Planckian values. However, even for very large values of the scalar field they can still remain in the sub-Planckian domain. In fact, considering a massive homogeneous field with negligible kinetic energy we infer that the energy density reaches the Planckian value for $\varphi \simeq m^{-1}$. Therefore, if $m \ll 1$, then for $m^{-1} > \varphi > 1$ we can safely disregard nonperturbative quantum gravity effects.

According to (5.33) the scalar field decreases linearly with time after joining the attractor. During the inflationary stage

$$p \simeq -\varepsilon + m^2/12\pi.$$

So when the potential energy density $\sim m^2\varphi^2$, which dominates the total energy density, drops to m^2, inflation is over. At this time the scalar field is of order unity (in Planckian units).

Let us determine the time dependence of the scale factor during inflation. Substituting (5.33) into (5.25) and neglecting the kinetic term, we obtain a simple equation which is readily integrated to yield

$$a(t) \simeq a_f \exp\left(-\frac{m^2}{6}(t_f - t)^2\right) \simeq a_i \exp\left(\frac{(H_i + H(t))}{2}(t - t_i)\right), \tag{5.35}$$

where a_i and H_i are the initial values of the scale factor and the Hubble parameter. Note that the Hubble constant $H(t) \simeq \sqrt{4\pi/3}\, m\varphi(t)$ also linearly decreases with time. It follows from (5.33) that inflation lasts for

$$\Delta t \simeq t_f - t_i \simeq \sqrt{12\pi}(\varphi_i/m). \tag{5.36}$$

During this time interval the scale factor increases

$$\frac{a_f}{a_i} \simeq \exp(2\pi\varphi_i^2) \tag{5.37}$$

times. The results obtained are in good agreement with the previous rough estimates (5.18) and (5.19). Inflation lasts more than 75 e-folds if the initial value of the scalar field, φ_i, is four times larger than the Planckian value. To obtain an estimate for the largest possible increase of the scale factor during inflation, let us consider a scalar field of mass 10^{13} GeV. The maximal possible value of the scalar field for which we still remain in the sub-Planckian domain is $\varphi_i \sim 10^6$, and hence

$$\left(\frac{a_f}{a_i}\right)_{\max} \sim \exp(10^{12}). \tag{5.38}$$

Thus, the actual duration of the inflationary stage can massively exceed the 75 e-folds needed. In this case our universe would constitute only a very tiny piece of an incredibly large homogeneous domain which originated from one causal region. The other important feature of inflation is that the Hubble constant decreases only by a factor 10^{-6}, while the scale factor grows by the tremendous amount given in (5.38), that is,

$$\frac{H_i}{H_f} \ll \frac{a_f}{a_i}.$$

Graceful exit and afterwards After the field drops below the Planckian value it begins to oscillate. To determine the attractor behavior in this regime we note that

$$\dot{\varphi}^2 + m^2\varphi^2 = \frac{3}{4\pi}H^2 \tag{5.39}$$

and use the Hubble parameter H and the angular variable θ, defined via

$$\dot{\varphi} = \sqrt{\frac{3}{4\pi}}H\sin\theta, \quad m\varphi = \sqrt{\frac{3}{4\pi}}H\cos\theta, \tag{5.40}$$

as the new independent variables. It is convenient to replace (5.27) by a system of two first order differential equations for H and θ:

$$\dot{H} = -3H^2\sin^2\theta, \tag{5.41}$$

$$\dot{\theta} = -m - \frac{3}{2}H\sin 2\theta, \tag{5.42}$$

where a dot denotes the derivative with respect to physical time t. The second term on the right hand side in (5.42) describes oscillations with *decaying* amplitude, as is evident from (5.41). Therefore, neglecting this term we obtain

$$\theta \simeq -mt + \alpha, \tag{5.43}$$

where the constant phase α can be set to zero. Thus, the scalar field oscillates with frequency $\omega \simeq m$. After substituting $\theta \simeq -mt$ into (5.41), we obtain a readily integrated equation with solution

$$H(t) \equiv \left(\frac{\dot{a}}{a}\right) \simeq \frac{2}{3t}\left(1 - \frac{\sin(2mt)}{2mt}\right)^{-1}, \tag{5.44}$$

where a constant of integration is removed by a time shift. This solution is applicable only for $mt \gg 1$. Therefore the oscillating term is small compared to unity and the expression on the right hand side in (5.44) can be expanded in powers of $(mt)^{-1}$. Substituting (5.43) and (5.44) into the second equation in (5.40), we obtain

$$\varphi(t) \simeq \frac{\cos(mt)}{\sqrt{3\pi}\,mt}\left(1 + \frac{\sin(2mt)}{2mt}\right) + O\left((mt)^{-3}\right). \tag{5.45}$$

The time dependence of the scale factor can easily be derived by integrating (5.44):

$$a \propto t^{2/3}\left(1 - \frac{\cos(2mt)}{6m^2t^2} - \frac{1}{24m^2t^2} + O\left((mt)^{-3}\right)\right). \tag{5.46}$$

Thus, in the leading approximation (up to decaying oscillating corrections), the universe expands like a matter-dominated universe with zero pressure. This is not surprising because an oscillating homogeneous field can be thought of as a condensate of massive scalar particles with zero momenta. Although the oscillating corrections are completely negligible in the expressions for $a(t)$ and $H(t)$, they must nevertheless be taken into account when we calculate the curvature invariants. For example, the scalar curvature is

$$R \simeq -\frac{4}{3t^2}\left(1 + 3\cos(2mt) + O\left((mt)^{-1}\right)\right) \tag{5.47}$$

(compare to $R = -4/3t^2$ in a matter-dominated universe).

We have shown that inflation with a smooth graceful exit occurs naturally in models with classical massive scalar fields. If the mass is small compared to the Planck mass, the inflationary stage lasts long enough and is followed by a cold-matter-dominated stage. This cold matter, consisting of heavy scalar particles, must finally be converted to radiation, baryons and leptons. We will see later that this can easily be achieved in a variety of ways.

5.4.2 General potential: slow-roll approximation

Equation (5.24) for a massive scalar field in an expanding universe coincides with the equation for a harmonic oscillator with a friction term proportional to the Hubble parameter H. It is well known that a large friction damps the initial velocities and enforces a slow-roll regime in which the acceleration can be neglected compared to the friction term. Because for a general potential $H \propto \sqrt{\varepsilon} \sim \sqrt{V}$, we expect that for large values of V the friction term can also lead to a slow-roll inflationary stage, where $\ddot{\varphi}$ is negligible compared to $3H\dot{\varphi}$. Omitting the $\ddot{\varphi}$ term and assuming that $\dot{\varphi}^2 \ll V$, (5.24) and (5.25) simplify to

$$3H\dot{\varphi} + V_{,\varphi} \simeq 0, \quad H \equiv \left(\frac{d \ln a}{dt}\right) \simeq \sqrt{\frac{8\pi}{3} V(\varphi)}. \tag{5.48}$$

Taking into account that

$$\frac{d \ln a}{dt} = \dot{\varphi} \frac{d \ln a}{d\varphi} \simeq -\frac{V_{,\varphi}}{3H} \frac{d \ln a}{d\varphi},$$

equations (5.48) give

$$-V_{,\varphi} \frac{d \ln a}{d\varphi} \simeq 8\pi V \tag{5.49}$$

and hence

$$a(\varphi) \simeq a_i \exp\left(8\pi \int_\varphi^{\varphi_i} \frac{V}{V_{,\varphi}} d\varphi\right). \tag{5.50}$$

This approximate solution is valid only if the slow-roll conditions

$$\left|\dot{\varphi}^2\right| \ll |V|, \quad |\ddot{\varphi}| \ll 3H\dot{\varphi} \sim \left|V_{,\varphi}\right|, \tag{5.51}$$

used to simplify (5.24) and (5.25), are satisfied. With the help of (5.48), they can easily be recast in terms of requirements on the derivatives of the potential itself:

$$\left(\frac{V_{,\varphi}}{V}\right)^2 \ll 1, \quad \left|\frac{V_{,\varphi\varphi}}{V}\right| \ll 1. \tag{5.52}$$

For a power-law potential, $V = (1/n)\lambda\varphi^n$, both conditions are satisfied for $|\varphi| \gg 1$. In this case the scale factor changes as

$$a(\varphi(t)) \simeq a_i \exp\left(\frac{4\pi}{n}\left(\varphi_i^2 - \varphi^2(t)\right)\right). \tag{5.53}$$

It is obvious that the bulk of the inflationary expansion takes place when the scalar field decreases by a factor of a few from its initial value. However, we are interested mainly in the last 50–70 e-folds of inflation because they determine the structure of the universe on present observable scales. The detailed picture of the expansion during these last 70 e-folds depends on the shape of the potential only within a rather narrow interval of scalar field values.

Problem 5.7 Find the time dependence of the scale factor for the power-law potential and estimate the duration of inflation.

Problem 5.8 Verify that for a general potential V the system of equations (5.24), (5.25) can be reduced to the following first order differential equation:

$$\frac{dy}{dx} = -3(1-y^2)\left(1 + \frac{V_{,x}}{6yV}\right), \tag{5.54}$$

where

$$x \equiv \sqrt{\frac{4\pi}{3}}\varphi; \quad y \equiv \sqrt{\frac{4\pi}{3}}\frac{d\varphi}{d\ln a}.$$

Assuming that $V_{,\varphi}/V \to 0$ as $|\varphi| \to \infty$, draw the phase diagram and analyze the behavior of the solutions in different asymptotic regions. Consider separately the case of the exponential potential. What is the physical meaning of the solutions in the regions corresponding to $y > 1$?

After the end of inflation the scalar field begins to oscillate and the universe enters the stage of deceleration. Assuming that the period of oscillation is smaller than the cosmological time, let us determine the effective equation of state. Neglecting the expansion and multiplying (5.24) by φ, we obtain

$$(\varphi\dot\varphi)\dot{} - \dot\varphi^2 + \varphi V_{,\varphi} \simeq 0. \tag{5.55}$$

As a result of averaging over a period, the first term vanishes and hence $\langle\dot\varphi^2\rangle \simeq \langle\varphi V_{,\varphi}\rangle$. Thus, the averaged effective equation of state for an oscillating scalar field is

$$w \equiv \frac{p}{\varepsilon} \simeq \frac{\langle\varphi V_{,\varphi}\rangle - \langle 2V\rangle}{\langle\varphi V_{,\varphi}\rangle + \langle 2V\rangle}. \tag{5.56}$$

It follows that for $V \propto \varphi^n$ we have $w \simeq (n-2)/(n+2)$. For an oscillating massive field ($n=2$) we obtain $w \simeq 0$ in agreement with our previous result. In the case of a quartic potential ($n=4$), the oscillating field mimics an ultra-relativistic fluid with $w \simeq 1/3$.

In fact, inflation can continue even after the end of slow-roll. Considering the potential which behaves as

$$V \sim \ln(|\varphi|/\varphi_c)$$

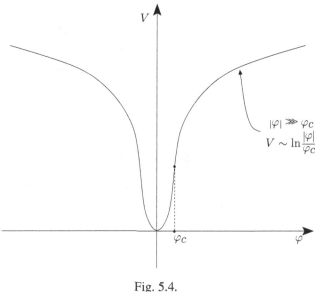

Fig. 5.4.

for $1 > |\varphi| \gg \varphi_c$ (see Figure 5.4), we infer from (5.56) that $w \to -1$. This is easy to understand. In the case of a convex potential, an oscillating scalar field spends most of the time near the potential walls where its kinetic energy is negligible and hence the main contribution to the equation of state comes from the potential term.

Problem 5.9 Which general conditions must a potential V satisfy to provide a stage of fast oscillating inflation? How long can such inflation last and why is it not very helpful for solving the initial conditions problem?

5.5 Preheating and reheating

The theory of reheating is far from complete. Not only the details, but even the overall picture of inflaton decay depend crucially on the underlying particle physics theory beyond the Standard Model. Because there are so many possible extensions of the Standard Model, it does not make much sense to study the particulars of the reheating processes in each concrete model. Fortunately we are interested only in the final outcome of reheating, namely, in the possibility of obtaining a thermal Friedmann universe. Therefore, to illustrate the physical processes which could play a major role we consider only simple toy models. The relative importance of the different reheating mechanisms cannot be clarified without an underlying particle theory. However, we will show that all of them lead to the desired result.

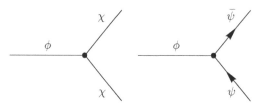

Fig. 5.5.

5.5.1 Elementary theory

We consider an inflaton field φ of mass m coupled to a scalar field χ and a spinor field ψ. Their simplest interactions are described by three-legged diagrams (Figure 5.5), which correspond to the following terms in the Lagrangian:

$$\Delta L_{int} = -g\varphi\chi^2 - h\varphi\bar{\psi}\psi. \tag{5.57}$$

We have seen that these kinds of couplings naturally arise in gauge theories with spontaneously broken symmetry, and they are enough for our illustrative purposes. To avoid a tachyonic instability we assume that $|g\varphi|$ is smaller than the squared "bare" mass m_χ^2. The decay rates of the inflaton field into $\chi\chi$ and $\bar{\psi}\psi$ pairs are determined by the coupling constants g and h respectively. They can easily be calculated and the corresponding results are cited in books on particle physics:

$$\Gamma_\chi \equiv \Gamma(\varphi \to \chi\chi) = \frac{g^2}{8\pi m}, \quad \Gamma_\psi \equiv \Gamma(\varphi \to \psi\psi) = \frac{h^2 m}{8\pi}. \tag{5.58}$$

Let us apply these results in order to calculate the decay rate of the inflaton. As we have noted, an *oscillating* homogeneous scalar field can be interpreted as a condensate of heavy particles of mass m "at rest," that is, their 3-momenta $\mathbf{k}$ are equal to zero. Keeping only the leading term in (5.45), we have

$$\varphi(t) \simeq \Phi(t)\cos(mt), \tag{5.59}$$

where $\Phi(t)$ is the slowly decaying amplitude of oscillations. The number density of φ particles can be estimated as

$$n_\varphi = \frac{\varepsilon_\varphi}{m} = \frac{1}{2m}\left(\dot{\varphi}^2 + m^2\varphi^2\right) \simeq \frac{1}{2}m\Phi^2. \tag{5.60}$$

This number is very large. For example, for $m \sim 10^{13}$ GeV, we have $n_\varphi \sim 10^{92}$ cm^{-3} immediately after the end of inflation, when $\Phi \sim 1$ in Planckian units.

One can show that quantum corrections do not significantly modify the interactions (5.57) only if $g < m$ and $h < m^{1/2}$. Therefore, for $m \ll m_{Pl}$, the highest decay rate into χ particles, $\Gamma_\chi \sim m$, is much larger than the highest possible rate for the decay into fermions, $\Gamma_\psi \sim m^2$. If $g \sim m$, then the lifetime of a φ particle is about $\Gamma_\chi^{-1} \sim m^{-1}$ and the inflaton decays after a few oscillations. Even if the coupling is not so large, the decay can still be very efficient. The reason is that the effective decay rate into bosons, Γ_{eff}, is equal to Γ_χ, given in (5.58), only if the phase space of χ particles is not densely populated by previously created χ particles. Otherwise Γ_{eff} can be made much larger by the effect of Bose condensation. This amplification of the inflaton decay is discussed in the next section.

Taking into account the expansion of the universe, the equations for the number densities of the φ and χ particles can be written as

$$\frac{1}{a^3}\frac{d(a^3 n_\varphi)}{dt} = -\Gamma_{\text{eff}} n_\varphi; \quad \frac{1}{a^3}\frac{d(a^3 n_\chi)}{dt} = 2\Gamma_{\text{eff}} n_\varphi, \qquad (5.61)$$

where the coefficient 2 in the second equation arises because *one* φ particle decays into *two* χ particles.

Problem 5.10 Substituting (5.60) into the first equation in (5.61), derive the *approximate* equation

$$\ddot{\varphi} + (3H + \Gamma_{\text{eff}})\dot{\varphi} + m^2 \varphi \simeq 0, \qquad (5.62)$$

which shows that the decay of the inflaton amplitude due to particle production may be roughly taken into account by introducing an extra friction term $\Gamma_{\text{eff}}\dot{\varphi}$. Why is this equation applicable only during the oscillatory phase?

5.5.2 Narrow resonance

The domain of applicability of elementary reheating theory is limited. Bose condensation effects become important very soon after the beginning of the inflaton decay. Because the inflaton particle is "at rest," the momenta of the two produced χ particles have the same magnitude k but opposite directions. If the corresponding states in the phase space of χ particles are already occupied, then the inflaton decay rate is enhanced by a Bose factor. The inverse decay process $\chi\chi \to \varphi$ can also take place. The rates of these processes are proportional to

$$\left|\langle n_\varphi - 1, n_{\mathbf{k}} + 1, n_{-\mathbf{k}} + 1 | \hat{a}_{\mathbf{k}}^+ \hat{a}_{-\mathbf{k}}^+ \hat{a}_\varphi^- | n_\varphi, n_{\mathbf{k}}, n_{-\mathbf{k}} \rangle\right|^2 = (n_{\mathbf{k}} + 1)(n_{-\mathbf{k}} + 1)n_\varphi$$

and

$$\left|\langle n_\varphi + 1, n_{\mathbf{k}} - 1, n_{-\mathbf{k}} - 1 | \hat{a}_\varphi^+ \hat{a}_{\mathbf{k}}^- \hat{a}_{-\mathbf{k}}^- | n_\varphi, n_{\mathbf{k}}, n_{-\mathbf{k}} \rangle\right|^2 = n_{\mathbf{k}} n_{-\mathbf{k}}(n_\varphi + 1)$$

respectively, where $\hat{a}_{\mathbf{k}}^\pm$ are the creation and annihilation operators for χ particles and $n_{\pm\mathbf{k}}$ are their occupation numbers. To avoid confusion the reader must always distinguish the occupation numbers from the number densities keeping in mind that the occupation number refers to a density per cell of volume $(2\pi)^3$ (in the Planckian units) in the phase space, while the number density is the number of particles per unit volume in the three-dimensional space. Taking into account that $n_{\mathbf{k}} = n_{-\mathbf{k}} \equiv n_k$ and $n_\varphi \gg 1$, we infer that the number densities n_φ and n_χ satisfy (5.61), where

$$\Gamma_{\text{eff}} \simeq \Gamma_\chi (1 + 2n_k). \tag{5.63}$$

Given a number density n_χ, let us calculate n_k. A φ particle "at rest" decays into two χ particles, both having energy $m/2$. Because of the interaction term (5.57), the effective squared mass of the χ particle depends on the value of the inflaton field and is equal to $m_\chi^2 + 2g\varphi(t)$. Therefore the corresponding 3-momentum of the produced χ particle is given by

$$k = \left(\left(\frac{m}{2}\right)^2 - m_\chi^2 - 2g\varphi(t) \right)^{1/2}, \tag{5.64}$$

where we assume that $m_\chi^2 + 2g\varphi \ll m^2$. The oscillating term,

$$g\varphi \simeq g\Phi \cos(mt),$$

leads to a "scattering" of the momenta in phase space. If $g\Phi \ll m^2/8$, then all particles are created within a thin shell of width

$$\Delta k \simeq m \left(\frac{4g\Phi}{m^2}\right) \ll m \tag{5.65}$$

located near the radius $k_0 \simeq m/2$ (Figure 5.6(a)). Therefore

$$n_{k=m/2} \simeq \frac{n_\chi}{(4\pi k_0^2 \Delta k)/(2\pi)^3} \simeq \frac{2\pi^2 n_\chi}{mg\Phi} = \frac{\pi^2 \Phi}{g} \frac{n_\chi}{n_\varphi}. \tag{5.66}$$

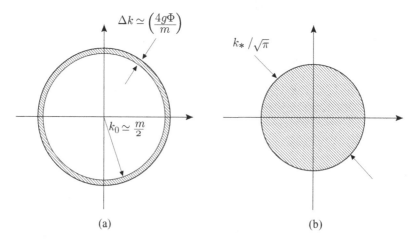

Fig. 5.6.

5.5 Preheating and reheating

The occupation numbers n_k exceed unity, and hence the Bose condensation effect is essential only if

$$n_\chi > \frac{g}{\pi^2 \Phi} n_\varphi. \tag{5.67}$$

Taking into account that at the end of inflation $\Phi \sim 1$, we infer that the occupation numbers begin to exceed unity as soon as the inflaton converts a fraction g of its energy to χ particles. The derivation above is valid only for $g\Phi \ll m^2/8$. Therefore, if $m \sim 10^{-6}$, then at most a fraction $g \sim m^2 \sim 10^{-12}$ of the inflaton energy can be transferred to χ particles in the regime where $n_k < 1$. Thus, the elementary theory of reheating, which is applicable for $n_k \ll 1$, fails almost immediately after the beginning of reheating. Given the result in (5.66), the effective decay rate (5.63) becomes

$$\Gamma_{\text{eff}} \simeq \frac{g^2}{8\pi m}\left(1 + \frac{2\pi^2 \Phi}{g}\frac{n_\chi}{n_\varphi}\right), \tag{5.68}$$

where we have used (5.58) for Γ_χ. Substituting this expression into the second equation in (5.61), we obtain

$$\frac{1}{a^3}\frac{d(a^3 n_\chi)}{dN} = \frac{g^2}{2m^2}\left(1 + \frac{2\pi^2 \Phi}{g}\frac{n_\chi}{n_\varphi}\right)n_\varphi, \tag{5.69}$$

where $N \equiv mt/2\pi$ is the number of inflaton oscillations. Let us neglect for a moment the expansion of the universe and disregard the decrease of the inflation amplitude due to particle production. In this case $\Phi = \text{const}$ and for $n_k \gg 1$ (5.69) can be easily integrated. The result is

$$n_\chi \propto \exp\left(\frac{\pi^2 g \Phi}{m^2} N\right) \propto \exp(2\pi \mu N), \tag{5.70}$$

where $\mu \equiv \pi g \Phi/(2m^2)$ is the parameter of instability.

Problem 5.11 Derive the following equation for the Fourier modes of the field χ in Minkowski space:

$$\ddot{\chi}_k + \left(k^2 + m_\chi^2 + 2g\Phi \cos mt\right)\chi_k = 0. \tag{5.71}$$

Reduce it to the well known Mathieu equation and, assuming that $m^2 \gg m_\chi^2 \geq 2|g\Phi|$, investigate the narrow parametric resonance. Determine the instability bands and the corresponding instability parameters. Compare the width of the first instability band with (5.65). Where is this band located? The minimal value of the initial amplitude of χ_k is due to vacuum fluctuations. The increase of χ_k with time can be interpreted as the production of χ particles by the external classical field φ, with

$n_\chi \propto |\chi_k|^2$. Show that in the center of the first instability band,

$$n_\chi \propto \exp\left(\frac{4\pi g \Phi}{m^2} N\right), \qquad (5.72)$$

where N is the number of oscillations. Compare this result with (5.70) and explain why they are different by a numerical factor in the exponent. Thus, Bose condensation can be interpreted as a narrow parametric resonance in the first instability band, and vice versa. Give a physical interpretation of the higher-order resonance bands in terms of particle production.

Using the results of this problem we can reduce the investigation of the inflaton decay due to the coupling

$$\Delta L_{int} = -\tfrac{1}{2}\tilde{g}^2 \varphi^2 \chi^2, \qquad (5.73)$$

to the case studied above. In fact, the equation for a massless scalar field χ, coupled to the inflaton $\varphi = \Phi \cos mt$, takes the form

$$\ddot{\chi}_k + \left(k^2 + \tilde{g}^2 \Phi^2 \cos^2 mt\right) \chi_k = 0, \qquad (5.74)$$

which coincides with (5.71) for $m_\chi^2 = 2g\Phi$ after the substitutions $\tilde{g}^2 \Phi^2 \to 4g\Phi$ and $m \to m/2$. Thus, the two problems are mathematically equivalent. Using this observation and making the corresponding replacements in (5.72), we immediately find that

$$n_\chi \propto \exp\left(\frac{\pi \tilde{g}^2 \Phi^2}{4m^2} N\right). \qquad (5.75)$$

The condition for narrow resonance is $\tilde{g}\Phi \ll m$ and the width of the first resonance band can be estimated from (5.65) as $\Delta k \sim m \left(\tilde{g}^2 \Phi^2 / m^2\right)$.

In summary, we have shown that even for a small coupling constant the elementary theory of reheating must be modified to take into account the Bose condensation effect, and that this can lead to an exponential increase of the reheating efficiency.

Problem 5.12 Taking a few concrete values for g and m, compare the results of the elementary theory with those obtained for narrow parametric resonance.

So far we have neglected the expansion of the universe, the back-reaction of the produced particles and their rescatterings. All these effects work to suppress the efficiency of the narrow parametric resonance. The expansion shifts the momenta of the previously created particles and takes them out of the resonance layer (Figure 5.6(a)). Thus, the occupation numbers relevant for Bose condensation are actually smaller than what one would expect according to the naive estimate (5.66). If the rate of supply of newly created particles in the resonance layer is smaller than the rate of their escape, then $n_k < 1$ and we can use the elementary theory of reheating. The other important effect is the decrease of the amplitude $\Phi(t)$ due to both the expansion of the universe and particle production. Because the width of the resonance layer is proportional to Φ, it becomes more and more narrow. As a result the particles can escape from this layer more easily and they do not stimulate the subsequent production of particles. The rescattering of the χ particles also suppresses the resonance efficiency by removing particles from the resonance layer. Another effect is the change of the effective inflaton mass due to the newly produced χ particles; this shifts the center of the resonance layer from its original location.

To conclude, narrow parametric resonance is very sensitive to the interplay of different complicating factors. It can be fully investigated only using numerical methods. From our analytical consideration we can only say that the inflaton field probably decays not as "slowly" as in the elementary theory, but not as "fast" as in the case of pure narrow parametric resonance.

5.5.3 Broad resonance

So far we have considered only the case of a small coupling constant. Quantum corrections to the Lagrangian are not very crucial if $g < m$ and $\tilde{g} < (m/\Phi)^{1/2}$. They can therefore be ignored when we consider inflaton decay in the strong coupling regime: $m > g > m^2/\Phi$ for the three-leg interaction and $(m/\Phi)^{1/2} > \tilde{g} > m/\Phi$ for the quartic interaction (5.73). In this case the condition for narrow resonance is not fulfilled and we cannot use the methods above. Perturbative methods fail because the higher-order diagrams, built from the elementary diagrams, give comparable contributions. Particle production can be treated only as a collective effect in which many inflaton particles participate simultaneously. We have to apply the methods of quantum field theory in an external classical background – as in Problem 5.11.

Let us consider quartic interaction (5.73). First, we neglect the expansion of the universe. For $\tilde{g}\Phi \gg m$ the mode equation (see (5.74)):

$$\ddot{\chi}_k + \omega^2(t)\chi_k = 0, \tag{5.76}$$

where

$$\omega(t) \equiv \left(k^2 + \tilde{g}^2\Phi^2\cos^2 mt\right)^{1/2}, \tag{5.77}$$

describes a broad parametric resonance. If the frequency $\omega(t)$ is a slowly varying function of time or, more precisely, $|\dot{\omega}| \ll \omega^2$, (5.76) can be solved in the

250 Inflation I: homogeneous limit

quasiclassical (WKB) approximation:

$$\chi_k \propto \frac{1}{\sqrt{\omega}} \exp\left(\pm i \int \omega dt\right). \tag{5.78}$$

In this case the number of particles, $n_\chi \sim \varepsilon_\chi/\omega$, is an adiabatic invariant and is conserved. For most of the time the condition $|\dot\omega| \ll \omega^2$ is indeed fulfilled. However, every time the oscillating inflaton vanishes at $t_j = m^{-1}(j + 1/2)\pi$, the effective mass of the χ field, proportional to $|\cos(mt)|$, vanishes. It is shortly before and after t_j that the adiabatic condition is strongly violated:

$$\frac{|\dot\omega|}{\omega^2} = \frac{m\tilde{g}^2\Phi^2 |\cos(mt)\sin(mt)|}{\left(k^2 + \tilde{g}^2\Phi^2 \cos^2(mt)\right)^{3/2}} \geq 1. \tag{5.79}$$

Considering a small time interval $\Delta t \ll m^{-1}$ in the vicinity of t_j, we can rewrite this condition as

$$\frac{\Delta t/\Delta t_*}{\left(k^2 \Delta t_*^2 + (\Delta t/\Delta t_*)^2\right)^{3/2}} \geq 1, \tag{5.80}$$

where

$$\Delta t_* \simeq (\tilde{g}\Phi m)^{-1/2} = \frac{1}{m}(\tilde{g}\Phi/m)^{-1/2}. \tag{5.81}$$

It follows that the adiabatic condition is broken only within short time intervals $\Delta t \sim \Delta t_*$ near t_j and only for modes with

$$k < k_* \simeq \Delta t_*^{-1} \simeq m(\tilde{g}\Phi/m)^{1/2}. \tag{5.82}$$

Therefore, we expect that χ particles with the corresponding momenta are created only during these time intervals. It is worth noting that the momentum of the created particle can be larger than the inflaton mass by the ratio $(\tilde{g}\Phi/m)^{1/2} > 1$; the χ particles are produced as a result of a collective process involving many inflaton particles. This is the reason why we cannot describe the broad resonance regime using the usual methods of perturbation theory.

To calculate the number of particles produced in a single inflaton oscillation we consider a short time interval in the vicinity of t_j and approximate the cosine in (5.76) by a linear function. Equation (5.76) then takes the form

$$\frac{d^2\chi_\kappa}{d\tau^2} + \left(\kappa^2 + \tau^2\right)\chi_\kappa = 0, \tag{5.83}$$

where the dimensionless wavenumber $\kappa \equiv k/k_*$ and time $\tau \equiv (t - t_j)/\Delta t_*$ have been introduced. In terms of the new variables the adiabaticity condition is broken at $|\tau| < 1$ and only for $\kappa < 1$. It is remarkable that the coupling constant $\tilde{g}$, the mass and the amplitude of the inflaton enter explicitly only in κ^2. The adiabaticity violation is largest for $k = 0$. In this case the parameters $\tilde{g}$, Φ and m drop from (5.83) and the amplitude $\chi_{\kappa=0}$ changes only by a numerical, parameter-independent factor as a result of passing through the nonadiabatic region at $|\tau| < 1$. Because the particle density n is proportional to $|\chi|^2$, its growth from one oscillation to the next can be written as

$$\left(\frac{n^{j+1}}{n^j}\right)_{k=0} = \exp(2\pi\mu_{k=0}), \tag{5.84}$$

where the instability parameter $\mu_{k=0}$ does not depend on $\tilde{g}$, Φ and m. For modes with $k \neq 0$, the parameter $\mu_{k\neq 0}$ is a function of $\kappa = k/k_*$. In this case the adiabaticity is not violated as strongly as for the $k = 0$ mode, and hence $\mu_{k\neq 0}$ is smaller than $\mu_{k=0}$. To calculate the instability parameters we have to determine the change of the amplitude χ in passing from the $\tau < -1$ region to the $\tau > 1$ region. This can be done using two independent WKB solutions of (5.83) in the asymptotic regions $|\tau| \gg 1$:

$$\chi_\pm = \frac{1}{(\kappa^2 + \tau^2)^{1/4}} \exp\left(\pm i \int \sqrt{\kappa^2 + \iota^2}\, d\iota\right) \simeq |\tau|^{-\frac{1}{2} \pm \frac{1}{2} i\kappa^2} \exp\left(\pm \frac{i\tau^2}{2}\right). \tag{5.85}$$

After passing through the nonadiabatic region the mode $A_+\chi_+$ becomes a mixture of the modes χ_+ and χ_-, that is,

$$A_+\chi_+ \to B_+\chi_+ + C_+\chi_-, \tag{5.86}$$

where A_+, B_+ and C_+ are the complex constant coefficients. Similarly, for the mode $A_-\chi_-$, we have

$$A_-\chi_- \to B_-\chi_- + C_-\chi_+. \tag{5.87}$$

Drawing an analogy with the scattering problem for the inverse parabolic potential, we note that the mixture arises due to an overbarrier reflection of the wave. The reflection is most efficient for the waves with $k = 0$ which "touch" the top of the barrier.

The quasi-classical solution is valid in the complex plane for $|\tau| \gg 1$. Traversing the appropriate contour $\tau = |\tau| e^{i\varphi}$ in the complex plane from $\tau \ll -1$ to $\tau \gg 1$, we infer from (5.85), (5.86) and (5.87) that

$$B_\pm = \mp i e^{-\frac{\pi}{2}\kappa^2} A_\pm. \tag{5.88}$$

The coefficients $C_\pm$ are not determined in this method. To find them we use the Wronskian

$$W \equiv \dot{\chi}\chi^* - \chi\dot{\chi}^*, \tag{5.89}$$

where χ is an arbitrary complex solution of (5.83). Taking the derivative of W and using (5.83) to express $\ddot{\chi}$ in terms of χ, we find

$$\dot{W} = 0 \tag{5.90}$$

and hence $W = \text{const}$. From this we infer that the coefficients A, B and C in (5.86) and (5.87) satisfy the "probability conservation" condition

$$|C_\pm|^2 - |B_\pm|^2 = |A_\pm|^2. \tag{5.91}$$

Substituting B from (5.88), we obtain

$$C_\pm = \sqrt{1 + e^{-\pi \kappa^2}}\, |A_\pm|\, e^{i\alpha_\pm}, \tag{5.92}$$

where the phases $\alpha_\pm$ remain undetermined.

At $|\tau| \gg 1$ the modes of field χ satisfy the harmonic oscillator equation with a slowly changing frequency $\omega \propto |\tau|$. In quantum field theory the occupation number n_k in the expression for the energy of the harmonic oscillator,

$$\varepsilon_k = \omega(n_k + 1/2), \tag{5.93}$$

is interpreted as the number of particles in the corresponding mode k. In the adiabatic regime ($|\tau| \gg 1$) this number is conserved and it changes only when the adiabatic condition is violated. Let us consider an arbitrary initial mixture of the modes χ_+ and χ_-. After passing through the nonadiabatic region at $t \sim t_j$, it changes as

$$\chi^j = A_+ \chi_+ + A_- \chi_- \to \chi^{j+1} = (B_+ + C_-)\chi_+ + (B_- + C_+)\chi_-. \tag{5.94}$$

Taking into account that

$$n + \frac{1}{2} = \frac{\varepsilon}{\omega} \simeq \omega |\chi|^2, \tag{5.95}$$

we see that as a result of this passage the number of particles in the mode k increases

$$\left(\frac{n^{j+1} + 1/2}{n^j + 1/2}\right)_k \simeq \frac{\omega |\chi^{j+1}|^2}{\omega |\chi^j|^2} \simeq \frac{|B_+ + C_-|^2 + |B_- + C_+|^2}{|A_+|^2 + |A_-|^2} \tag{5.96}$$

times, where we have averaged $|\chi|^2$ over the time interval $m^{-1} > t > \omega^{-1}$. With B and C from (5.88) and (5.92), this expression becomes

$$\left(\frac{n^{j+1} + 1/2}{n^j + 1/2}\right)_k \simeq \left(1 + 2e^{-\pi \kappa^2}\right) + \frac{4|A_-||A_+|}{|A_+|^2 + |A_-|^2} \cos\theta\, e^{-\frac{\pi}{2}\kappa^2} \sqrt{1 + e^{-\pi \kappa^2}}. \tag{5.97}$$

Problem 5.13 Verify (5.97) and explain the origin of the phase θ. (*Hint* Derive and use the relation Re $B_+C_- =$ Re B_-C_+, which follows from the "probability conservation" condition for (5.94).)

In the vacuum initial state $n_k = 0$ but the amplitude of the field χ does not vanish because of the existence of vacuum fluctuations; we have $|A_+^0|^2 \neq 0$ and $|A_-^0|^2 = 0$. It follows from the "probability conservation" condition that

$$|A_+|^2 - |A_-|^2 = |A_+^0|^2 \qquad (5.98)$$

at every moment of time. This means that as a result of particle production the coefficients $|A_+|^2$ and $|A_-|^2$ grow by the same amount. When $|A_+|$ becomes much larger than $|A_+^0|$ we have $|A_+| \simeq |A_-|$. Taking this into account and beginning in the vacuum state, we find from (5.97) that after $N \gg 1$ inflaton oscillations the particle number in mode k is

$$n_k \simeq \tfrac{1}{2} \exp(2\pi \mu_k N), \qquad (5.99)$$

where the instability parameter is given by

$$\mu_k \simeq \frac{1}{2\pi} \ln\!\left(1 + 2e^{-\pi\kappa^2} + 2\cos\theta\, e^{-\frac{\pi}{2}\kappa^2} \sqrt{1 + e^{-\pi\kappa^2}}\right). \qquad (5.100)$$

This parameter takes its maximal value

$$\mu_k^{\max} = \pi^{-1} \ln\!\left(1 + \sqrt{2}\right) \simeq 0.28$$

for $k = 0$ and $\theta = 0$. In the interval $-\pi < \theta < \pi$ we find that $\mu_{k=0}$ is positive if $3\pi/4 > \theta > -3\pi/4$ and negative otherwise. Thus, assuming random θ, we conclude that the particle number in every mode changes stochastically. However, if all θ are equally probable, then the number of particles increases three quarters of the time and therefore it also increases on average, in agreement with entropic arguments. The net instability parameter, characterizing the average growth in particle number, is obtained by skipping the $\cos\theta$ term in (5.100):

$$\bar{\mu}_k \simeq \frac{1}{2\pi} \ln\!\left(1 + 2e^{-\pi\kappa^2}\right). \qquad (5.101)$$

With slight modifications the results above can be applied to an expanding universe. First of all we note that the expansion randomizes the phases θ and hence the effective instability parameter is given by (5.101). For particles with physical momenta $k < k_*/\sqrt{\pi}$, the instability parameter $\bar{\mu}_k$ can be roughly estimated by its value at the center of the instability region, $\bar{\mu}_{k=0} = (\ln 3)/2\pi \simeq 0.175$. To understand how the expansion can influence the efficiency of broad resonance, it is again helpful to use the phase space picture. The particles created in the broad resonance regime occupy the entire sphere of radius $k_*/\sqrt{\pi}$ in phase space (see

Figure 5.6(b)). During the passage through the nonadiabatic region the number of particles in every cell of the sphere, and hence the total number density, increases on average $\exp(2\pi \times 0.175) \simeq 3$ times. At the stage when inflaton energy is still dominant, the physical momentum of the created particle decreases in inverse proportion to the scale factor $(k \propto a^{-1})$, while the radius of the sphere shrinks more slowly, namely, as $\Phi^{1/2} \propto t^{-1/2} \propto a^{-3/4}$. As a result, the created particles move away from the boundary of the sphere towards its center where they participate in the next "act of creation," enhancing the probability by a Bose factor. Furthermore, expansion also makes broad resonance less sensitive to rescattering and back-reaction effects. These two effects influence the resonance efficiency by removing those particles which are located near the boundary of the resonance sphere. Because expansion moves particles away from this region, the impact of these effects is diminished. Thus, in contrast to the narrow resonance case, expansion stabilizes broad resonance and at the beginning of reheating it can be realized in its pure form.

Taking into account that the initial volume of the resonance sphere is about

$$k_*^3 \simeq m^3(\tilde{g}\Phi_0/m)^{3/2},$$

we obtain the following estimate for the ratio of the particle number densities after N inflaton oscillations:

$$\frac{n_\chi}{n_\varphi} \sim \frac{k_*^3 \exp(2\pi \bar{\mu}_{k=0} N)}{m\Phi_0^2} \sim m^{1/2}\tilde{g}^{3/2} \cdot 3^N, \quad (5.102)$$

where $\Phi_0 \sim O(1)$ is the value of the inflaton amplitude after the end of inflation. Since in the adiabatic regime the effective mass of the χ particles is of order $\tilde{g}\Phi$, where Φ decreases in inverse proportion to N, we also obtain an estimate for the ratio of the energy densities:

$$\frac{\varepsilon_\chi}{\varepsilon_\varphi} \sim \frac{m_\chi n_\chi}{m n_\varphi} \sim m^{-1/2} g^{5/2} N^{-1} 3^N. \quad (5.103)$$

The formulae above fail when the energy density of the created particles begins to exceed the energy density stored in the inflaton field. In fact, at this time, the amplitude $\Phi(t)$ begins to decrease very quickly because of the very efficient energy transfer from the inflaton to the χ particles. Broad resonance is certainly over when $\Phi(t)$ drops to the value $\Phi_r \sim m/\tilde{g}$, and we enter the narrow resonance regime. For the coupling constant $\sqrt{m} > \tilde{g} > O(1)m$, the number of the inflaton oscillation N_r in the broad resonance regime can be roughly estimated using the condition $\varepsilon_\chi \sim \varepsilon_\varphi$:

$$N_r \sim (0.75–2)\log_3 m^{-1}. \quad (5.104)$$

As an example, if $m \simeq 10^{13}$ GeV, we have $N_r \simeq 10\text{–}25$ for a wide range of the coupling constants $10^{-3} > \tilde{g} > 10^{-6}$. Taking into account that the total energy decays as $m^2(\Phi_0/N)^2$, we obtain

$$\frac{\varepsilon_\varphi}{\varepsilon_\chi + \varepsilon_\varphi} \sim \frac{m^2 \Phi_r^2}{m^2(\Phi_0/N_r)^2} \sim N_r^2 \left(\frac{m}{\tilde{g}\Phi_0}\right)^2, \qquad (5.105)$$

that is, the energy still stored in the inflaton field at the end of broad resonance is only a small fraction of the total energy. In particular, for $m \simeq 10^{13}$ GeV, this ratio varies in the range 10^{-6}–$O(1)$ depending on the coupling constant $\tilde{g}$.

Problem 5.14 Investigate inflaton decay due to the three-leg interaction in the strong coupling regime: $m > g > m^2/\Phi$.

5.5.4 Implications

It follows from the above considerations that broad parametric resonance can play a very important role in the preheating phase. During only 15–25 oscillations of the inflaton, it can convert most of the inflaton energy into other scalar particles. The most interesting aspect of this process is that the effective mass and the momenta of the particles produced can exceed the inflaton mass. For example, for $m \simeq 10^{14}$ GeV, the effective mass $m_\chi^{\text{eff}} = \tilde{g}\Phi |\cos(mt)|$ can be as large as 10^{16} GeV. Therefore, if the χ particles are coupled to bosonic and fermionic fields heavier than the inflaton, then the inflaton may indirectly decay into these heavy particles. This brings Grand Unification scales back into play. For instance, even if the inflation ends at low energy scales, preheating may rescue the GUT baryogenesis models. Another potential outcome of the above mechanism is the far-from-equilibrium production of topological defects after inflation. Obviously their numbers must not conflict with observations and this leads to cosmological bounds on admissible theories.

If, after the period of broad resonance, the slightest amount of the inflaton remained – given by (5.105) – it would be a cosmological disaster. Since the inflaton particles are nonrelativistic, if they were present in any substantial amount, they would soon dominate and leave us with a cold universe. Fortunately, these particles should easily decay in the subsequent narrow resonance regime or as a result of elementary particle decay. These decay channels thus become necessary ingredients of the reheating theory.

The considerations of this section do not constitute a complete theory of reheating. We have studied only elementary processes which could play a role in producing a hot Friedmann universe. The final outcome of reheating must be matter in thermal equilibrium. The particles which are produced in the preheating processes are initially in a highly nonequilibrium state. Numerical calculations show that as a result of their scatterings they quickly reach local thermal equilibrium. Parameterizing the total preheating and reheating time in terms of the inflaton oscillations number N_T, we obtain the following estimate for the reheating temperature:

$$T_R \sim \frac{m^{1/2}}{N_T^{1/2} N^{1/4}}, \qquad (5.106)$$

where N is the effective number degrees of freedom of the light fields at $T \sim T_R$. Assuming that $N_T \sim 10^6$, and taking $N \sim 10^2$ and $m \simeq 10^{13}$ GeV, we obtain $T_R \sim 10^{12}$ GeV. This does not mean, however, that we can ignore physics beyond this scale. As we have already pointed out, nonequilibrium preheating processes can play a nontrivial role.

Reheating is an important ingredient of inflationary cosmology. We have seen that there is no general obstacle to arranging successful reheating. A particle theory should be tested on its ability to realize reheating in combination with baryogenesis. In this way, cosmology enables us to preselect realistic particle physics theories beyond the Standard Model.

5.6 "Menu" of scenarios

All we need for successful inflation is a scalar condensate satisfying the slow-roll conditions. Building concrete scenarios then becomes a "technical" problem. Involving two or more scalar condensates, and assuming them to be equally relevant during inflation, extends the number of possibilities, but simultaneously diminishes the predictive power of inflation. This especially concerns cosmological perturbations, which are among the most important robust predictions of inflation. Because inflation can be falsified experimentally (or more accurately, observationally) only if it makes such predictions, we consider only simple scenarios with a single inflaton component. Fortunately all of them lead to very similar predictions which differ only slightly in the details. This makes a unique scenario, the one actually realized in nature, less important. The situation here is very different from particle physics, where the concrete models are as important as the ideas behind them. This does not mean we do not need the correct scenario; if one day it becomes available, we will be able to verify more delicate predictions of inflation. However, even in the absence of the true scenario, we can nonetheless verify observationally the most important predictions of the stage of cosmic acceleration. The purpose of this section is to give the reader a very brief guide to the "menu of scenarios" discussed in the literature.

Inflaton candidates The first question which naturally arises is "what is the most realistic candidate for the inflaton field?". There are many because the only requirement is that this candidate imitates a scalar condensate in the slow-roll regime. This

5.6 "Menu" of scenarios

can be achieved by a fundamental scalar field or by a fermionic condensate described in terms of an effective scalar field. This, however, does not exhaust all possibilities. The scalar condensate can also be imitated entirely within the theory of gravity itself. Einstein gravity is only a low curvature limit of some more complicated theory whose action contains higher powers of the curvature invariants, for example,

$$S = -\frac{1}{16\pi} \int \left(R + \alpha R^2 + \beta R_{\mu\nu} R^{\mu\nu} + \gamma R^3 + \cdots \right) \sqrt{-g} d^4x. \tag{5.107}$$

The quadratic and higher-order terms can be either of fundamental origin or they can arise as a result of vacuum polarization. The corresponding dimensional coefficients in front of these terms are likely of Planckian size. The theory with (5.107) can provide us with inflation. This can easily be understood. Einstein gravity is the only metric theory in four dimensions where the equations of motion are second order. Any modification of the Einstein action introduces higher-derivative terms. This means that, in addition to the gravitational waves, the gravitational field has extra degrees of freedom including, generically, a spin 0 field.

Problem 5.15 Consider a gravity theory with metric $g_{\mu\nu}$ and action

$$S = \frac{1}{16\pi} \int f(R) \sqrt{-g} d^4x, \tag{5.108}$$

where $f(R)$ is an arbitrary function of the scalar curvature R. Derive the following equations of motion:

$$\frac{\partial f}{\partial R} R^\mu_\nu - \frac{1}{2} \delta^\mu_\nu f + \left(\frac{\partial f}{\partial R} \right)^{;\alpha}_{;\alpha} \delta^\mu_\nu - \left(\frac{\partial f}{\partial R} \right)^{;\mu}_{;\nu} = 0. \tag{5.109}$$

Verify that under the conformal transformation $g_{\mu\nu} \to \tilde{g}_{\mu\nu} = F g_{\mu\nu}$, the Ricci tensor and the scalar curvature transform as

$$R^\mu_\nu \to \tilde{R}^\mu_\nu = F^{-1} R^\mu_\nu - F^{-2} F^{;\mu}_{;\nu} - \frac{1}{2} F^{-2} F^{;\alpha}_{;\alpha} \delta^\mu_\nu + \frac{3}{2} F^{-3} F_{;\nu} F^{;\mu}, \tag{5.110}$$

$$R \to \tilde{R} = F^{-1} R - 3 F^{-2} F^{;\alpha}_{;\alpha} + \frac{3}{2} F^{-3} F_{;\alpha} F^{;\alpha}. \tag{5.111}$$

Introduce the "scalar field"

$$\varphi \equiv \sqrt{\frac{3}{16\pi}} \ln F(R), \tag{5.112}$$

and show that the equations

$$\tilde{R}^\mu_\nu - \frac{1}{2} \tilde{R} \delta^\mu_\nu = 8\pi \tilde{T}^\mu_\nu(\varphi), \tag{5.113}$$

coincide with (5.109) if we set $F = \partial f/\partial R$ and take the following potential for the scalar field:

$$V(\varphi) = \frac{1}{16\pi} \frac{f - R\partial f/\partial R}{(\partial f/\partial R)^2}. \tag{5.114}$$

Problem 5.16 Study the inflationary solutions in R^2 gravity:

$$S = -\frac{1}{16\pi} \int \left(R - \frac{1}{6M^2} R^2 \right) \sqrt{-g} d^4x. \tag{5.115}$$

What is the physical meaning of the constant M?

Thus, the higher derivative gravity theory is conformally equivalent to Einstein gravity with an extra scalar field. If the scalar field potential satisfies the slow-roll conditions, then we have an inflationary solution in the conformal frame for the metric $\tilde{g}_{\mu\nu}$. However, one should not confuse the conformal metric with the original physical metric. They generally describe manifolds with different geometries and the final results must be interpreted in terms of the original metric. In our case the use of the conformal transformation is a mathematical tool which simply allows us to reduce the problem to one we have studied before. The conformal metric is related to the physical metric $g_{\mu\nu}$ by a factor F, which depends on the curvature invariants; it does not change significantly during inflation. Therefore, we also have an inflationary solution in the original physical frame.

So far we have been considering inflationary solutions due to the potential of the scalar field. However, inflation can be realized even without a potential term. It can occur in Born–Infeld-type theories, where the action depends nonlinearly on the kinetic energy of the scalar field. These theories do not have higher-derivative terms, but they have some other peculiar properties.

Problem 5.17 Consider a scalar field with action

$$S = \int p(X, \varphi) \sqrt{-g} d^4x, \tag{5.116}$$

where p is an arbitrary function of φ and $X \equiv \frac{1}{2}(\partial_\mu \varphi \partial^\mu \varphi)$. Verify that the energy–momentum tensor for this field can be written in the form

$$T^\mu_\nu = (\varepsilon + p) u^\mu u_\nu - p \delta^\mu_\nu, \tag{5.117}$$

where the Lagrangian p plays the role of the effective pressure and

$$\varepsilon = 2X \frac{\partial p}{\partial X} - p, \quad u_\nu = \frac{\partial_\nu \varphi}{\sqrt{2X}}. \tag{5.118}$$

If the Lagrangian p satisfies the condition $X \partial p/\partial X \ll p$ for some range of X and φ, then the equation of state is $p \approx -\varepsilon$ and we have an inflationary solution. Why

5.6 "Menu" of scenarios

is inflation not satisfactory if p depends only on X? Consider a general $p(X, \varphi)$ without an explicit potential term, that is, $p \to 0$ when $X \to 0$. Formulate the conditions which this function must satisfy to provide us with a slow-roll inflationary stage and a graceful exit. The inflationary scenario based on the nontrivial dependence of the Lagrangian on the kinetic term is called k inflation.

Scenarios The simplest inflationary scenarios can be subdivided into three classes. They correspond to the usual scalar field with a potential, higher-derivative gravity and k inflation. The cosmological consequences of scenarios from the different classes are almost indistinguishable — they can exactly imitate each other. Within each class, however, we can try to make further distinctions by addressing the questions: (a) what was before inflation and (b) how does a graceful exit to a Friedmann stage occur? For our purpose it will be sufficient to consider only the simplest case of a scalar field with canonical kinetic energy. The potential can have different shapes, as shown in Figure 5.7. The three cases presented correspond to the so-called old, new and chaotic inflationary scenarios. The first two names refer to their historical origins.

Old inflation (see Figure 5.7(a)) assumes that the scalar field arrives at the local minimum of the potential at $\varphi = 0$ as a result of a supercooling of the initially hot universe. After that the universe undergoes a stage of accelerated expansion with a subsequent graceful exit via bubble nucleation. It was clear from the very beginning that this scenario could not provide a successful graceful exit because all the energy released in a bubble is concentrated in its wall and the bubbles have no chance to collide. This difficulty was avoided in the new inflationary scenario, a scenario similar to a successful model in higher-derivative gravity which had previously been invented.

New inflation is based on a Coleman–Weinberg type potential (Figure 5.7(b)). Because the potential is very flat and has a maximum at $\varphi = 0$, the scalar field

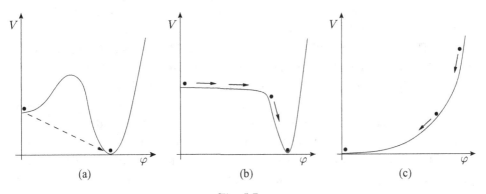

Fig. 5.7.

escapes from the maximum not via tunneling, but due to the quantum fluctuations. It then slowly rolls towards the global minimum where the energy is released homogeneously in the whole space. Originally the pre-inflationary state of the universe was taken to be thermal so that the symmetry was restored due to thermal corrections. This was a justification for the initial conditions of the scalar field. Later it was realized that the thermal initial state of the universe is quite unlikely, and so now the original motivation for the initial conditions in the new inflationary model seems to be false. Instead, the universe might be in a "self-reproducing" regime (for more details see Section 8.5).

Chaotic inflation is the name given to the broadest possible class of potentials satisfying the slow-roll conditions (Figure 5.7(c)). We have considered it in detail in the previous sections. The name *chaotic* is related to the possibility of having almost arbitrary initial conditions for the scalar field. To be precise, this field must initially be larger than the Planckian value but it is otherwise arbitrary. Indeed, it could have varied from one spatial region to another and, as a result, the universe would have a very complicated global structure. It could be very inhomogeneous on scales much larger than the present horizon and extremely homogeneous on "small" scales corresponding to the observable domain. We will see in Section 8.5 that in the case of chaotic inflation, quantum fluctuations lead to a self-reproducing universe.

Since chaotic inflation encompasses so many potentials, one might think it worthwhile to consider special cases, for example, an exponential potential. For an exponential potential, if the slow-roll conditions are satisfied once, they are always satisfied. Therefore, it describes (power-law) inflation without a graceful exit. To arrange a graceful exit we have to "damage" the potential. For two or more scalar fields the number of options increases. Thus it is not helpful here to go into the details of the different models.

In the absence of the underlying fundamental particle theory, one is free to play with the potentials and invent more new scenarios. In this sense the situation has changed since the time the importance of inflation was first realized. In fact, in the 1980s many people considered inflation a useful application of the Grand Unified Theory that was believed to be known. Besides solving the initial conditions problem, inflation also explained why we do not have an overabundance of the monopoles that are an inevitable consequence of a Grand Unified Theory. Either inflation ejects all previously created monopoles, leaving less than one monopole per present horizon volume, or the monopoles are never produced. The same argument applies to the heavy stable particles that could be overproduced in the state of thermal equilibrium at high temperatures. Many authors consider the solution of the monopole and heavy particle problems to be as important as a solution of the initial conditions problem. We would like to point out, however, that the initial conditions problem is posed to us by nature, while the other problems are, at present, not more than internal problems of theories beyond the Standard Model. By solving these extra problems, inflation opens the door to theories that would otherwise be prohibited by cosmology. Depending on one's attitude, this is either a useful or damning achievement of inflation.

De Sitter solution and inflation The last point we would like to make concerns the role of a cosmological constant and a pure de Sitter solution for inflation. We have already said that the pure de Sitter solution cannot provide us with a model with a graceful exit. Even the notion of expansion is not unambiguously defined in de Sitter space. We saw in Section 1.3.6 that this space has the same symmetry group as Minkowski space. It is spatially homogeneous and time-translation-invariant. Therefore any space-like surface is a hypersurface of constant energy. To characterize an expansion we can use not only $k = 0, \pm 1$ Friedmann coordinates but also, for example, "static coordinates" (see Problem 2.7), which describe an expanding space outside the event horizon. In all these cases the 3-geometries of constant time hypersurfaces are very different. These differences, however, simply characterize the different slicings of the perfectly symmetrical space and there is no obvious preferable choice for the coordinates.

It is important therefore that inflation is never realized by a pure de Sitter solution. There must be deviations from the vacuum equation of state, which finally determine the "hypersurface" of transition to the hot universe. The de Sitter universe is still, however, a very useful zeroth order approximation for nearly all inflationary models. In fact, the effective equation of state must satisfy the condition $\varepsilon + 3p < 0$ for at least 75 e-folds. This is generally possible only if during most of the time we have $p \approx -\varepsilon$ to a rather high accuracy. Therefore, one can use the language of constant time hypersurfaces defined in various coordinate systems in de Sitter space. Our earlier considerations show that the transition from inflation to the Friedmann universe occurs along a hypersurface of constant time in the expanding isotropic a coordinates ($\eta = $ const), but not along a $r = $ const hypersurface in the "static coordinates." The next question is out of the three possible isotropic coordinate systems ($k = 0, \pm 1$), which must be used to match the de Sitter space to the Friedmann universe? Depending on the answer to this question, we obtain flat, open or closed Friedmann universes. It turns out, however, that this answer seems not to be relevant for the observable domain of the universe. In fact, if inflation lasts more than 75 e-folds, the observable part of the universe corresponds only to a tiny piece of the matched global conformal diagrams for de Sitter and Friedmann universes. This piece is located near the upper border of the conformal diagram for

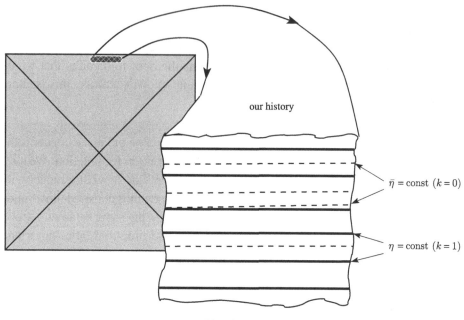

Fig. 5.8.

de Sitter space and the lower border for the flat, open or closed Friedmann universes (Figure 5.8), where the difference between the hypersurfaces of constant time for flat, open or closed cases is negligibly small. After a graceful exit we obtain a very large domain of the Friedmann universe with incredibly small flatness and this domain covers all present observable scales. The global structure of the universe on scales much larger than the present horizon is not relevant for an observer — at least not for the next 100 billion years. In Part II we will see that the issue of the global structure is complicated by quantum fluctuations. These fluctuations are amplified during inflation and as a result the hypersurface of transition has "wrinkles." The wrinkles are rather small on scales corresponding to the observable universe but they become huge on the very large scales. Hence, globally the universe is very different from the Friedmann space and the question about the spatial curvature of the whole universe no longer makes sense. It also follows that the global properties of an exact de Sitter solution have no relevance for the real physical universe.

Part II

Inhomogeneous universe

6
Gravitational instability in Newtonian theory

Measurements of the cosmic microwave background tell us that the universe was very homogeneous and isotropic at the time of recombination. Today, however, the universe has a well developed nonlinear structure. This structure takes the form of galaxies, clusters and superclusters of galaxies, and, on larger scales, of voids, sheets and filaments of galaxies. Deep redshift surveys show, however, that when averaged over a few hundred megaparsecs, the inhomogeneities in the density distribution remain small. The simple explanation as to how nonlinear structure could develop from small initial perturbations is based on the fact of gravitational instability.

Gravitational instability is a natural property of gravity. Matter is attracted to high-density regions, thus amplifying already existing inhomogeneities. To ensure that the small initial inhomogeneities present at recombination produce the nonlinear structure observed today, we have to study how fast they grow in an expanding universe. The complete general relativistic analysis of gravitational instability is rather involved and the physical interpretation of the results is not always straightforward. For this reason we develop the theory of gravitational instability in several steps.

In this chapter we consider gravitational instability in the Newtonian theory of gravity. The results derived in this theory are applicable only to nonrelativistic matter on scales not exceeding the Hubble horizon. First, we find out how small inhomogeneities grow in a nonexpanding universe (Jeans theory). The main purpose here is to determine which types of perturbations can exist in homogeneous, isotropic media, and to introduce methods to analyze them. Although the formulae describing the rate of instability in a nonexpanding universe are not very useful, the results obtained help us gain a solid intuitive understanding of the behavior of perturbations. Next, we consider linear perturbations in an expanding universe. This is not only a useful exercise, but a realistic theory describing the growth of inhomogeneities on subhorizon scales after recombination. We apply this theory to study the rate of instability in a matter-dominated universe and then see how a smooth unclustered energy component, like radiation or vacuum energy, influences the growth of inhomogeneities in the cold matter component. Finally, we derive a few exact solutions which describe the behavior of perturbations with certain spatial geometrical symmetries into the nonlinear regime. Based on these solutions, we are able to explain the general features of the matter distribution on nonlinear scales.

6.1 Basic equations

On large scales matter can be described in a perfect fluid approximation. This means that at any given moment of time it can be completely characterized by the energy density distribution $\varepsilon(\mathbf{x}, t)$, the entropy per unit mass $S(\mathbf{x}, t)$, and the vector field of 3-velocities $\mathbf{V}(\mathbf{x}, t)$. These quantities satisfy the hydrodynamical equations and we begin with a brief reminder of their derivation.

Continuity equation If we consider a *fixed volume element* ΔV in Euler (noncomoving) coordinates $\mathbf{x}$, then the rate of change of its mass can be written as

$$\frac{dM(t)}{dt} = \int_{\Delta V} \frac{\partial \varepsilon(\mathbf{x}, t)}{\partial t} dV. \tag{6.1}$$

On the other hand, this rate is entirely determined by the flux of matter through the surface surrounding the volume:

$$\frac{dM(t)}{dt} = -\oint \varepsilon \mathbf{V} \cdot d\boldsymbol{\sigma} = -\int_{\Delta V} \nabla(\varepsilon \mathbf{V}) dV. \tag{6.2}$$

These two expressions are consistent only if

$$\frac{\partial \varepsilon}{\partial t} + \nabla(\varepsilon \mathbf{V}) = 0. \tag{6.3}$$

Euler equations The acceleration $\mathbf{g}$ of a small *matter element* of mass ΔM is determined by the gravitational force

$$\mathbf{F}_{gr} = -\Delta M \cdot \nabla \phi, \tag{6.4}$$

where ϕ is the gravitational potential, and by the pressure p:

$$\mathbf{F}_{pr} = -\oint p \cdot d\boldsymbol{\sigma} = -\int_{\Delta V} \nabla p \, dV \simeq -\nabla p \cdot \Delta V. \tag{6.5}$$

With

$$\mathbf{g} \equiv \frac{d\mathbf{V}(\mathbf{x}(t), t)}{dt} = \left(\frac{\partial \mathbf{V}}{\partial t}\right)_x + \frac{dx^i(t)}{dt}\left(\frac{\partial \mathbf{V}}{\partial x^i}\right) = \frac{\partial \mathbf{V}}{\partial t} + (\mathbf{V} \cdot \nabla)\mathbf{V}, \tag{6.6}$$

Newton's force law

$$\Delta M \cdot \mathbf{g} = \mathbf{F}_{gr} + \mathbf{F}_{pr} \tag{6.7}$$

becomes the Euler equations

$$\frac{\partial \mathbf{V}}{\partial t} + (\mathbf{V} \cdot \boldsymbol{\nabla}) \mathbf{V} + \frac{\boldsymbol{\nabla} p}{\varepsilon} + \boldsymbol{\nabla} \phi = 0. \tag{6.8}$$

Conservation of entropy Neglecting dissipation, the entropy of a *matter element* is conserved:

$$\frac{dS(\mathbf{x}(t), t)}{dt} = \frac{\partial S}{\partial t} + (\mathbf{V} \cdot \boldsymbol{\nabla}) S = 0. \tag{6.9}$$

Poisson equation Finally, the equation which determines the gravitational potential is the well known Poisson equation,

$$\Delta \phi = 4\pi G \varepsilon. \tag{6.10}$$

Equations (6.3), (6.8)–(6.10), taken together with the equation of state

$$p = p(\varepsilon, S), \tag{6.11}$$

form a complete set of seven equations which, in principle, allows us to determine the seven unknown functions ε, $\mathbf{V}$, S, ϕ, p. Note that only the first five equations contain first time derivatives. Hence the most general solution of these equations should depend on five constants of integration which in our case are five arbitrary functions of the spatial coordinates $\mathbf{x}$. The hydrodynamical equations are nonlinear and in general it is not easy to find their solutions. However, to study the behavior of *small* perturbations around a homogeneous, isotropic background, it is appropriate to linearize them.

6.2 Jeans theory

Let us first consider a static nonexpanding universe, assuming the homogeneous, isotropic background with constant, time-independent matter density: $\varepsilon_0(t, \mathbf{x}) = $ const. This assumption is in obvious contradiction with the hydrodynamical equations. In fact, the energy density remains unchanged only if the matter is at rest and the gravitational force, $F \propto \boldsymbol{\nabla} \phi$, vanishes. But then the Poisson equation $\Delta \phi = 4\pi G \varepsilon_0$ is not satisfied. This inconsistency can, in principle, be avoided if we consider a static Einstein universe, where the gravitational force of the matter is compensated by the "antigravitational" force of an appropriately chosen cosmological constant.

Slightly disturbing the matter distribution, we have:

$$\varepsilon(\mathbf{x}, t) = \varepsilon_0 + \delta\varepsilon(\mathbf{x}, t), \quad \mathbf{V}(\mathbf{x}, t) = \mathbf{V}_0 + \delta\mathbf{v} = \delta\mathbf{v}(\mathbf{x}, t),$$
$$\phi(\mathbf{x}, t) = \phi_0 + \delta\phi(\mathbf{x}, t), \quad S(\mathbf{x}, t) = S_0 + \delta S(\mathbf{x}, t), \quad (6.12)$$

where $\delta\varepsilon \ll \varepsilon_0$, etc. The pressure is equal to

$$p(\mathbf{x}, t) = p(\varepsilon_0 + \delta\varepsilon, S_0 + \delta S) = p_0 + \delta p(\mathbf{x}, t), \quad (6.13)$$

and in linear approximation its perturbation δp can be expressed in terms of the energy density and entropy perturbations as

$$\delta p = c_s^2 \delta\varepsilon + \sigma \delta S. \quad (6.14)$$

Here $c_s^2 \equiv (\partial p/\partial \varepsilon)_S$ is the square of the speed of sound and $\sigma \equiv (\partial p/\partial S)_\varepsilon$. For nonrelativistic matter ($p \ll \varepsilon$), the speed of sound as well as the velocities $\delta\mathbf{v}$ are much less than the speed of light.

Substituting (6.12) and (6.14) into (6.3), (6.8)–(6.10) and keeping only the terms which are linear in the perturbations, we obtain:

$$\frac{\partial \delta\varepsilon}{\partial t} + \varepsilon_0 \nabla(\delta\mathbf{v}) = 0, \quad (6.15)$$

$$\frac{\partial \delta\mathbf{v}}{\partial t} + \frac{c_s^2}{\varepsilon_0}\nabla\delta\varepsilon + \frac{\sigma}{\varepsilon_0}\nabla\delta S + \nabla\delta\phi = 0, \quad (6.16)$$

$$\frac{\partial \delta S}{\partial t} = 0, \quad (6.17)$$

$$\Delta\delta\phi = 4\pi G \delta\varepsilon. \quad (6.18)$$

Equation (6.17) has a simple general solution

$$\delta S(\mathbf{x}, t) = \delta S(\mathbf{x}), \quad (6.19)$$

which states that the entropy is an arbitrary time-independent function of the spatial coordinates.

Taking the divergence of (6.16) and using the continuity and Poisson equations to express $\nabla\delta\mathbf{v}$ and $\Delta\delta\phi$ in terms of $\delta\varepsilon$, we obtain

$$\frac{\partial^2 \delta\varepsilon}{\partial t^2} - c_s^2 \Delta\delta\varepsilon - 4\pi G\varepsilon_0 \delta\varepsilon = \sigma \Delta\delta S(\mathbf{x}). \quad (6.20)$$

This is a closed, linear equation for $\delta\varepsilon$, where the entropy perturbation serves as a given source.

6.2.1 Adiabatic perturbations

First we will assume that entropy perturbations are absent, that is, $\delta S = 0$. The coefficients in (6.20) do not depend on the spatial coordinates, so upon taking the Fourier transform,

$$\delta\varepsilon(\mathbf{x}, t) = \int \delta\varepsilon_{\mathbf{k}}(t) \exp(i\mathbf{k}\mathbf{x}) \frac{d^3k}{(2\pi)^{3/2}}, \quad (6.21)$$

we obtain a set of independent ordinary differential equations for the time-dependent Fourier coefficients $\delta\varepsilon_{\mathbf{k}}(t)$:

$$\delta\ddot{\varepsilon}_{\mathbf{k}} + \left(k^2 c_s^2 - 4\pi G\varepsilon_0\right)\delta\varepsilon_{\mathbf{k}} = 0, \quad (6.22)$$

where a dot denotes the derivative with respect to time t and $k = |\mathbf{k}|$.

Equation (6.22) has two independent solutions

$$\delta\varepsilon_{\mathbf{k}} \propto \exp(\pm i\omega(k)t), \quad (6.23)$$

where

$$\omega(k) = \sqrt{k^2 c_s^2 - 4\pi G\varepsilon_0}.$$

The behavior of these so-called adiabatic perturbations depends crucially on the sign of the expression under the square root. Defining the Jeans length as

$$\lambda_J = \frac{2\pi}{k_J} = c_s \left(\frac{\pi}{G\varepsilon_0}\right)^{1/2}, \quad (6.24)$$

so that $\omega(k_J) = 0$, we conclude that if $\lambda < \lambda_J$, the solutions describe sound waves

$$\delta\varepsilon \propto \sin(\omega t + \mathbf{k}\mathbf{x} + \alpha), \quad (6.25)$$

propagating with phase velocity

$$c_{phase} = \frac{\omega}{k} = c_s \sqrt{1 - \frac{k_J^2}{k^2}}. \quad (6.26)$$

In the limit $k \gg k_J$, or on very small scales ($\lambda \ll \lambda_J$) where gravity is negligible compared to the pressure, we have $c_{phase} \to c_s$, as it should be.

On large scales gravity dominates and if $\lambda > \lambda_J$, we have

$$\delta\varepsilon_{\mathbf{k}} \propto \exp(\pm |\omega| t). \quad (6.27)$$

One of these solutions describes the exponentially fast growth of inhomogeneities, while the other corresponds to a decaying mode. When $k \to 0$, $|\omega| t \to t/t_{gr}$, where $t_{gr} \equiv (4\pi G\varepsilon_0)^{-1/2}$. We interpret t_{gr} as the characteristic collapse time for a region with initial density ε_0.

270   *Gravitational instability in Newtonian theory*

The Jeans length $\lambda_J \sim c_s t_{gr}$ is the "sound communication" scale over which the pressure can still react to changes in the energy density due to gravitational instability. Gravitational instability is very efficient in a static universe. Even if the adiabatic perturbation is initially extremely small, say 10^{-100}, gravity needs only a short time $t \sim 230 t_{gr}$ to amplify it to order unity.

Problem 6.1 Find and analyze the expression for $\delta v_\mathbf{k}$ and $\delta\phi_\mathbf{k}$ for sound waves and for the perturbations on scales larger than the Jeans wavelength.

6.2.2 Vector perturbations

The trivial solution of (6.20) with $\delta\varepsilon = 0$ and $\delta S = 0$ can correspond to a nontrivial solution of the complete system of the hydrodynamical equations. In this case, in fact, (6.15)–(6.18) reduce to

$$\nabla \delta \mathbf{v} = 0, \quad \frac{\partial \delta \mathbf{v}}{\partial t} = 0. \tag{6.28}$$

From the second equation it follows that $\delta \mathbf{v}$ can be an arbitrary time-independent function of the spatial coordinates, $\delta \mathbf{v} = \delta \mathbf{v}(\mathbf{x})$. The first equation tells us that for the plane wave perturbation, $\delta \mathbf{v} = \mathbf{w_k} \exp(i \mathbf{k}\mathbf{x})$, the velocity is perpendicular to the wave vector $\mathbf{k}$:

$$\mathbf{w_k} \cdot \mathbf{k} = 0. \tag{6.29}$$

These *vector perturbations* describe shear motions of the media which do not disturb the energy density. Because there are two independent directions perpendicular to $\mathbf{k}$, there exist two independent vector modes for a given $\mathbf{k}$.

6.2.3 Entropy perturbations

In the presence of entropy inhomogeneities ($\delta S \neq 0$), the Fourier transform of (6.20) is

$$\delta\ddot{\varepsilon}_\mathbf{k} + \left(k^2 c_s^2 - 4\pi G \varepsilon_0\right)\delta\varepsilon_\mathbf{k} = -\sigma k^2 \delta S_\mathbf{k}. \tag{6.30}$$

The general solution of this equation can be written as the sum of its particular solution and a general solution of the homogeneous equation with $\delta S_\mathbf{k} = 0$. The particular time-independent solution of (6.30),

$$\delta\varepsilon_\mathbf{k} = -\frac{\sigma k^2 \delta S_\mathbf{k}}{\left(k^2 c_s^2 - 4\pi G \varepsilon_0\right)}, \tag{6.31}$$

is called the *entropy perturbation*. Note, that in the short distance limit $k \to \infty$, when gravity is unimportant, $\delta\varepsilon_\mathbf{k} \to -\sigma \delta S_\mathbf{k}/c_s^2$. In this case the contribution to the pressure due to the energy density inhomogeneities is exactly compensated by the corresponding contribution from the entropy perturbations, so that $\delta p_{\mathbf{k}} = c_s^2 \delta \varepsilon_{\mathbf{k}} + \sigma \delta S_{\mathbf{k}}$ vanishes.

Entropy perturbations can occur only in multi-component fluids. For example, in a fluid consisting of baryons and radiation, the baryons can be distributed inhomogeneously on a homogeneous background of radiation. In such a case, the entropy, which is equal to the number of photons per baryon, varies from place to place.

Thus we have found the complete set of modes — two adiabatic modes, two vector modes and one entropy mode — describing perturbations in a gravitating homogeneous non-expanding medium. The most interesting is the exponentially growing adiabatic mode which is responsible for the origin of structure in the universe.

6.3 Instability in an expanding universe

Background In an expanding homogeneous and isotropic universe, the background energy density is a function of time, and the background velocities obey the Hubble law:

$$\varepsilon = \varepsilon_0(t), \quad \mathbf{V} = \mathbf{V}_0 = H(t) \cdot \mathbf{x}. \tag{6.32}$$

Substituting these expressions into (6.3), we obtain the familiar equation

$$\dot{\varepsilon}_0 + 3H\varepsilon_0 = 0, \tag{6.33}$$

which states that the total mass of nonrelativistic matter is conserved. The divergence of the Euler equations (6.8) together with the Poisson equation (6.10) leads to the Friedmann equation:

$$\dot{H} + H^2 = -\frac{4\pi G}{3}\varepsilon_0. \tag{6.34}$$

Perturbations Ignoring entropy perturbations and substituting the expressions

$$\varepsilon = \varepsilon_0 + \delta\varepsilon(\mathbf{x}, t), \quad \mathbf{V} = \mathbf{V}_0 + \delta\mathbf{v}, \quad \phi = \phi_0 + \delta\phi,$$
$$p = p_0 + \delta p = p_0 + c_s^2 \delta\varepsilon, \tag{6.35}$$

into (6.3), (6.8), (6.10), we derive the following set of linearized equations for small perturbations:

$$\frac{\partial \delta\varepsilon}{\partial t} + \varepsilon_0 \nabla \delta\mathbf{v} + \nabla(\delta\varepsilon \cdot \mathbf{V}_0) = 0, \tag{6.36}$$

$$\frac{\partial \delta\mathbf{v}}{\partial t} + (\mathbf{V}_0 \cdot \nabla)\delta\mathbf{v} + (\delta\mathbf{v} \cdot \nabla)\mathbf{V}_0 + \frac{c_s^2}{\varepsilon_0}\nabla\delta\varepsilon + \nabla\delta\phi = 0, \tag{6.37}$$

$$\Delta\delta\phi = 4\pi G \delta\varepsilon. \tag{6.38}$$

The Hubble velocity $\mathbf{V}_0$ depends explicitly on $\mathbf{x}$ and therefore the Fourier transform with respect to the Eulerian coordinates $\mathbf{x}$ does not reduce these equations to a decoupled set of ordinary differential equations. This is why it is more convenient to use the Lagrangian (*comoving with the Hubble flow*) coordinates $\mathbf{q}$, which are related to the Eulerian coordinates via

$$\mathbf{x} = a(t)\mathbf{q}, \tag{6.39}$$

where $a(t)$ is the scale factor. The partial derivative with respect to time taken at constant $\mathbf{x}$ is different from the partial derivative taken at constant $\mathbf{q}$. For a general function $f(\mathbf{x}, t)$ we have

$$\left(\frac{\partial f(\mathbf{x}=a\mathbf{q},t)}{\partial t}\right)_\mathbf{q} = \left(\frac{\partial f}{\partial t}\right)_\mathbf{x} + \dot{a}q^i\left(\frac{\partial f}{\partial x^i}\right)_t \tag{6.40}$$

and therefore

$$\left(\frac{\partial}{\partial t}\right)_\mathbf{x} = \left(\frac{\partial}{\partial t}\right)_\mathbf{q} - (\mathbf{V}_0 \cdot \nabla_\mathbf{x}). \tag{6.41}$$

The spatial derivatives are more simply related:

$$\nabla_\mathbf{x} = \frac{1}{a}\nabla_\mathbf{q}. \tag{6.42}$$

Replacing the derivatives in (6.36)–(6.38) and introducing the fractional amplitude of the density perturbations $\delta \equiv \delta\varepsilon/\varepsilon_0$, we finally obtain

$$\left(\frac{\partial \delta}{\partial t}\right) + \frac{1}{a}\nabla\delta\mathbf{v} = 0, \tag{6.43}$$

$$\left(\frac{\partial \delta\mathbf{v}}{\partial t}\right) + H\delta\mathbf{v} + \frac{c_s^2}{a}\nabla\delta + \frac{1}{a}\nabla\delta\phi = 0, \tag{6.44}$$

$$\Delta\delta\phi = 4\pi G a^2 \varepsilon_0 \delta, \tag{6.45}$$

where $\nabla \equiv \nabla_\mathbf{q}$ and Δ are now the derivatives with respect to the Lagrangian coordinates $\mathbf{q}$ and the time derivatives are taken at constant $\mathbf{q}$. In deriving (6.43) we have used (6.33) for the background and noted that $\nabla_\mathbf{x}\mathbf{V}_0 = 3H$ and $(\delta\mathbf{v} \cdot \nabla_\mathbf{x})\mathbf{V}_0 = H\delta\mathbf{v}$. Taking the divergence of (6.44) and using the continuity and Poisson equations to express $\nabla\delta\mathbf{v}$ and $\Delta\delta\phi$ in terms of δ, we derive the closed form equation

$$\ddot{\delta} + 2H\dot{\delta} - \frac{c_s^2}{a^2}\Delta\delta - 4\pi G\varepsilon_0\delta = 0, \tag{6.46}$$

which describes gravitational instability in an expanding universe.

6.3.1 Adiabatic perturbations

Taking the Fourier transform of (6.46) with respect to the comoving coordinates **q**, we obtain the ordinary differential equation

$$\ddot{\delta}_{\mathbf{k}} + 2H\dot{\delta}_{\mathbf{k}} + \left(\frac{c_s^2 k^2}{a^2} - 4\pi G \varepsilon_0\right)\delta_{\mathbf{k}} = 0 \qquad (6.47)$$

for every Fourier mode $\delta = \delta_{\mathbf{k}}(t)\exp(i\mathbf{kq})$. The behavior of each perturbation depends crucially on its spatial size; the critical lengthscale is the Jeans length

$$\lambda_J^{ph} = \frac{2\pi a}{k_J} = c_s\sqrt{\frac{\pi}{G\varepsilon_0}}. \qquad (6.48)$$

Here λ^{ph} is the physical wavelength (measured, for example, in centimeters), related to the comoving wavelength $\lambda = 2\pi/k$ via $\lambda^{ph} = a \cdot \lambda$. In a flat, matter-dominated universe $\varepsilon_0 = (6\pi G t^2)^{-1}$ and hence

$$\lambda_J^{ph} \sim c_s t, \qquad (6.49)$$

that is, the Jeans length is of order the sound horizon. Sometimes instead of the Jeans length, one uses the Jeans mass, defined as $M_J \equiv \varepsilon_0(\lambda_J^{ph})^3$.

Perturbations on scales much smaller than the Jeans length ($\lambda \ll \lambda_J$) are sound waves. If c_s changes adiabatically, then the solution of (6.47) is

$$\delta_{\mathbf{k}} \propto \frac{1}{\sqrt{c_s a}} \exp\left(\pm k \int \frac{c_s dt}{a}\right). \qquad (6.50)$$

Problem 6.2 Derive the solution in (6.50). Explain why the amplitude of sound waves decays with time. (*Hint* Using conformal time $\eta \equiv \int dt/a$ instead of the physical time t, derive the equation for the rescaled amplitude $\sqrt{a}\delta_{\mathbf{k}}$ and solve it in the WKB approximation.)

On scales much larger than the Jeans scale ($\lambda \gg \lambda_J$), gravity dominates and we can neglect the k-dependent term in (6.47). Then one of the solutions is simply proportional to the Hubble constant $H(t)$. In fact, substituting $\delta_d = H(t)$ in (6.47), where we set $c_s^2 k^2 = 0$, one finds that the resulting equation coincides with the time derivative of the Friedmann equation (6.34). Note that $\delta_d = H(t)$ is the decaying solution of the perturbation equation (H decreases with time) in a matter-dominated universe with *arbitrary* curvature.

Actually, one could have guessed this solution using the following simple argument. Both the background energy density $\varepsilon_0(t)$ and the time-shifted energy density $\varepsilon_0(t + \tau)$, where $\tau = $ const, satisfy (6.33), (6.34). Indeed, using (6.33) to express H in terms of ε_0 and substituting this into (6.34), we obtain an equation for $\varepsilon_0(t)$ in which the time does not explicitly appear. Hence its solution is time-translational-invariant. For small τ, the time-shifted solution $\varepsilon_0(t+\tau)$ can be considered as a perturbation of the background $\varepsilon_0(t)$ with amplitude

$$\delta_d = \frac{\varepsilon_0(t+\tau) - \varepsilon_0(t)}{\varepsilon_0(t)} \approx \frac{\dot{\varepsilon}_0 \tau}{\varepsilon_0} \propto H(t).$$

Once we know one solution of the second order differential equation, δ_d, then the other independent solution δ_i can easily be found with the help of the Wronskian,

$$W \equiv \dot{\delta}_d \delta_i - \delta_d \dot{\delta}_i. \tag{6.51}$$

Taking the derivative of the Wronskian and using (6.47) to express $\ddot{\delta}$ in terms of $\dot{\delta}$ and δ, we find that W satisfies the equation

$$\dot{W} = -2HW, \tag{6.52}$$

which has the obvious solution

$$W \equiv \dot{\delta}_d \delta_i - \delta_d \dot{\delta}_i = \frac{C}{a^2}, \tag{6.53}$$

where C is a constant of integration. Substituting the ansatz $\delta_i = \delta_d f(t)$ into (6.53), we obtain an equation for f that is readily integrated:

$$f = -C \int \frac{dt}{a^2 \delta_d^2}. \tag{6.54}$$

Thus the most general longwave solution of (6.47) is

$$\delta = C_1 H \int \frac{dt}{a^2 H^2} + C_2 H. \tag{6.55}$$

In a flat, matter-dominated universe, $a \propto t^{2/3}$ and $H \propto t^{-1}$. In this case, we have

$$\delta = C_1 t^{2/3} + C_2 t^{-1}. \tag{6.56}$$

Hence, we see that in an expanding universe, gravitational instability is *much less efficient* and the perturbation amplitude increases only as a power of time. In the important case of a flat, matter-dominated universe, the growing mode is proportional to the scale factor. Therefore, if we want to obtain large inhomogeneities ($\delta \gtrsim 1$) today, we have to assume that at early times (for example, at redshifts $z = 1000$) the inhomogeneities were already substantial ($\delta \gtrsim 10^{-3}$). This imposes rather strong constraints on the initial spectrum of perturbations. We will see in Chapter 8 that the required initial spectrum can be explained naturally in inflationary cosmology.

Problem 6.3 Calculate the peculiar velocities and gravitational potential for the long-wavelength perturbations. Analyze their behavior and give the physical interpretation of the behavior of the gravitational potential for the growing mode.

6.3 Instability in an expanding universe

Problem 6.4 If we use the redshift z as a time parameter, then the integral in (6.55) can be calculated explicitly for a matter-dominated universe with arbitrary Ω_0. Find the corresponding solution $\delta(z)$ and show that for $\Omega_0 \ll 1$ the perturbation amplitude freezes out at $z \sim 1/\Omega_0$.

6.3.2 Vector perturbations

With $\delta = 0$, (6.43)–(6.45) reduce to

$$\nabla \delta \mathbf{v} = 0, \quad \frac{\partial \delta \mathbf{v}}{\partial t} + H \delta \mathbf{v} = 0. \tag{6.57}$$

From the first equation it follows that for a plane wave perturbation, $\delta \mathbf{v} \propto \delta \mathbf{v_k}(t) \exp(i \mathbf{kq})$, the peculiar velocity $\delta \mathbf{v}$ is perpendicular to the wavenumber $\mathbf{k}$. The second equation becomes

$$\delta \dot{\mathbf{v}}_\mathbf{k} + \frac{\dot{a}}{a} \delta \mathbf{v_k} = 0, \tag{6.58}$$

and has the obvious solution $\delta \mathbf{v}_k \propto 1/a$. Thus, the vector perturbations decay as the universe expands. These perturbations can have significant amplitudes at present only if their initial amplitudes were so large that they completely spoiled the isotropy of the very early universe. In an inflationary universe there is no room for such large primordial vector perturbations and they do not play any role in the formation of the large-scale structure of the universe. Vector perturbations, however, can be generated at late times, after nonlinear structure has been formed, and can explain the rotation of galaxies.

6.3.3 Self-similar solution

For large-scale perturbations we can neglect the pressure, and the spatial derivatives drop out of (6.46). In this case the solution for perturbations can be written directly in coordinate space

$$\delta(\mathbf{q}, t) = A(\mathbf{q}) \delta_i(t) + B(\mathbf{q}) \delta_d(t), \tag{6.59}$$

where δ_i and δ_d are growing and decaying modes respectively. Without losing generality we can set $\delta_i(t_0) = \delta_d(t_0) = 1$ at some initial moment of time t_0. If the density distribution at this time is described by the function $\delta(\mathbf{q}, t_0)$ and the matter is at rest with respect to the Hubble flow ($\delta \mathbf{v} \propto \dot{\delta}(\mathbf{q}, t_0) = 0$), then, expressing $A(\mathbf{q})$ and $B(\mathbf{q})$ in terms of $\delta(\mathbf{q}, t_0)$, we obtain

$$\delta(\mathbf{q}, t) = \delta(\mathbf{q}, t_0) \left(\frac{\delta_i(t)}{1 - (\dot{\delta}_i/\dot{\delta}_d)_{t_0}} + \frac{\delta_d(t)}{1 - (\dot{\delta}_d/\dot{\delta}_i)_{t_0}} \right). \tag{6.60}$$

In this particular case the perturbation preserves its initial spatial shape as it develops. Such a solution is said to be self-similar.

Generically the shape of inhomogeneity changes. However, at late times ($t \gg t_0$) when the growing mode dominates, we can omit the second term in (6.59) and the *linear* perturbation grows in a self-similar way.

6.3.4 Cold matter in the presence of radiation or dark energy

There is convincing evidence that along with the cold matter in the universe there exists a smooth dark energy component. This dark energy changes the expansion rate and, as a result, influences the growth of inhomogeneities in the cold matter.

To study the gravitational instability in the presence of relativistic matter, in principle we need the full relativistic theory. However, on scales smaller than the Jeans length for relativistic matter, which is comparable to the horizon scale, the inhomogeneities in the cold matter distribution do not disturb the relativistic component and it remains practically homogeneous. As a result we can still apply modified Newtonian theory to *the perturbations in the cold matter itself* on scales smaller than the horizon. In the following we consider the growth of the perturbations in the presence of a homogeneous relativistic energy component. This can be radiation, with equation of state $w = 1/3$, or dark matter with $w < -1/3$.

It is easy to verify that the equation for the perturbation in the cold component alone, $\delta \equiv \delta\varepsilon_d/\varepsilon_d$, coincides with (6.46), but the Hubble constant is now determined by the *total* energy density

$$\varepsilon_{tot} = \frac{\varepsilon_{eq}}{2}\left(\left(\frac{a_{eq}}{a}\right)^3 + \left(\frac{a_{eq}}{a}\right)^{3(1+w)}\right), \tag{6.61}$$

via the usual relation, which for a *flat universe* is

$$H^2 = \frac{8\pi G}{3}\varepsilon_{tot}. \tag{6.62}$$

Here a_{eq} is the scale factor at "equality" when the energy densities of both components are equal. To find the explicit solutions of (6.46), it is convenient to rewrite it using as a time variable the normalized scale factor $x \equiv a/a_{eq}$ instead of t. Taking into account that ε_0 entering (6.46) is the cold matter density alone, equal to

$$\varepsilon_d = \frac{\varepsilon_{eq}}{2}\left(\frac{a_{eq}}{a}\right)^3, \tag{6.63}$$

and using (6.62) to express the Hubble parameter in terms of x, (6.46) becomes

$$x^2\left(1 + x^{-3w}\right)\frac{d^2\delta}{dx^2} + \frac{3}{2}x\left(1 + (1-w)x^{-3w}\right)\frac{d\delta}{dx} - \frac{3}{2}\delta = 0. \tag{6.64}$$

6.3 Instability in an expanding universe

We have skipped in (6.64) the term proportional to c_s^2 because it is determined by the pressure of the cold matter alone and hence is negligible. The general solution of (6.64) for an arbitrary $w = $ const is given by a linear combination of hypergeometric functions. However, at least in two important cases, they reduce to simple elementary functions.

Cosmological constant ($w = -1$) One can easily verify that in this case,

$$\delta_1(x) = \sqrt{1 + x^{-3}} \tag{6.65}$$

satisfies (6.64). The other solution can be obtained using the properties of the Wronskian.

Problem 6.5 Verify that if $\delta_1(x)$ is a solution of (6.64), then the second independent solution is given by

$$\delta_2(x) = \delta_1(x) \int^x \frac{dy}{y^{3/2}(1 + y^{-3w})^{1/2} \delta_1^2(y)}. \tag{6.66}$$

For $w = -1$, the general solution of (6.64) is thus

$$\delta(x) = C_1\sqrt{1 + x^{-3}} + C_2\sqrt{1 + x^{-3}} \int_0^x \left(\frac{y}{1 + y^3}\right)^{3/2} dy, \tag{6.67}$$

where C_1 and C_2 are constants of integration. At early times when the cold matter dominates ($x \ll 1$), the perturbation grows as

$$\delta(x) = C_1 x^{-3/2} + \tfrac{2}{5} C_2 x + O(x^{3/2}), \tag{6.68}$$

in complete agreement with our previous result. Subsequently the cosmological constant becomes dominant and in the limit $x \gg 1$ we have

$$\delta(x) = (C_1 + I C_2) - \tfrac{1}{2} C_2 x^{-2} + O(x^{-3}), \tag{6.69}$$

where

$$I = \int_0^\infty \left(\frac{y}{1 + y^3}\right)^{3/2} dy \simeq 0.57. \tag{6.70}$$

Thus, when the cosmological constant overtakes the matter density the growth ceases and the amplitude of the perturbation is frozen. According to (6.45), the induced gravitational potential decays in inverse proportion to the scale factor since $\varepsilon_d \propto a^{-3}$. The results obtained are not surprising because the cosmological constant acts as "antigravity" and tries to prevent the growth of the perturbations.

Problem 6.6 Verify that if $w = -1$ or $w = -1/3$, then $\delta \propto H$ is the solution of (6.46) in the universe with arbitrary curvature. Using (6.55), find the general solutions in these cases and analyze their behavior in open and closed universes.

Radiation background The Jeans length for radiation is comparable to the horizon size because the speed of sound in the radiation component is of order the speed of light ($c_s^2 = 1/3$). Therefore cold dark matter, which interacts only gravitationally with radiation, does not induce significant perturbations in radiation and to study the growth of inhomogeneities in cold matter alone we can still use (6.64), setting $w = 1/3$. In this case,

$$\delta_1(x) = 1 + \frac{3}{2}x \tag{6.71}$$

satisfies (6.64). The other independent solution can be found by substituting (6.71) into (6.66). The integral can be calculated explicitly and the general solution of (6.64) is

$$\delta(x) = C_1\left(1 + \frac{3}{2}x\right) + C_2\left[\left(1 + \frac{3}{2}x\right)\ln\frac{\sqrt{1+x}+1}{\sqrt{1+x}-1} - 3\sqrt{1+x}\right]. \tag{6.72}$$

At early times, during the radiation-dominated stage ($x \ll 1$), the amplitude of perturbations grows as

$$\delta(x) = (C_1 - 3C_2) - C_2 \ln(x/4) + O(x), \tag{6.73}$$

that is, logarithmically at most. Thus, by influencing the rate of expansion, the radiation suppresses the growth of inhomogeneities in the cold component. After matter–radiation equality, matter overtakes radiation and at $x \gg 1$, the amplitude of perturbation is

$$\delta(x) = C_1\left(1 + \frac{3}{2}x\right) + \frac{4}{15}C_2 x^{-3/2} + O\left(x^{-5/2}\right), \tag{6.74}$$

that is, it grows proportionally to the scale factor. Since the perturbations cannot be amplified significantly during the radiation epoch, small initial perturbations can produce nonlinear structure only if the cold matter starts to dominate early enough. This imposes a lower bound on the amount of cold matter. In particular, if the amplitude of the initial inhomogeneities is of order 10^{-4}, as favored by the observed CMB fluctuations, we can explain the nonlinear structure seen only if the initial perturbations started to grow before recombination. This is possible if there exists a cold *dark* matter component of *non-baryonic* origin, which interacts only gravitationally with radiation. We can reconcile the small initial perturbations with the observed large-scale structure only if this dark matter constitutes about 30% of the present critical density. It is clear that baryons cannot substantially contribute to the dark matter. They are tightly coupled to radiation before recombination and perturbations in the baryon component can start to grow only after recombination, when the dark matter must already have begun to cluster. Furthermore, a high baryon density would spoil nucleosynthesis.

6.4 Beyond linear approximation

The Hubble flow stretches *linear* inhomogeneities; their spatial size is proportional to the scale factor. The relative amplitude of the linear perturbation grows, while its energy density, equal to

$$\varepsilon = \varepsilon_0 \left(1 + \delta + O(\delta^2)\right),$$

decays only slightly more slowly than the background energy density ε_0. It is obvious that when the perturbation amplitude reaches unity ($\delta \sim 1$), the neglected nonlinear terms $\sim \delta^2$ etc., become important. At this time the gravitational field created by the perturbation leads to a contraction which overwhelms the Hubble expansion. As a result the inhomogeneity drops out of the Hubble flow, reaches its maximal size and recollapses to form a stable nonlinear structure.

Even for pressureless matter, exact solutions describing nonlinear evolution can be obtained only in a few particular cases where the spatial shape of the inhomogeneity possesses a special symmetry. To build intuition about the nonlinear behavior of perturbations we will derive exact solutions in two special cases: for a spherically symmetric perturbation and for an anisotropic one-dimensional inhomogeneity. The behavior of realistic nonsymmetric perturbations can then be qualitatively understood on the basis of these two limiting cases.

Let us first recast the hydrodynamical equations (6.3), (6.8), (6.10) in a slightly different form, which is more convenient for finding their nonlinear solutions. The continuity equation (6.3) can be written as

$$\left(\frac{\partial}{\partial t}\bigg|_x + V^i \nabla_i \right) \varepsilon + \varepsilon \nabla_i V^i = 0, \tag{6.75}$$

where $\nabla_i \equiv \partial/\partial x^i$. Taking the divergence of the Euler equations (6.8) and using the Poisson equation (6.10), we obtain

$$\left(\frac{\partial}{\partial t}\bigg|_x + V^i \nabla_i \right)(\nabla_j V^j) + (\nabla_j V^i)(\nabla_i V^j) + 4\pi G \varepsilon = 0, \tag{6.76}$$

where we have assumed that the pressure is equal to zero.

In the next step we replace the Eulerian coordinates **x** with the comoving Lagrangian coordinates **q**, enumerating the *matter elements*:

$$\mathbf{x} = \mathbf{x}(\mathbf{q}, t). \tag{6.77}$$

These new coordinates can be used until the trajectories of the matter elements start to cross one other. The velocity of a matter element with the given Lagrangian coordinates **q** is equal to

$$V^i \equiv \frac{dx^i}{dt} = \left.\frac{\partial x^i(\mathbf{q},t)}{\partial t}\right|_q. \tag{6.78}$$

The derivatives of the velocity field with respect to the Eulerian coordinates can then be written as

$$\nabla_j V^i = \frac{\partial q^k}{\partial x^j} \frac{\partial}{\partial q^k}\left(\frac{\partial x^i(\mathbf{q},t)}{\partial t}\right) = \frac{\partial q^k}{\partial x^j} \frac{\partial J_k^i}{\partial t}, \tag{6.79}$$

where we have introduced the strain tensor

$$J_k^i(\mathbf{q},t) \equiv \frac{\partial x^i(\mathbf{q},t)}{\partial q^k}. \tag{6.80}$$

Taking into account that

$$\left(\left.\frac{\partial}{\partial t}\right|_x + V^i \nabla_i\right) = \left.\frac{\partial}{\partial t}\right|_x + \left.\frac{\partial x^i(q,t)}{\partial t}\right|_q \frac{\partial}{\partial x^i} = \left.\frac{\partial}{\partial t}\right|_q, \tag{6.81}$$

and substituting (6.79) into (6.75) and (6.76), we obtain

$$\frac{\partial \varepsilon}{\partial t} + \varepsilon \frac{\partial q^k}{\partial x^i} \frac{\partial J_k^i}{\partial t} = 0, \tag{6.82}$$

$$\frac{\partial}{\partial t}\left(\frac{\partial q^k}{\partial x^i} \frac{\partial J_k^i}{\partial t}\right) + \left(\frac{\partial q^k}{\partial x^i} \frac{\partial J_k^j}{\partial t}\right)\left(\frac{\partial q^l}{\partial x^j} \frac{\partial J_l^i}{\partial t}\right) + 4\pi G\varepsilon = 0, \tag{6.83}$$

where the time derivatives are taken at constant **q**. The elements of the strain tensor (6.80) form a 3×3 matrix $\mathbf{J} \equiv \|J_k^i\|$, and since

$$\frac{\partial q^k}{\partial x^j} \cdot \frac{\partial x^i}{\partial q^k} = \frac{\partial x^i}{\partial x^j} = \delta_j^i, \tag{6.84}$$

the derivatives $\partial q^k/\partial x^i$ are the elements of the inverse matrix $\mathbf{J}^{-1}$. Consequently, we can rewrite (6.82) and (6.83) in matrix notation as

$$\dot{\varepsilon} + \varepsilon \operatorname{tr}\left(\dot{\mathbf{J}} \cdot \mathbf{J}^{-1}\right) = 0, \tag{6.85}$$

$$\left[\operatorname{tr}\left(\dot{\mathbf{J}} \cdot \mathbf{J}^{-1}\right)\right]^{\cdot} + \operatorname{tr}\left[\left(\dot{\mathbf{J}} \cdot \mathbf{J}^{-1}\right)^2\right] + 4\pi G\varepsilon = 0, \tag{6.86}$$

where a dot denotes the partial time derivative.

Problem 6.7 Prove that

$$\operatorname{tr}\left(\dot{\mathbf{J}} \cdot \mathbf{J}^{-1}\right) = (\ln J)^{\cdot}, \tag{6.87}$$

where $J(\mathbf{q}, t) \equiv \det \mathbf{J}$.

After substitution of (6.87) into (6.85), the resulting equation can easily be integrated to give

$$\varepsilon(\mathbf{q}, t) = \frac{\varrho_0(\mathbf{q})}{J(\mathbf{q}, t)}, \tag{6.88}$$

where $\varrho_0(\mathbf{q})$ is an arbitrary time-independent function of the Lagrangian coordinates. With (6.87) and (6.88), (6.86) simplifies to

$$(\ln J)^{\cdot\cdot} + \operatorname{tr}\left[\left(\dot{\mathbf{J}} \cdot \mathbf{J}^{-1}\right)^2\right] + 4\pi G \varrho_0 J^{-1} = 0. \tag{6.89}$$

This resulting equation for J can be solved exactly for a few interesting cases.

6.4.1 Tolman solution

Let us consider a spherically symmetric inhomogeneity. In this case one can always find a coordinate system where the strain tensor is proportional to the unit tensor:

$$J^i_k = a(R, t)\delta^i_k, \tag{6.90}$$

where $R \equiv |\mathbf{q}|$ is the radial Lagrangian coordinate. Substituting

$$J = a^3, \quad \operatorname{tr}\left[\left(\dot{\mathbf{J}} \cdot \mathbf{J}^{-1}\right)^2\right] = 3\left(\frac{\dot{a}}{a}\right)^2 \tag{6.91}$$

into (6.89), we obtain

$$\ddot{a}(R, t) = -\frac{4\pi G \varrho_0(R)}{3a^2(R, t)}. \tag{6.92}$$

Multiplying this equation by $\dot{a}$, we easily derive its first integral

$$\dot{a}^2(R, t) - \frac{8\pi G \varrho_0(R)}{3a(R, t)} = F(R), \tag{6.93}$$

where $F(R)$ is a constant of integration. Note that for a homogeneous matter distribution ϱ_0, a and F do not depend on R and (6.93) coincides with the Friedmann equation for a matter-dominated universe.

Problem 6.8 Verify that the solution of (6.93) can be written in the following parametric form:

$$a(R, \eta) = \frac{4\pi G \varrho_0}{3|F|}(1 - \cos \eta), \quad t(R, \eta) = \frac{4\pi G \varrho_0}{3|F|^{3/2}}(\eta - \sin \eta) + t_0, \quad (6.94)$$

for $F < 0$, and

$$a(R, \eta) = \frac{4\pi G \varrho_0}{3F}(\cosh \eta - 1), \quad t(R, \eta) = \frac{4\pi G \varrho_0}{3F^{3/2}}(\sinh \eta - \eta) + t_0, \quad (6.95)$$

for $F > 0$. Here $t_0 \equiv t_0(R)$ is a further integration constant. Note that the same "conformal time" η generally corresponds to different values of physical time t for different R. Assuming that the initial singularity ($a \to 0$) occurs at the same moment of physical time $t = 0$ everywhere in space, we can set $t_0 = 0$.

Let us consider the evolution of a spherically symmetric overdense region in a flat, matter-dominated universe. Far away from the center of this region the matter remains undisturbed and hence $\varrho_0(R \to \infty) \to \varrho_\infty = $ const. The condition of flatness requires $F \to 0$ as $R \to \infty$. Taking the limit $|F| \to 0$ so that the ratio $\eta/\sqrt{|F|}$ remains fixed, we immediately obtain from (6.94)

$$a(R \to \infty, t) = (6\pi G \varrho_\infty)^{1/3} t^{2/3}. \quad (6.96)$$

The energy density is consequently

$$\varepsilon(R \to \infty, t) = \frac{\varrho_\infty}{a^3} = \frac{1}{6\pi G t^2}, \quad (6.97)$$

in complete agreement with what one would expect for a flat dust-dominated universe. Inside the overdense region, F is negative and the energy density does not continually decrease. Because $\varepsilon \propto a^{-3}$, the density at some point R takes its minimal value ε_m when $a(R, t)$ reaches its maximal value

$$a_m = \frac{8\pi G \varrho_0}{3|F|} \quad (6.98)$$

at $\eta = \pi$ (see (6.94)). This happens at the moment of physical time

$$t_m = \frac{4\pi^2 G \varrho_0}{3|F|^{3/2}}, \quad (6.99)$$

when the energy density is equal to

$$\varepsilon_m(R) = \frac{\varrho_0(R)}{a_m^3(R)} = \frac{27|F|^3}{(8\pi G)^3 \varrho_0^2} = \frac{3\pi}{32 G t_m^2}. \quad (6.100)$$

Comparing this result with the averaged density at $t = t_m$, given by (6.97), we find that in those places where the energy density exceeds the averaged density by a

factor of
$$\frac{\varepsilon_m}{\varepsilon(R \to \infty)} = \frac{9\pi^2}{16} \simeq 5.55, \qquad (6.101)$$

the matter detaches from the Hubble flow and begins to collapse.

Formally the energy density becomes infinite at $t = 2t_m$; in reality, however, this does not happen because there always exist deviations from exact spherical symmetry. As a result a spherical cloud of particles virializes and forms a stationary spherical object.

Problem 6.9 Consider a homogeneous spherical cloud of particles at rest and, using the virial theorem, verify that after virialization its size is halved. Assuming that virialization is completed at $t = 2t_m$, compare the density inside the cloud with the average density in the universe at this time. (*Hint* The virial theorem states that at equilibrium, $U = -2K$, where U and K are the total potential and kinetic energies respectively.)

Problem 6.10 Assuming that $\eta \ll 1$ and expanding the expressions in (6.94) in powers of η, derive the following expansion for the energy density in powers of $(t/t_m)^{2/3} \ll 1$:

$$\varepsilon = \frac{1}{6\pi t^2}\left[1 + \frac{3}{20}\left(\frac{6\pi t}{t_m}\right)^{2/3} + O\left(\left(\frac{t}{t_m}\right)^{4/3}\right)\right], \qquad (6.102)$$

where t_m is defined in (6.99). The second term inside the square brackets is obviously the amplitude of the linear perturbation δ. Thus, when the actual density exceeds the averaged density by a factor of 5.5, according to the linearized theory

$$\delta(t_m) = 3(6\pi)^{2/3}/20 \simeq 1.06.$$

Later, at $t = 2t_m$, the Tolman solution formally gives $\varepsilon \to \infty$, while the linear perturbation theory predicts $\delta(2t_m) \simeq 1.69$.

6.4.2 Zel'dovich solution

The geometrical shapes of realistic inhomogeneities are typically far from spherical and their collapse is strongly anisotropic. To build intuition about the main features of anisotropic collapse we consider the Zel'dovich solution. This solution describes the nonlinear behavior of a *one-dimensional* perturbation, superimposed on three-dimensional Hubble flow. In this case the relation between the Eulerian and Lagrangian coordinates can be written as

$$x^i = a(t)\left(q^i - f^i(q^j, t)\right). \qquad (6.103)$$

If we ignore vector perturbations then $f^i = \partial \psi/\partial q^i$, where ψ is the potential for the peculiar velocities. For a one-dimensional perturbation, ψ depends only on one of the coordinates, say q^1. Then the strain tensor takes the form

$$\mathbf{J} = a(t) \begin{pmatrix} 1 - \lambda(q^1, t) & 0 & 0 \\ 0 & 1 & 0 \\ 0 & 0 & 1 \end{pmatrix}, \quad (6.104)$$

and hence

$$J = a^3(1 - \lambda), \quad \text{tr}\left[\left(\dot{\mathbf{J}} \cdot \mathbf{J}^{-1}\right)^2\right] = \left(H - \frac{\dot{\lambda}}{1 - \lambda}\right)^2 + 2H^2, \quad (6.105)$$

where $\lambda(q^1, t) \equiv \partial f^1/\partial q^1$. Substituting (6.105) into (6.89), we find that for $\varrho_0(\mathbf{q}) = $ const this equation reduces to two independent equations:

$$\dot{H} + H^2 = -\frac{4\pi G}{3}\varepsilon_0, \quad (6.106)$$

$$\ddot{\lambda} + 2H\dot{\lambda} - 4\pi G \varepsilon_0 \lambda = 0, \quad (6.107)$$

where $\varepsilon_0(t) \equiv \varrho_0/a^3$. The first equation is the familiar Friedmann equation for the homogeneous background. The second equation *coincides* with (6.46) for linear perturbations in pressureless matter. However, it must be stressed that in deriving (6.107) we did not assume that the perturbations were small, and hence its solutions are valid in both the linear and the nonlinear regime.

According to (6.88), the energy density is equal to

$$\varepsilon(q, t) = \frac{\varepsilon_0(t)}{\left(1 - \lambda(q^1, t)\right)}, \quad (6.108)$$

and $\lambda(q^1, t)$ can be written as (see (6.59))

$$\lambda(q^1, t) = \alpha(q^1)\,\delta_i(t) + \varkappa(q^1)\,\delta_d(t). \quad (6.109)$$

Here $\delta_i(t)$ and $\delta_d(t)$ are the growing and decaying modes from the linearized theory. For example, in a flat matter-dominated universe $\delta_i \propto t^{2/3}$ and $\delta_d \propto t^{-1}$. For $\lambda(q^1, t) \ll 1$, the exact solution in (6.108) obviously reproduces the results of the linearized theory.

Problem 6.11 How must (6.89) be modified in the presence of a homogeneous relativistic component? Find and analyze the corresponding Zel'dovich solutions in a flat universe with a cosmological constant.

6.4 Beyond linear approximation

The decaying mode soon becomes negligible and does not influence the evolution even in the nonlinear phase. Ignoring this mode we have

$$\varepsilon(q, t) = \frac{\varepsilon_0(t)}{(1 - \alpha(q^1)\delta_i(t))}. \tag{6.110}$$

In those places where $\alpha(q^1)$ is positive, the energy density exceeds the averaged density $\varepsilon_0(t)$ and the relative density contrast grows. However, during the linear stage, when $\alpha\delta_i \ll 1$, the energy density itself decays. Only after the perturbation enters the nonlinear regime ($\alpha\delta_i \sim 1$) does the inhomogeneous region drop out of the Hubble flow and start to collapse. To estimate when this turnaround happens we have to find when $\dot\varepsilon(q, t) = 0$. The time derivative of the expression in (6.110) vanishes when

$$\frac{\varepsilon(q, t)}{\varepsilon_0(t)} = 1 + 3\frac{H}{(\ln \delta_i)^\cdot}. \tag{6.111}$$

In a flat matter-dominated universe, $\delta_i \propto a$ and, according to (6.111), as soon as the energy density exceeds the averaged density by a factor of 4, the region detaches from the Hubble flow and begins to collapse (compare to (6.101)). The collapse is one-dimensional and produces a two-dimensional structure known as a Zel'dovich "pancake." According to (6.110), at some moment of time the energy density of the pancake becomes infinite. However, in contrast with spherical collapse, the gravitational force and velocities at this moment remain finite. Once the matter trajectories cross, the solution in (6.110) becomes invalid.

In places where $\alpha(q^1)$ is negative, the energy density always decreases. Matter "escapes" from these regions and they eventually become empty.

In reality, the situation is more complicated because a typical inhomogeneity is neither spherical nor one-dimensional. To describe the evolution of a perturbation with an arbitrary shape, Zel'dovich suggested generalizing the solution in (6.110) to

$$\varepsilon(q, t) = \frac{\varepsilon_0(t)}{(1 - \alpha\delta_i(t))(1 - \beta\delta_i(t))(1 - \gamma\delta_i(t))}, \tag{6.112}$$

where α, β, γ characterize the deformation along the three principle axes of the strain tensor and they now depend on all coordinates q^i. The corresponding strain tensor

$$\mathbf{J} = a\mathbf{I} - a\delta_i \begin{pmatrix} \alpha & 0 & 0 \\ 0 & \beta & 0 \\ 0 & 0 & \gamma \end{pmatrix}, \tag{6.113}$$

where $\mathbf{I}$ is the unit matrix, satisfies (6.89) only to leading order. Therefore, the approximate solution in (6.112) has a very limited range of applicability. In fact, substituting (6.113) into (6.86), we find

$$\varepsilon(q,t) = \frac{\varepsilon_0\left[1 - \left((\alpha\beta + \alpha\gamma + \beta\gamma)\delta_i^2 - 2\alpha\beta\gamma\delta_i^3\right)\right]}{(1 - \alpha\delta_i)(1 - \beta\delta_i)(1 - \gamma\delta_i)}. \qquad (6.114)$$

On the other hand, it follows from (6.88) that ε should be given by the expression in (6.112). Hence the expected error of the *Zel'dovich approximation* is of order the disagreement between the results in (6.112) and (6.114), that is, $\sim O\left((\alpha\beta + \alpha\gamma + \beta\gamma)\delta_i^2, \alpha\beta\gamma\delta_i^3\right)$. When the perturbations are small, $\alpha\delta_i, \beta\delta_i, \gamma\delta_i \ll 1$, the Zel'dovich approximation reproduces the results of the linear perturbation theory. However, in the nonlinear regime, it is not very reliable. If, for example, $\alpha \gg \beta, \gamma$, then we can trust only the leading term, given by (6.110), which describes one-dimensional contraction along α-axis. The linear corrections $\beta\delta_i, \gamma\delta_i \ll 1$ become unreliable when $\alpha\delta_i$ reaches a value of order unity. If $\alpha \sim \beta$, then the formula in (6.112) fails to reproduce even the basic feature of the nonlinear collapse.

6.4.3 Cosmic web

The strain tensor is very useful when we try to understand the nonlinear large-scale structure of the universe. The initial inhomogeneities can be characterized completely by the strain tensor, or equivalently by three functions $\alpha(q^i)$, $\beta(q^i)$, $\gamma(q^i)$. How an inhomogeneity will grow in a particular region depends on the relation between the values of α, β, γ. Based on the results above, we see that the collapse is one-dimensional and produces two-dimensional pancakes (*walls*) in those regions where $\alpha \gg \beta, \gamma$. In the places where $\alpha \sim \beta \gg \gamma$, we expect two-dimensional collapse, leading to formation of one-dimensional filaments. For $\alpha \sim \beta \sim \gamma$ the collapse is nearly spherical.

For initial Gaussian perturbations the probability distribution of the strain tensor eigenvalues can be calculated exactly. A helpful *lower dimensional* visualization of the initial density field is a mountainous landscape in which the mountain peaks represent local maxima in the density and valleys correspond to local minima. In the concordance model (cold dark matter plus inflationary perturbation spectrum), inhomoheneities with significant amplitudes are present in nearly all scales and hence mountains with nearly all base sizes are superimposed on each other. If we are interested in the structure on scales exceeding some particular size, we have to smear the inhomogeneities on smaller scales, in other words, remove the mountains with small base sizes.

The first nonlinear structures are obviously formed near the tallest peaks where the energy density takes its maximal value. Typically, the two curvature scales are comparable near the mountain peak. Therefore, we expect that at a peak in the

6.4 Beyond linear approximation

density, which is a higher-dimensional analog of a mountain peak, $\alpha \sim \beta \sim \gamma$. Hence the surrounding region collapses in a nearly spherical way resulting in a *spherical or somewhat elliptical* object. Usually neighbouring peaks are connected by a saddleback, which is lower in height than the peaks. Along the saddleback one curvature scale is significantly smaller than the other. Similarly, density peaks are connected by higher-dimensional saddlebacks, where $\alpha \sim \beta$ and $\gamma \ll \alpha, \beta$. The collapse in these regions is two-dimensional and produces *filaments*. The filaments connect the spherical objects formed earlier, resulting in a web-like structure (see Figure 6.1). There are also regions that have no analog in the lower-dimensional landscape, where only one of the eigenvalues α, say, has a local maximum and $\beta, \gamma \ll \alpha$. In these regions the collapse is one-dimensional and *walls* (pancakes)

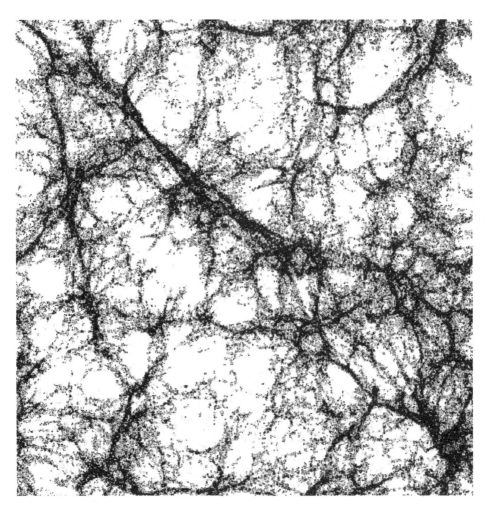

Fig. 6.1.

are formed. The walls connect the filaments. It is clear that in those places where α, β, γ are all negative (valleys) the expansion will continue forever. Matter will be diluted in these regions and empty voids, occupying most of the volume, will be produced.

Over time these anisotropic structures simply become elements of the structure forming on even larger scales. Finally the anisotropic substructure of the larger structural units becomes virialized and disappears. That is, the web of filaments and walls is only an intermediate stage of evolution between an initial, nearly homogeneous state and a final, virialized isotropic structure.

At present we observe filaments and pancakes on scales which entered the non-linear stage not very long ago. Therefore their average density is only a few times greater than the average density in the universe. The filaments connect quasi-spherical structures, in which the density is larger than in filaments.

Baryons fall into the potential wells created by the cold dark matter and form luminous galaxies. Most of the galaxies are concentrated in the quasi-spherical inhomogeneities corresponding to clusters and superclusters of galaxies with scales of order of a few Mpc. Some of them reside in filaments which have scales of 10–30 Mpc, and others are yet in pancakes.

The complete quantitative theory of the structure formation is a challenging numerical problem and represents a frontier of current research.

7

Gravitational instability in General Relativity

The Newtonian analysis of gravitational instability has limitations. It clearly fails for perturbations on scales larger than the Hubble radius. In the case of a relativistic fluid we have to use General Relativity for both short-wavelength and long-wavelength perturbations. This theory gives us a unified description for any matter on all scales. Unfortunately, the physical interpretation of the results obtained is less transparent in General Relativity than in Newtonian theory. The main problem is the freedom in the choice of coordinates used to describe the perturbations. In contrast to the homogeneous and isotropic universe, where the preferable coordinate system is fixed by the symmetry properties of the background, there are no obvious preferable coordinates for analyzing perturbations. The freedom in the coordinate choice, or gauge freedom, leads to the appearance of fictitious perturbation modes. These fictitious modes do not describe any real inhomogeneities, but reflect only the properties of the coordinate system used.

To demonstrate this point let us consider an *undisturbed* homogeneous isotropic universe, where $\varepsilon(\mathbf{x}, t) = \varepsilon(t)$. In General Relativity any coordinate system is allowed, and we can in principle decide to use a "new" time coordinate $\tilde{t}$, related to the "old" time t via $\tilde{t} = t + \delta t(\mathbf{x}, t)$. Then the energy density $\tilde{\varepsilon}(\tilde{t}, \mathbf{x}) \equiv \varepsilon(t(\tilde{t}, \mathbf{x}))$ on the hypersurface $\tilde{t} = $ const depends, in general, on the spatial coordinates $\mathbf{x}$ (Figure 7.1). Assuming that $\delta t \ll t$, we have

$$\varepsilon(t) = \varepsilon(\tilde{t} - \delta t(\mathbf{x}, t)) \simeq \varepsilon(\tilde{t}) - \frac{\partial \varepsilon}{\partial t}\delta t \equiv \varepsilon(\tilde{t}) + \delta\varepsilon(\mathbf{x}, \tilde{t}). \quad (7.1)$$

The first term on the right hand side must be interpreted as the background energy density in the new coordinate system, while the second describes a linear perturbation. This perturbation is nonphysical and entirely due to the choice of the new "disturbed" time. Thus we can "produce" *fictitious* perturbations simply by perturbing the coordinates. Moreover, we can "remove" a *real* perturbation in the energy density by choosing the hypersurfaces of constant time to be the same as

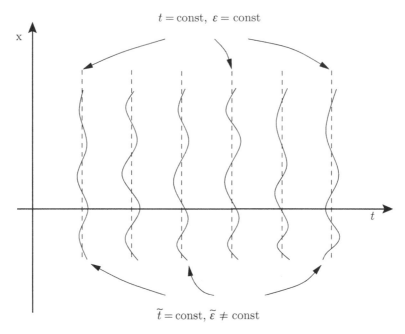

Fig. 7.1.

the hypersurfaces of constant energy: in this case $\delta\varepsilon = 0$ in spite of the presence of the *real* inhomogeneities.

To resolve real and fictitious perturbation modes in General Relativity, it is necessary to have a full set of variables. To be precise we need both the matter field perturbations and the metric perturbations.

In this chapter we introduce gauge-invariant variables, which do not depend on the particular choice of coordinates and have a clear physical interpretation. We apply the formalism developed to study the behavior of relativistic perturbations in a few interesting cases. To simplify the formulae we consider only a spatially flat universe. The generalization of the results obtained to nonflat universes is largely straightforward.

7.1 Perturbations and gauge-invariant variables

Inhomogeneities in the matter distribution induce metric perturbations which can be decomposed into irreducible pieces. In the linear approximation different types of perturbations evolve independently and therefore can be analyzed separately. In this section we first classify metric perturbations, then determine how they transform under general coordinate (gauge) transformations and finally construct the gauge-invariant variables. The relation between the different coordinate systems prevalent in the literature is also discussed.

7.1.1 Classification of perturbations

The metric of a flat Friedmann universe with small perturbations can be written as

$$ds^2 = \left[{}^{(0)}g_{\alpha\beta} + \delta g_{\alpha\beta}(x^\gamma)\right]dx^\alpha dx^\beta, \tag{7.2}$$

where $|\delta g_{\alpha\beta}| \ll |{}^{(0)}g_{\alpha\beta}|$. Using conformal time, the background metric becomes

$${}^{(0)}g_{\alpha\beta}dx^\alpha dx^\beta = a^2(\eta)\left(d\eta^2 - \delta_{ij}dx^i dx^j\right). \tag{7.3}$$

The metric perturbations $\delta g_{\alpha\beta}$ can be categorized into three distinct types: *scalar*, *vector* and *tensor* perturbations. This classification is based on the symmetry properties of the homogeneous, isotropic background, which at a given moment of time is obviously invariant with respect to the group of spatial rotations and translations. The δg_{00} component behaves as a scalar under these rotations and hence

$$\delta g_{00} = 2a^2\phi, \tag{7.4}$$

where ϕ is a 3-scalar.

The spacetime components δg_{0i} can be decomposed into the sum of the spatial gradient of some scalar B and a vector S_i with zero divergence:

$$\delta g_{0i} = a^2\left(B_{,i} + S_i\right). \tag{7.5}$$

Here a comma with index denotes differentiation with respect to the corresponding spatial coordinate, e.g. $B_{,i} = \partial B/\partial x^i$. The vector S_i satisfies the constraint $S^i_{,i} = 0$ and therefore has two independent components. *From now on the spatial indices are always raised and lowered with the unit metric δ_{ij} and we assume summation over repeated spatial indices.*

In a similar way, the components δg_{ij}, which behave as a tensor under 3-rotations, can be written as the sum of the irreducible pieces:

$$\delta g_{ij} = a^2\left(2\psi\delta_{ij} + 2E_{,ij} + F_{i,j} + F_{j,i} + h_{ij}\right). \tag{7.6}$$

Here ψ and E are scalar functions, vector F_i has zero divergence ($F^i_{,i} = 0$) and the 3-tensor h_{ij} satisfies the four constraints

$$h^i_i = 0, \quad h^i_{j,i} = 0, \tag{7.7}$$

that is, it is traceless and transverse. Counting the number of independent functions used to form $\delta g_{\alpha\beta}$, we find we have four functions for the scalar perturbations, four functions for the vector perturbations (two 3-vectors with one constraint each), and two functions for the tensor perturbations (a symmetric 3-tensor has six independent components and there are four constraints). Thus we have ten functions altogether. This number coincides with the number of independent components of $\delta g_{\alpha\beta}$.

Scalar perturbations are characterized by the four scalar functions ϕ, ψ, B, E. They are induced by energy density inhomogeneities. These perturbations are most important because they exhibit gravitational instability and may lead to the formation of structure in the universe.

Vector perturbations are described by the two vectors S_i and F_i and are related to the rotational motions of the fluid. As in Newtonian theory, they decay very quickly and are not very interesting from the point of view of cosmology.

Tensor perturbations h_{ij} have no analog in Newtonian theory. They describe *gravitational waves*, which are the degrees of freedom of the gravitational field itself. In the linear approximation the gravitational waves do not induce any perturbations in the perfect fluid.

Scalar, vector and tensor perturbations are decoupled and thus can be studied separately.

7.1.2 Gauge transformations and gauge-invariant variables

Let us consider the coordinate transformation

$$x^\alpha \to \tilde{x}^\alpha = x^\alpha + \xi^\alpha, \tag{7.8}$$

where ξ^α are infinitesimally small functions of space and time. At *a given point of the spacetime manifold* the metric tensor in the coordinate system $\tilde{x}$ can be calculated using the usual transformation law

$$\tilde{g}_{\alpha\beta}(\tilde{x}^\rho) = \frac{\partial x^\gamma}{\partial \tilde{x}^\alpha} \frac{\partial x^\delta}{\partial \tilde{x}^\beta} g_{\gamma\delta}(x^\rho) \approx {}^{(0)}g_{\alpha\beta}(x^\rho) + \delta g_{\alpha\beta} - {}^{(0)}g_{\alpha\delta}\xi^\delta_{,\beta} - {}^{(0)}g_{\gamma\beta}\xi^\gamma_{,\alpha}, \tag{7.9}$$

where we have kept only the terms linear in δg and ξ. In the new coordinates $\tilde{x}$ the metric can also be split into background and perturbation parts,

$$\tilde{g}_{\alpha\beta}(\tilde{x}^\rho) = {}^{(0)}g_{\alpha\beta}(\tilde{x}^\rho) + \delta \tilde{g}_{\alpha\beta}, \tag{7.10}$$

where ${}^{(0)}g_{\alpha\beta}$ is the Friedmann metric (7.3), which now depends on $\tilde{x}$. Comparing the expressions in (7.9) and (7.10) and taking into account that

$${}^{(0)}g_{\alpha\beta}(x^\rho) \approx {}^{(0)}g_{\alpha\beta}(\tilde{x}^\rho) - {}^{(0)}g_{\alpha\beta,\gamma}\xi^\gamma, \tag{7.11}$$

we infer the following gauge transformation law:

$$\delta g_{\alpha\beta} \to \delta \tilde{g}_{\alpha\beta} = \delta g_{\alpha\beta} - {}^{(0)}g_{\alpha\beta,\gamma}\xi^\gamma - {}^{(0)}g_{\gamma\beta}\xi^\gamma_{,\alpha} - {}^{(0)}g_{\alpha\delta}\xi^\delta_{,\beta}. \tag{7.12}$$

7.1 Perturbations and gauge-invariant variables

Problem 7.1 Consider a 4-scalar $q(x^\rho) = {}^{(0)}q(x^\rho) + \delta q$, where ${}^{(0)}q$ is its background value, and verify that the perturbation δq transforms under (7.8) as

$$\delta q \to \delta \tilde{q} = \delta q - {}^{(0)}q_{,\alpha}\xi^\alpha. \tag{7.13}$$

Similarly, show that for a *covariant* 4-vector,

$$\delta u_\alpha \to \delta \tilde{u}_\alpha = \delta u_\alpha - {}^{(0)}u_{\alpha,\gamma}\xi^\gamma - {}^{(0)}u_\gamma \xi^\gamma_{,\alpha}. \tag{7.14}$$

Of course the value of a 4-scalar q at a given point of the manifold does not change as a result of the coordinate transformation, but the way we split it into a background value and a perturbation depends on the coordinates used.

Let us write the spatial components of the infinitesimal vector $\xi^\alpha \equiv (\xi^0, \xi^i)$ as

$$\xi^i = \xi^i_\perp + \zeta^{,i}, \tag{7.15}$$

where $\xi^i_\perp$ is a 3-vector with zero divergence $\left(\xi^i_{\perp,i} = 0\right)$ and ζ is a scalar function. Since in the Friedmann universe ${}^{(0)}g_{00} = a^2(\eta)$ and ${}^{(0)}g_{ij} = -a^2(\eta)\delta_{ij}$, we obtain from (7.12)

$$\begin{aligned}
\delta \tilde{g}_{00} &= \delta g_{00} - 2a\left(a\xi^0\right)', \\
\delta \tilde{g}_{0i} &= \delta g_{0i} + a^2\left[\xi'_{\perp i} + \left(\zeta' - \xi^0\right)_{,i}\right], \\
\delta \tilde{g}_{ij} &= \delta g_{ij} + a^2\left[2\frac{a'}{a}\delta_{ij}\xi^0 + 2\zeta_{,ij} + \left(\xi_{\perp i,j} + \xi_{\perp j,i}\right)\right],
\end{aligned} \tag{7.16}$$

where $\xi_{\perp i} \equiv \xi^i_\perp$ and a prime denotes the derivative with respect to conformal time η. Combining these results with (7.4)–(7.6), we immediately derive the transformation laws for the different types of perturbations.

Scalar perturbations For scalar perturbations the metric takes the form

$$ds^2 = a^2\left[(1+2\phi)\,d\eta^2 + 2B_{,i}dx^i d\eta - \left((1-2\psi)\delta_{ij} - 2E_{,ij}\right)dx^i dx^j\right]. \tag{7.17}$$

Under the change of coordinates we have

$$\begin{aligned}
\phi \to \tilde{\phi} &= \phi - \frac{1}{a}\left(a\xi^0\right)', & B \to \tilde{B} &= B + \zeta' - \xi^0, \\
\psi \to \tilde{\psi} &= \psi + \frac{a'}{a}\xi^0, & E \to \tilde{E} &= E + \zeta.
\end{aligned} \tag{7.18}$$

Thus, only ξ^0 and ζ contribute to the transformations of scalar perturbations and by choosing them appropriately we can make any two of the four functions ϕ, ψ, B, E vanish. The simplest *gauge-invariant* linear combinations of these functions, which span the two-dimensional space of the physical perturbations, are

$$\Phi \equiv \phi - \frac{1}{a}\left[a(B - E')\right]', \quad \Psi \equiv \psi + \frac{a'}{a}(B - E'). \tag{7.19}$$

It is easy to see that they do not change under the coordinate transformations and if Φ and Ψ vanish in one particular coordinate system, they will be zero in any coordinate system. This means we can immediately distinguish physical inhomogeneities from fictitious perturbations; if both Φ and Ψ are equal to zero, then the metric perturbations (if they are present) are fictitious and can be removed by a change of coordinates.

Of course one can construct an infinite number of gauge-invariant variables, since any combination of Φ and Ψ will also be gauge-invariant. Our choice of these variables is justified only by reason of convenience. As with the electric and magnetic fields in electrodynamics, the potentials Φ and Ψ are the simplest possible combinations and satisfy simple equations of motion (see the following section).

Problem 7.2 Using the results in Problem 7.1, verify that

$$\overline{\delta\varepsilon} = \delta\varepsilon - \varepsilon'_0(B - E') \tag{7.20}$$

is the gauge-invariant variable characterizing the energy density perturbations.

Taking into account that the 4-velocity of a fluid in a homogeneous universe is $^{(0)}u_\alpha = (a, 0, 0, 0)$, show that

$$\overline{\delta u_0} = \delta u_0 - \left[a(B - E')\right]', \quad \overline{\delta u_i} = \delta u_i - a(B - E')_{,i} \tag{7.21}$$

are the gauge-invariant variables for the *covariant* components of the velocity perturbations δu_α.

Vector perturbations For vector perturbations the metric is

$$ds^2 = a^2\left[d\eta^2 + 2S_i dx^i d\eta - (\delta_{ij} - F_{i,j} - F_{j,i})dx^i dx^j\right], \tag{7.22}$$

and the variables S_i and F_i transform as

$$S_i \to \tilde{S}_i = S_i + \xi'_{\perp i}, \quad F_i \to \tilde{F}_i = F_i + \xi_{\perp i}. \tag{7.23}$$

It is obvious that

$$\overline{V}_i = S_i - F'_i \tag{7.24}$$

is gauge-invariant. Only two of the four independent functions S_i, F_i characterize physical perturbations; the other two reflect the coordinate freedom. The variables (7.24) span the two-dimensional space of physical perturbations and describe rotational motions. The corresponding *covariant* components of the rotational velocity $\delta u_{\perp i}$, satisfying the condition $(\delta u_{\perp i})^{,i} = 0$, are also gauge-invariant.

Tensor perturbations For tensor perturbations,

$$ds^2 = a^2\left[d\eta^2 - (\delta_{ij} - h_{ij})\, dx^i dx^j\right] \tag{7.25}$$

and h_{ij} does not change under coordinate transformations. It already describes the gravitational waves in a gauge-invariant manner.

7.1.3 Coordinate systems

Gauge freedom has its most important manifestation in scalar perturbations. We can use it to impose two conditions on the functions $\phi, \psi, B, E, \delta\varepsilon$ and the potential velocity perturbations $\delta u_{\parallel i} = \varphi_{,i}$. This is possible since we are free to choose the two functions ξ^0 and ζ. Imposing the gauge conditions is equivalent to fixing the (class of) coordinate system(s). In the following we consider several choices of gauge and show how, knowing the solution for the gauge-invariant variables, one can calculate the metric and density perturbations in any particular coordinate system in a simple way.

Longitudinal (conformal-Newtonian) gauge Longitudinal gauge is defined by the conditions $B_l = E_l = 0$. From (7.18), it follows that these conditions fix the coordinate system uniquely. In fact, the condition $E_l = 0$ is violated by any $\zeta \neq 0$, and using this result we see that any time transformation with $\xi^0 \neq 0$ destroys the condition $B_l = 0$. Hence there is no extra coordinate freedom which preserves $B_l = E_l = 0$. In the corresponding coordinate system the metric takes the form

$$ds^2 = a^2\left[(1 + 2\phi_l)d\eta^2 - (1 - 2\psi_l)\delta_{ij}dx^i dx^j\right]. \tag{7.26}$$

If the spatial part of the energy–momentum tensor is diagonal, that is, $\delta T^i_j \propto \delta^i_j$, we have $\phi_l = \psi_l$ (see the following section) and there remains only one variable characterizing scalar metric perturbations. The variable ϕ_l is a generalization of the Newtonian potential, which explains the choice of the name "conformal-Newtonian" for this coordinate system. As can be seen from (7.19)–(7.21), the gauge-invariant variables have a very simple physical interpretation: they are the amplitudes of the metric, density and velocity perturbations in the conformal-Newtonian coordinate system, in particular, $\Phi = \phi_l$, $\Psi = \psi_l$.

Synchronous gauge Synchronous coordinates, where $\delta g_{0\alpha} = 0$, have been used most widely in the literature. In our notation, they correspond to the gauge choice $\phi_s = 0$ and $B_s = 0$. This does not fix the coordinates uniquely; there exists a whole class of synchronous coordinate systems. From (7.18), it follows that if the conditions $\phi_s = 0$ and $B_s = 0$ are satisfied in some coordinate system $x^\alpha \equiv (\eta, \mathbf{x})$, then they will also be satisfied in another coordinate system $\tilde{x}^\alpha$, related to x^α by

$$\tilde{\eta} = \eta + \frac{C_1}{a}, \quad \tilde{x}^i = x^i + C_{1,i} \int \frac{d\eta}{a} + C_{2,i}, \tag{7.27}$$

where $C_1 \equiv C_1(x^j)$ and $C_2 \equiv C_2(x^j)$ are arbitrary functions of the spatial coordinates. This residual coordinate freedom leads to the appearance of unphysical gauge modes, which render the interpretation of the results difficult, especially on scales larger than the Hubble radius.

If we know a solution for perturbations in terms of the gauge-invariant variables or, equivalently, in the conformal-Newtonian coordinate system, then the behavior of perturbations in the synchronous coordinate system can easily be determined without needing to solve the Einstein equations again. Using the definitions in (7.19) we have

$$\Phi = \frac{1}{a}\left[a E_s'\right]', \quad \Psi = \psi_s - \frac{a'}{a} E_s'. \tag{7.28}$$

These two equations can easily be resolved to express ψ_s and E_s in terms of the gauge-invariant potentials:

$$E_s = \int \frac{1}{a}\left(\int^\eta a\Phi\, d\tilde{\eta}\right) d\eta, \quad \psi_s = \Psi + \frac{a'}{a^2}\int a\Phi\, d\eta. \tag{7.29}$$

Similarly, from (7.20) it follows that the energy density perturbations are

$$\delta\varepsilon_s = \overline{\delta\varepsilon} - \frac{\varepsilon_0'}{a}\int a\Phi\, d\eta. \tag{7.30}$$

The constants of integration arising in these formulae correspond to unphysical, fictitious modes.

Problem 7.3 Impose the *comoving* gauge conditions

$$\phi = 0, \quad \delta u_\parallel^i = -\frac{1}{a^2}\delta u_{\parallel i} + \frac{1}{a} B_{,i} = 0, \tag{7.31}$$

where $\delta u_\parallel^i$ are the *contravariant* spatial components of the potential 4-velocity, and find the metric perturbations in the comoving coordinate system in terms of the gauge-invariant variables. Do these conditions fix the coordinates uniquely?

7.2 Equations for cosmological perturbations

To derive the equations for the perturbations we have to linearize the Einstein equations

$$G^\alpha_\beta \equiv R^\alpha_\beta - \tfrac{1}{2}\delta^\alpha_\beta R = 8\pi G T^\alpha_\beta,$$

for small inhomogeneities about a Friedmann universe. The calculation of the Einstein tensor for the background metric (7.3) is very simple and the result is

$$^{(0)}G^0_0 = \frac{3\mathcal{H}^2}{a^2}, \quad {}^{(0)}G^0_i = 0, \quad {}^{(0)}G^i_j = \frac{1}{a^2}(2\mathcal{H}' + \mathcal{H}^2)\delta_{ij}, \qquad (7.32)$$

where $\mathcal{H} \equiv a'/a$. It is clear that in order to satisfy the background Einstein equations, the energy–momentum tensor for the matter, ${}^{(0)}T^\alpha_\beta$, must have the following symmetry properties:

$$^{(0)}T^0_i = 0, \quad {}^{(0)}T^i_j \propto \delta_{ij}. \qquad (7.33)$$

For a metric with small perturbations, the Einstein tensor can be written as $G^\alpha_\beta = {}^{(0)}G^\alpha_\beta + \delta G^\alpha_\beta + \cdots$, where δG^α_β denote terms linear in metric fluctuations. The energy–momentum tensor can be split in a similar way and the linearized equations for perturbations are

$$\delta G^\alpha_\beta = 8\pi G \delta T^\alpha_\beta. \qquad (7.34)$$

Neither δG^α_β nor δT^α_β are gauge-invariant. Combining them with the metric perturbations, however, we can construct corresponding gauge-invariant quantities.

Problem 7.4 Derive the transformation laws for δT^α_β and verify that

$$\begin{aligned}
\overline{\delta T}^0_0 &= \delta T^0_0 - \left({}^{(0)}T^0_0\right)'(B - E'), \\
\overline{\delta T}^0_i &= \delta T^0_i - \left({}^{(0)}T^0_0 - {}^{(0)}T^k_k/3\right)(B - E')_{,i}, \\
\overline{\delta T}^i_j &= \delta T^i_j - \left({}^{(0)}T^i_j\right)'(B - E'),
\end{aligned} \qquad (7.35)$$

where T^k_k is the trace of the spatial components, are gauge-invariant.

In a similar manner to (7.35), we can construct

$$\overline{\delta G}^0_0 = \delta G^0_0 - \left({}^{(0)}G^0_0\right)'(B - E'), \text{ etc.} \qquad (7.36)$$

and rewrite (7.34) in the form

$$\overline{\delta G}^\alpha_\beta = 8\pi G \overline{\delta T}^\alpha_\beta. \qquad (7.37)$$

The components of $\overline{\delta T}^\alpha_\beta$ can also be decomposed into scalar, vector and tensor pieces; each piece contributes only to the evolution of the corresponding perturbation.

Scalar perturbations The left hand side in (7.37) is gauge-invariant and depends only on the metric perturbations. Therefore it can be expressed entirely in terms of the potentials Φ and Ψ. The direct calculation of $\overline{\delta G}^\alpha_\beta$ for the metric (7.17) gives the equations:

$$\Delta\Psi - 3\mathcal{H}(\Psi' + \mathcal{H}\Phi) = 4\pi G a^2 \overline{\delta T}^0_{\ 0}, \tag{7.38}$$

$$\left(\Psi' + \mathcal{H}\Phi\right)_{,i} = 4\pi G a^2 \overline{\delta T}^0_{\ i}, \tag{7.39}$$

$$\left[\Psi'' + \mathcal{H}(2\Psi + \Phi)' + \left(2\mathcal{H}' + \mathcal{H}^2\right)\Phi + \tfrac{1}{2}\Delta(\Phi - \Psi)\right]\delta_{ij} \\ - \tfrac{1}{2}(\Phi - \Psi)_{,ij} = -4\pi G a^2 \overline{\delta T}^i_{\ j}. \tag{7.40}$$

We have to stress that these equations can be derived without imposing any gauge conditions and they are valid in an arbitrary coordinate system. To obtain the explicit form of the equations for the metric perturbations in a particular coordinate system, we simply have to express Φ and Ψ in (7.38)–(7.40) through these perturbations. For example, in the synchronous coordinate system, we would use the expressions in (7.28).

Problem 7.5 Write down the equations for ψ_s and E_s in the synchronous coordinate system. (*Hint* Do not forget that E_s enters the definition of $\overline{\delta T}^\alpha_\beta$.)

The equations for the metric perturbations in the conformal-Newtonian coordinate system obviously coincide with (7.38)–(7.40). Therefore calculations in these coordinates are identical to calculations in terms of the gauge-invariant potentials, with the advantage that we need not carry B and E through the intermediate formulae.

Problem 7.6 Derive (7.38)–(7.40). (*Hint*: The direct calculation of $\overline{\delta G}^\alpha_\beta$ in terms of the gauge-invariant potentials is rather tedious. However, it can be significantly simplified if we take into account that these potentials coincide with the metric perturbations in the conformal-Newtonian coordinate system. Therefore, calculate the Einstein tensor for the metric (7.26) and then replace ϕ_l, ψ_l with Φ and Ψ respectively. It is convenient to calculate δG^α_β in two steps: (a) set $a = 1$ in (7.26) and find the Einstein tensor in this case, (b) make a conformal transformation to an arbitrary $a(t)$ and calculate δG^α_β using formulae (5.110), (5.111), where $F = a^2$.)

Vector perturbations The equations for the vector perturbations take the forms

$$\Delta \overline{V}_i = 16\pi G a^2 \overline{\delta T}^0_{i(V)}, \tag{7.41}$$

$$\left(\overline{V}_{i,j} + \overline{V}_{j,i}\right)' + 2\mathcal{H}\left(\overline{V}_{i,j} + \overline{V}_{j,i}\right) = -16\pi G a^2 \overline{\delta T}^i_{j(V)}, \tag{7.42}$$

where $\overline{V}_i$ is defined in (7.24) and $\overline{\delta T}^\alpha_{\beta(V)}$ is the vector part of the energy–momentum tensor.

Tensor perturbations For the gravitational waves we obtain

$$\left(h''_{ij} + 2\mathcal{H} h'_{ij} - \Delta h_{ij}\right) = 16\pi G a^2 \overline{\delta T}^i_{j(T)}, \tag{7.43}$$

where $\overline{\delta T}^i_{j(T)}$ is that part of the energy–momentum tensor which has the same structural form as h_{ij}.

Problem 7.7 Derive (7.41), (7.42) and (7.43).

7.3 Hydrodynamical perturbations

Let us consider a perfect fluid with energy–momentum tensor

$$T^\alpha_\beta = (\varepsilon + p) u^\alpha u_\beta - p \delta^\alpha_\beta. \tag{7.44}$$

One can easily verify that its gauge-invariant perturbations, defined in (7.35), can be written as

$$\overline{\delta T}^0_0 = \overline{\delta \varepsilon}, \quad \overline{\delta T}^0_i = \frac{1}{a}(\varepsilon_0 + p_0)\left(\overline{\delta u}_{\|i} + \delta u_{\perp i}\right), \quad \overline{\delta T}^i_j = -\overline{\delta p} \delta^i_j, \tag{7.45}$$

where $\overline{\delta \varepsilon}$, $\overline{\delta u}_{\|i}$ and $\overline{\delta p}$ are defined in (7.20), (7.21). The only term, which contributes to the vector perturbations is proportional to $\delta u_{\perp i}$; all other terms have the same structural form as the scalar metric perturbations.

7.3.1 Scalar perturbations

Since $\delta T^i_j = 0$ for $i \neq j$, (7.40) reduces to

$$(\Phi - \Psi)_{,ij} = 0 \quad (i \neq j). \tag{7.46}$$

The only solution consistent with Φ and Ψ being perturbations is $\Psi = \Phi$. Then substituting (7.45) into (7.38)–(7.40) we arrive at the following set of equations for the scalar perturbations:

$$\Delta \Phi - 3\mathcal{H}(\Phi' + \mathcal{H}\Phi) = 4\pi G a^2 \overline{\delta \varepsilon}, \tag{7.47}$$

$$(a\Phi)'_{,i} = 4\pi G a^2 (\varepsilon_0 + p_0) \overline{\delta u}_{\|i}, \tag{7.48}$$

$$\Phi'' + 3\mathcal{H}\Phi' + (2\mathcal{H}' + \mathcal{H}^2)\Phi = 4\pi G a^2 \overline{\delta p}. \tag{7.49}$$

In a nonexpanding universe $\mathcal{H} = 0$, and the first equation exactly coincides with the usual Poisson equation for the gravitational potential. In an expanding universe, the second and third terms on the left hand side in (7.47) are suppressed on sub-Hubble scales by a factor $\sim \lambda/H^{-1}$, and hence can be neglected. Thus (7.47) generalizes the Poisson equation and supports the interpretation of Φ as the relativistic generalization of the Newtonian gravitational potential. Note that, on scales smaller than the Hubble radius, (7.47) can be applied even to nonlinear inhomogeneities, because it requires only $|\Phi| \ll 1$ but not necessarily $|\delta\varepsilon/\varepsilon_0| \ll 1$. From (7.48) it follows that the time derivative of $(a\Phi)'$ serves as the velocity potential.

Given $p(\varepsilon, S)$, the pressure fluctuations $\overline{\delta p}$ can be expressed in terms of the energy density and entropy perturbations,

$$\overline{\delta p} = c_s^2 \overline{\delta \varepsilon} + \tau \delta S, \tag{7.50}$$

where $c_s^2 \equiv (\partial p/\partial \varepsilon)_S$ is the square of the speed of sound and $\tau \equiv (\partial p/\partial S)_\varepsilon$. Taking this into account and combining (7.47) and (7.49), we obtain the closed form equation for the gravitational potential

$$\Phi'' + 3\left(1 + c_s^2\right)\mathcal{H}\Phi' - c_s^2 \Delta\Phi + \left(2\mathcal{H}' + \left(1 + 3c_s^2\right)\mathcal{H}^2\right)\Phi = 4\pi G a^2 \tau \delta S. \tag{7.51}$$

Below we begin by finding the exact solutions of this equation in two particular cases: (a) for nonrelativistic matter with zero pressure, and (b) for relativistic fluid with constant equation of state $p = w\varepsilon$. Then we analyze the behavior of adiabatic perturbations ($\delta S = 0$) for a general equation of state $p(\varepsilon)$, and finally consider entropy perturbations.

Nonrelativistic matter ($p = 0$) In a flat matter-dominated universe $a \propto \eta^2$ and $\mathcal{H} = 2/\eta$. In this case (7.51) simplifies to

$$\Phi'' + \frac{6}{\eta}\Phi' = 0, \tag{7.52}$$

and has the solution

$$\Phi = C_1(\mathbf{x}) + \frac{C_2(\mathbf{x})}{\eta^5}, \tag{7.53}$$

where $C_1(\mathbf{x})$ and $C_2(\mathbf{x})$ are arbitrary functions of the comoving spatial coordinates. From (7.47) we find the gauge-invariant density perturbations

$$\frac{\overline{\delta\varepsilon}}{\varepsilon_0} = \frac{1}{6}\left[\left(\Delta C_1 \eta^2 - 12 C_1\right) + \left(\Delta C_2 \eta^2 + 18 C_2\right)\frac{1}{\eta^5}\right]. \tag{7.54}$$

The nondecaying mode of the gravitational potential, C_1, remains constant regardless of the relative size of the lengthscale of the inhomogeneity relative to the

7.3 Hydrodynamical perturbations

Hubble radius. The behavior of the energy density perturbations, however, depends crucially on the scale.

Let us consider a plane wave perturbation with comoving wavenumber $k \equiv |\mathbf{k}|$, for which $C_{1,2} \propto \exp(i\mathbf{k}\mathbf{x})$. If the physical scale $\lambda_{ph} \sim a/k$ is much larger than the Hubble scale $H^{-1} \sim a\eta$ or, equivalently, $k\eta \ll 1$, then to leading order

$$\frac{\overline{\delta\varepsilon}}{\varepsilon_0} \simeq -2C_1 + \frac{3C_2}{\eta^5}. \tag{7.55}$$

Neglecting the decaying mode, the relation between the energy density fluctuations and the gravitational potential on superhorizon scales becomes $\overline{\delta\varepsilon}/\varepsilon_0 \simeq -2\Phi$.

For shortwave perturbations with $k\eta \gg 1$, we have

$$\frac{\overline{\delta\varepsilon}}{\varepsilon_0} \simeq -\frac{k^2}{6}(C_1\eta^2 + C_2\eta^{-3}) = \tilde{C}_1 t^{2/3} + \tilde{C}_2 t^{-1}, \tag{7.56}$$

in agreement with the Newtonian result (6.56).

Problem 7.8 Determine the behavior of the peculiar velocity for nonrelativistic matter.

Problem 7.9 Substituting (7.53) and (7.54) in (7.29) and (7.30), calculate the metric and energy density perturbations in the synchronous coordinate system. Analyze the behavior of the long- and short-wavelength perturbations. Is the Newtonian limit explicit in this coordinate system?

Ultra-relativistic matter Let us now study the behavior of adiabatic perturbations ($\delta S = 0$) in the universe dominated by relativistic matter with equation of state $p = w\varepsilon$, where w is a positive constant. In this case the scale factor increases as $a \propto \eta^{2/(1+3w)}$ (see Problem 1.18). With $c_s^2 = w$, and for a plane wave perturbation $\Phi = \Phi_\mathbf{k}(\eta)\exp(i\mathbf{k}\mathbf{x})$, (7.51) becomes

$$\Phi_\mathbf{k}'' + \frac{6(1+w)}{1+3w}\frac{1}{\eta}\Phi_\mathbf{k}' + wk^2\Phi_\mathbf{k} = 0. \tag{7.57}$$

The solution of this equation is

$$\Phi_\mathbf{k} = \eta^{-\nu}\left[C_1 J_\nu(\sqrt{w}k\eta) + C_2 Y_\nu(\sqrt{w}k\eta)\right], \quad \nu \equiv \frac{1}{2}\left(\frac{5+3w}{1+3w}\right), \tag{7.58}$$

where J_ν and Y_ν are Bessel functions of order ν.

Considering long-wavelength inhomogeneities, for which $\sqrt{w}k\eta \ll 1$, and using the small-argument expansion of the Bessel functions, we see that in this limit the nondecaying mode of Φ is constant. It follows from (7.47) that

$$\overline{\delta\varepsilon}/\varepsilon_0 \simeq -2\Phi. \tag{7.59}$$

Perturbations on scales smaller than the Jeans length $\lambda_J \sim c_s t$, for which $\sqrt{w}k\eta \gg 1$, behave as sound waves with decaying amplitude

$$\Phi_k \propto \eta^{-\nu-\frac{1}{2}} \exp(\pm i\sqrt{w}k\eta). \tag{7.60}$$

In a *radiation-dominated* universe ($w = 1/3$), the order of the Bessel functions is $\nu = 3/2$ and they can be expressed in terms of elementary functions. We have

$$\Phi_k = \frac{1}{x^2}\left[C_1\left(\frac{\sin x}{x} - \cos x\right) + C_2\left(\frac{\cos x}{x} + \sin x\right)\right], \tag{7.61}$$

where $x \equiv k\eta/\sqrt{3}$. The corresponding energy density perturbations are

$$\frac{\overline{\delta\varepsilon}}{\varepsilon_0} = 2C_1\left[\left(\frac{2-x^2}{x^2}\right)\left(\frac{\sin x}{x} - \cos x\right) - \frac{\sin x}{x}\right]$$
$$+ 4C_2\left[\left(\frac{1-x^2}{x^2}\right)\left(\frac{\cos x}{x} + \sin x\right) + \frac{\sin x}{2}\right]. \tag{7.62}$$

General case Unfortunately, (7.51) cannot be solved exactly for an arbitrary equation of state $p(\varepsilon)$. However, it turns out to be possible to derive asymptotic solutions for both long-wavelength and short-wavelength perturbations. To do this, it is convenient to recast the equation in a slightly different form. The "friction term" proportional to Φ' can be eliminated if we introduce the new variable

$$u \equiv \exp\left(\frac{3}{2}\int(1+c_s^2)\mathcal{H}d\eta\right)\Phi \tag{7.63}$$

$$= \exp\left(-\frac{1}{2}\int\left(1+\frac{p_0'}{\varepsilon_0'}\right)\frac{\varepsilon_0'}{(\varepsilon_0+p_0)}d\eta\right)\Phi = \frac{\Phi}{(\varepsilon_0+p_0)^{1/2}},$$

where we have used $c_s^2 = p_0'/\varepsilon_0'$ and expressed $\mathcal{H}$ in terms of ε and p via the conservation law $\varepsilon' = -3\mathcal{H}(\varepsilon + p)$. After some tedious calculations, and using the background equations (see (1.67), (1.68))

$$\mathcal{H}^2 = \frac{8\pi G}{3}a^2\varepsilon_0, \quad \mathcal{H}^2 - \mathcal{H}' = 4\pi Ga^2(\varepsilon_0 + p_0), \tag{7.64}$$

the equation for u can be written in the form

$$u'' - c_s^2\Delta u - \frac{\theta''}{\theta}u = 0, \tag{7.65}$$

where

$$\theta \equiv \frac{1}{a}\left(1+\frac{p_0}{\varepsilon_0}\right)^{-1/2} = \frac{1}{a}\left(\frac{2}{3}\left(1-\frac{\mathcal{H}'}{\mathcal{H}^2}\right)\right)^{-1/2}. \tag{7.66}$$

For a plane wave perturbation, $u \propto \exp(i\mathbf{kx})$, the solutions of (7.65) can easily be found in two limiting cases: on scales much larger and scales much smaller than the Jeans length.

Long-wavelength perturbations When $c_s k\eta \ll 1$, we can omit the spatial derivative term in (7.65) and then $u \propto \theta$ is obviously the solution of this equation. The second solution is derived using the Wronskian,

$$u \simeq C_1 \theta + C_2 \theta \int_{\eta_0} \frac{d\eta}{\theta^2} = C_2 \theta \int_{\bar{\eta}_0} \frac{d\eta}{\theta^2}, \qquad (7.67)$$

where the latter equality is obtained by changing the lower limit of integration from η_0 to $\bar{\eta}_0$ and, thus, absorbing the C_1 mode. Using the definition in (7.66) and integrating by parts, we reduce the integral in (7.67) to

$$\int \frac{d\eta}{\theta^2} = \frac{2}{3} \int a^2 \left[1 + \left(\frac{1}{\mathcal{H}}\right)'\right] d\eta = \frac{2}{3}\left(\frac{a^2}{\mathcal{H}} - \int a^2 d\eta\right). \qquad (7.68)$$

With this result the gravitational potential becomes

$$\Phi = (\varepsilon_0 + p_0)^{1/2} u = A\left(1 - \frac{\mathcal{H}}{a^2}\int a^2 d\eta\right) = A\frac{d}{dt}\left(\frac{1}{a}\int a\, dt\right), \qquad (7.69)$$

where $t = \int a\, d\eta$.

Let us apply this result to find the behavior of long-wavelength, adiabatic perturbations ($\delta S = 0$) in a universe with a mixture of radiation and cold baryons. In this case the scale factor increases as

$$a(\eta) = a_{eq}\left(\xi^2 + 2\xi\right), \qquad (7.70)$$

where $\xi \equiv \eta/\eta_\star$ (see (1.81)) Substituting (7.70) into (7.69), we obtain

$$\Phi = \frac{\xi+1}{(\xi+2)^3}\left[A\left(\frac{3}{5}\xi^2 + 3\xi + \frac{1}{\xi+1} + \frac{13}{3}\right) + B\frac{1}{\xi^3}\right], \qquad (7.71)$$

where A and B are constants of integration multiplying the nondecaying and decaying modes respectively.

Problem 7.10 Calculate the energy density fluctuations.

The behavior of the gravitational potential and the energy density perturbation for an inhomogeneity which enters the Hubble horizon after equality, is shown in Figure 7.2, where we have neglected the decaying mode. We see that Φ and $\overline{\delta\varepsilon}/\varepsilon_0$ are constant at times both early and late compared to $\eta_{eq} \sim \eta_\star$. After the transition from the radiation- to the matter-dominated epoch, the amplitude of both Φ and $\overline{\delta\varepsilon}/\varepsilon_0$ decreases by a factor of 9/10. During the period of matter domination, Φ remains constant while $\overline{\delta\varepsilon}/\varepsilon_0$ begins to increase after the perturbation enters horizon

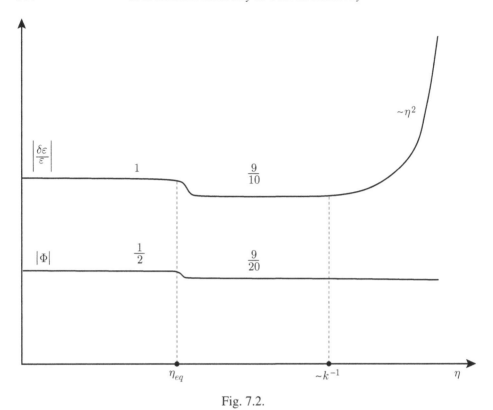

Fig. 7.2.

at $\eta \sim k^{-1}$. It follows from (7.47) that for a constant potential Φ, the amplitude of $\overline{\delta\varepsilon}/\varepsilon_0$ is always equal to -2Φ on superhorizon scales.

The change in the amplitude of Φ can also be inferred from a widely used "conservation law" for the quantity

$$\zeta \equiv \frac{2}{3}\left(\frac{8\pi G}{3}\right)^{1/2}\theta^2\left(\frac{u}{\theta}\right)'. \tag{7.72}$$

Substituting the long-wavelength solution (7.67) into (7.72), we see that ζ remains constant (is conserved) even if $w \equiv p/\varepsilon$ is changing. Recalling the definitions of u, θ and using the background equations of motion, ζ becomes

$$\zeta = \frac{2}{3}\frac{\mathcal{H}^{-1}\Phi' + \Phi}{1+w} + \Phi. \tag{7.73}$$

Let us assume the equation of state w is initially a constant w_i, and later on changes to another constant w_f. In this case the initial and final values of Φ are also constants

7.3 Hydrodynamical perturbations

and it immediately follows from (7.73) that

$$\Phi_f = \left(\frac{1+w_f}{1+w_i}\right)\left(\frac{5+3w_i}{5+3w_f}\right)\Phi_i. \tag{7.74}$$

For a matter–radiation universe, $w_i = 1/3$ and $w_f = 0$, and we obtain the familiar result $\Phi_f = (9/10)\,\Phi_i$.

Problem 7.11 Verify that for a mode with wavenumber k, equation (7.65) can be rewritten in the following integral form:

$$u_k(\eta) = C_1\theta + C_2\theta \int \frac{d\eta}{\theta^2} - k^2\theta \int^\eta \left(\int^{\tilde{\eta}} c_s^2 \theta u_k d\bar{\eta}\right) \frac{1}{\theta^2(\tilde{\eta})} d\tilde{\eta}. \tag{7.75}$$

Using this equation, calculate the subleading k^2-corrections to the long-wavelength solution (7.67) and determine the violation of the "conserved" quantity ζ.

Short-wavelength perturbations When $c_s k \eta \gg 1$, the last term in (7.65) can be neglected. The resulting equation,

$$u'' + c_s^2 k^2 u \simeq 0, \tag{7.76}$$

can easily be solved in the WKB approximation for a slowly varying speed of sound. Its solution describes sound waves with the time-dependent amplitude.

Matching conditions Sometimes it is convenient to approximate the continuous change of the equation of state by a sharp jump. In this case the pressure $p(\varepsilon)$ is discontinuous on the hypersurface of transition Σ, $\varepsilon_T = $ const, and its derivatives become singular. Therefore we cannot directly use the equation for the gravitational potential and, instead, must derive matching conditions for Φ and Φ' on Σ. These conditions can be obtained if we recast (7.65) in the following form:

$$\left[\theta^2\left(\frac{u}{\theta}\right)'\right]' = c_s^2 \theta^2 \Delta\left(\frac{u}{\theta}\right). \tag{7.77}$$

Evidently u/θ should be continuous. Because the scale factor a and the energy density ε are both continuous, the gravitational potential Φ does not jump during the transition, or equivalently, the 3-metric induced on Σ is continuous.

To determine the jump in the derivative of u/θ let us integrate (7.77) within an infinitesimally thin layer near Σ:

$$\left[\theta^2\left(\frac{u}{\theta}\right)'\right]_\pm = \int_{\Sigma-0}^{\Sigma+0} c_s^2 \theta^2 \Delta\left(\frac{u}{\theta}\right) d\eta, \tag{7.78}$$

where $[X]_\pm \equiv X_+ - X_-$ denotes the jump of a variable X on Σ. The integrand in (7.78) is singular and to perform the integration we note that

$$c_s^2 \theta^2 = \left(\frac{p_0'}{\varepsilon_0'}\right)\left(\frac{1}{a^2}\frac{\varepsilon_0}{\varepsilon_0 + p_0}\right) = \frac{\varepsilon_0}{3a^2\mathcal{H}}\left(\frac{1}{\varepsilon_0 + p_0}\right)' - \frac{\varepsilon_0}{a^2(\varepsilon_0 + p_0)}.$$

The second term in the latter equality does not contribute to the integral, while the first term gives

$$\int_{\Sigma-0}^{\Sigma+0} c_s^2 \theta^2 \Delta\left(\frac{u}{\theta}\right) d\eta = \left[\frac{\varepsilon_0}{3a^2\mathcal{H}}\left(\frac{1}{\varepsilon_0 + p_0}\right)\Delta\left(\frac{u}{\theta}\right)\right]_\pm, \quad (7.79)$$

taking into account that a, ε and u/θ are continuous. Substituting (7.79) into (7.78) and expressing u in terms of the gravitational potential, we finally arrive at the following matching conditions:

$$[\Phi]_\pm = 0, \quad \left[\zeta - \frac{2}{9\mathcal{H}^2}\frac{\Delta\Phi}{1+w}\right]_\pm = 0, \quad (7.80)$$

where ζ is defined in (7.72), (7.73). For long-wavelength perturbations the term proportional to $\Delta\Phi$ can be neglected. Hence on superhorizon scales the matching conditions reduce to the continuity of Φ and to the conservation law for ζ.

Problem 7.12 Assuming a sharp transition from the radiation- to the matter-dominated epoch, determine the amplitude of metric perturbations after the transition for both short- and long-wavelength perturbations.

Problem 7.13 Write down the matching conditions explicitly in terms of the metric perturbations in the synchronous coordinate system.

Entropy perturbations Until now we have been considering adiabatic perturbations in an isentropic fluid where the pressure depends only on the energy density. In a multi-component media both adiabatic and entropy perturbations can arise. Generally speaking, the analysis of perturbations in this case is rather complicated because of extra instability modes due to the relative motion of the components. We will consider the problem of cold matter mixed with a baryon–radiation plasma in Section 7.4. Here we content ourselves with studying a fluid of cold baryons tightly coupled to radiation. The baryons do not move with respect to the radiation and this simplifies our task enormously. In particular, we can still use the one-component perfect fluid approximation. However, in this case the pressure depends not only on the energy density, but also on the baryon-to-radiation distribution, characterized by the entropy per baryon, $S \sim T_\gamma^3/n_b$, where n_b is the number density of baryons. Consequently, entropy perturbations can arise. We can study the evolution of these perturbations with the help of (7.51). Here δS is constant because the entropy per baryon is conserved.

First we need to determine the parameter $\tau \equiv (\partial p / \partial S)_\varepsilon$ entering (7.51). The cold baryons do not contribute to the pressure and hence the fluctuations of the total pressure are entirely due to the radiation:

$$\delta p = \delta p_\gamma = \tfrac{1}{3} \delta \varepsilon_\gamma. \tag{7.81}$$

In turn, $\delta \varepsilon_\gamma$ can be expressed in terms of the total energy density perturbation

$$\delta \varepsilon = \delta \varepsilon_\gamma + \delta \varepsilon_b \tag{7.82}$$

and the entropy fluctuation δS. Because the energy density of the radiation, ε_γ, is proportional to T_γ^4 and $\varepsilon_b \propto n_b$, we have $S \propto \varepsilon_\gamma^{3/4} / \varepsilon_b$ and therefore

$$\frac{\delta S}{S} = \frac{3}{4} \frac{\delta \varepsilon_\gamma}{\varepsilon_\gamma} - \frac{\delta \varepsilon_b}{\varepsilon_b}. \tag{7.83}$$

Solving (7.82) and (7.83) for $\delta \varepsilon_\gamma$ in terms of $\delta \varepsilon$ and δS, and substituting the result into (7.81) we obtain

$$\delta p = \frac{1}{3}\left(1 + \frac{3}{4}\frac{\varepsilon_b}{\varepsilon_\gamma}\right)^{-1} \delta \varepsilon + \frac{1}{3}\varepsilon_b \left(1 + \frac{3}{4}\frac{\varepsilon_b}{\varepsilon_\gamma}\right)^{-1} \frac{\delta S}{S}. \tag{7.84}$$

Comparing this expression with (7.50), we can read off the speed of sound c_s and τ:

$$c_s^2 = \frac{1}{3}\left(1 + \frac{3}{4}\frac{\varepsilon_b}{\varepsilon_\gamma}\right)^{-1}, \quad \tau = \frac{c_s^2 \varepsilon_b}{S}. \tag{7.85}$$

For $\delta S \neq 0$, the general solution of (7.51) is the sum of a particular solution and a general solution of the homogeneous equation ($\delta S = 0$). To find a particular solution, we note that

$$2\mathcal{H}' + \left(1 + 3c_s^2\right)\mathcal{H}^2 = 8\pi G a^2 \left(c_s^2 \varepsilon - p\right) = 2\pi G c_s^2 \varepsilon_b.$$

Substituting this expression and τ from (7.85) into (7.51), we immediately see that for the *long-wavelength perturbations*, for which the $\Delta \Phi$ term can be neglected,

$$\Phi = 2\frac{\delta S}{S} = \text{const} \tag{7.86}$$

is a particular solution of this equation.

Physically, the general solution of (7.51), when $\delta S \neq 0$, describes a mixture of adiabatic and entropy modes. How to distinguish between them is a matter of definition. Based on the intuitive idea that in the very early universe the entropy mode should describe inhomogeneities in the baryon distribution on a nearly homogeneous radiation background, we define the entropy mode to have the initial condition

$$\Phi \to 0 \text{ as } \eta \to 0. \tag{7.87}$$

Obviously, the particular solution (7.86) does not satisfy this condition. The solution which does is obtained by adding to (7.86) a general solution (7.71) and choosing the integration constants so that (7.87) is satisfied. The result is

$$\Phi = \frac{1}{5} \xi \frac{\xi^2 + 6\xi + 10}{(\xi + 2)^3} \frac{\delta S}{S}. \tag{7.88}$$

Problem 7.14 Calculate $\delta\varepsilon/\varepsilon$ and $\delta\varepsilon_b/\varepsilon_b$.

In Figure 7.3 we plot the time dependence of Φ, $\delta\varepsilon/\varepsilon$ and $\delta\varepsilon_b/\varepsilon_b$ for the entropy mode. The amplitudes of Φ and $\delta\varepsilon/\varepsilon$ increase linearly until η_{eq}, whereas they are constant for adiabatic perturbations. The fluctuations in the cold-matter density $\delta\varepsilon_b/\varepsilon_b$ are somewhat frozen before η_{eq} and decrease to 2/5 of their initial values

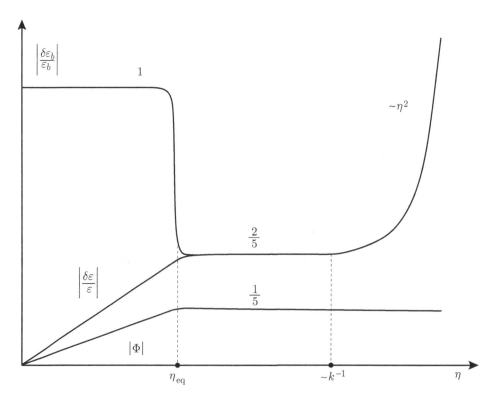

Fig. 7.3.

by the time of equality. For $\eta > \eta_{eq}$, the entropy perturbations evolve like the non-decaying mode of the adiabatic perturbations. There is a key difference, however: it follows from (7.83) that for adiabatic perturbations ($\delta S = 0$) we always have

$$\frac{\delta \varepsilon_\gamma}{\varepsilon_\gamma} = \frac{4}{3}\frac{\delta \varepsilon_b}{\varepsilon_b}, \tag{7.89}$$

whereas for entropy perturbations

$$\frac{\delta \varepsilon_\gamma}{\varepsilon_\gamma} \simeq -2\frac{\delta \varepsilon_b}{\varepsilon_b} \tag{7.90}$$

after equality.

We can also define the isocurvature mode of perturbations by imposing the conditions $\Phi_i = 0$ and $\Phi'_i = 0$ at some initial moment of time $\eta_i \neq 0$. One can easily verify that this mode soon approaches the entropy mode.

7.3.2 Vector and tensor perturbations

For a perfect fluid, the only nonvanishing vector components of δT^α_β are $\delta T^0_i = a^{-1}(\varepsilon_0 + p_0)\delta u_{\perp i}$. Equations (7.45) become

$$\Delta \overline{V}_i = 16\pi G a(\varepsilon_0 + p_0)\delta u_{\perp i}, \tag{7.91}$$

$$\left(\overline{V}_{i,j} + \overline{V}_{j,i}\right)' + 2\mathcal{H}\left(\overline{V}_{i,j} + \overline{V}_{j,i}\right) = 0. \tag{7.92}$$

The solution of the second equation is

$$\overline{V}_i = \frac{C_{\perp i}}{a^2}, \tag{7.93}$$

where $C_{\perp i}$ is a constant of integration. Taking into account that the *physical* velocity is $\delta v^i = a(dx^i/ds) = -a^{-1}\delta u_{\perp i}$, we obtain

$$\delta v^i \propto \frac{1}{a^4(\varepsilon_0 + p_0)}. \tag{7.94}$$

Thus, in a matter-dominated universe, the rotational velocities decay in inverse proportion to the scale factor, in agreement with the result of Newtonian theory. In a radiation-dominated universe $\delta v = $ const. In both cases the metric perturbations given by (7.93) decay very quickly and the primordial vector fluctuations have significant amplitudes at present only if they were originally very large. There is no reason to expect such large primordial vector perturbations and from now on we will completely ignore them.

Tensor perturbations are more interesting, and as we will see in the following chapter, they can be generated during an inflationary stage. In a hydrodynamical universe with perfect fluid $\overline{\delta T}^i_{j(T)} = 0$, (7.43) simplifies to

$$\left(h''_{ij} + 2\mathcal{H} h'_{ij} - \Delta h_{ij}\right) = 0. \tag{7.95}$$

Introducing the rescaled variable v via

$$h_{ij} = \frac{v}{a} e_{ij}, \tag{7.96}$$

where e_{ij} is a time-independent polarization tensor, and considering a plane wave perturbation with the wavenumber k, (7.95) becomes

$$v'' + \left(k^2 - \frac{a''}{a}\right) v = 0. \tag{7.97}$$

In a *radiation-dominated* universe $a \propto \eta$, hence $a'' = 0$ and $v \propto \exp(\pm ik\eta)$. In this case the exact solution of (7.95) is

$$h_{ij} = \frac{1}{\eta}(C_1 \sin(k\eta) + C_2 \cos(k\eta)) e_{ij}. \tag{7.98}$$

The nondecaying mode of the gravitational wave with wavelength larger than the Hubble scale ($k\eta \ll 1$) is constant. After the wavelength becomes smaller than the Hubble radius, the amplitude decays in inverse proportion to the scale factor. This is a general result valid for any equation of state. In fact, for long-wavelength perturbations with $k\eta \ll 1$, we can neglect the k^2 term in (7.97) and its solution becomes

$$v \simeq C_1 a + C_2 a \int \frac{d\eta}{a^2}. \tag{7.99}$$

Hence,

$$h_{ij} = \left(C_1 + C_2 \int \frac{d\eta}{a^2}\right) e_{ij}. \tag{7.100}$$

For $p < \varepsilon$ the second term describes the decaying mode.

For short-wavelength perturbations ($k\eta \gg 1$), we have $k^2 \gg a''/a$ and $h \propto \exp(\pm ik\eta)/a$.

Problem 7.15 Find the exact solution of (7.95) for an arbitrary constant equation of state $p = w\varepsilon$ and analyze the behavior of the short- and long-wavelengh gravitational waves. Consider separately the cases $p = \pm \varepsilon$.

7.4 Baryon–radiation plasma and cold dark matter

Understanding the perturbations in a multi-component medium consisting of a mixture of a baryon–radiation plasma and cold dark matter is important both to analyze the anisotropy of the cosmic microwave background and to determine the transfer function relating the primordial spectrum of density inhomogeneities created during inflation to the spectrum after matter–radiation equality. As a prelude to analysis of the microwave background anisotropies, we consider in this section the calculation of the gravitational potential and the radiation fluctuations at recombination.

Before recombination, baryons are strongly coupled to radiation and the baryon–radiation component can be treated in the hydrodynamical approximation as a single *imperfect* fluid. The other component consists of heavy cold particles, which we assume interact only gravitationally with the baryon–photon plasma and are otherwise free to move with respect to it.

We assume that the number of photons per cold dark matter particle is initially spatially uniform on supercurvature scales (wavelengths greater than H^{-1}) but that the matter and radiation densities vary in space. In other words, we consider *adiabatic* perturbations. As the universe expands and the inhomogeneity scale becomes smaller than the curvature scale, the components move with respect to one another and the entropy (number of photons) per *cold dark matter particle* varies spatially. In contrast, the entropy per *baryon* remains spatially uniform on all scales until the baryons decouple from the radiation.

7.4.1 Equations

Because the baryon–radiation and cold dark matter components interact only gravitationally, their energy–momentum tensors satisfy the conservation laws, $T^\alpha_{\beta;\alpha} = 0$, separately. The *cold* particles have negligible relative velocities and hence the cold dark matter can be described as a dust-like perfect fluid with zero pressure. In the baryon–radiation plasma, the photons can efficiently transfer energy from one region of the fluid to another over distances determined by the mean free path of the photons (e.g. through diffusion). Shear viscosity and heat conduction play an important role in this case and lead to the dissipation of perturbations on small scales (Silk damping). In the limit of low baryon density, corresponding to current observations, heat conduction is not as important as shear viscosity, and so we will only consider the latter. The derivation of the energy–momentum tensor for an imperfect fluid is given in many books and we will not repeat it here. The energy–momentum tensor is found to be

$$T^\alpha_\beta = (\varepsilon + p) u^\alpha u_\beta - p \delta^\alpha_\beta - \eta \left(P^\alpha_\gamma u_\beta{}^{;\gamma} + P^\gamma_\beta u^\alpha{}_{;\gamma} - \tfrac{2}{3} P^\alpha_\beta u^\gamma{}_{;\gamma} \right), \quad (7.101)$$

where η is the shear viscosity coefficient and

$$P^\alpha_\beta \equiv \delta^\alpha_\beta - u^\alpha u_\beta$$

is the projection operator. We will analyze small perturbations using the conformal Newtonian coordinate system in which the metric takes the form

$$ds^2 = a^2(\eta)\left[(1+2\Phi)\,d\eta^2 - (1+2\Psi)\,\delta_{ij}dx^i dx^j\right]. \tag{7.102}$$

The potentials Φ and Ψ are equal only if the nondiagonal spatial components of the energy–momentum tensor are equal to zero. This is obviously not true for an imperfect fluid. However, one can easily check that even when the main contribution to the gravitational potential comes from the imperfect fluid, the difference $\Phi - \Psi$ is suppressed compared to Φ, at least by the ratio of the photon mean free path to the perturbation scale. After equality, the gravitational potential is mainly due to the cold dark matter and the contribution of the nondiagonal components of the energy–momentum tensor can be completely neglected. Therefore we set $\Psi = \Phi$. In this case the Christoffel symbols are

$$\Gamma^0_{00} = \mathcal{H} + \Phi'; \quad \Gamma^0_{0i} = \Gamma^i_{00} = \Phi_{,i}; \quad \Gamma^0_{ij} = \left((1-4\Phi)\mathcal{H} - \Phi'\right)\delta_{ij};$$
$$\Gamma^i_{j0} = (\mathcal{H}-\Phi')\,\delta_{ij}; \quad \Gamma^j_{ik} = \Phi_{,j}\delta_{ik} - \Phi_{,i}\delta_{jk} - \Phi_{,k}\delta_{ij}, \tag{7.103}$$

and the zero component of the 4-velocity to first order in perturbations is equal to

$$u^0 = \frac{1}{a}(1-\Phi), \quad u_0 = a(1+\Phi). \tag{7.104}$$

Given these relations, it is easy to show that, to zeroth order in the perturbations, the relation $T^\alpha_{0;\alpha} = 0$ reduces to the homogeneous energy–momentum conservation law and $T^\alpha_{i;\alpha} = 0$ is trivially satisfied. To the next order, the equation $T^\alpha_{0;\alpha} = 0$ leads to

$$\delta\varepsilon' + 3\mathcal{H}(\delta\varepsilon + \delta p) - 3(\varepsilon+p)\Phi' + a(\varepsilon+p)\,u^i_{,i} = 0. \tag{7.105}$$

Note that the shear viscosity does not appear in this relation. As for the remaining equations, $T^\alpha_{i;\alpha} = 0$, if we are only interested in scalar perturbations, it suffices to take the spatial divergence. We find

$$\frac{1}{a^4}\left(a^5(\varepsilon+p)u^i_{,i}\right)' - \frac{4}{3}\eta\Delta u^i_{,i} + \Delta\delta p + (\varepsilon+p)\Delta\Phi = 0. \tag{7.106}$$

As already noted, the two equations above are separately valid for the dark matter and the baryon–radiation plasma components.

Problem 7.16 Derive (7.105) and (7.106).

Dark matter For dark matter, the pressure p and the shear viscosity η are both equal to zero. Taking into account that $\varepsilon_d a^3 = \text{const}$, we obtain from (7.105) that the fractional perturbation in the energy of dark matter component, $\delta_d \equiv \delta\varepsilon_d/\varepsilon_d$, satisfies the equation

$$(\delta_d - 3\Phi)' + au^i_{,i} = 0. \tag{7.107}$$

Solving for $u^i_{,i}$ in terms of δ_d and Φ and substituting into (7.106), the resulting equation takes the form

$$\left(a(\delta_d - 3\Phi)'\right)' - a\Delta\Phi = 0. \tag{7.108}$$

Baryon–radiation plasma The baryons and radiation are tightly coupled before recombination and, therefore, their energy and momentum are not conserved separately. Nevertheless, when the baryons are nonrelativistic, (7.105), in contrast to (7.106), is valid for the baryon and radiation components separately, because the energy conservation law for the baryons, $T^\alpha_{0;\alpha} = 0$, reduces to the conservation law for total baryon number. (Specifically, if $T^\alpha_0 = m_b n_b u^\alpha u_0$, where m_b is the baryon mass, then $T^\alpha_{0;\alpha} = 0$ is equivalent to $(n_b u^\alpha)_{;\alpha} = 0$ up to linear order in perturbations.) Hence, the fractional baryon density fluctuation, $\delta_b \equiv \delta\varepsilon_b/\varepsilon_b$, satisfies an equation similar to (7.107):

$$(\delta_b - 3\Phi)' + au^i_{,i} = 0. \tag{7.109}$$

The corresponding equation for the perturbations in the radiation component, $\delta_\gamma \equiv \delta\varepsilon_\gamma/\varepsilon_\gamma$, is, according to (7.105),

$$(\delta_\gamma - 4\Phi)' + \tfrac{4}{3}au^i_{,i} = 0. \tag{7.110}$$

Since the photons and baryons are tightly coupled, they move together, and hence the velocities entering both of these equations are the same. Multiplying (7.110) by 3/4, subtracting (7.109), and integrating, we obtain

$$\frac{\delta S}{S} \equiv \frac{3}{4}\delta_\gamma - \delta_b = \text{const}, \tag{7.111}$$

where $\delta S/S$ is the fractional entropy fluctuation in the baryon–radiation plasma (see also (7.83)). Equation (7.111) states that $\delta S/S$ is conserved *on all scales*. In the case of adiabatic perturbations, $\delta S = 0$ and therefore

$$\delta_b = \tfrac{3}{4}\delta_\gamma. \tag{7.112}$$

Expressing $u^i_{,i}$ in terms of δ_γ and Φ and substituting into (7.106), we obtain

$$\left(\frac{\delta'_\gamma}{c_s^2}\right)' - \frac{3\eta}{\varepsilon_\gamma a}\Delta\delta'_\gamma - \Delta\delta_\gamma = \frac{4}{3c_s^2}\Delta\Phi + \left(\frac{4\Phi'}{c_s^2}\right)' - \frac{12\eta}{\varepsilon_\gamma a}\Delta\Phi', \tag{7.113}$$

where Δ is the Laplacian operator and c_s is the speed of sound in the baryon–radiation plasma given in (7.85). In deriving (7.113), we used the relations

$$\varepsilon + p = \varepsilon_b + \frac{4}{3}\varepsilon_\gamma = \frac{4}{9c_s^2}\varepsilon_\gamma \tag{7.114}$$

and $\varepsilon_\gamma a^4 =$ const. Neglecting polarization effects, the shear viscosity coefficient entering (7.113) is given by

$$\eta = \frac{4}{15}\varepsilon_\gamma \tau_\gamma, \tag{7.115}$$

where τ_γ is the mean free time for photon scattering.

We have, thus far, two perturbation equations, (7.108) and (7.113), for three unknown variables, δ_d, δ_γ and Φ. To these equations, we add the $0-0$ component of the Einstein equations (see (7.38)), which in the case under consideration becomes

$$\Delta\Phi - 3\mathcal{H}\Phi' - 3\mathcal{H}^2\Phi = 4\pi Ga^2\big(\delta\varepsilon_d + \delta\varepsilon_b + \delta\varepsilon_\gamma\big)$$
$$= 4\pi Ga^2\left(\varepsilon_d\delta_d + \frac{1}{3c_s^2}\varepsilon_\gamma\delta_\gamma\right), \tag{7.116}$$

where (7.112) has been used to express δ_b in terms of δ_γ.

Using (7.110), we obtain a useful relation for *the radiation contribution* to the divergence of the $0-i$ components of the energy–momentum tensor:

$$T^i_{0,i} = \tfrac{4}{3}\varepsilon_\gamma u_0 u^i_{,i} = \big(4\Phi - \delta_\gamma\big)' \varepsilon_\gamma, \tag{7.117}$$

which will be used in Section 9.3.

7.4.2 Evolution of perturbations and transfer functions

If the density fluctuations are decomposed into modes with comoving wavenumber k, then their behavior for a given k depends on whether $k\eta < 1$ or $k\eta > 1$. The crossover from $k\eta < 1$ to $k\eta > 1$ corresponds to the transition in which a mode changes from having a wavelength exceeding the curvature scale to having a wavelength less than the curvature scale. In a decelerating universe, as time evolves and η grows, the curvature scale $H^{-1} = a/\dot{a}$ increases faster than the physical scale of the perturbation, $\lambda_{ph} \simeq a/k$, and encompasses modes with smaller and smaller k. We shall refer to modes for which $k\eta < 1$ as supercurvature modes and those with $k\eta > 1$ as subcurvature ones.

The initial perturbation spectrum produced in inflation can be characterized by the "frozen" amplitudes of the metric fluctuations Φ_k^0 on supercurvature scales during the radiation stage (see the following chapter for details). After the perturbation enters the curvature scale it evolves in a nontrivial way. Our goal here is to determine

7.4 Baryon–radiation plasma and cold dark matter

the amplitude of the gravitational potential Φ and the radiation fluctuations δ_γ at recombination for a given initial Φ_k^0 or, equivalently, to find the *transfer functions* relating the initial spectrum of perturbations with the resulting one at recombination. We will see in Chapter 9 that Φ and δ_γ determine the anisotropies in the background radiation.

Long-wavelength perturbations ($k\eta_r < 1$) We first consider the long-wavelength perturbations which are supercurvature modes at recombination. They are described by solution (7.71) for adiabatic perturbations in a two-component medium consisting of cold matter and radiation. Although the dark matter particles are not tightly coupled to the radiation, the entropy per cold dark matter particle is nevertheless conserved on supercurvature scales because, as can be seen intuitively, there is insufficient time to move matter distances greater than the Hubble scale. One can formally arrive at this conclusion by noting that, for long-wavelength perturbations, the $u^i_{,i}$-terms in (7.107) and (7.110) are negligible and, consequently, the steps which lead to the entropy conservation law per baryon (see (7.111)) can be repeated here.

Knowing the gravitational potential, which is given by (7.71), we can easily find δ_γ. Skipping the velocity term in (7.110), which is negligible for the long-wavelength perturbations, and integrating, we obtain

$$\delta_\gamma - 4\Phi = C, \tag{7.118}$$

where C is a constant of integration. To determine C we note that, during the radiation-dominated epoch, the gravitational potential is mainly due to the fluctuations in the radiation component and stays constant on supercurvature scales. At these early times,

$$\delta_\gamma \simeq -2\Phi(\eta \ll \eta_{eq}) \equiv -2\Phi^0; \tag{7.119}$$

hence $C = -6\Phi^0$. After equality, when the cold matter overtakes the radiation, the gravitational potential Φ changes its value by a factor of $9/10$ and remains constant afterwards, that is,

$$\Phi(\eta \gg \eta_{eq}) \simeq \tfrac{9}{10}\Phi^0. \tag{7.120}$$

Therefore, assuming that cold dark matter dominates at recombination, we obtain from (7.118)

$$\delta_\gamma(\eta_r) = -6\Phi^0 + 4\Phi(\eta_r) = -\tfrac{8}{3}\Phi(\eta_r) = -\tfrac{8}{3}\left(\tfrac{9}{10}\Phi^0\right). \tag{7.121}$$

One arrives at the same conclusion by noting that, for adiabatic perturbations, $\delta_\gamma = 4\delta_d/3$ and $\delta_d \simeq -2\Phi(\eta_r)$ at recombination.

Standard inflation predicts adiabatic fluctuations. In principle, one can imagine alternative possibilities for the initial inhomogeneities, such as *entropy perturbations*. For example, the dark matter might initially be distributed inhomogeneously on a homogeneous radiation background. It is clear that, at early times when radiation dominates, δ_γ and Φ both vanish and the constant of integration on the right hand side of (7.118) is equal to zero. After equality, the dark matter inhomogeneities induce the gravitational potential. Then, it follows from (7.118) that the fluctuations in the radiation component on supercurvature scales are equal to $\delta_\gamma = 4\Phi$, where Φ is mainly due to the cold dark matter fluctuations (see also (7.90)). The differences between this case with entropic perturbations and the adiabatic case give rise to distinctive cosmic microwave background anisotropies.

Short-wavelength perturbations $(k\eta_r > 1)$ We next consider the subcurvature modes which enter the horizon before recombination. These perturbations are especially interesting since they are responsible for the acoustic peaks in the cosmic microwave background spectrum.

To simplify the consideration, we neglect the contribution of baryons to the gravitational potential. This approximation is valid in realistic models where the baryons constitute only a small fraction of the total matter density. Although we neglect the contribution of the baryons to the gravitational potential, they cannot be completely ignored since they substantially affect the speed of sound after equality.

In general, there exist four independent instability modes in the two-component medium. The set of equations for the perturbations is rather complicated and they cannot be solved analytically without making further approximations. Let us consider the evolution of perturbations after equality, at $\eta > \eta_{eq}$. In this case the problem is greatly simplified if one notes that at $\eta > \eta_{eq}$ the gravitational potential is mainly due to the perturbations in the cold dark matter and is therefore time-independent for both the long-wavelength and the short-wavelength perturbations. Thus, at $\eta > \eta_{eq}$, the potential Φ can be considered as an external source in (7.113) and the general solution of this equation can be written as a sum of the general solution of the homogeneous equation (with $\Phi = 0$) and a particular solution for δ_γ. Using the variable x defined by $dx = c_s^2 d\eta$ and taking into account that the time derivatives of the potential on the right hand side of (7.113) are zero ($\Phi = $ const), (7.113) becomes

$$\frac{d^2\delta_\gamma}{dx^2} - \frac{4\tau_\gamma}{5a}\Delta\frac{d\delta_\gamma}{dx} - \frac{1}{c_s^2}\Delta\delta_\gamma = \frac{4}{3c_s^4}\Delta\Phi, \qquad (7.122)$$

where the second term is due to the viscosity. If the speed of sound is slowly varying, (7.122) has an obvious approximate particular solution:

$$\delta_\gamma \simeq -\frac{4}{3c_s^2}\Phi. \qquad (7.123)$$

7.4 Baryon–radiation plasma and cold dark matter

To find the general solution of the homogeneous equation (7.122) we employ the WKB approximation. Let us consider a plane wave perturbation with wavenumber k and introduce the new variable

$$y_k(x) \equiv \delta_\gamma(k, x) \exp\left(\frac{2}{5}k^2 \int \frac{\tau_\gamma}{a} dx\right). \tag{7.124}$$

Then, it follows from (7.122) that the variable y_k satisfies the equation

$$\frac{d^2 y_k}{dx^2} + \frac{k^2}{c_s^2}\left[1 - \frac{4c_s^2}{25}\left(\frac{k\tau_\gamma}{a}\right)^2 - \frac{2c_s^2}{5}\left(\frac{\tau_\gamma}{a}\right)'\right] y_k = 0. \tag{7.125}$$

First note that, for a perturbation whose physical wavelength ($\lambda_{ph} \sim a/k$) is much larger than the mean free path of the photons ($\sim \tau_\gamma$), the second term in the square brackets is negligible. The third term is roughly $\tau_\gamma/a\eta \sim \tau_\gamma/t \ll 1$ and can also be skipped. With these simplifications, the WKB solution for y is

$$y_k \simeq A_k \sqrt{c_s} \cos\left(k \int \frac{dx}{c_s}\right), \tag{7.126}$$

where A_k is a constant of integration and the indefinite integral implies that the argument of the cosine function includes an arbitrary phase. Given the definition of y_k in (7.124) and combining this solution with (7.123), we obtain the general approximate solution of (7.122), valid at $\eta > \eta_{eq}$:

$$\delta_\gamma(k, \eta) \simeq -\frac{4}{3c_s^2}\Phi_k(\eta > \eta_{eq}) + A_k \sqrt{c_s} \cos\left(k \int c_s d\eta\right) e^{-(k/k_D)^2}. \tag{7.127}$$

Here we have converted back from x to conformal time η and introduced the dissipation scale corresponding to the comoving wave number:

$$k_D(\eta) \equiv \left(\frac{2}{5}\int_0^\eta c_s^2 \frac{\tau_\gamma}{a} d\tilde\eta\right)^{-1/2}. \tag{7.128}$$

In the limit of constant speed of sound and vanishing viscosity, the solution is exact and valid not only for the short-wavelength perturbations but also for the perturbations with $k\eta \ll 1$.

Problem 7.17 Find the corrections to (7.127) and determine when the WKB solution is applicable.

Silk damping From (7.127), it is clear that the viscosity efficiently damps perturbations on comoving scales $\lambda \leq 1/k_D$. Viscous damping is due to the scattering and mixing of the photons. Therefore, the scale where the viscous damping is important is of order the photon diffusion scale at a given cosmological time t. To estimate the diffusion scale, we note that the photons undergo about $N \sim t/\tau_\gamma$ scatterings during time t. After every scattering the direction of propagation is completely random, so the photon trajectory is similar to that of a "drunken sailor." Therefore, after N steps of length τ_γ, the typical distance travelled (the diffusion scale) is about $\tau_\gamma \sqrt{N} \sim \sqrt{\tau_\gamma t}$. Consequently, the ratio of the physical damping scale to the horizon scale is

$$\frac{\lambda_D^{ph}}{t} \sim (k_D \eta)^{-1} \sim \sqrt{\frac{\tau_\gamma}{t}}. \tag{7.129}$$

This simple estimate is in agreement with the more rigorous result (7.128).

Before recombination, the mean free path of the photons is determined by Thomson scattering on free electrons:

$$\tau_\gamma = \frac{1}{\sigma_T n_e}, \tag{7.130}$$

where $\sigma_T \simeq 6.65 \times 10^{-25}$ cm^2 is the Thomson cross-section and n_e is the number density of free electrons. We are mainly interested in the dissipation scale at recombination time when the universe is dominated by cold dark matter. Taking into account that $t_r \propto (\Omega_m h^2)^{-1/2} z_r^{-3/2}$ and $n_e \propto (\Omega_b h^2) z_r^3$, we infer from (7.129) that

$$(k_D \eta_r)^{-1} \sim (\sigma_T n_e t_r)^{-1/2} \propto (\Omega_m h^2)^{1/4} (\Omega_b h_{75}^2)^{-1/2} z_r^{-3/4}. \tag{7.131}$$

Problem 7.18 Using the exact formula (7.128) with $c_s^2 = 1/3$ and assuming instantaneous recombination, calculate k_D and show that

$$(k_D \eta_r)^{-1} \simeq 0.6 (\Omega_m h^2)^{1/4} (\Omega_b h^2)^{-1/2} z_r^{-3/4}. \tag{7.132}$$

The dissipation scale can never exceed the curvature scale $H^{-1} \approx t$, because there is insufficient time for radiation to rearrange itself on those scales. This imposes a limit on the range of validity of (7.132), namely, $(k_D \eta_r)^{-1} < 1$. If τ_γ grows and begins to exceed the cosmological time t, then we have to use the kinetic description for photons. Our analysis of viscous damping is also not valid on scales smaller than the mean free path of the photons since the hydrodynamical description fails in this limit. On scales smaller than the mean free path, another effect, *free streaming*, becomes important. Free streaming refers to the propagation of photons without scattering. On scales smaller than the mean free path, photons coming from different directions with different temperatures intermingle, smearing spatial inhomogeneities in the radiation energy density distribution. However, in contrast with viscous damping, free streaming does not remove the angular temperature anisotropy of the radiation at a given point (see Problem 9.2). As with viscous damping, free streaming has no effect on scales larger than the horizon scale.

Transfer functions For long-wavelength perturbations with $k\eta_r \ll 1$, the amplitudes of the metric and radiation fluctuations after equality in terms of Φ^0 are given in (7.120) and (7.121). To find the transfer functions for short-wavelength inhomogeneities we have to express $\Phi(\eta > \eta_{eq})$ and A_k, entering (7.127), in terms of Φ_k^0. This can be done analytically in two limiting cases: for perturbations which enter the horizon *long enough after* and *well before* equality, that is, for the modes with $k\eta_{eq} \ll 1$ and $k\eta_{eq} \gg 1$.

The perturbation with $k\eta_{eq} \ll 1$ enters the horizon when the cold matter already dominates and determines the gravitational potential. Therefore the gravitational potential does not change and it is given by

$$\Phi_k(\eta > \eta_{eq}) = \tfrac{9}{10}\Phi_k^0. \qquad (7.133)$$

The solution (7.127) for δ_γ is applicable even when the wavelength of the perturbation exceeds the curvature scale (see Problem 7.17). After equality at $\eta \gg \eta_{eq}$, the amplitude of δ_γ for the supercurvature modes with $k\eta \ll 1$, according to (7.127), should be equal to $\delta_\gamma \simeq -8\Phi_k/3 = $ const. Assuming that the baryons have a negligible effect on the speed of sound at this time, so that $c_s^2 \to 1/3$, we find that $A = 4\Phi_k/3^{3/4}$. As a result, after equality but before recombination, we have

$$\delta_\gamma(\eta) = \left[-\frac{4}{3c_s^2} + \frac{4\sqrt{c_s}}{3^{3/4}}\cos\left(k\int_0^\eta c_s d\tilde\eta\right)e^{-(k/k_D)^2}\right]\left(\frac{9}{10}\Phi_k^0\right) \qquad (7.134)$$

for the modes with $\eta_{eq}^{-1} \gg k \gg \eta_r^{-1}$.

Now we consider perturbations with $k\eta_{eq} \gg 1$, which enter the horizon well before equality. At $\eta \ll \eta_{eq}$, radiation dominates over dark matter and baryons. Therefore, Φ and δ_γ are well approximated by (7.61) and (7.62), describing perturbations in the radiation-dominated universe. Neglecting the decaying mode on supercurvature scales and expressing C_1 in terms of Φ_k^0, we find that at $\eta_{eq} \gg \eta \gg k^{-1}$

$$\delta_\gamma \simeq 6\Phi_k^0 \cos(k\eta/\sqrt{3}), \quad \Phi_k(\eta) \simeq -\frac{9\Phi_k^0}{(k\eta)^2}\cos(k\eta/\sqrt{3}). \qquad (7.135)$$

To determine the fluctuations in the cold dark matter component, we integrate (7.108) to obtain

$$\delta_d(\eta) = 3\Phi(\eta) + \int^\eta \frac{d\tilde\eta}{a}\int^{\tilde\eta} a\Delta\Phi d\bar\eta. \qquad (7.136)$$

This is an exact relation which is always valid for any k. During the radiation-dominated epoch, the main contribution to the gravitational potential is due to radiation, and, therefore, we can treat Φ in (7.136) as an external source given by (7.61). The two constants of integration can be fixed by substituting (7.61) in (7.136) and noting that, at earlier times when the wavelength of the mode exceeds the curvature scale, one should match the result for the long-wavelength perturbations:

$$\delta_d \simeq 3\delta_\gamma/4 \simeq -3\Phi_k^0/2. \quad (7.137)$$

Problem 7.19 Determine the constants of integration in (7.136) and show that after the perturbation enters the Hubble scale, but before equality,

$$\delta_d \simeq -9\left[\mathbf{C}-\frac{1}{2}+\ln\left(k\eta/\sqrt{3}\right)+O\left((k\eta)^{-1}\right)\right]\Phi_k^0, \quad (7.138)$$

where $\mathbf{C} = 0.577\ldots$ is the Euler constant.

It is easy to see from (7.116) that, before equality, the contribution of the dark matter perturbations to the gravitational potential is suppressed by a factor $\varepsilon_d/\varepsilon_\gamma$ compared to the fluctuations in the radiation component. At equality, the dark matter contribution begins to dominate and the density perturbation δ_d starts to grow $\propto \eta^2$, as shown in Section 7.3.1. The gravitational potential "freezes" at the value

$$\Phi_k(\eta > \eta_{eq}) \sim -\frac{4\pi G a^2 \varepsilon}{k^2}\delta_d\bigg|_{\eta_{eq}} \sim O(1)\frac{\ln(k\eta_{eq})}{(k\eta_{eq})^2}\Phi_k^0 \quad (7.139)$$

and stays constant until recombination.

More work is required to obtain the exact coefficients in (7.139).

Problem 7.20 For short-wavelength perturbations, the time derivatives of the gravitational potential in (7.108) and (7.116) can be neglected compared to the spatial derivatives. From these relations, it follows that

$$(a\delta_d')' - 4\pi G a^3\left(\varepsilon_d\delta_d + \frac{1}{3c_s^2}\varepsilon_\gamma\delta_\gamma\right) = 0. \quad (7.140)$$

Show that the second term here induces corrections to (7.138) that become significant only near equality. These corrections are mainly due to the $\varepsilon_d\delta_d$ contribution. Show that the $\varepsilon_\gamma\delta_\gamma$ term remains negligible throughout and, hence, can be omitted in (7.140). Then (7.140) coincides with the equation describing the instability in the nonrelativistic cold matter component on a homogeneous radiation background and its solution is given in (6.72). At $x \ll 1$ this solution should coincide with (7.138). Considering this limit, show that the integration constants in (6.72) are

$$C_1 \simeq -9\left(\ln\left(\frac{2k\eta_*}{\sqrt{3}}\right)+\mathbf{C}-\frac{7}{2}\right)\Phi_k^0, \quad C_2 \simeq 9\Phi_k^0, \quad (7.141)$$

where $\eta_* = \eta_{eq}/(\sqrt{2} - 1)$. Neglecting the decaying mode and using the relation between the gravitational potential and δ_d, verify that

$$\Phi_k(\eta > \eta_{eq}) \simeq \frac{\ln(0.15k\eta_{eq})}{(0.27k\eta_{eq})^2} \Phi_k^0. \quad (7.142)$$

The fluctuations in the radiation component after equality continue to behave as sound waves in the external gravitational potential given by (7.142). The integration constant A in (7.127) can be fixed by comparing the oscillating part of this solution to the result in (7.135) at $\eta \sim \eta_{eq}$. Then we find that at $\eta > \eta_{eq}$,

$$\delta_\gamma \simeq \left[-\frac{4}{3c_s^2} \frac{\ln(0.15k\eta_{eq})}{(0.27k\eta_{eq})^2} + 3^{5/4}\sqrt{4c_s} \cos\left(k \int_0^\eta c_s d\eta\right) e^{-(k/k_D)^2} \right] \Phi_k^0 \quad (7.143)$$

for modes with $k\eta_{eq} \gg 1$.

It follows from (7.134) and (7.143) that, for $k > \eta_r^{-1}$, the spectrum of δ_γ at recombination is partially modulated by the cosine. This is because all sound waves with the same $k = |\mathbf{k}|$ enter the horizon and begin to oscillate simultaneously. As we will see in the next chapter, this leads to peaks and valleys in the spectrum of the temperature fluctuations of background radiation.

In summary, the results obtained in this section allow us to express the gravitational potential and the radiation energy density fluctuations at recombination in terms of the basic cosmic parameters and the primordial perturbation spectrum. The primordial spectrum is described by the gravitational potential Φ_k^0, characterizing a perturbation with comoving wavenumber k at very early times when its size still exceeds the curvature scale. For modes whose wavelength exceeds the curvature scale at recombination, the spectrum of Φ remains unchanged except that its amplitude drops by a factor of 9/10 after matter–radiation equality, and the amplitude of radiation fluctuations is given in (7.121). For perturbations whose wavelength is less than the Hubble scale, we have derived asymptotic expressions for the modes which enter the curvature scale *well before* equality (see (7.142), (7.143)), and *long enough after* equality (see (7.133), (7.134)). For these perturbations the initial spectrum is substantially changed as a result of evolution.

8
Inflation II: origin of the primordial inhomogeneities

One of the central issues of contemporary cosmology is the explanation of the origin of primordial inhomogeneities, which serve as the seeds for structure formation. Before the advent of inflationary cosmology the initial perturbations were *postulated* and their spectrum was designed to fit observational data. In this way practically any observation could be "explained", or more accurately *described, by* arranging the appropriate initial conditions. In contrast, inflationary cosmology *truly explains* the origin of primordial inhomogeneities and *predicts* their spectrum. Thus it becomes possible to test this theory by comparing its predictions with observations.

According to cosmic inflation, primordial perturbations originated from quantum fluctuations. These fluctuations have substantial amplitudes only on scales close to the Planckian length, but during the inflationary stage they are stretched to galactic scales with nearly unchanged amplitudes. Thus, inflation links the large-scale structure of the universe to its microphysics. The resulting spectrum of inhomogeneities is not very sensitive to the details of any particular inflationary scenario and has nearly universal shape. This leads to concrete predictions for the spectrum of cosmic microwave background anisotropies.

In the previous chapter we studied gravitational instability in a universe filled with hydrodynamical matter. To understand the generation of primordial fluctuations we have to extend our analysis to the case of a scalar field condensate and quantize the cosmological perturbations. In this chapter we study the behavior of perturbations during an inflationary stage and calculate their resulting spectrum. We first consider a simple inflationary model and use the slow-roll approximation to solve the perturbation equations. Then the rigorous quantum theory is developed and applied to a general inflationary scenario.

8.1 Characterizing perturbations

At a given moment in time small inhomogeneities can be characterized by the spatial distribution of the gravitational potential Φ or by the energy density fluctuations $\delta\varepsilon/\varepsilon_0$. It turns out to be convenient to treat them as *random fields*, for which we will use the common notation $f(\mathbf{x})$. Subdividing an infinite universe into a set of large spatial regions we can consider a particular configuration $f(\mathbf{x})$ within some region as a realization of a random process. This means that the relative number of regions where a given configuration $f(\mathbf{x})$ occurs can be described by a probability distribution function. Then averaging over the statistical ensemble is equivalent to averaging over the volume of the whole infinite universe.

It is convenient to describe the random process using Fourier methods. The Fourier expansion of the function $f(\mathbf{x})$ in a given region of volume V can be written as

$$f(\mathbf{x}) = \frac{1}{\sqrt{V}} \sum_{\mathbf{k}} f_{\mathbf{k}} e^{i\mathbf{k}\mathbf{x}}. \tag{8.1}$$

In the case of a dimensionless function f the complex Fourier coefficients, $f_{\mathbf{k}} = a_{\mathbf{k}} + i b_{\mathbf{k}}$, have dimension cm$^{3/2}$. The reality of f requires $f_{-\mathbf{k}} = f_{\mathbf{k}}^*$ and hence the real and imaginary parts of $f_{\mathbf{k}}$ must satisfy the constraints $a_{-\mathbf{k}} = a_{\mathbf{k}}$ and $b_{-\mathbf{k}} = -b_{\mathbf{k}}$. The coefficients $a_{\mathbf{k}}$ and $b_{\mathbf{k}}$ take different values in different spatial regions. Given a very large number N of such regions, the definition of the probability distribution function $p(a_{\mathbf{k}}, b_{\mathbf{k}})$ tells us that in

$$dN = N p(a'_{\mathbf{k}}, b'_{\mathbf{k}}) \, da_{\mathbf{k}} db_{\mathbf{k}} \tag{8.2}$$

of them the value of $a_{\mathbf{k}}$ lies between $a'_{\mathbf{k}}$ and $a'_{\mathbf{k}} + da_{\mathbf{k}}$ and that of $b_{\mathbf{k}}$ between $b'_{\mathbf{k}}$ and $b'_{\mathbf{k}} + db_{\mathbf{k}}$. Inflation predicts only *homogeneous and isotropic Gaussian* processes, for which:

$$p(a_{\mathbf{k}}, b_{\mathbf{k}}) = \frac{1}{\pi \sigma_k^2} \exp\left(-\frac{a_{\mathbf{k}}^2}{\sigma_k^2}\right) \exp\left(-\frac{b_{\mathbf{k}}^2}{\sigma_k^2}\right), \tag{8.3}$$

where the variance depends only on $k = |\mathbf{k}|$; it is the same for both *independent* variables $a_{\mathbf{k}}$ and $b_{\mathbf{k}}$ and is equal to $\sigma_k^2/2$. This variance characterizes the corresponding Gaussian process entirely and all correlation functions can be expressed in terms of σ_k^2. For example, for the expectation value of the product of Fourier coefficients one finds

$$\langle f_{\mathbf{k}} f_{\mathbf{k}'} \rangle = \langle a_{\mathbf{k}} a_{\mathbf{k}'} \rangle + i(\langle a_{\mathbf{k}} b_{\mathbf{k}'} \rangle + \langle a_{\mathbf{k}'} b_{\mathbf{k}} \rangle) - \langle b_{\mathbf{k}} b_{\mathbf{k}'} \rangle = \sigma_k^2 \delta_{\mathbf{k},-\mathbf{k}'}, \tag{8.4}$$

where we have taken into account that $a_{-\mathbf{k}} = a_{\mathbf{k}}$ and $b_{-\mathbf{k}} = -b_{\mathbf{k}}$. Here $\delta_{\mathbf{k},-\mathbf{k}'} = 1$ for $\mathbf{k} = -\mathbf{k}'$ and is otherwise equal to zero.

In the continuous limit, as $V \to \infty$, the sum in (8.1) is replaced by the integral

$$f(\mathbf{x}) = \int f_\mathbf{k} e^{i\mathbf{k}\mathbf{x}} \frac{d^3k}{(2\pi)^{3/2}} \tag{8.5}$$

and (8.4) becomes

$$\langle f_\mathbf{k} f_{\mathbf{k}'} \rangle = \sigma_k^2 \delta(\mathbf{k} + \mathbf{k}'), \tag{8.6}$$

where $\delta(\mathbf{k} + \mathbf{k}')$ is the Dirac delta function. Note that the Fourier coefficients in (8.5) are related to the Fourier coefficients in (8.1) by a factor of $\sqrt{V}$ and have dimension cm^3. In contrast to the dimensionless $\delta_{\mathbf{k},-\mathbf{k}'}$, the Dirac delta function has dimension cm^3. The dimension of σ_k^2 does not change in the transition to the continuous limit and this quantity does not acquire any extra volume factors.

Alternatively, a Gaussian random field can be characterized by the spatial two-point correlation function

$$\xi_f(\mathbf{x} - \mathbf{y}) \equiv \langle f(\mathbf{x}) f(\mathbf{y}) \rangle. \tag{8.7}$$

This function tells us how large the field fluctuations are on different scales. In the homogeneous and isotropic case, the correlation function depends only on the distance between the points $\mathbf{x}$ and $\mathbf{y}$, that is, $\xi_f = \xi_f(|\mathbf{x} - \mathbf{y}|)$. Substituting (8.5) into (8.7) and averaging over the ensemble with the help of (8.6), we find

$$\xi_f(|\mathbf{x} - \mathbf{y}|) = \int \frac{\sigma_k^2 k^3}{2\pi^2} \frac{\sin(kr)}{kr} \frac{dk}{k}, \tag{8.8}$$

where $r \equiv |\mathbf{x} - \mathbf{y}|$. In deriving this relation we have taken into account isotropy and performed the integration over angles. The *dimensionless variance*

$$\delta_f^2(k) \equiv \frac{\sigma_k^2 k^3}{2\pi^2} \tag{8.9}$$

is roughly the typical squared amplitude of the fluctuations on scales $\lambda \sim 1/k$.

Problem 8.1 Verify that the typical fluctuations of f, averaged over the volume $V \sim \lambda^3$, can be estimated as

$$\left\langle \left(\frac{1}{V} \int_V f d^3x \right)^2 \right\rangle^{1/2} \sim O(1) \delta_f(k \sim \lambda). \tag{8.10}$$

When does this estimate fail?

Problem 8.2 Show that different variances for $a_\mathbf{k}$ and $b_\mathbf{k}$ contradict the assumption of homogeneity and that the dependence of $\sigma_\mathbf{k}^2$ on the direction of $\mathbf{k}$ is in conflict with isotropy.

8.2 Perturbations on inflation (slow-roll approximation)

Thus, in the case of the Gaussian random process we need to know only σ_k^2 or, equivalently, $\delta_f(k)$. For small perturbations Fourier modes evolve independently. Therefore, the spatial distribution of inhomogeneities remains Gaussian and only their spectrum changes with time. When the perturbations enter the nonlinear regime, different Fourier modes start to "interact." As a result the statistical analysis of nonlinear structure becomes very complicated.

In this chapter we consider only small inhomogeneities and our main task is to derive the initial perturbation spectrum generated in inflation. This spectrum will be characterized by the variance of the gravitational potential $\sigma_k^2 \equiv |\Phi_k|^2$ or, equivalently, by the dimensionless variance

$$\delta_\Phi^2(k) \equiv \frac{|\Phi_k|^2 k^3}{2\pi^2}. \tag{8.11}$$

In the following we refer to $\delta_\Phi^2(k)$ as the power spectrum. Given δ_Φ^2 the corresponding power spectrum for the energy density fluctuations can easily be calculated.

8.2 Perturbations on inflation (slow-roll approximation)

To aid our intuition we begin with a *nonrigorous* derivation of the inflationary spectrum in a simple model with a scalar field. Let us consider the universe filled by a scalar field φ with potential $V(\varphi)$, and see how small inhomogeneities $\delta\varphi(\mathbf{x}, \eta)$, superimposed on a homogeneous component $\varphi_0(\eta)$, evolve during the inflationary stage. In curved spacetime the scalar field satisfies the Klein–Gordon equation,

$$\frac{1}{\sqrt{-g}} \frac{\partial}{\partial x^\alpha} \left(\sqrt{-g} g^{\alpha\beta} \frac{\partial \varphi}{\partial x^\beta} \right) + \frac{\partial V}{\partial \varphi} = 0, \tag{8.12}$$

which follows immediately from the action

$$S = \int \left(\tfrac{1}{2} g^{\gamma\delta} \varphi_{,\gamma} \varphi_{,\delta} - V \right) \sqrt{-g} d^4 x. \tag{8.13}$$

A small perturbation $\delta\varphi(\mathbf{x}, \eta)$ induces scalar metric perturbations and as a result the metric takes the form (7.17). Substituting

$$\varphi = \varphi_0(\eta) + \delta\varphi(\mathbf{x}, \eta)$$

together with (7.17) into (8.12), we find that the Klein–Gordon equation for the homogeneous component reduces to

$$\varphi_0'' + 2\mathcal{H}\varphi_0' + a^2 V_{,\varphi} = 0 \tag{8.14}$$

(compare with (5.24)). To linear order in metric perturbations and $\delta\varphi$ it becomes

$$\delta\varphi'' + 2\mathcal{H}\delta\varphi' - \Delta\left(\delta\varphi - \varphi_0'(B - E')\right) + a^2 V_{,\varphi\varphi}\delta\varphi$$
$$-\varphi_0'(3\psi + \phi)' + 2a^2 V_{,\varphi}\phi = 0. \quad (8.15)$$

This equation is valid in any coordinate system. Using the background equation (8.14), we can easily recast it in terms of the gauge-invariant variables Φ and Ψ, defined in (7.19), and the gauge-invariant scalar field perturbation

$$\overline{\delta\varphi} \equiv \delta\varphi - \varphi_0'(B - E'). \quad (8.16)$$

The result is

$$\overline{\delta\varphi}'' + 2\mathcal{H}\overline{\delta\varphi}' - \Delta\overline{\delta\varphi} + a^2 V_{,\varphi\varphi}\overline{\delta\varphi} - \varphi_0'(3\Psi + \Phi)' + 2a^2 V_{,\varphi}\Phi = 0. \quad (8.17)$$

Problem 8.3 Derive (8.15) and (8.17). (*Hint* As previously noted, the quickest way to derive gauge-invariant equations is to use the longitudinal gauge. After the equations are obtained in this gauge, we simply have to replace the perturbation variables by the corresponding gauge-invariant quantities: $\phi_l \to \Phi$, $\psi_l \to \Psi$ and $\delta\varphi_l \to \overline{\delta\varphi}$. Then using the explicit expressions for Φ, Ψ and $\overline{\delta\varphi}$ we can write the equations in an arbitrary coordinate system.)

Equation (8.17) contains three unknown variables, $\overline{\delta\varphi}$, Φ and Ψ, and should be supplemented by the Einstein equations. For these we need the energy–momentum tensor for the scalar field, which follows from the action in (8.13) upon variation with respect to the metric $g_{\alpha\beta}$:

$$T^\alpha_\beta = g^{\alpha\gamma}\varphi_{,\gamma}\varphi_{,\beta} - \left(g^{\gamma\delta}\varphi_{,\gamma}\varphi_{,\delta} - V(\varphi)\right)\delta^\alpha_\beta. \quad (8.18)$$

It is convenient to use (7.39). For the energy–momentum tensor (8.18), the perturbed gauge-invariant components $\overline{\delta T}^0_i$, defined in (7.35), are

$$\overline{\delta T}^0_i = \frac{1}{a^2}\varphi_0'\delta\varphi_{,i} - \frac{1}{a^2}\varphi_0'^2(B - E')_{,i} = \frac{1}{a^2}\left(\varphi_0'\overline{\delta\varphi}\right)_{,i}. \quad (8.19)$$

Equation (7.39) becomes

$$\Psi' + \mathcal{H}\Phi = 4\pi\varphi_0'\overline{\delta\varphi}, \quad (8.20)$$

where we have set $G = 1$. Finally, we note that the nondiagonal spatial components of the energy–momentum tensor are equal to zero, that is, $\delta T^i_k = 0$ for $i \neq k$, and hence $\Psi = \Phi$.

We will solve (8.17) and (8.20) in two limiting cases: for perturbations with a physical wavelength λ_{ph} much smaller than the curvature scale H^{-1} and for long-wavelength perturbations with $\lambda_{ph} \gg H^{-1}$. The curvature scale does not change very much during inflation. On the other hand the physical scale of a perturbation,

8.2 Perturbations on inflation (slow-roll approximation)

$\lambda_{ph} \sim a/k$, grows. For the modes we will be interested in, the physical wavelength starts smaller than the Hubble radius but eventually exceeds it.

Our strategy will be the following. We start with a short-wavelength perturbation and fix its amplitude at the minimal possible level allowed by the uncertainty principle (*vacuum fluctuations*). We then study how the perturbation evolves after it crosses the Hubble radius.

We occasionally adopt the widespread custom of referring to the curvature (Hubble) radius as the (event) horizon scale. To avoid confusion the reader should be sure to distinguish the curvature radius from the *particle* horizon scale, which grows exponentially during inflation. What is relevant for the dynamics of the perturbations is the curvature scale, not the particle horizon size which has a kinematical origin.

8.2.1 Inside the Hubble scale

The gravitational field is not crucial to the evolution of the short-wavelength perturbations with $\lambda_{ph} \ll H^{-1}$ or, equivalently, with $k \gg Ha \sim |\eta|^{-1}$. In fact, for very large $k|\eta|$ the spatial derivative term dominates in (8.17) and its solution behaves as $\exp(\pm ik\eta)$ to leading order. The gravitational field also oscillates, so that $\Phi' \sim k\Phi$, and can be estimated from (8.20) as $\Phi \sim k^{-1}\varphi_0' \overline{\delta\varphi}$. Using this estimate and taking into account that during inflation $V_{,\varphi\varphi} \ll V \sim H^2$, we find that only the first three terms in (8.17) are relevant. Thus, for a plane wave perturbation with comoving wavenumber k, this equation reduces to

$$\overline{\delta\varphi}_k'' + 2\mathcal{H}\overline{\delta\varphi}_k' + k^2 \overline{\delta\varphi}_k \simeq 0, \tag{8.21}$$

and with the substitution $\overline{\delta\varphi}_k = u_k/a$, it becomes

$$u_k'' + \left(k^2 - \frac{a''}{a}\right) u_k = 0. \tag{8.22}$$

For $k|\eta| \gg 1$ the last term in (8.22) can be neglected and the resulting solution for $\overline{\delta\varphi}_k$ is

$$\overline{\delta\varphi}_k \simeq \frac{C_k}{a} \exp(\pm ik\eta), \tag{8.23}$$

where C_k is a constant of integration which has to be fixed by the initial conditions. The physical ingredient is that the initial scalar field modes arise as vacuum quantum fluctuations.

Quantum fluctuations To make a rough estimate for the typical amplitude of the vacuum quantum fluctuations $\delta\varphi_L$ on physical scales L, we consider a finite volume

$V \sim L^3$. Assuming that the field is nearly homogeneous within this volume, we can write its action (see (8.13)) in the form

$$S \simeq \tfrac{1}{2} \int \left(\dot{X}^2 + \cdots\right) dt,$$

where $X \equiv \delta\varphi_L L^{3/2}$ and the dot denotes the derivative with respect to the physical time t. It is clear that X plays the role of the canonical quantization variable and the corresponding conjugated momentum is $P = \dot{X} \sim X/L$; in the latter estimate we have assumed that the mass of the field is negligible and hence the field propagates with the speed of light. The variables X and P satisfy the uncertainty relation $\Delta X \Delta P \sim 1$ ($\hbar = 1$) and it follows that the minimal amplitude of the quantum fluctuations is $X_m \sim \sqrt{L}$ or $\delta\varphi_L \sim L^{-1}$. Thus the amplitude of the minimal fluctuations of a massless scalar field is inversely proportional to the physical scale. Taking into account that $\delta\varphi_L \sim |\delta\varphi_k| k^{3/2}$, where $k \sim a/L$ is the *comoving* wavenumber, we obtain

$$|\delta\varphi_k| \sim \frac{k^{-1/2}}{a}. \tag{8.24}$$

Comparing this result with (8.23), we infer that $|C_k| \sim k^{-1/2}$. The evolution of the mode according to (8.23) preserves the vacuum spectrum.

The result obtained is not surprising and has a simple physical interpretation. On scales smaller than the curvature scale one can always use the local inertial frame in which spacetime can be well approximated by the Minkowski metric. Therefore the short-wavelength fluctuations "think" they are in Minkowski space and the vacuum is preserved. Above, we have simply described this vacuum in an expanding coordinate system, where the perturbations with given comoving wavenumbers are continuously being stretched by the expansion. As a result, for a given physical scale, they are replaced by perturbations which were initially on sub-Planckian scales. This does not mean, however, that for a consistent treatment of quantum perturbations we need nonperturbative quantum gravity. Given a *physical scale*, which is larger than the Planckian length, vacuum fluctuations with amplitudes given above will always be present, irrespective of whether they are formally described as "being stretched from sub-Planckian scales" in the expanding coordinate system or as always existing at the given scale in the nonexpanding local inertial frame.

We noted in Chapter 5 that inflation "washes out" all pre-existing classical inhomogeneities by stretching them to very large scales. It is sometimes said that inflation removes the "classical hairs." However, it cannot remove quantum fluctuations ("quantum hairs"). In place of the stretched quantum fluctuations, new ones "are generated" via the Heisenberg uncertainty relation. For a given comoving wavenumber k, the typical amplitude of fluctuations is of order

$$\delta_\varphi(k) \sim |\delta\varphi_k| k^{3/2} \sim \frac{k}{a_k} \sim H_{k\sim Ha}, \tag{8.25}$$

at the moment of the horizon crossing, $Ha_k \sim k$ (or $k\eta_k \sim 1$). During the inflationary stage $Ha = \dot{a}$ increases and a perturbation with a given k eventually leaves the horizon. To see whether it will remain large enough after being stretched to galactic scales, we have to find out how it will behave on supercurvature scales.

8.2.2 The spectrum of generated perturbations

To determine the behavior of long-wavelength perturbations we use the slow-roll approximation. In Chapter 5 we saw that for the homogeneous mode this approximation means that in the equation

$$\ddot{\varphi}_0 + 3H\dot{\varphi}_0 + V_{,\varphi} = 0 \tag{8.26}$$

we can neglect the second derivative with respect to the physical time t and it simplifies to

$$3H\dot{\varphi}_0 + V_{,\varphi} \simeq 0. \tag{8.27}$$

To take advantage of the slow-roll approximation for the perturbations, we have to recast (8.17) and (8.20) in terms of the physical time t:

$$\delta\ddot{\varphi} + 3H\delta\dot{\varphi} - \Delta\delta\varphi + V_{,\varphi\varphi}\delta\varphi - 4\dot{\varphi}_0\dot{\Phi} + 2V_{,\varphi}\Phi = 0, \tag{8.28}$$

$$\dot{\Phi} + H\Phi = 4\pi\dot{\varphi}_0\delta\varphi, \tag{8.29}$$

where $\delta\varphi \equiv \overline{\delta\varphi}$ and we have taken into account that $\Psi = \Phi$. First of all we note that the spatial derivative term $\Delta\delta\varphi$ can be neglected for long-wavelength inhomogeneities. To find the nondecaying slow-roll mode we next omit terms proportional to $\delta\ddot{\varphi}$ and $\dot{\Phi}$. (After finding the solution of the simplified equations one can check that the omitted terms are actually negligible.) The equations for the perturbations become

$$3H\delta\dot{\varphi} + V_{,\varphi\varphi}\delta\varphi + 2V_{,\varphi}\Phi \simeq 0, \quad H\Phi \simeq 4\pi\dot{\varphi}_0\delta\varphi. \tag{8.30}$$

Introducing the new variable

$$y \equiv \delta\varphi/V_{,\varphi}$$

and using (8.27), they further simplify to

$$3H\dot{y} + 2\Phi = 0, \quad H\Phi = 4\pi\dot{V}y. \tag{8.31}$$

Since $3H^2 \simeq 8\pi V$ during inflation, we obtain the equation

$$\frac{d(yV)}{dt} = 0, \tag{8.32}$$

which is readily integrated to give

$$y = A/V, \tag{8.33}$$

where A is a constant of integration. The final result for the nondecaying mode is

$$\delta\varphi_k = A_k \frac{V_{,\varphi}}{V}, \quad \Phi_k = 4\pi A_k \frac{\dot{\varphi}_0}{H} \frac{V_{,\varphi}}{V} = -\frac{1}{2} A_k \left(\frac{V_{,\varphi}}{V}\right)^2. \tag{8.34}$$

The behavior of $\delta\varphi_k(a)$ is shown in Figure 8.1. For $a < a_k \sim k/H$ the perturbation is still inside the horizon and its amplitude decreases in inverse proportion to the scale factor. After horizon crossing, for $a > a_k$, the perturbation amplitude slowly increases since $V_{,\varphi}/V$ grows towards the end of inflation. In particular, for a power-law potential, $V \propto \varphi^n$, we have $\delta\varphi_k \propto \varphi^{-1}$. The integration constant A_k in (8.34) can be fixed by requiring that $\delta\varphi_k$ has the minimal vacuum amplitude at the moment of horizon crossing. Comparing (8.34) to (8.25), we find

$$A_k \sim \frac{k^{-1/2}}{a_k} \left(\frac{V}{V_{,\varphi}}\right)_{k \sim Ha},$$

where the index $k \sim Ha$ means that the corresponding quantity is estimated at the moment of horizon crossing. At the end of inflation $(t \sim t_f)$, the slow-roll condition is violated and $V_{,\varphi}/V$ becomes of order unity. Therefore, it follows from (8.34) that at this time the typical amplitude of the metric fluctuations on supercurvature

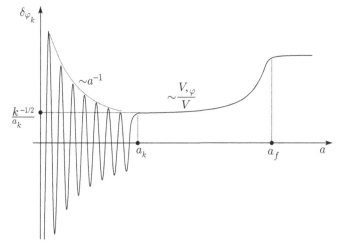

Fig. 8.1.

scales is

$$\delta_\Phi(k, t_f) \sim A_k k^{3/2} \sim \left(H\frac{V}{V_{,\varphi}}\right)_{k\sim Ha} \sim \left(\frac{V^{3/2}}{V_{,\varphi}}\right)_{k\sim Ha}. \tag{8.35}$$

In particular, for the power-law potential $V = \lambda \varphi^n/n$ we obtain

$$\delta_\Phi(k, t_f) \sim \lambda^{1/2}\left(\varphi^2_{k\sim Ha}\right)^{\frac{n+2}{4}} \sim \lambda^{1/2}(\ln \lambda_{ph} H_k)^{\frac{n+2}{4}}, \tag{8.36}$$

where (5.53) has been used in $k \sim aH \equiv a_k H_k$ to express $\varphi^2_{k\sim Ha}$ in terms of the physical wavelength $\lambda_{ph} \sim a(t_f) k^{-1}$. The spectrum (8.36) is shown in Figure 8.2. The effect of H_k in the logarithm in (8.36) is not very significant; we make only a slight error by taking $H_k \sim H(t_f)$.

For a massive scalar field $V = m^2 \varphi^2/2$ and the amplitude of the metric fluctuations is

$$\delta_\Phi \sim m \ln(\lambda_{ph} H_k). \tag{8.37}$$

We will show in the next section that perturbations present at the end of inflation survive the subsequent reheating phase practically unchanged. Since galactic scales correspond to $\ln(\lambda_{ph} H_k) \sim 50$ and we require the amplitude of the gravitational potential to be $O(1) \times 10^{-5}$, the mass of the scalar field should be about 10^{-6} in Planck units or $m \sim 10^{13}$ GeV. This determines the energy scale at the end of inflation to be $\varepsilon \sim m^2 \sim 10^{-12} \varepsilon_{Pl}$. In the absence of a fundamental particle theory we cannot predict the amplitude of the perturbations; it is a free parameter of the theory. However, the shape of the spectrum is predicted: it has logarithmic deviations from a flat spectrum with the amplitude growing slightly towards larger scales. We will see that this is a rather generic and robust prediction of inflation.

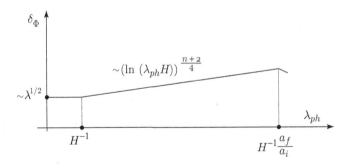

Fig. 8.2.

Problem 8.4 Show that the scalar field perturbations in the synchronous gauge are given by

$$\delta\varphi_s = \overline{\delta\varphi} - \dot\varphi_0 \int \Phi dt \qquad (8.38)$$

(compare with (7.30)). Substituting (8.34) into (8.38), verify that $\delta\varphi_s = C_1 \dot\varphi_0$, where C_1 is the integration constant in (8.38) corresponding to the purely coordinate mode. It is easy to understand why this mode is fictitious by simply considering the homogeneous field $\varphi_0(\eta)$ and making the coordinate transformation (7.27) which preserves the synchronous gauge. As we will see in Problem 8.8, the long-wavelength physical perturbations of the scalar fields are suppressed in the synchronous gauge by a factor $(k\eta)^2$.

Problem 8.5 It is clear that large-scale metric fluctuations after inflation can depend only on the few parameters characterizing them during the inflationary stage. Give arguments for why the most natural candidates for these parameters are $\delta\varphi$, $\dot\varphi_0$ and H. Out of these parameters, build the reasonable dimensionless combination which could describe metric fluctuations after inflation. Substituting for $\delta\varphi$ the amplitude of the quantum fluctuations, compare the estimate obtained with (8.35). Which questions remain open in this dimensional-based approach?

Problem 8.6 Consider two slow-roll fields φ_1 and φ_2 with potential $V(\varphi_1, \varphi_2) = V(\varphi_1) + V(\varphi_2)$ and verify that the nondecaying mode of the long-wavelength perturbations is given by

$$\Phi = A \frac{\dot H}{H^2} + B \frac{1}{H} \frac{V_1 \dot V_2 - \dot V_1 V_2}{V_1 + V_2}. \qquad (8.39)$$

The first term on the right hand side here is similar to (8.34) and it can be interpreted as the adiabatic mode. The second term describes the entropic contribution which is present when we have more than one field. When two or more scalar fields play an important role during inflation we can get a variety of different spectra and inflation, to a large extent, loses its predictive power. Therefore, we will not consider this case any further. (*Hint* Introduce the new variables $y_1 \equiv \delta\varphi_1 / V_{1,\varphi}$ and $y_2 \equiv \delta\varphi_2 / V_{2,\varphi}$.)

8.2.3 Why do we need inflation?

It is natural to ask whether quantum metric fluctuations can be substantially amplified in an expanding universe without an inflationary stage. Let us explain why this is impossible.

Quantum metric fluctuations are large only near the Planckian scale. For example, in Minkowski space the typical amplitude of vacuum metric fluctuations corresponding to gravitational waves can be estimated on dimensional grounds as $h \sim l_{Pl}/L$, where $l_{Pl} \sim 10^{-33}$ cm. It is incredibly small: on galactic scales $L \sim 10^{25}$ cm, so $h \sim 10^{-58}$. Scalar metric perturbations due to vacuum fluctuations of the scalar field are even smaller. Thus the only way to get the required amplitude $\Phi \sim 10^{-5}$ on large scales from initial quantum fluctuations is by stretching the very short-wavelength fluctuations. During this stretching, the mode must not lose its amplitude. Let us consider a scalar field perturbation, which determines the metric fluctuations, and find out what generally happens to its amplitude when the spatial size of the perturbation is stretched. As we have seen, the amplitude decays in inverse proportion to the spatial size until the perturbation starts to "feel" the curvature of the universe. This happens when its size begins to exceed the curvature scale H^{-1}. Therefore, if during expansion the perturbation size always remains smaller than the curvature scale, then its amplitude continuously decreases; it "arrives" at large scales with negligible vacuum amplitude. In a decelerating universe, the curvature scale $H^{-1} = a/\dot{a}$ grows faster than the physical wavelength of the perturbation $(\lambda_{ph} \propto a)$ because $\dot{a}$ is decreasing (see Figure 8.3). Hence, if a perturbation is initially inside the horizon, it remains there and decays. Perturbations on those scales which were initially a little larger than the Hubble radius will soon enter the horizon and also decay. Thus, in a decelerating expanding universe, quantum fluctuations can never be significantly amplified to become relevant for large-scale structure.

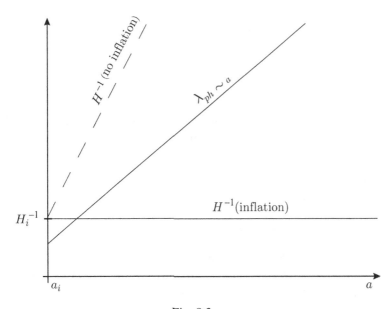

Fig. 8.3.

In a universe which undergoes a stage of accelerated expansion, the Hubble scale $H^{-1} = a/\dot{a}$ grows more slowly than the scale factor a because the rate of the expansion $\dot{a}$ increases. Hence a perturbation which was initially inside the horizon soon leaves it (see Figure 8.3) and starts to "feel" the curvature effects which preserve its amplitude from decay. The amplitude even grows slightly. We will see later that this growth of the amplitude is a rather general property of inflationary scenarios and it results in the deviation of the spectrum from a flat one. Thus, the initial amplitude of the subcurvature perturbation decays only until the moment of horizon crossing. After that it freezes out and the perturbation is stretched to galactic scales with *nearly* unchanged amplitude. Because the curvature scale does not change significantly during inflation, the freeze-out amplitude is nearly the same for different scales and this leads to a nearly flat spectrum for the produced inhomogeneities.

The initial quantum fluctuations are Gaussian. Subsequent evolution influences only their power spectrum and preserves the statistical properties of the fluctuations. As a consequence, simple inflation predicts *Gaussian adiabatic* perturbations.

8.3 Quantum cosmological perturbations

In this section we develop a consistent quantum theory of cosmological perturbations. We consider a flat universe filled by a scalar field condensate described by the action

$$S = \int p(X, \varphi) \sqrt{-g} d^4 x, \tag{8.40}$$

where

$$X \equiv \tfrac{1}{2} g^{\alpha\beta} \varphi_{,\alpha} \varphi_{,\beta}. \tag{8.41}$$

The Lagrangian $p(X, \varphi)$ plays the role of pressure. Indeed, by varying action (8.40) with respect to the metric, we obtain the energy–momentum tensor in the form of an ideal hydrodynamical fluid (see Problem 5.17):

$$T^\alpha_\beta = (\varepsilon + p) u^\alpha u_\beta - p \delta^\alpha_\beta. \tag{8.42}$$

Here $u_\nu \equiv \varphi_{,\nu}/\sqrt{2X}$ and the energy density ε is given by the expression

$$\varepsilon \equiv 2X p_{,X} - p, \tag{8.43}$$

where $p_{,X} \equiv \partial p/\partial X$. Thus, a scalar field can be used to describe potential flow of an ideal fluid. Conversely, hydrodynamics provides a useful analogy for a scalar field with an arbitrary Lagrangian. Action (8.40) is enough to describe all single-field inflationary scenarios, including k inflation. If p depends only on X, then $\varepsilon = \varepsilon(X)$, and in many cases (8.43) can be rearranged to give $p = p(\varepsilon)$, the equation of state for an isentropic fluid. For $p \propto X^n$ we have $p = \varepsilon/(2n-1)$, so, for example, the Lagrangian $p \propto X^2$ describes an "ultra-relativistic fluid" with equation of state $p = \varepsilon/3$. In the general case, $p = p(X, \varphi)$, the pressure cannot be expressed only in terms of ε since X and φ are independent. However, even in this case, the hydrodynamical analogy is still useful. For a canonical scalar field we have $p = X - V(\varphi)$ and, correspondingly, $\varepsilon = X + V$.

8.3.1 Equations

Here we derive the equations for perturbations and recast them in a simple, convenient form. The reader interested only in the final result can go directly to (8.56)–(8.58).

Background The state of a flat, homogeneous universe is characterized completely by the scale factor $a(\eta)$ and the homogeneous field $\varphi_0(\eta)$, which satisfy the familiar equations

$$\mathcal{H}^2 = \frac{8\pi}{3} a^2 \varepsilon, \tag{8.44}$$

and

$$\varepsilon' = \varepsilon_{,X} X_0' + \varepsilon_{,\varphi} \varphi_0' = -3\mathcal{H}(\varepsilon + p), \tag{8.45}$$

where $X_0 = \varphi_0'^2/(2a^2)$ and we have set $G = 1$. Substituting ε from (8.44) into the left hand side of (8.45), we obtain the relation

$$\mathcal{H}' - \mathcal{H}^2 = -4\pi a^2 (\varepsilon + p), \tag{8.46}$$

which is useful in what follows.

Perturbations To derive the equations for inhomogeneities we must first express the gauge-invariant perturbations of the energy–momentum tensor $\overline{\delta T}^\alpha_\beta$ in terms of the scalar field and metric perturbations. The calculation can easily be done in the longitudinal gauge, where the metric takes the form

$$ds^2 = a^2(\eta) \left[(1 + 2\Phi) d\eta^2 - (1 - 2\Psi) \delta_{ik} dx^i dx^k \right]. \tag{8.47}$$

To linear order in perturbation we have

$$\delta X = \frac{1}{2} \delta g^{00} \varphi_0'^2 + g^{00} \varphi_0' \delta\varphi' = 2X_0 \left(-\Phi + \frac{\delta\varphi'}{\varphi_0'} \right), \tag{8.48}$$

and the δT_0^0 component is

$$\delta T_0^0 = \delta\varepsilon = \varepsilon_{,X}\delta X + \varepsilon_{,\varphi}\delta\varphi = \varepsilon_{,X}\left(\delta X - X_0'\frac{\delta\varphi}{\varphi_0'}\right) - 3\mathcal{H}(\varepsilon+p)\frac{\delta\varphi}{\varphi_0'}$$

$$= \frac{\varepsilon+p}{c_s^2}\left(\left(\frac{\delta\varphi}{\varphi_0'}\right)' + \mathcal{H}\frac{\delta\varphi}{\varphi_0'} - \Phi\right) - 3\mathcal{H}(\varepsilon+p)\frac{\delta\varphi}{\varphi_0'}. \qquad (8.49)$$

We have used here the second equality in (8.45) to express $\varepsilon_{,\varphi}$ in terms of $\varepsilon_{,X}$, ε and p, and introduced the "speed of sound"

$$c_s^2 \equiv \frac{p_{,X}}{\varepsilon_{,X}} = \frac{\varepsilon+p}{2X\varepsilon_{,X}}. \qquad (8.50)$$

For a canonical scalar field the "speed of sound" is always equal to the speed of light, $c_s = 1$. The components δT_i^0 are readily calculated and the result is

$$\delta T_i^0 = (\varepsilon+p)u^0\delta u_i = (\varepsilon+p)g^{00}\frac{\varphi_0'}{\sqrt{2X_0}}\frac{\delta\varphi_{,i}}{\sqrt{2X_0}} = (\varepsilon+p)\left(\frac{\delta\varphi}{\varphi_0'}\right)_{,i}. \qquad (8.51)$$

Replacing $\delta\varphi$ by $\overline{\delta\varphi}$, defined in (8.16), and substituting (8.49) and (8.51) into (7.38) and (7.39), one obtains the for the gauge-invariant variables Ψ, Φ and $\overline{\delta\varphi}$:

$$\Delta\Psi - 3\mathcal{H}(\Psi' + \mathcal{H}\Phi)$$
$$= 4\pi a^2(\varepsilon+p)\left[\frac{1}{c_s^2}\left(\left(\frac{\overline{\delta\varphi}}{\varphi_0'}\right)' + \mathcal{H}\frac{\overline{\delta\varphi}}{\varphi_0'} - \Phi\right) - 3\mathcal{H}\frac{\overline{\delta\varphi}}{\varphi_0'}\right], \qquad (8.52)$$

$$(\Psi' + \mathcal{H}\Phi) = 4\pi a^2(\varepsilon+p)\left(\frac{\overline{\delta\varphi}}{\varphi_0'}\right). \qquad (8.53)$$

Since $\delta T_k^i = 0$ for $i \neq k$, we have $\Psi = \Phi$; the two equations above are sufficient to determine the gravitational potential and the perturbation of the scalar field. It is useful, however, to recast them in a slightly different, more convenient form. Using (8.53) to express Φ in terms of Ψ' and $\overline{\delta\varphi}$ and substituting the result into (8.52), we obtain

$$\Delta\Psi = \frac{4\pi a^2(\varepsilon+p)}{c_s^2\mathcal{H}}\left(\mathcal{H}\frac{\overline{\delta\varphi}}{\varphi_0'} + \Psi\right)', \qquad (8.54)$$

where the background equations (8.44) and (8.46) have also been used. Because $\Phi = \Psi$, (8.53) can be rewritten as

$$\left(a^2\frac{\Psi}{\mathcal{H}}\right)' = \frac{4\pi a^4(\varepsilon+p)}{\mathcal{H}^2}\left(\mathcal{H}\frac{\overline{\delta\varphi}}{\varphi_0'} + \Psi\right). \qquad (8.55)$$

Finally, in terms of the new variables

$$u \equiv \frac{\Psi}{4\pi(\varepsilon + p)^{1/2}}, \quad v \equiv \sqrt{\varepsilon_{,X}} a\left(\overline{\delta\varphi} + \frac{\varphi_0'}{\mathcal{H}}\Psi\right), \tag{8.56}$$

(8.54) and (8.55) become

$$c_s \Delta u = z\left(\frac{v}{z}\right)', \quad c_s v = \theta\left(\frac{u}{\theta}\right)' \tag{8.57}$$

where

$$z \equiv \frac{a^2(\varepsilon + p)^{1/2}}{c_s \mathcal{H}}, \quad \theta \equiv \frac{1}{c_s z} = \sqrt{\frac{8\pi}{3}} \frac{1}{a}\left(1 + \frac{p}{\varepsilon}\right)^{-1/2}. \tag{8.58}$$

8.3.2 Classical solutions

Substituting v from the second equation in (8.57) into the first gives a closed form, second order differential equation for u:

$$u'' - c_s^2 \Delta u - \frac{\theta''}{\theta} u = 0. \tag{8.59}$$

The variables u and θ coincide (up to irrelevant numerical factors) with the corresponding quantities defined in (7.63) and (7.66) for the hydrodynamical fluid. However, now they describe the perturbations in the homogeneous scalar condensate.

The solutions of (8.59) were discussed in the previous chapter. Considering a *short-wavelength* plane wave perturbation with a wavenumber k ($c_s^2 k^2 \gg |\theta''/\theta|$), we obtain in the WKB approximation

$$u \simeq \frac{C}{\sqrt{c_s}} \exp\left(\pm ik \int c_s d\eta\right), \tag{8.60}$$

where C is a constant of integration. The *long-wavelength* solution, valid for $c_s^2 k^2 \ll |\theta''/\theta|$, is

$$u = C_1 \theta + C_2 \theta \int_{\eta_0} \frac{d\eta}{\theta^2} + O\left((k\eta)^2\right). \tag{8.61}$$

Given u, the gravitational potential can be inferred from the definition in (8.56):

$$\Phi = \Psi = 4\pi(\varepsilon + p)^{1/2} u \tag{8.62}$$

and a perturbation of the scalar field is calculated using (8.53):

$$\overline{\delta\varphi} = \varphi_0' \frac{(a\Phi)'}{4\pi a^3(\varepsilon + p)} = \dot{\varphi}_0 \frac{(\dot{\Phi} + H\Phi)}{4\pi(\varepsilon + p)}. \tag{8.63}$$

Taking into account that

$$\varepsilon + p = 2X p_{,X} = \frac{1}{a^2}\varphi_0'^2 p_{,X} \qquad (8.64)$$

and substituting (8.60) into (8.62) and (8.63), we have

$$\Phi \simeq 4\pi C \dot\varphi_0 \sqrt{\frac{p_{,X}}{c_s}} \exp\left(\pm ik\int \frac{c_s}{a}dt\right), \qquad (8.65)$$

$$\overline{\delta\varphi} \simeq C \sqrt{\frac{1}{c_s p_{,X}}}\left(\pm i c_s \frac{k}{a} + H + \cdots\right)\exp\left(\pm ik\int\frac{c_s}{a}dt\right), \qquad (8.66)$$

for a *short-wavelength* perturbation.

In the *long-wavelength* limit the calculation is identical to that done in deriving (7.69), and the result is

$$\Phi \simeq A\frac{d}{dt}\left(\frac{1}{a}\int a\,dt\right) = A\left(1 - \frac{H}{a}\int a\,dt\right), \qquad (8.67)$$

$$\overline{\delta\varphi} \simeq A\dot\varphi_0\left(\frac{1}{a}\int a\,dt\right), \qquad (8.68)$$

where A is a constant of integration. (A second constant of integration corresponding to the decaying mode can always be shifted to the lower limit of integration.)

Let us first find out how a perturbation behaves during inflation. It follows from (8.65) and (8.66) that in the short-wavelength regime both metric and scalar field perturbations oscillate. The amplitude of the metric perturbation is proportional to $\dot\varphi_0$ and it grows only slightly towards the end of inflation, while the amplitude of scalar field perturbation decays in inverse proportion to the scale factor. After a perturbation enters the long-wavelength regime it is described by (8.67) and (8.68). These formulae are simplified during slow-roll. Integrating by parts, we obtain the following asymptotic expansion:

$$\frac{1}{a}\int a\,dt = \frac{1}{a}\int\frac{da}{H} = H^{-1} - \frac{1}{a}\int\frac{da}{H}(H^{-1})^{\cdot}$$
$$= H^{-1}\left[1 - (H^{-1})^{\cdot} + \left(H^{-1}(H^{-1})^{\cdot}\right)^{\cdot} - \cdots\right] + \frac{B}{a}, \qquad (8.69)$$

where B is a constant of integration corresponding to the decaying mode. Neglecting this mode we find that to leading order

$$\Phi \simeq A(H^{-1})^{\cdot} = -A\frac{\dot H}{H^2}, \qquad \overline{\delta\varphi} \simeq A\frac{\dot\varphi_0}{H}. \qquad (8.70)$$

It is easy to see that for standard slow-roll inflation these formulae are in agreement with (8.34). Result (8.70) is applicable only during inflation. After the slow-roll

8.3 Quantum cosmological perturbations

stage is over we must use (8.67) and (8.68) directly. Inflation is usually followed by an oscillatory stage where the scale factor grows as some power of time, $a \propto t^p$, with p depending on the scalar field potential. We have found that for the quadratic potential $p = 2/3$ and for the quartic potential $p = 1/2$. Neglecting the decaying mode we obtain from (8.67) and (8.68)

$$\Phi \simeq \frac{A}{p+1}, \quad \overline{\delta\varphi} \simeq \frac{At\dot{\varphi}_0}{p+1}, \tag{8.71}$$

that is, the amplitude of the gravitational potential freezes out after inflation.

The scalar field finally converts its energy into ultra-relativistic matter corresponding to $p = 1/2$. This influences the perturbations only via the change of the effective equation of state and the resulting amplitude is

$$\Phi \simeq \tfrac{2}{3}A. \tag{8.72}$$

Using (8.70), we can express A in terms of $\overline{\delta\varphi}$, $\dot{\varphi}_0$ and H at the moment of sound horizon crossing, when $c_s k \sim H a$. For those perturbations which leave the horizon during inflation the final result is

$$\Phi \simeq \frac{2}{3}\left(H\frac{\overline{\delta\varphi}}{\dot{\varphi}_0}\right)_{c_s k \sim H a}. \tag{8.73}$$

Given initial quantum fluctuations, this is consistent with the estimate in (8.35) and we infer that the amplitude of the perturbation in the radiation-dominated epoch differs from its amplitude at the end of inflation only by a numerical factor of order unity. Note that (8.73) can also be applied to calculate the perturbations in theories with a non-minimal kinetic term.

Problem 8.7 Using the integral representation of (8.59) (see (7.75)), calculate the k^2-corrections to the long-wavelength solution (8.61). Verify that the "conserved" quantity $\zeta \propto \theta^2(u/\theta)'$ (see also (7.72)) blows up during an oscillatory stage. Hence, contrary to the claims often made in the literature, it cannot be used to trace the evolution of perturbation through this stage.

Problem 8.8 *Synchronous coordinate system.* (a) Verify that a scalar field perturbation in the synchronous coordinate system can be expressed through the gravitational potential as

$$\delta\varphi_s = \overline{\delta\varphi} - \dot{\varphi}_0 \int \Phi dt = F_s \dot{\varphi}_0 - \dot{\varphi}_0 \int \frac{c_s^2}{\dot{H}a^2} \Delta\Phi dt, \tag{8.74}$$

where F_s is a constant of integration corresponding to the fictitious mode. The relation above is exact. In the long-wavelength limit the physical mode of $\delta\varphi_s$ is of order k^2. Considering a long-wavelength perturbation and using (8.70), show that in an inflationary phase

$$\delta\varphi_s \simeq F_s\dot\varphi_0 + \frac{1}{2}A\frac{\dot\varphi_0}{H}\left(\frac{kc_s}{Ha}\right)^2. \qquad (8.75)$$

Skipping the fictitious mode, express A in terms of $\delta\varphi_s$, H and $\dot\varphi_0$ at the moment of the sound horizon crossing. Compare the result obtained with the expression previously derived for A in terms of $\overline{\delta\varphi}$. (*Hint* To derive the second equality in (8.74), use (8.52) to express $\overline{\delta\varphi}$ in terms of Φ and $\Delta\Phi$.)

(b) Substituting (8.67) into (7.29), verify that for a long-wavelength perturbation

$$\psi_s \simeq A + F_1 H, \quad E_s \simeq A\int\frac{1}{a^3}\left(\int^t a d\tilde t\right)dt + F_1\int\frac{dt}{a^2} + F_2, \qquad (8.76)$$

where F_1 and F_2 are constants of integration corresponding to fictitious modes. Find the relation between F_s and F_1, F_2. Write down the metric components δg_{ik} in the synchronous coordinate system.

Starting with quantum fluctuations, the resulting amplitude of perturbations in the post-inflationary epoch can be fixed if we know $\delta\varphi$ at horizon crossing. The natural question arises: which $\delta\varphi$ plays the role of a canonical quantization variable? We found in Problem 8.8 that one can get results differing by a numerical factor depending on whether we relate the quantum perturbation to $\delta\varphi_s$ or $\overline{\delta\varphi}$. This is not surprising because at the moment of horizon crossing the metric fluctuations may become relevant and the Minkowski space approximation fails. To resolve the gauge ambiguity and derive the exact numerical coefficients we need a rigorous quantum theory.

8.3.3 Quantizing perturbations

Action In order to construct a canonical quantization variable and properly normalize the amplitude of quantum fluctuations, we need the action for the cosmological perturbations. To obtain it one expands the action for the gravitational and scalar fields to second order in perturbations. After use of the constraints, the result is reduced to an expression containing only the physical degrees of freedom. The steps are very cumbersome but fortunately they can be avoided. This is because the action for the perturbations can be unambiguously inferred directly from the equations of motion (8.57) up to an overall time-independent factor. This factor can then be fixed by calculating the action in some simple limiting case. The first order action reproducing the equations of motion (8.57) is

$$S = \int\left[\left(\frac{v}{z}\right)'\hat O\left(\frac{u}{\theta}\right) - \frac{1}{2}c_s^2(\Delta u)\hat O u + \frac{1}{2}c_s^2 v\hat O v\right]d\eta d^3x, \qquad (8.77)$$

where $\hat{O} \equiv \hat{O}(\Delta)$ is a time-independent operator to be determined. Using the first equation in (8.57) to express u in terms of $(v/z)'$, we obtain

$$S = \frac{1}{2} \int \left[z^2 \left(\frac{v}{z}\right)' \frac{\hat{O}}{\Delta} \left(\frac{v}{z}\right)' + c_s^2 v \hat{O} v \right] d\eta d^3 x. \tag{8.78}$$

Problem 8.9 Write down the action for a massless scalar field in a flat de Sitter universe. Comparing action (8.78) in the limiting case $\dot{\varphi}_0/H \to 0$ to the action for a free scalar field in the de Sitter universe, verify that $\hat{O} = \Delta$.

With the result of the above problem, (8.78) becomes

$$S \equiv \int \mathcal{L} d\eta d^3 x = \frac{1}{2} \int \left(v'^2 + c_s^2 v \Delta v + \frac{z''}{z} v^2 \right) d\eta d^3 x, \tag{8.79}$$

after we drop the total derivative terms. Varying this action with respect to v we obtain

$$v'' - c_s^2 \Delta v - \frac{z''}{z} v = 0. \tag{8.80}$$

Note that this equation also follows from the second equation in (8.57) after substituting u in terms of v.

Problem 8.10 The long-wavelength solution of (8.80) can be written in a similar manner to (8.61):

$$v = C_1^{(v)} z + C_2^{(v)} z \int_{\eta_0} \frac{d\eta}{z^2} + O((k\eta)^2), \tag{8.81}$$

where $C_1^{(v)}$ and $C_2^{(v)}$ are constants of integration. Because u and v satisfy a system of two first order differential equations, there are only two independent constants of integration. Therefore $C_1^{(v)}$ and $C_2^{(v)}$ can be expressed in terms of the C_1 and C_2 in (8.61). Verify that $C_1^{(v)} = C_2$ and $C_2^{(v)} = -k^2 C_1$.

Quantization The quantization of cosmological perturbations with action (8.79) is thus formally equivalent to the quantization of a "free scalar field" v with time-dependent "mass" $m^2 = -z''/z$ in Minkowski space. The time dependence of the "mass" is due to the interaction of the perturbations with the homogeneous expanding background. The energy of the perturbations is not conserved and they can be excited by borrowing energy from the Hubble expansion.

The canonical quantization variable

$$v = \sqrt{\varepsilon_{,X}} a \left(\overline{\delta\varphi} + \frac{\varphi_0'}{\mathcal{H}} \Psi \right) = \sqrt{\varepsilon_{,X}} a \left(\delta\varphi + \frac{\varphi_0'}{\mathcal{H}} \psi \right) \tag{8.82}$$

is a gauge-invariant combination of the scalar field and metric perturbations.

Problem 8.11 Considering only the physical (nonfictitious) mode of a long-wavelength perturbatioun in the synchronous coordinate system, verify that the second term in the second equality in (8.82) dominates over the first.

The first step in quantizing (8.79) is to define the momentum π canonically conjugated to v,

$$\pi \equiv \frac{\partial \mathcal{L}}{\partial v'} = v'. \tag{8.83}$$

In quantum theory, the variables v and π become operators $\hat{v}$ and $\hat{\pi}$, which at any moment of time η satisfy the standard commutation relations:

$$[\hat{v}(\eta, \mathbf{x}), \hat{v}(\eta, \mathbf{y})] = [\hat{\pi}(\eta, \mathbf{x}), \hat{\pi}(\eta, \mathbf{y})] = 0,$$

$$[\hat{v}(\eta, \mathbf{x}), \hat{\pi}(\eta, \mathbf{y})] = [\hat{v}(\eta, \mathbf{x}), \hat{v}'(\eta, \mathbf{y})] = i\delta(\mathbf{x} - \mathbf{y}), \tag{8.84}$$

where we have set $\hbar = 1$. The operator $\hat{v}$ obeys the same equation as the corresponding classical variable v,

$$\hat{v}'' - c_s^2 \Delta \hat{v} - \frac{z''}{z}\hat{v} = 0, \tag{8.85}$$

and its general solution can be written as

$$\hat{v}(\eta, \mathbf{x}) = \frac{1}{\sqrt{2}} \int \left[v_\mathbf{k}^*(\eta) e^{i\mathbf{k}\mathbf{x}} \hat{a}_\mathbf{k}^- + v_\mathbf{k}(\eta) e^{-i\mathbf{k}\mathbf{x}} \hat{a}_\mathbf{k}^+ \right] \frac{d^3k}{(2\pi)^{3/2}}, \tag{8.86}$$

where the temporal mode functions $v_\mathbf{k}(\eta)$ satisfy

$$v_k'' + \omega_k^2(\eta) v_\mathbf{k} = 0, \quad \omega_k^2(\eta) \equiv c_s^2 k^2 - z''/z. \tag{8.87}$$

We are free to impose the bosonic commutation relations for the creation and annihilation operators on the conjugated operator-valued constants of integration $\hat{a}_\mathbf{k}^-$ and $\hat{a}_{\mathbf{k}'}^+$:

$$[\hat{a}_\mathbf{k}^-, \hat{a}_{\mathbf{k}'}^-] = [\hat{a}_\mathbf{k}^+, \hat{a}_{\mathbf{k}'}^+] = 0, \quad [\hat{a}_\mathbf{k}^-, \hat{a}_{\mathbf{k}'}^+] = \delta(\mathbf{k} - \mathbf{k}'). \tag{8.88}$$

Substituting (8.86) into (8.84), we find that they are consistent with commutation relations (8.84) only if the mode functions $v_\mathbf{k}(\eta)$ obey the normalization condition

$$v_\mathbf{k}' v_\mathbf{k}^* - v_\mathbf{k} v_\mathbf{k}^{*\prime} = 2i. \tag{8.89}$$

The expression on the left hand side is a Wronskian of (8.87) built from two independent solutions $v_\mathbf{k}$ and $v_\mathbf{k}^*$; therefore it does not depend on time. It follows from (8.89) that $v_\mathbf{k}(\eta)$ is a *complex* solution of the second order differential equation (8.87). To specify it fully and thus determine the physical meaning of the operators $\hat{a}_\mathbf{k}^\pm$ we need the initial conditions for $v_\mathbf{k}$ and $v_\mathbf{k}'$ at some initial time $\eta = \eta_i$.

8.3 Quantum cosmological perturbations

Substituting

$$v_{\mathbf{k}} = r_{\mathbf{k}} \exp(i\alpha_{\mathbf{k}})$$

into (8.89) we infer that the real functions $r_{\mathbf{k}}$ and $\alpha_{\mathbf{k}}$ obey the condition

$$r_{\mathbf{k}}^2 \alpha_{\mathbf{k}}' = 1. \tag{8.90}$$

Next we note that (8.87) describes a harmonic oscillator with energy

$$E_{\mathbf{k}} = \frac{1}{2}\left(|v_{\mathbf{k}}'|^2 + \omega_k^2 |v_{\mathbf{k}}|^2\right)$$
$$= \frac{1}{2}(r_{\mathbf{k}}'^2 + r_{\mathbf{k}}^2 \alpha_{\mathbf{k}}'^2 + \omega_k^2 r_{\mathbf{k}}^2) = \frac{1}{2}\left(r_{\mathbf{k}}'^2 + \frac{1}{r_{\mathbf{k}}^2} + \omega_k^2 r_{\mathbf{k}}^2\right). \tag{8.91}$$

We want to consider the minimal possible fluctuations allowed by the uncertainty relations. The energy is minimized when $r_{\mathbf{k}}'(\eta_i) = 0$ and $r_{\mathbf{k}}(\eta_i) = \omega_k^{-1/2}$. We thus obtain

$$v_{\mathbf{k}}(\eta_i) = \frac{1}{\sqrt{\omega_k}} e^{i\alpha_{\mathbf{k}}(\eta_i)}, \quad v_{\mathbf{k}}'(\eta_i) = i\sqrt{\omega_k} e^{i\alpha_{\mathbf{k}}(\eta_i)}. \tag{8.92}$$

Although the phase factors $\alpha_{\mathbf{k}}(\eta_i)$ remain undetermined, they are irrelevant and we can set them to zero. Note that the above considerations are valid only if $\omega_k^2 > 0$, that is, for modes with $c_s^2 k^2 > (z''/z)_i$.

The next step in quantization is to define the "vacuum" state $|0\rangle$ as the state annihilated by operators $\hat{a}_{\mathbf{k}}^-$:

$$\hat{a}_{\mathbf{k}}^- |0\rangle = 0. \tag{8.93}$$

We further assume that a complete set of independent states in the corresponding Hilbert space can be obtained by acting with the products of creation operators on the vacuum state $|0\rangle$. If the ω_k do not depend on time, then the vector $|0\rangle$ corresponds to the familiar Minkowski vacuum. Assuming c_s changes adiabatically, we find that modes with $c_s^2 k^2 \gg (z''/z)$ remain unexcited and minimal fluctuations are well defined. On the other hand, for modes with $c_s^2 k^2 < (z''/z)_i$ we have $\omega_k^2(\eta_i) < 0$, and the initial minimal fluctuations on corresponding scales cannot be unambiguously determined. These scales exceed the Hubble scale at the beginning of inflation and are subsequently stretched to huge unobservable scales; therefore the question of initial fluctuations here is fortunately moot. The inhomogeneities responsible for the observable structure originate from quantum fluctuations on scales where the minimal fluctuations are unambiguously defined.

Spectrum Our final task is to calculate the correlation function, or equivalently, the power spectrum of the gravitational potential. Taking into account (8.56), we have the following expansion for the operator $\hat{\Phi}$:

$$\hat{\Phi}(\eta, \mathbf{x}) = \frac{4\pi(\varepsilon + p)^{1/2}}{\sqrt{2}} \int \left[u_\mathbf{k}^*(\eta) e^{i\mathbf{k}\mathbf{x}} \hat{a}_\mathbf{k}^- + u_\mathbf{k}(\eta) e^{-i\mathbf{k}\mathbf{x}} \hat{a}_\mathbf{k}^+ \right] \frac{d^3k}{(2\pi)^{3/2}}, \tag{8.94}$$

where the mode functions $u_\mathbf{k}(\eta)$ obey (8.59) and are related to the mode functions $v_\mathbf{k}(\eta)$ via (8.57). For the initial vacuum state $|0\rangle$ the correlation function at $\eta > \eta_i$ is

$$\langle 0 | \hat{\Phi}(\eta, \mathbf{x}) \hat{\Phi}(\eta, \mathbf{y}) | 0 \rangle = \int 4(\varepsilon + p) |u_k|^2 k^3 \frac{\sin kr}{kr} \frac{dk}{k}, \tag{8.95}$$

where $r \equiv |\mathbf{x} - \mathbf{y}|$. According to the definition of the power spectrum in (8.8) and (8.9), we have

$$\delta_\Phi^2(k, \eta) = 4(\varepsilon + p) |u_\mathbf{k}(\eta)|^2 k^3. \tag{8.96}$$

Given $v_\mathbf{k}(\eta_i)$ and $v'_\mathbf{k}(\eta_i)$, the initial conditions for $u_\mathbf{k}$ can be inferred from (8.57). Let us consider a short-wavelength perturbation with $c_s^2 k^2 \gg (z''/z)_i$ for which $\omega_k(\eta_i) \simeq c_s k$. In this case the initial conditions (8.92) can be rewritten in terms of $u_\mathbf{k}$ as

$$u_\mathbf{k}(\eta_i) \simeq -\frac{i}{\sqrt{c_s} k^{3/2}}, \quad u'_\mathbf{k}(\eta_i) \simeq \frac{\sqrt{c_s}}{k^{1/2}}, \tag{8.97}$$

where we have neglected higher-order terms, which are suppressed by powers of $(c_s k \eta_i)^{-1} \ll 1$. The corresponding short-wavelength WKB solution, valid for $c_s^2 k^2 \gg |\theta''/\theta|$, is

$$u_\mathbf{k}(\eta) \simeq -\frac{i}{\sqrt{c_s} k^{3/2}} \exp\left(ik \int_{\eta_i}^{\eta} c_s d\tilde{\eta} \right). \tag{8.98}$$

During inflation the ratio $|\theta''/\theta|$ can be estimated roughly as $\eta^{-2}|\dot{H}/H^2|$. Because $|\dot{H}/H^2| \ll 1$, (8.98) is still applicable within the short time interval

$$\frac{1}{c_s k} > |\eta| > \frac{1}{k} |\dot{H}/H^2|^{1/2} \tag{8.99}$$

after the sound horizon crossing. At this time the argument in the exponent is almost constant and $u_\mathbf{k}$ freezes out. After a perturbation enters the long-wavelength regime the time evolution of the gravitational potential is described by (8.67), and hence

$$u_\mathbf{k}(\eta) \equiv \frac{\Phi}{4\pi(\varepsilon + p)^{1/2}} = \frac{A_\mathbf{k}}{4\pi(\varepsilon + p)^{1/2}} \left(1 - \frac{H}{a} \int a \, dt \right). \tag{8.100}$$

We can use (8.69) to simplify this expression during inflation:

$$u_{\mathbf{k}}(\eta) \simeq -\frac{A_{\mathbf{k}}}{4\pi(\varepsilon+p)^{1/2}}\left(\frac{\dot{H}}{H^2}\right) = A_{\mathbf{k}}\frac{(\varepsilon+p)^{1/2}}{H^2}. \quad (8.101)$$

Taking into account that within the time interval (8.99) the ratio

$$\frac{(\varepsilon+p)^{1/2}}{H^2}$$

is almost constant and comparing (8.98) and (8.101), we obtain

$$A_{\mathbf{k}} \simeq -\frac{i}{k^{3/2}}\left(\frac{H^2}{\sqrt{c_s}(\varepsilon+p)^{1/2}}\right)_{c_s k \simeq Ha}. \quad (8.102)$$

Substituting (8.98) into (8.96) gives the scale-independent power spectrum

$$\delta_\Phi^2(k, t) \simeq \frac{4(\varepsilon+p)}{c_s} \quad (8.103)$$

for short-wavelength perturbations with $k > Ha(t)/c_s$. Using (8.100) with $A_{\mathbf{k}}$ as given in (8.102), we obtain

$$\delta_\Phi^2(k, t) \simeq \frac{16}{9}\left(\frac{\varepsilon}{c_s(1+p/\varepsilon)}\right)_{c_s k \simeq Ha}\left(1 - \frac{H}{a}\int a\, dt\right)^2 \quad (8.104)$$

for long-wavelength perturbations with $Ha(t)/c_s > k > Ha_i/c_s$, where $a_i \equiv a(t_i)$.

Problem 8.12 Verify that for a massive scalar field of mass m the spectrum $\delta_\Phi(\lambda_{ph}, t)$ as a function of the physical scale $\lambda_{ph} \sim a(t)/k$ is given by

$$\delta_\Phi \simeq \frac{m}{\sqrt{3\pi}}\begin{cases} 1, & \lambda_{ph} < H^{-1}, \\ \left(1 + \frac{\ln(\lambda_{ph} H)}{\ln(a_f/a(t))}\right), & H^{-1}\frac{a(t)}{a_i} > \lambda_{ph} > H^{-1}, \end{cases} \quad (8.105)$$

for $a_f > a(t) > a_i$, where a_i and a_f are the values of the scale factor at the beginning and at the end of inflation respectively. The evolution of the spectrum (8.105) is sketched in Figure 8.4.

It follows from (8.104) that in the post-inflationary, radiation-dominated epoch the resulting power spectrum is

$$\delta_\Phi^2 \simeq \frac{64}{81}\left(\frac{\varepsilon}{c_s(1+p/\varepsilon)}\right)_{c_s k \simeq Ha}. \quad (8.106)$$

This formula is applicable only on scales corresponding to $(c_s^{-1}Ha)_f > k > (c_s^{-1}Ha)_i$. This range surely encompasses the observable universe. The

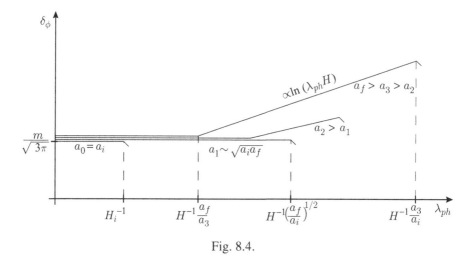

Fig. 8.4.

supercurvature perturbations are frozen during the radiation-dominated stage and they survive unchanged until recombination. Only for those scales which re-enter the horizon does the evolution proceed in a nontrivial way.

Problem 8.13 Verify that for an inflationary model with potential $V = (\lambda/n)\varphi^n$,

$$\delta_\Phi^2 \simeq \frac{128}{27} \frac{n^{\frac{n}{2}-2}}{(4\pi)^{n/2}} \lambda \left[\ln\left(\lambda_{ph}/\lambda_\gamma\right)\right]^{\frac{n+2}{2}}, \tag{8.107}$$

where λ_γ is the typical wavelength of the background radiation.

Spectral tilt It follows from (8.106) that the amplitude of the metric perturbation on a given comoving scale is determined by the energy density and by deviation of the equation of state from the vacuum equation of state at the time of horizon crossing. On galactic scales, δ_Φ^2 is of order 10^{-10} and $(1 + p/\varepsilon)$ can be estimated as $\sim 10^{-2}$; therefore we conclude that $\varepsilon \sim 10^{-12}$ of the Planckian density at this time. This is a rather robust and generic estimate for inflation during the last 70 e-folds. Only if $c_s \ll 1$, for instance in k inflation, can we avoid this conclusion.

Since inflation must have a graceful exit, the energy density and the equation of state slowly change during inflation. As a consequence the amplitude of the perturbations generated depends slightly on the lengthscale. The energy density always decreases and it is natural to expect that the deviation of the equation of state from that for the vacuum should increase towards the end of inflation. It follows then from (8.106) that the amplitude of those perturbations which crossed the horizon earlier must be larger than the amplitude of perturbations which crossed later. Within a narrow range of scales, one can always approximate the spectrum by the power law, $\delta_\Phi^2(k) \propto k^{n_S-1}$, and thus characterize it by the spectral index n_S. A flat spectrum corresponds to $n_S = 1$.

The expression for the spectral index follows from (8.106):

$$n_S - 1 \equiv \frac{d \ln \delta_\Phi^2}{d \ln k} \simeq -3\left(1 + \frac{p}{\varepsilon}\right) - \frac{1}{H}\left(\ln\left(1 + \frac{p}{\varepsilon}\right)\right)^{\cdot} - \frac{(\ln c_s)^{\cdot}}{H}, \qquad (8.108)$$

where the quantities on the right hand side must be calculated at the time of horizon crossing. In deriving this formula we have taken into account that $d \ln k \simeq d \ln a_k$. This relation follows from the condition determining horizon crossing, $c_s k \simeq H a_k$, if we neglect the change in c_s and H. All terms on the right hand side of (8.108) are negative for a generic inflationary scenario. Therefore, inflation does not predict a flat spectrum, as is quite often mistakenly stated. Instead, it predicts a red-tilted spectrum: $n_S < 1$ so that the amplitude grows slightly towards the larger scales. The physical reason for this tilt is the necessity for a smooth graceful exit. To obtain an estimate for the tilt we note that the galactic scales cross the horizon around 50–60 e-folds before the end of inflation. At this time $(1 + p/\varepsilon)$ is larger than 10^{-2}. The second term in (8.108) is about the same order of magnitude and the spectral index can thus be estimated as $n_S \simeq 0.96$. The concrete value of n_S depends on a particular inflationary scenario. Even without knowing this scenario, however, one could expect that $n_S \leq 0.97$. By inspection of the variety of scenarios, one infers that it is rather difficult to get a very large deviation from the flat spectrum and that it is likely $n_S > 0.92$.

Problem 8.14 Consider inflation in a model with potential V and verify that

$$n_S - 1 \simeq -\frac{3}{8\pi}\left(\frac{V_{,\varphi}}{V}\right)^2 + \frac{1}{4\pi}\frac{V_{,\varphi\varphi}}{V}. \qquad (8.109)$$

Check that for the power-law potential, $V \propto \varphi^n$,

$$n_S - 1 \simeq -\frac{n(n+2)}{8\pi \varphi^2_{k \simeq Ha}} \simeq -\frac{n+2}{2N}, \qquad (8.110)$$

where N is the number of e-folds before the end of inflation when the corresponding perturbation crosses the horizon. In the case of a massive scalar field, $n = 2$, and $n_S \simeq 0.96$ on galactic scales for which $N \simeq 50$. For the quartic potential $n = 4$ and $n_S \simeq 0.94$. How much does the spectral index "run" when the scale changes by one decade?

How do quantum fluctuations become classical? When we look at the sky we see the galaxies in certain positions. If these galaxies originated from initial quantum fluctuations, a natural question arises: how does a galaxy, e.g. Andromeda, find itself at a *particular* place if the initial vacuum state was translational-invariant with no preferred position in space? Quantum mechanical unitary evolution does not destroy translational invariance and hence the answer to this question must lie in the transition from quantum fluctuations to classical inhomogeneities. Decoherence is a necessary condition for the emergence of classical inhomogeneities and can easily be justified for amplified cosmological perturbations. However, decoherence is not sufficient to explain the breaking of translational invariance. It can be shown that as a result of unitary evolution we obtain a state which is a superposition of many macroscopically different states, each corresponding to a particular realization of galaxy distribution. Most of these realizations have the same statistical properties. Such a state is a close cosmic analog of the "Schrödinger cat." Therefore, to pick an observed macroscopic state from the superposition we have to appeal either to Bohr's reduction postulate or to Everett's many-worlds interpretation of quantum mechanics. The first possibility does not look convincing in the cosmological context. The reader who would like to pursue this issue can consult the corresponding references in "Bibliography" (Everett, 1957; De Witt and Graham, 1973).

8.4 Gravitational waves from inflation

Quantizing gravitational waves In a similar manner to scalar perturbations, long-wavelength gravitational waves are also generated in inflation. The calculations are not very different from those presented in the previous section. First of all we need the action for the gravitational waves. This action can be derived by expanding the Einstein action up to the second order in transverse, traceless metric perturbations h_{ik}. The result is

$$S = \frac{1}{64\pi} \int a^2 \left(h_j^{i\prime} h_i^{j\prime} - h_{j,l}^i h_i^{j,l} \right) d\eta d^3x, \qquad (8.111)$$

where the spatial indices are raised and lowered with the help of the unit tensor δ_{ik}.

Problem 8.15 Derive (8.111). (*Hint* Calculate the curvature tensor R considering small perturbations around Minkowski space and then use (5.111) to make the appropriate conformal transformation to an expanding universe.)

Substituting the expansion

$$h_j^i(\mathbf{x}, \eta) = \int h_\mathbf{k}(\eta) \, e_j^i(\mathbf{k}) \, e^{i\mathbf{k}\mathbf{x}} \frac{d^3k}{(2\pi)^{3/2}}, \qquad (8.112)$$

where $e_j^i(\mathbf{k})$ is the polarization tensor, into (8.111), we obtain

$$S = \frac{1}{64\pi} \int a^2 e_j^i e_i^j \left(h_\mathbf{k}' h_{-\mathbf{k}}' - k^2 h_\mathbf{k} h_{-\mathbf{k}} \right) d\eta d^3k. \qquad (8.113)$$

8.4 Gravitational waves from inflation

Rewritten in terms of the new variable

$$v_{\mathbf{k}} = \sqrt{\frac{e_j^i e_i^j}{32\pi}} a h_{\mathbf{k}}, \qquad (8.114)$$

the action becomes

$$S = \frac{1}{2} \int \left(v'_{\mathbf{k}} v'_{-\mathbf{k}} - \left(k^2 - \frac{a''}{a} \right) v_{\mathbf{k}} v_{-\mathbf{k}} \right) d\eta d^3 k. \qquad (8.115)$$

It describes a real scalar field in terms of its Fourier components. The resulting equations of motion are

$$v''_{\mathbf{k}} + \omega_k^2(\eta) v_{\mathbf{k}} = 0, \quad \omega_k^2(\eta) \equiv k^2 - a''/a. \qquad (8.116)$$

There is no need to repeat the quantization procedure for this case. Taking into account (8.114) and (8.112), we immediately find the correlation function

$$\langle 0 | h^i_j(\eta, \mathbf{x}) h^j_i(\eta, \mathbf{y}) | 0 \rangle = \frac{8}{\pi a^2} \int |v_{\mathbf{k}}|^2 k^3 \frac{\sin kr}{kr} \frac{dk}{k}, \qquad (8.117)$$

where $v_{\mathbf{k}}$ is the solution of (8.116) with initial conditions

$$v_{\mathbf{k}}(\eta_i) = \frac{1}{\sqrt{\omega_k}}, \quad v'_{\mathbf{k}}(\eta_i) = i\sqrt{\omega_k}. \qquad (8.118)$$

These initial conditions make sense only if $\omega_k > 0$, that is, for gravitational waves with $k^2 > (a''/a)_{\eta_i}$. The power spectrum, characterizing the strength of a gravitational wave with comoving wavenumber k, is correspondingly

$$\delta_h^2(k, \eta) = \frac{8 |v_{\mathbf{k}}|^2 k^3}{\pi a^2}. \qquad (8.119)$$

Inflation In contrast to scalar perturbations, the deviation of the equation of state from the vacuum equation of state is not so crucial to the evolution of gravitational waves. Therefore, we first consider a pure de Sitter universe where $a = -(H_\Lambda \eta)^{-1}$. In this case (8.116) simplifies to

$$v''_{\mathbf{k}} + \left(k^2 - \frac{2}{\eta^2} \right) v_{\mathbf{k}} = 0 \qquad (8.120)$$

and has the exact solution

$$v_{\mathbf{k}}(\eta) = \frac{1}{\eta} \{ C_1 [k\eta \cos(k\eta) - \sin(k\eta)] + C_2 [k\eta \sin(k\eta) + \cos(k\eta)] \}. \qquad (8.121)$$

Let us consider gravitational waves with $k|\eta_i| \gg 1$ for which $\omega_k \simeq k$. Taking into account the initial conditions in (8.118), we can determine the constants of

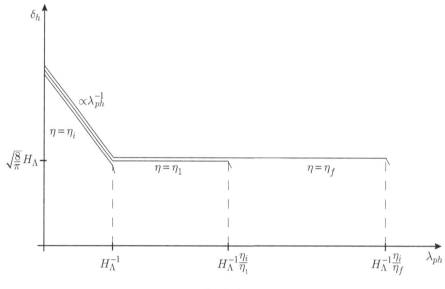

Fig. 8.5.

integration C_1 and C_2 and the solution becomes

$$v_{\mathbf{k}}(\eta) = \frac{1}{\sqrt{k}}\left(1 + \frac{i}{k\eta}\right)\exp(ik(\eta - \eta_i)). \tag{8.122}$$

Substituting this into (8.119), we obtain

$$\delta_h^2 = \frac{8H_\Lambda^2}{\pi}\left[1 + (k\eta)^2\right] = \frac{8H_\Lambda^2}{\pi}\left[1 + \left(\frac{k_{ph}}{H_\Lambda}\right)^2\right], \tag{8.123}$$

where $k_{ph} \equiv k/a$ is the physical wavelength. This formula is applicable only for $k_{ph} \gg H_\Lambda(\eta/\eta_i)$. The amplitude δ_h as a function of the physical wavelength $\lambda_{ph} \sim k_{ph}^{-1}$ is sketched in Figure 8.5. Long-wavelength gravitational waves with $H_\Lambda^{-1}(\eta_i/\eta) > \lambda_{ph} > H_\Lambda^{-1}$ have a flat spectrum with amplitude proportional to H_Λ.

The above consideration refers to a pure de Sitter universe where H_Λ is exactly constant. In realistic inflationary models the Hubble constant slowly changes with time. Recalling that the nondecaying mode of a gravitational wave is frozen on supercurvature scales (see Section 7.3.2), we obtain

$$\delta_h^2 \simeq \frac{8H_{k\simeq Ha}^2}{\pi} = \frac{64}{3}\varepsilon_{k\simeq Ha}. \tag{8.124}$$

The tensor spectral index is then equal to

$$n_T \equiv \frac{d\ln\delta_h^2}{d\ln k} \simeq -3\left(1 + \frac{p}{\varepsilon}\right)_{k\simeq Ha}, \tag{8.125}$$

8.4 Gravitational waves from inflation

and hence the spectrum of the gravitational waves is also slightly tilted to the red. (Note that the tensor and scalar spectral indices are defined differently – see (8.108).) The ratio of tensor to scalar power spectrum amplitudes on supercurvature scales during the post-inflationary, radiation-dominated epoch is

$$\frac{\delta_h^2}{\delta_\Phi^2} \simeq 27 \left[c_s \left(1 + \frac{p}{\varepsilon} \right) \right]_{k \simeq Ha}. \tag{8.126}$$

For a canonical scalar field ($c_s = 1$), this ratio is between 0.2 and 0.3. However, in k inflation, where $c_s \ll 1$, it can be strongly suppressed. Thus, at least in principle, k inflation is phenomenologically distinguishable from inflation based on a scalar field potential.

Post-inflationary epoch We found in Section 7.3.2 that the amplitude of a gravitational wave stays constant on supercurvature scales irrespective of changes in the equation of state. When the gravitational wave re-enters the horizon, however, its amplitude begins to decay in inverse proportion to the scale factor. Hence the spectrum remains unchanged only on large scales and is altered within the Hubble horizon. Neglecting the dark energy component, we can express the Hubble constant at earlier times in terms of its present value H_0 as

$$H(a) \simeq H_0 \begin{cases} (a_0/a)^{3/2}, & z < z_{eq}, \\ z_{eq}^{-1/2}(a_0/a)^2, & z > z_{eq}, \end{cases} \tag{8.127}$$

where z_{eq} is the redshift at matter–radiation equality. For a gravitational wave with a comoving wavenumber k, the value of the scale factor at horizon crossing, a_k, is determined from the condition $k \simeq H(a_k) a_k$. After that the amplitude decreases by a factor a_k/a_0 and we obtain the following spectrum at the present time:

$$\delta_h \sim H_\Lambda \begin{cases} z_{eq}^{-1/2}(\lambda_{ph} H_0), & \lambda_{ph} < H_0^{-1} z_{eq}^{-1/2}, \\ (\lambda_{ph} H_0)^2, & H_0^{-1} > \lambda_{ph} > H_0^{-1} z_{eq}^{-1/2}, \\ 1, & \lambda_{ph} > H_0^{-1}, \end{cases} \tag{8.128}$$

where H_Λ is the value of the Hubble constant during inflation and $\lambda_{ph} \sim a_0/k$ is the physical wavelength. This spectrum is sketched in Figure 8.6. On scales of several light years the typical amplitude of the primordial gravitational waves can be estimated as roughly 10^{-17} for a realistic model of inflation. This amplitude drops linearly towards smaller scales and so the prospects of direct detection of the primordial gravitational background are not very promising. However, as we shall see in the next chapter, these gravitational waves influence the CMB temperature fluctuations and therefore may be detectable indirectly.

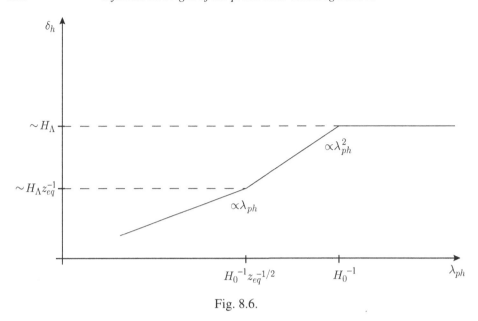

Fig. 8.6.

8.5 Self-reproduction of the universe

The amplitude of scalar perturbations takes its maximum value on scales corresponding to $k \simeq H_i a_i$, that is, those which left the horizon at the very beginning of inflation. For a massive scalar field at the end of inflation, this maximal amplitude can be estimated from (8.37) and is equal to

$$\delta_\Phi^{\max} \sim m \ln(a_f/a_i) \sim m\varphi_i^2. \tag{8.129}$$

If the initial value φ_i is larger than $m^{-1/2}$, then inhomogeneities on scales $\lambda_{ph} \sim H_i^{-1} a/a_i$ become very large ($\delta_\Phi > 1$) before inflation ends. Therefore, for large initial values of the scalar field, the initial homogeneity is completely spoiled by amplified quantum fluctuations on scales exceeding $\lambda_{ph}(t_f) \sim H^{-1} \exp(m^{-1})$. For realistic values of m these scales are enormous. For example, if $m \sim 10^{-6}$, they are larger than $H^{-1} \exp(10^6)$ and exceed the observable scales, $\sim H^{-1} \exp(70)$, by many orders of magnitude. On scales smaller than $H^{-1} \exp(m^{-1})$ the universe remains quasi-homogeneous. Thus, if inflation begins at $m^{-1} > \varphi_i > m^{-1/2}$, then, on one hand, it produces a very homogeneous, isotropic piece of space which is large enough to encompass the observed universe while, on the other hand, quantum fluctuations induce a large inhomogeneity on scales much larger than the observable scale.

Futhermore, if $\varphi_i > m^{-1/2}$ in one causally connected region, then inflation never ends but continues eternally somewhere in space. To see why this happens let us consider a causal domain of size H^{-1}. In a typical Hubble time, $\Delta t_H \sim H^{-1}$,

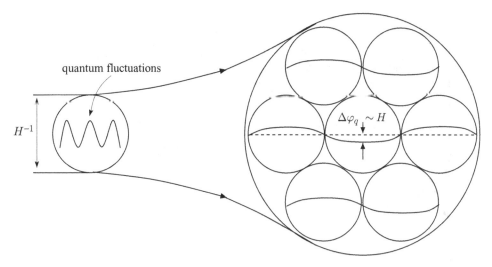

Fig. 8.7.

the size of this domain grows to $H^{-1} \exp(H \Delta t) \sim e H^{-1}$ and hence gives rise to $e^3 \simeq 20$ new domains of size H^{-1}. Now consider the averaged value of the scalar field in each of these new domains. During a Hubble time the classical scalar field decreases by an amount

$$\Delta \varphi_{cl} \simeq -\frac{V_{,\varphi}}{3H} \Delta t_H \sim -\varphi^{-1}. \tag{8.130}$$

Simultaneously, quantum fluctuations stretched from sub-Hubble scales begin to contribute to the mean value of the scalar field in each domain of size H^{-1}. Quantum fluctuations with wavelength of order H^{-1} and amplitude $\Delta \varphi_q \sim H \sim m\varphi$ are superimposed on the classical field; in half of the regions they decrease the value of the scalar field still further (Figure 8.7), while in the other half they increase the field value. The overall change of φ in these latter domains is about

$$\Delta \varphi_{tot} = \Delta \varphi_{cl} + \Delta \varphi_q \sim -\varphi^{-1} + m\varphi. \tag{8.131}$$

It is clear that if $\varphi > m^{-1/2}$, the field grows and inflation always produces regions where the scalar field exceeds its "initial" value. In Figure 8.8 we sketch the typical trajectories describing the evolution of the scalar field within a typical Hubble domain. For $\varphi \ll m^{-1/2}$, quantum fluctuations only slightly disturb the classical slow-roll trajectory, while for $\varphi \gg m^{-1/2}$ they dominate and induce a "random walk." Because each domain of size H^{-1} in turn produces other domains at an exponential rate, the physical volume of space where the scalar field is larger than its initial value grows exponentially. Thus, inflation continues forever and the universe is said to be "self-reproducing." In those regions where the field drops below the

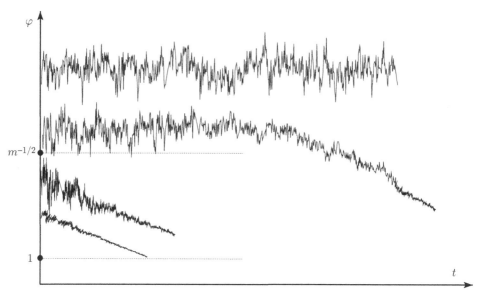

Fig. 8.8.

self-reproduction scale, $\varphi_{rep} \sim m^{-1/2}$, quantum fluctuations are no longer relevant. The size of these regions grows exponentially and eventually they produce very large homogeneous domains where the universe is hot.

Problem 8.16 Determine the self-reproduction scale for the power-law potential $V = (\lambda/n)\varphi^n$.

To conclude, we have found that inflation generically leads to a self-reproducing universe and induces complicated global structure on very large scales. This global structure may become relevant for an observer but only in *many many* billions of years. However, for a better understanding of the initial condition problem, the self-reproduction regime is very important. There is no complete and reliable description of the self-reproducing universe at present and much more work has to be done to clarify the question of the global structure of the universe.

8.6 Inflation as a theory with predictive power

Assuming a stage of cosmic acceleration – inflation – we are able to make robust predictions even in the absence of the actual inflationary scenario. The most important among them are:

(*i*) *the flatness of the universe;*
(*ii*) *Gaussian scalar metric perturbations with a slightly red-tilted spectrum;*
(*iii*) *long-wavelength gravitational waves.*

8.6 Inflation as a theory with predictive power

The condition of flatness is not as "natural" as it might appear at first glance. We recall that $\Omega_0 = 1$ was strongly disfavored by observations not so long ago. If gravity were always an attractive force, it is absolutely unclear why the current value of Ω_0 could not be, for instance, 0.01 or 0.2. Only inflation gives a natural justification for $\Omega_0 = 1$. The deuterium abundance clearly indicates that baryons cannot contribute more than a small percentage of the critical energy density. Therefore, inflation also predicts the existence of a dark component. It can be dark matter, dark energy or a combination of the two. In the absence of the actual inflationary scenario, we cannot make any prediction about the composition of the dark component. In spite of the tremendous progress made, we are still far from understanding the true nature of dark matter and dark energy. The current data on CMB fluctuations favor the critical density and, combined with the results from high-redshift supernovae, make it almost impossible to doubt the existence of dark matter and dark energy.

The predicted spectrum for the scalar perturbations is also in good agreement with the current data. However, the accuracy of the observations is not yet sufficient to determine a small spectral tilt. The deviation of the spectrum from flat is an inevitable consequence of simple inflation and therefore it is extremely important to detect it. The amplitude of the power spectrum is a free parameter of the theory.

The production of a significant number of long-wavelength gravitational waves is another generic prediction of a broad class of simple inflationary scenarios. While their detection would strongly support inflation, the absence of gravitational waves would not allow us to exclude simple inflation since their production can be avoided in k inflation.

Since we do not know which concrete scenario was realized in nature, the question of the robustness of the predictions of inflation is of particular importance. Simple inflation does not leave much room for ambiguities. However, it is clear that by introducing extra parameters and by fine-tuning, one can alter the robust predictions of the simple inflationary models. For example, by designing specifically fine-tuned potentials one can avoid the flatness constraint. Similarly, by involving many scalar fields, or by studying models with several different stages of inflation, one can obtain practically any spectrum of cosmological perturbations and induce nongaussianity. In our point of view, an increase of complexity of the models simultaneously increases the "price-to-performance" ratio; the theory gradually loses its predictive power and becomes less attractive. Only observations confirming the robust predictions of inflation can completely assure us that we are on the right track in understanding our universe.

9
Cosmic microwave background anisotropies

After recombination, the primordial radiation freely streams through the universe without any further scattering. An observer today detects the photons that last interacted with matter at redshift $z \approx 1000$, far beyond the stars and galaxies. The pattern of the angular temperature fluctuations gives us a direct snapshot of the distribution of radiation and energy at the moment of recombination, which is representative of what the universe looked like when it was a thousand times smaller and a hundred thousand times younger than today.

The first striking feature is that the variations in intensity across the sky are tiny, less than 0.01% on average. We can conclude from this that the universe was extremely homogeneous at that time, in contrast to the lumpy, highly inhomogeneous distribution of matter seen today. The second striking feature is that the average amplitude of the inhomogeneities is just what is required in a universe composed of cold dark matter and ordinary matter to explain the formation of galaxies and large-scale structure. Moreover, the temperature autocorrelation function indicates that the inhomogeneities have statistical properties in perfect accordance with what is predicted by inflationary models of the universe.

In a map showing the microwave background temperature across the sky, the features subtending a given angle are associated with physics on a spatial scale that can be computed from the angle and the angular diameter distance to the last scattering surface. The latter depends on the cosmological model. The angular scale $\theta \sim 1°$ corresponds to the Hubble radius at recombination, which is the dividing line between the large-scale inhomogeneities that have not changed much since inflation and the small-scale perturbations that have entered the horizon before recombination and been substantially modified by gravitational instability. Hence, observations of temperature fluctuations on large angular scales give us direct information about the primordial spectrum of perturbations, and observations of the small-scale fluctuations enable us to determine the values of the cosmological parameters that control the change of perturbation amplitudes after they enter the Hubble scale.

The purpose of this chapter is to derive the spectrum of microwave background fluctuations, assuming a nearly scale-invariant spectrum of primordial inhomogeneities, as occurs in inflationary models. Today, sophisticated computer programs are used to obtain numerically precise predictions. Here, though, our purpose is to understand from first principles the physics behind the characteristic features of the spectrum and to determine how they depend on fundamental parameters. We are willing to sacrifice a little accuracy to obtain a solid, analytic insight.

We first use an approximation of instantaneous recombination in which the radiation behaves as a perfect fluid before recombination and as an ensemble of the free photons immediately afterwards. This approximation is very good for large angular fluctuations arising from inhomogeneities on scales larger than the Hubble radius at recombination. In fact, recombination is a more gradual process that extends over a finite range of redshifts and this substantially influences the temperature fluctuations on small angular scales. The computations are then more complicated, but the problem is still treatable analytically.

Throughout this chapter we consider a spatially flat universe predicted by inflation and favored by the current observations. The modifications of the most important features of the CMB spectrum induced by the spatial curvature are rather obvious and will be briefly discussed.

9.1 Basics

Before recombination, radiation is strongly coupled to ordinary matter and it is well approximated as a perfect fluid. When sufficient neutral hydrogen has been formed, the photons cease interacting with the matter and, therefore, they must be described by a kinetic equation.

Phase volume and Liouville's theorem The state of a single photon with a given polarization at (conformal) time η can be completely characterized by its position in the space $x^i(\eta)$ and its 3-momentum $p_i(\eta)$, where $i = 1, 2, 3$ is the spatial index. Since the 4-momentum p_α satisfies the equation $g^{\alpha\beta} p_\alpha p_\beta = 0$, the "energy" p_0 can be expressed in terms of the metric $g_{\alpha\beta}$ and p_i.

The *one-particle* phase volume element is a product of the differentials of spatial coordinates and *covariant* components of the momentum:

$$d^3x d^3p \equiv dx^1 dx^2 dx^3 dp_1 dp_2 dp_3. \tag{9.1}$$

It is *invariant* under general coordinate transformations. To prove this, let us go to another coordinate system

$$\tilde{\eta} = \tilde{\eta}(\eta, x^i), \quad \tilde{x}^i = \tilde{x}^i(\eta, x^j).$$

The phase volume $d^3\tilde{x} d^3\tilde{p}$ calculated *at the hypersurface* $\tilde{\eta} = $ const in this new coordinate system is related to (9.1) by

$$d^3\tilde{x} d^3\tilde{p} = J d^3x d^3p, \tag{9.2}$$

where

$$J = \frac{\partial(\tilde{x}^1, \tilde{x}^2, \tilde{x}^3, \tilde{p}_1, \tilde{p}_2, \tilde{p}_3)}{\partial(x^1, x^2, x^3, p_1, p_2, p_3)}, \tag{9.3}$$

is the Jacobian of the transformation

$$x^i \to \tilde{x}^i = \tilde{x}^i(x^j, \tilde{\eta}), \quad p_i \to \tilde{p}_i = (\partial x^\alpha / \partial \tilde{x}^i)_{\tilde{x}} \, p_\alpha. \tag{9.4}$$

Note that the new coordinates $\tilde{x}^i$ should be considered here as functions of the old coordinates x^j and the new time $\tilde{\eta}$. Since $(\partial \tilde{x}/\partial p) = 0$, we have

$$J = \det\left(\frac{\partial \tilde{x}^i}{\partial x^j}\bigg|_{\tilde{\eta}=\text{const}}\right) \det\left(\frac{\partial x^j}{\partial \tilde{x}^k}\bigg|_{\tilde{\eta}=\text{const}}\right) = \det\left(\delta^j_k\right) = 1, \tag{9.5}$$

and therefore the phase volume is invariant.

Problem 9.1 Verify that $dx^1 dx^2 dx^3 dp^1 dp^2 dp^3$ is not invariant under coordinate transformations.

Liouville's theorem, which can easily be proved in flat spacetime, states that the phase volume of the Hamiltonian system is *invariant* under canonical transformations, or in other words, it is *conserved* along the trajectory of the particle. Considering an infinitesimal volume element, one can always go to a local inertial coordinate system (the Einstein elevator) in the vicinity of any point along the particle trajectory where the same theorem will obviously be valid. Yet, since the phase volume (9.1) is independent of the particular coordinate system, it must also be conserved as we move along a trajectory in curved spacetime. Hence, Liouville's theorem must continue to hold in General Relativity.

The Boltzmann equation Let us consider an ensemble of *noninteracting* identical particles. If dN is the number of particles per volume element $d^3x d^3p$, then the distribution function f, characterizing the number density in one-particle phase space, is defined by

$$dN = f(x^i, p_j, t) d^3x d^3p. \tag{9.6}$$

9.1 Basics

Since phase volume is invariant under coordinate transformations, f is a spacetime scalar. In the absence of particle interactions (scatterings), the particle number within the conserved phase volume does not change. As a result, the distribution function obeys the collisionless Boltzmann equation

$$\frac{Df(x^i(\eta), p_i(\eta), \eta)}{D\eta} \equiv \frac{\partial f}{\partial \eta} + \frac{dx^i}{d\eta}\frac{\partial f}{\partial x^i} + \frac{dp_i}{d\eta}\frac{\partial f}{\partial p_i} = 0, \quad (9.7)$$

where $dx^i/d\eta$ and $dp_i/d\eta$ are the derivatives calculated along the geodesics.

Temperature and its transformation properties Let us consider a nearly homogeneous isotropic universe filled by slightly perturbed thermal radiation. The frequency (energy) of the photon measured by an observer is equal to the time component of the photon's 4-momentum in the *comoving local inertial frame* of the observer. Therefore in an arbitrary coordinate system where an observer has 4-velocity u^α and the photon 4-momentum is p_α, this frequency can be expressed as $\omega = p_\alpha u^\alpha$. If the radiation coming to an observer from different directions

$$l^i \equiv -\frac{p_i}{\Sigma p_i^2}$$

has the Planckian spectra, then the distribution function is

$$f = \bar{f}\left(\frac{\omega}{T}\right) \equiv \frac{2}{\exp(\omega/T(x^\alpha, l^i)) - 1}. \quad (9.8)$$

The effective temperature $T(x^\alpha, l^i)$ depends not only on the direction l^i but also on the observer's location x^i and on the moment of time η. The factor of 2 in the numerator accounts for the two possible polarizations of the photons.

In a nearly isotropic universe, this temperature can be written as

$$T(x^\alpha, l^i) = T_0(\eta) + \delta T(x^\alpha, l^i), \quad (9.9)$$

where $\delta T \ll T_0$. To understand how the fluctuations δT depend on the coordinate system, let us consider two observers O and $\tilde{O}$, who are at rest with respect to two different frames related by the coordinate trasformation $\tilde{x}^\alpha = x^\alpha + \xi^\alpha$. In the rest frame of each observer, the zeroth component of 4-velocity can be expressed through the metric by using the relation $g_{\alpha\beta}u^\alpha u^\beta = g_{00}(u^0)^2 = 1$. Thus we conclude that the frequency of the same photon differs as measured by different observers. They are equal to

$$\omega = p_\alpha u^\alpha = p_0/\sqrt{g_{00}} \quad \text{and} \quad \tilde{\omega} = \tilde{p}_0/\sqrt{\tilde{g}_{00}}$$

respectively, where the photon's momentum and metric components in different frames are related by coordinate transformation laws. Using these laws together with the relation $p_\alpha p^\alpha = 0$ we get

$$\tilde{\omega} = \omega\left(1 + \frac{\partial \xi^i}{\partial \eta} l^i\right), \tag{9.10}$$

where we have kept only the first order terms in ξ^α and in the metric perturbations around the homogeneous isotropic universe. Since the distribution function (9.8) is a scalar,

$$\omega/T(x^\alpha) = \tilde{\omega}/\tilde{T}(\tilde{x}^\alpha),$$

and hence the temperature fluctuations measured by two observers are related as

$$\widetilde{\delta T} = \delta T - T_0' \xi^0 + T_0 \frac{\partial \xi^i}{\partial \eta} l^i. \tag{9.11}$$

We can see from this expression that the monopole (l^i-independent) and dipole (proportional to l^i) components of the temperature fluctuations depend on the particular coordinate system in which an observer is at rest. If we can only observe the radiation from one vantage point, then the monopole term can always be removed by redefinition of the background temperature. The dipole component depends on the motion of the observer with respect to the "preferred frame" determined by the background radiation. For these reasons, neither the monopole nor the dipole components are very informative about the initial fluctuations. We have to look to the quadrupole and higher-order multipoles instead, which do not depend on the motion of the particular observer and coordinate system we use to calculate them.

9.2 Sachs–Wolfe effect

In this section we will solve the Boltzmann equation for freely propagating radiation in a flat universe using the conformal Newtonian coordinate system where the metric takes the form

$$ds^2 = a^2 \left\{(1 + 2\Phi) d\eta^2 - (1 - 2\Phi) \delta_{ik} dx^i dx^k\right\}. \tag{9.12}$$

Here $\Phi \ll 1$ is the gravitational potential of the scalar metric perturbations. We discount gravitational waves for the moment and consider them later.

Geodesics The geodesic equations describing the propagation of the radiation in an arbitrary curved spacetime can be rewritten as (see Problem 2.13)

$$\frac{dx^\alpha}{d\lambda} = p^\alpha, \quad \frac{dp_\alpha}{d\lambda} = \frac{1}{2} \frac{\partial g_{\gamma\delta}}{\partial x^\alpha} p^\gamma p^\delta, \tag{9.13}$$

9.2 Sachs–Wolfe effect

where λ is an affine parameter along the geodesic. Since the photons have zero mass, the first integral of these equations is $p^\alpha p_\alpha = 0$. Using this relation, we can express the time component of the photon 4-momentum in terms of the spatial components. Up to first order in the metric perturbations, one obtains

$$p^0 = \frac{1}{a^2}\left(\Sigma p_i^2\right)^{1/2} \equiv \frac{p}{a^2} \qquad p_0 = (1+2\Phi)\, p. \tag{9.14}$$

Then, from the first equation in (9.13) we can see that

$$\frac{dx^i}{d\eta} = \frac{p^i}{p^0} = \frac{-\frac{1}{a^2}(1+2\Phi)\, p_i}{p^0} = l^i(1+2\Phi), \tag{9.15}$$

where $l^i \equiv -p_i/p$. Expressing p^0 and p^i in terms of p_i and substituting the metric (9.12) into the second equation (9.13), we obtain

$$\frac{dp_\alpha}{d\eta} = \frac{1}{2}\frac{\partial g_{\gamma\delta}}{\partial x^\alpha}\frac{p^\gamma p^\delta}{p^0} = 2p\frac{\partial \Phi}{\partial x^\alpha}. \tag{9.16}$$

Equation for the temperature fluctuations Using the geodesic equations (9.15) and (9.16), the Boltzmann equation (9.7) takes the form

$$\frac{\partial f}{\partial \eta} + l^i(1+2\Phi)\frac{\partial f}{\partial x^i} + 2p\frac{\partial \Phi}{\partial x^j}\frac{\partial f}{\partial p_j} = 0. \tag{9.17}$$

Since f is the function of the single variable

$$y \equiv \frac{\omega}{T} = \frac{p_0}{T\sqrt{g_{00}}} \simeq \frac{p}{T_0 a}\left(1 + \Phi - \frac{\delta T}{T_0}\right), \tag{9.18}$$

the Boltzmann equation to zeroth order in the perturbations reduces to

$$(T_0 a)' = 0, \tag{9.19}$$

and to linear order becomes

$$\left(\frac{\partial}{\partial \eta} + l^i \frac{\partial}{\partial x^i}\right)\left(\frac{\delta T}{T} + \Phi\right) = 2\frac{\partial \Phi}{\partial \eta}. \tag{9.20}$$

Solutions The zeroth order equation (9.19) informs us that the temperature of the background radiation in a homogeneous universe is inversely proportional to the scale factor, while (9.20) determines the temperature fluctuations of the microwave background. In the case of practical interest, the universe is matter-dominated after recombination and the main mode in Φ is constant. Therefore, the right hand side of (9.20) vanishes. The operator on the left hand side is just a total time derivative and we obtain the result that

$$\left(\frac{\delta T}{T} + \Phi\right) = \text{const}, \tag{9.21}$$

along null geodesics. The influence of the gravitational potential on the microwave background fluctuations is known as the Sachs–Wolfe effect.

In reality, radiation is a small but not completely negligible fraction of the energy density immediately after recombination, and so Φ is actually slowly time-varying. Consequently, the combination $(\delta T/T + \Phi)$, according to (9.20), varies by an amount proportional to the integral of $\partial\Phi/\partial\eta$ along the geodesic of the photon. The contribution to the temperature fluctuations induced by the change of the gravitational potential due to the residual radiation after equality is called the early integrated Sachs–Wolfe effect. When dark energy, either quintessence or vacuum density, overtakes the matter density in more recent epochs, the gravitational potential begins to vary once again, causing a further contribution to the temperature fluctuations. This phenomenon is referred to as the late integrated Sachs–Wolfe effect. To simplify the final formulae, we neglect both integrated Sachs–Wolfe effects which in any case never contribute more than 10–20% to the resulting amplitudes of the fluctuations. The reader can easily generalize the formulae derived below to include these effects.

Free-streaming Let us consider an initial distribution of free photons with temperature $T + \delta T$ where

$$\frac{\delta T}{T}(\eta_i, \mathbf{x}, \mathbf{l}) = A_k \sin(\mathbf{kx})\, g(\mathbf{l}). \tag{9.22}$$

In this case, the radiation energy density is inhomogeneous and its spatial variation is proportional to

$$\langle \delta T/T \rangle_\mathbf{l} = A_k \sin(\mathbf{kx})\, \langle g \rangle_\mathbf{l},$$

where $\langle \rangle_\mathbf{l}$ denotes the average over directions $\mathbf{l}$.

Problem 9.2 Neglecting the gravitational potential Φ in (9.20), show that at later times

$$\left\langle \frac{\delta T}{T}(\eta) \right\rangle_\mathbf{l} = \frac{\sin(k(\eta - \eta_i))}{k(\eta - \eta_i)} \left\langle \frac{\delta T}{T}(\eta_{in}) \right\rangle_\mathbf{l}. \tag{9.23}$$

Thus, at $\eta \gg \eta_i$, the initial spatial inhomogeneities of the energy density of free photons will be suppressed by the ratio of the inhomogeneity scale to the Hubble radius. This damping effect is known as free-streaming. The suppression occurs as a result of the mixing of photons arriving at a given point from regions with different temperatures.

Unlike Silk damping, free-streaming causes the spatial variation (the x-dependence) of the photon distribution to decrease as a power law rather than exponentially with k. Furthermore, free-streaming does not make the distribution function isotropic. Although the spatial variation (the x-dependence) of the photon distribution is damped, the initial angular anisotropy (the l-dependence) is preserved. Note that the perturbations on superhorizon scales are unaffected because the photons have no chance to mix.

9.3 Initial conditions

As follows from (9.15), the photons arriving from direction l^i seen at the present time η_0 by an observer located at x_0^i propagate along geodesics

$$x^i(\eta) \simeq x_0^i + l^i(\eta - \eta_0). \tag{9.24}$$

Therefore, from (9.21), we find that $\delta T/T$ in the direction l^i on the sky today is equal to

$$\frac{\delta T}{T}(\eta_0, x_0^i, l^i) = \frac{\delta T}{T}(\eta_r, x^i(\eta_r), l^i) + \Phi(\eta_r, x^i(\eta_r)) - \Phi(\eta_0, x_0^i), \tag{9.25}$$

where η_r is the conformal time of recombination and $x^i(\eta_r)$ is given by (9.24). Since we can observe from only one vantage point in the universe, we are only interested in the l^i-dependence of the temperature fluctuations. Hence the last term, which only contributes to the monopole component, can be ignored. The angular dependence of $(\delta T/T)_0$ is set by two contributions: (a) the "initial" temperature fluctuations on the last scattering surface; and (b) the value of the gravitational potential Φ at this same location. The first contribution, $(\delta T/T)_r$, can be expressed in terms of the gravitational potential and the fluctuations of the photon energy density $\delta_\gamma \equiv \delta\varepsilon_\gamma/\varepsilon_\gamma$ on the last scattering surface. For this purpose, we use matching conditions for the *hydrodynamic* energy–momentum tensor, which describes the radiation before decoupling, and the *kinetic* energy–momentum tensor, which characterizes the gas of free photons after decoupling,

$$T_\beta^\alpha = \frac{1}{\sqrt{-g}} \int f \frac{p^\alpha p_\beta}{p^0} d^3 p. \tag{9.26}$$

Substituting the metric (9.12) into (9.26) and assuming a Planckian distribution (9.8), we get an expression to linear order for the $0-0$ component of the kinetic energy–momentum tensor

$$T_0^0 \simeq \frac{1}{a^4(1-2\Phi)} \int \bar{f}\left(\frac{\omega}{T}\right) p_0 d^3 p \simeq (T_0)^4 \int \left(1 + 4\frac{\delta T}{T_0}\right) \bar{f}(y) y^3 dy d^2 l, \tag{9.27}$$

where $y \equiv \omega/T$ and p_0 and p have been expressed in terms of ω using (9.14) and (9.18). The integral over y can be calculated explicitly and simply gives a numerical factor that, combined with $4\pi T_0^4$, represents the energy density of the unperturbed radiation. This expression, describing the gas of photons immediately after recombination, should continuously match the $0-0$ component of the hydrodynamic energy–momentum tensor which characterizes radiation before recombination: $T_0^0 = \varepsilon_\gamma (1 + \delta_\gamma)$. The matching condition implies

$$\delta_\gamma = 4 \int \frac{\delta T}{T} \frac{d^2 l}{4\pi}. \tag{9.28}$$

Similarly, one can derive from (9.26) that the other components of the kinetic energy–momentum tensor are:

$$T_0^i \simeq 4\varepsilon_\gamma \int \frac{\delta T}{T} l^i \frac{d^2 l}{4\pi}. \tag{9.29}$$

Taking the divergence of this expression and comparing it to the divergence of the hydrodynamical energy–momentum tensor for radiation before recombination, which is given by (7.117), we get the second matching condition

$$\delta'_\gamma = -4 \int l^i \nabla_i \left(\frac{\delta T}{T}\right) \frac{d^2 l}{4\pi}, \tag{9.30}$$

where we have neglected the radiation contribution to the gravitational potential and therefore set $\Phi'(\eta_r) = 0$.

It is straightforward to show that to satisfy both (9.28) and (9.30) the spatial Fourier component of the temperature fluctuations should be related to the energy density inhomogeneities in radiation as

$$\left(\frac{\delta T}{T}\right)_{\mathbf{k}} (\mathbf{l}, \eta_r) = \frac{1}{4}\left(\delta_{\mathbf{k}} + \frac{3i}{k^2}(k_m l^m) \delta'_{\mathbf{k}}\right). \tag{9.31}$$

Here and throughout this chapter we drop the subscript γ, keeping in mind that the notation δ is always used for the fractional energy fluctuations in the radiation component itself. Substituting (9.31) in the Fourier expansion of (9.25), we obtain the final expression for the temperature fluctuations in the direction $\mathbf{l} \equiv (l^1, l^2, l^3)$ as observed at location $\mathbf{x}_0 \equiv (x^1, x^2, x^3)$:

$$\frac{\delta T}{T}(\eta_0, \mathbf{x}_0, \mathbf{l}) = \int \left[\left(\Phi + \frac{\delta}{4}\right)_{\mathbf{k}} - \frac{3\delta'_{\mathbf{k}}}{4k^2} \frac{\partial}{\partial \eta_0}\right]_{\eta_r} e^{i\mathbf{k}(\mathbf{x}_0 + \mathbf{l}(\eta_r - \eta_0))} \frac{d^3 k}{(2\pi)^{3/2}}, \tag{9.32}$$

where $k \equiv |\mathbf{k}|$, $\mathbf{k} \cdot \mathbf{l} \equiv k_m l^m$ and $\mathbf{k} \cdot \mathbf{x}_0 \equiv k_n x_0^n$. Because η_r/η_0 is less than $1/30$, we can neglect η_r in favor of η_0 in this expression. The first term in square brackets represents the combined result from the initial inhomogeneities in the radiation energy density itself and the Sachs–Wolfe effect, and the second term is related to the velocities of the baryon–radiation plasma at recombination. The latter term is therefore often referred to in the literature as the Doppler contribution to the fluctuations. The characteristic peaks in the temperature anisotropy power spectrum (described below) are sometimes called "Doppler peaks," but we will see that the Doppler term is not the dominant cause of these peaks.

9.4 Correlation function and multipoles

A sky map of the cosmic microwave background temperature fluctuations can be fully characterized in terms of an infinite sequence of correlation functions. If the spectrum of fluctuations is Gaussian, as predicted by inflation and as current data suggest, then only the even order correlation functions are nonzero and all of them can be directly expressed through the two-point correlation function (also known as the temperature autocorrelation function):

$$C(\theta) \equiv \left\langle \frac{\delta T}{T_0}(\mathbf{l}_1) \frac{\delta T}{T_0}(\mathbf{l}_2) \right\rangle, \tag{9.33}$$

where the brackets $\langle \rangle$ denote averaging over all directions $\mathbf{l}_1$ and $\mathbf{l}_2$, satisfying the condition $\mathbf{l}_1 \cdot \mathbf{l}_2 = \cos(\theta)$. The squared temperature difference between two directions separated by angle θ, averaged over the sky, is related to $C(\theta)$ by

$$\left\langle \left(\frac{\delta T}{T_0}(\theta) \right)^2 \right\rangle \equiv \left\langle \left(\frac{T(\mathbf{l}_1) - T(\mathbf{l}_2)}{T_0} \right)^2 \right\rangle = 2(C(0) - C(\theta)). \tag{9.34}$$

The temperature autocorrelation function is a detailed fingerprint that can be used first to discriminate among cosmological models and then, once the model is fixed, to determine the values of its fundamental parameters. The three-point function, also known as the *bispectrum*, is a sensitive test for a non-Gaussian contribution to the fluctuation spectrum since it is precisely zero in the Gaussian limit.

The cosmic microwave background is also polarized. An expanded set of n-point correlation functions can be constructed which quantifies the correlation of the polarization over long distances and the correlation between polarization and temperature fluctuations. For the purpose of this primer, however, we focus on computing the temperature autocorrelation function, since this has proven to be the most useful to date. The generalization to other correlation functions is straightforward.

The universe is homogeneous and isotropic on large scales. Consequently, averaging over all directions on the sky from a single vantage point (e.g., the Earth) should be close to the average of the results obtained by other observers in many points in space for given directions. The latter average corresponds to the *cosmic mean* and is determined by correlation functions of the random field of inhomogeneities. The root-mean-square difference between a local measurement and the cosmic mean is known as *cosmic variance*. This difference is due to the poorer statistics of a single observer and depends on the number of appropriate representatives of the random inhomogeneities within an horizon. The variance is tiny at small angular scales but substantial for angular separations of 10 degrees or more.

Cosmic variance is an unavoidable uncertainty, but experiments usually introduce additional uncertainty by measuring only a finite fraction of the full sky. The total difference from the cosmic mean, called *sample variance*, is inversely proportional to the square root of the fractional area measured, approaching cosmic variance as the covered-by-measurements area approaches the full sky.

Because of the homogeneous isotropic nature of the random fluctuations one can calculate the cosmic mean of the angular correlation function $C(\theta)$ by simply averaging over the observer position $\mathbf{x}_0$ and keeping the directions $\mathbf{l}_1$ and $\mathbf{l}_2$ fixed. For the Gaussian field, this is equivalent to an ensemble average of the appropriate random Fourier components, $\langle \Phi_\mathbf{k} \Phi_{\mathbf{k}'} \rangle = |\Phi_k|^2 \delta(\mathbf{k}+\mathbf{k}')$, etc. Keeping this remark in mind and substituting (9.32) into (9.33), after integration over the angular part of $\mathbf{k}$, the *cosmic mean* of temperature autocorrelation function can be written as

$$C(\theta) = \int \left(\Phi_k + \frac{\delta_k}{4} - \frac{3\delta'_k}{4k^2}\frac{\partial}{\partial \eta_1} \right) \left(\Phi_k + \frac{\delta_k}{4} - \frac{3\delta'_k}{4k^2}\frac{\partial}{\partial \eta_2} \right)^*$$
$$\times \frac{\sin(k|\mathbf{l}_1\eta_1 - \mathbf{l}_2\eta_2|)}{k|\mathbf{l}_1\eta_1 - \mathbf{l}_2\eta_2|} \frac{k^2 dk}{2\pi^2}, \quad (9.35)$$

where after differentiation with respect to η_1 and η_2, we set $\eta_1 = \eta_2 = \eta_0$. Using the well known expansion

$$\frac{\sin(k|\mathbf{l}_1\eta_1 - \mathbf{l}_2\eta_2|)}{k|\mathbf{l}_1\eta_1 - \mathbf{l}_2\eta_2|} = \sum_{l=0}^{\infty}(2l+1)\, j_l(k\eta_1)\, j_l(k\eta_2)\, P_l(\cos\theta), \quad (9.36)$$

where $P_l(\cos\theta)$ and $j_l(k\eta)$ are the Legendre polynomials and spherical Bessel functions of order l, respectively, we can rewrite the expression for $C(\theta)$ as a discrete sum over *multipole moments* C_l:

$$C(\theta) = \frac{1}{4\pi}\sum_{l=2}^{\infty}(2l+1)\, C_l P_l(\cos\theta). \quad (9.37)$$

The monopole and dipole components ($l = 0, 1$) have been excluded here and

$$C_l = \frac{2}{\pi}\int \left| \left(\Phi_k(\eta_r) + \frac{\delta_k(\eta_r)}{4}\right) j_l(k\eta_0) - \frac{3\delta'_k(\eta_r)}{4k}\frac{dj_l(k\eta_0)}{d(k\eta_0)} \right|^2 k^2 dk. \quad (9.38)$$

If $\delta T/T$ is expanded in terms of spherical harmonics,

$$\frac{\delta T(\theta, \phi)}{T_0} = \sum_{l,m} a_{lm} Y_{lm}(\theta, \phi), \qquad (9.39)$$

then the complex coefficients a_{lm}, in a homogeneous and isotropic universe, satisfy the condition

$$\langle a^*_{l'm'} a_{lm} \rangle = \delta_{ll'} \delta_{mm'} C_l, \qquad (9.40)$$

where the brackets refer to a cosmic mean. The multipole moments, $C_l = \langle |a_{lm}|^2 \rangle$, receive their main contribution from fluctuations on angular scale $\theta \sim \pi/l$ and $l(l+1) C_l$ is about typical squared temperature fluctuations on this scale.

Problem 9.3 Generalize the formula (9.38) for $\Phi'_k(\eta_r) \neq 0$, thereby incorporating the integrated Sachs–Wolfe effect.

Inflation predicts a flat universe with a nearly scale-invariant, adiabatic spectrum of Gaussian fluctuations. As we shall show, these lead to certain *qualitative* features in the temperature anisotropy power spectrum: a flat plateau for large angular scales, a sequence of peaks and valleys with a first peak at $l \approx 200$, and a steady damping of the oscillation amplitude as l increases. Once these features are confirmed, then a precise measurement of the power spectrum can be used to constrain many of the cosmological parameters which inflation does not fix uniquely. First, there are the amplitude B and spectral index n of the primordial density inhomogeneities generated by inflation. The rather generic prediction of inflation is that $|\Phi_k^2 k^3| = B k^{n_S - 1}$, with $1 - n_S \sim 0.03$–0.08. The amplitude B is not predicted by inflation. Its value is chosen to fit the observations. The other parameters involved in defining the shape of the temperature power spectrum are the Hubble constant h_{75}, the fraction of the critical density today due to the baryon density Ω_b, the total matter (baryonic plus cold dark matter) density Ω_m and the vacuum (or quintessence) energy density Ω_Λ.

The present data are consistent with inflation, and suggest a flat universe that consists approximately of 5% baryonic matter, 25% cold dark matter, and 70% dark energy. We will take these values for our fiducial model, also called the *concordance model*, and compute the temperature fluctuation spectrum for a range of parameters around this model.

9.5 Anisotropies on large angular scales

The fluctuations on large angular scales ($\theta \gg 1°$) are induced by inhomogeneities with wavelengths which exceed the Hubble radius at recombination and have not had a chance to evolve significantly since the end of inflation. Thus their spectrum represents pristine information about the primordial inhomogeneities. In this section we show that for perturbations predicted by inflation, the spectrum of temperature fluctuations on large angular scales has a nearly flat plateau, the height and slope of which are mainly determined by the amplitude and spectral index characterizing the primordial inhomogeneities and practically independent of the other cosmological parameters.

The Hubble scale at recombination $H_r^{-1} = 3t_r/2$ spans $0.87°$ on the sky today (see (2.73)). Therefore, the results derived in this section refer to the angles $\theta \gg 1°$, or to the multipoles $l \ll \pi/\theta_H \sim 200$.

As shown in Section 7.4, for adiabatic perturbations with $k\eta_r \ll 1$, the relative energy density fluctuations in the radiation component itself can be expressed through the gravitational potential as

$$\delta_k(\eta_r) \simeq -\frac{8}{3}\Phi_k(\eta_r), \quad \delta_k'(\eta_r) \simeq 0. \tag{9.41}$$

According to (9.32), the resulting temperature fluctuation due to large-scale imhomogeneities is

$$\frac{\delta T}{T}(\eta_0, \mathbf{x}_0, \mathbf{l}) \simeq \frac{1}{3}\Phi(\eta_r, \mathbf{x}_0 - \mathbf{l}\eta_0). \tag{9.42}$$

That is, the fluctuation amplitude is equal to one third of the gravitational potential at the point on the last scattering surface from which the photons emanated. In this estimate, we neglect the contribution of radiation to the gravitational potential at recombination and both integrated Sachs–Wolfe effects, which are subdominant.

After matter–radiation equality, the potential on superhorizon scales drops by a factor of 9/10. Taking this into account, substituting (9.41) into (9.38), and calculating the integral with the help of the identity

$$\int_0^\infty s^{m-1} j_l^2(s)\, ds = 2^{m-3}\pi \frac{\Gamma(2-m)\,\Gamma\!\left(l+\frac{m}{2}\right)}{\Gamma^2\!\left(\frac{3-m}{2}\right)\Gamma\!\left(l+2-\frac{m}{2}\right)}, \tag{9.43}$$

we find for a scale-invariant initial spectrum with $|(\Phi_k^0)^2 k^3| = B$, the plateau:

$$l(l+1)\, C_l \simeq \frac{9B}{100\pi} = \text{const}, \tag{9.44}$$

on large angular scales or for $l \ll 200$. Since the main contribution to large angular scales comes from superhorizon inhomogeneities, we have neglected here the modification of the spectrum for subhorizon modes. In actuality, each C_l is a weighted integral over all k, including near horizon and subhorizon scales, where the fluctuation amplitude rises and falls. The above result is a good approximation for l up to 20 or so. For $l > 20$, the neglected effects become essential, leading first to the rise in amplitude of the temperature fluctuations and then to the acoustic peaks.

Problem 9.4 Find the correction to (9.44) if the initial spectrum is not scale-invariant, $|(\Phi_k^0)^2 k^3| = Bk^{n_S-1}$, assuming that $|n_S - 1| \ll 1$.

Problem 9.5 Determine how (9.42) is modified for the entropy perturbations considered in Section 7.3.

Unfortunately, the information about statistical properties of the primordial spectrum gathered from a *single* vantage point is limited by cosmic variance. Since there are only $2l + 1$ independent a_{lm}, the variance is

$$\frac{\Delta C_l}{C_l} \simeq (2l+1)^{-1/2}. \tag{9.45}$$

The typical fluctuation is about 50% for the quadrupole ($l = 2$) and 15% for $l \sim 20$. Therefore, we are forced to go to smaller angular scales to obtain precise constraints on the spectrum of primordial inhomogeneities. The bad news is that, for these scales, we can no longer ignore evolution. On the other hand, if we can deconvolve the effects of evolution, we gain information about both the primordial spectrum and the parameters that control cosmic evolution.

9.6 Delayed recombination and the finite thickness effect

On small angular scales, recombination can no longer be approximated as instantaneous. The finite duration of recombination introduces uncertainty as to the precise moment and position when a given photon last scatters. As a result, photons arriving from a given direction yield only "smeared out" information. In turn this leads to a suppression of the temperature fluctuations on small angular scales known as the finite thickness effect. The spread in the time of last scattering also increases the Silk damping scale, changing the conditions in the region from which the photon last scatters.

We first consider the finite thickness effect. A photon arriving from direction **l** might last scatter at any value of the redshift in the interval $1200 > z > 900$. If the last scattering occurs at conformal time η_L, the photon carries information about conditions at position

$$\mathbf{x}(\eta_L) = \mathbf{x}_0 + \mathbf{l}(\eta_L - \eta_0).$$

Since the total flux of radiation arriving from direction **l** consists of photons that last scattered over a range of times, the information it carries represents a weighted average over a scale $\Delta x \sim \Delta \eta_r$, where $\Delta \eta_r$ is roughly the duration of recombination. Clearly, the contribution of inhomogeneities with scales smaller than $\Delta \eta_r$ to the temperature fluctuations will be smeared out and therefore strongly suppressed.

Let us calculate the probability that the photon was scattered within the time interval Δt_L at physical time t_L (corresponding to the conformal time η_L) and then avoided further scattering until the present time t_0. We can divide the time interval $t_0 > t > t_L$ into N small intervals of duration Δt, where the jth interval begins at time $t_j = t_L + j\Delta t$ and $N > j > 1$. The probability is then

$$\Delta P = \frac{\Delta t_L}{\tau(t_L)}\left(1 - \frac{\Delta t}{\tau(t_1)}\right)\cdots\left(1 - \frac{\Delta t}{\tau(t_j)}\right)\cdots\left(1 - \frac{\Delta t}{\tau(t_N)}\right), \quad (9.46)$$

where

$$\tau(t_j) = \frac{1}{\sigma_T n_t(t_j) X(t_j)}$$

is the mean free time for Thomson scattering, n_t is the number density of all (bound and free) electrons and X is the ionization fraction. Taking the limit $N \to \infty$ ($\Delta t \to 0$), and converting from physical time t to conformal time η, we obtain

$$dP(\eta_L) = \mu'(\eta_L)\exp[-\mu(\eta_L)]\,d\eta_L, \quad (9.47)$$

where the prime denotes the derivative with respect to conformal time and $\mu(\eta_L)$ is the optical depth:

$$\mu(\eta_L) \equiv \int_{t_L}^{t_0} \frac{dt}{\tau(t)} = \int_{\eta_L}^{\eta_0} \sigma_T n_t X_e a(\eta)\,d\eta. \quad (9.48)$$

The uncertainty in the last scattering time causes us to modify our expression for the temperature fluctuation (9.32), replacing the recombination moment η_r by η_L, and integrating over η_L with the probability weight (9.47):

$$\frac{\delta T}{T} = \int \left[\Phi + \frac{\delta}{4} - \frac{3\delta'}{4k^2}\frac{\partial}{\partial \eta_0}\right]_{\eta_L} e^{i\mathbf{k}\cdot(\mathbf{x}_0 + \mathbf{l}(\eta_L - \eta_0))}\mu' e^{-\mu} d\eta_L \frac{d^3k}{(2\pi)^{3/2}}. \quad (9.49)$$

Unlike in (9.32), here we cannot neglect η_L compared to η_0 because for $k > (\Delta \eta_r)^{-1}$ an oscillating exponential factor in (9.49) changes significantly during the time interval $\Delta \eta_r$ when the visibility function

$$\mu'(\eta_L)\exp[-\mu(\eta_L)]$$

is substantially different from zero. The visibility function vanishes at very small η_L (because $\mu \gg 1$) and at large η_L (where $\mu' \to 0$), and reaches its maximum at

9.6 Delayed recombination and the finite thickness effect

η_r determined by the condition

$$\mu'' = \mu'^2. \tag{9.50}$$

Since recombination is really spread over an interval of time, we will use η_r henceforth to represent the conformal time when the visibility function takes its maximum value. This maximum is located within the narrow redshift range $1200 > z > 900$. During this short time interval, the scale factor and the total number density n_t do not vary substantially, so we can use their values at $\eta = \eta_r$. On the other hand, the ionization fraction X changes by several orders of magnitude over this same interval of time, so its variation cannot be ignored. Substituting (9.48) in (9.50), we can re-express the condition determining η_r as

$$X'_r \simeq -(\sigma_T n_t a)_r X_r^2, \tag{9.51}$$

where the subscript r refers to the value at η_r. For redshifts $1200 > z > 900$, X is well described by (3.202). The time variation of X is mainly due to the exponential factor, and hence

$$X' \simeq -\frac{1.44 \times 10^4}{z} \mathcal{H} X, \tag{9.52}$$

where $\mathcal{H} \equiv a'/a$. Substituting this relation in (9.51) we obtain

$$X_r \simeq \mathcal{H}_r \kappa (\sigma_T n_t a)_r^{-1}, \tag{9.53}$$

where $\kappa \equiv 14400/z_r$. Together with (3.202), this equation determines that the visibility function reaches its maximum at $z_r \simeq 1050$, irrespective of the values of cosmological parameters. At this time, the ionization fraction X_r is $\kappa \simeq 13.7$ times larger than the ionization fraction at decoupling, as defined by the condition $t \sim \tau_\gamma$ (see (3.206)). Near its maximum, the visibility function can be well approximated as a Gaussian:

$$\mu' \exp(-\mu) \propto \exp\left(-\frac{1}{2}(\mu - \ln \mu')'_r (\eta_L - \eta_r)^2\right). \tag{9.54}$$

Calculating the derivatives with the help of (9.52) and (9.53), we obtain

$$\mu' \exp(-\mu) \simeq \frac{(\kappa \mathcal{H} \eta)_r}{\sqrt{2\pi} \eta_r} \exp\left(-\frac{1}{2}(\kappa \mathcal{H} \eta)_r^2 \left(\frac{\eta_L}{\eta_r} - 1\right)^2\right), \tag{9.55}$$

where the prefactor has been chosen to satisfy the normalization condition

$$\int \mu' \exp(-\mu) \, d\eta_L = 1.$$

The expression inside the square brackets in (9.49) does not change as much as the oscillating exponent and we obtain a good estimate just taking its value at

$\eta_L = \eta_r$. Then, substituting (9.55) in (9.49) and performing an explicit integration over η_L, one gets

$$\frac{\delta T}{T} = \int \left(\Phi + \frac{\delta}{4} - \frac{3\delta'}{4k^2} \frac{\partial}{\partial \eta_0} \right)_{\eta_r} e^{-(\sigma k \eta_r)^2} e^{i\mathbf{k}(\mathbf{x}_0 + \mathbf{l}(\eta_r - \eta_0))} \frac{d^3 k}{(2\pi)^{3/2}} \quad (9.56)$$

where

$$\sigma \equiv \frac{1}{\sqrt{6}(\kappa \mathcal{H} \eta)_r}. \quad (9.57)$$

In deriving (9.56) we replaced $(\mathbf{k} \cdot \mathbf{l})^2$ with $k^2/3$, using the fact that the perturbations field is isotropic. Now it is safe to neglect η_r compared to η_0.

To find out how σ depends on cosmological parameters, we have to calculate $(\mathcal{H} \eta)_r$. At recombination, the dark energy contribution can be ignored and the scale factor is well described by (1.81); hence,

$$(\mathcal{H} \eta)_r = 2 \times \frac{1 + (\eta_r/\eta_*)}{2 + (\eta_r/\eta_*)}. \quad (9.58)$$

The ratio (η_r/η_*) can be expressed through the ratio of the redshifts at equality and recombination using an obvious relation

$$\left(\frac{\eta_r}{\eta_*} \right)^2 + 2 \left(\frac{\eta_r}{\eta_*} \right) \simeq \frac{z_{eq}}{z_r}. \quad (9.59)$$

Taking this into account and substituting (9.58) in (9.57) we obtain

$$\sigma \simeq 1.49 \times 10^{-2} \left[1 + \left(1 + \frac{z_{eq}}{z_r} \right)^{-1/2} \right]. \quad (9.60)$$

The exact value of z_{eq} depends on the matter contribution to the total energy density and the number of ultra-relativistic species present in the early universe. For three types of light neutrinos we have

$$\frac{z_{eq}}{z_r} \simeq 12.8 \left(\Omega_m h_{75}^2 \right). \quad (9.61)$$

The parameter σ is only weakly sensitive to the amount of cold matter and number of light neutrinos: for $\Omega_m h_{75}^2 \simeq 0.3$, $\sigma \simeq 2.2 \times 10^{-2}$, and for $\Omega_m h_{75}^2 \simeq 1$, $\sigma \simeq 1.9 \times 10^{-2}$.

Problem 9.6 Find how σ depends on the number of light neutrinos for a given cold matter density.

Next let us consider how noninstantaneous recombination influences the Silk dissipation scale. As mentioned above, the ionization fraction at $\eta = \eta_r$ is $\kappa \approx 13.7$ times bigger than at decoupling and the mean free path is consequently κ times

9.6 Delayed recombination and the finite thickness effect

smaller than the horizon scale. Therefore, one can try to use the result (7.128), obtained in the imperfect fluid approximation, to estimate the extra dissipation during non-instantaneous recombination.

Problem 9.7 Using (3.202) for the ionization fraction (which is valid when the ionization fraction drops below unity), calculate the dissipation scale and show that

$$(k_D \eta)_r^{-2} \simeq 4.9 \times 10^{-3} c_s^2 \frac{\sqrt{\Omega_m h_{75}^2}}{\eta_{10}} + \frac{12}{5} c_s^2 \sigma^2. \qquad (9.62)$$

The first term here is what was obtained for instantaneous recombination (equation (7.132)), where we have expressed $\Omega_m h^2$ in terms of $\eta_{10} \equiv 10^{10} n_b/n_\gamma$ (see (3.121)). This term accounts for the dissipation before recombination starts. The second term accounts for the additional dissipation during recombination.

Note that the second term in (9.62) corresponds to a scale which is smaller than the mean free path τ_γ at η_r, and so the imperfect fluid approximation cannot be trusted. However, within the time interval $\Delta \eta \sim \eta_r \sigma$, when the visibility function is different from zero, free photons have only enough time to propagate the comoving distance $\lambda \sim \eta_r \sigma$, which is roughly the second term in (9.62). The inhomogeneities in the radiation can be smeared (mainly because of free streaming) only within these scales but not on larger scales. Therefore, the result (9.62) can be still used as a reasonable rough estimate for the scale below which the inhomogeneities will be suppressed.

At very low baryon density the first term in (9.62) dominates, and most of the dissipation happens before ionization significantly drops. However, for realistic values of the dark matter and baryon densities, $\Omega_m h_{75}^2 \simeq 0.3$ and $\eta_{10} \simeq 5$, the second term can be nearly twice the first term. Thus, the corrections to Silk dissipation due to noninstantaneous recombination can be important.

Problem 9.8 If the baryon density is too low the approximations used above are invalid. What is the minimal value of η_{10} (or $\Omega_b h_{75}^2$) for which the derived results can still be trusted? For smaller values, how would the results be modified?

In summary we have found that the finite duration of recombination produces two effects. First, the damping scale can be essentially greater than if recombination were instantaneous. Second, the uncertainty in the time of decoupling results in an extra suppression of temperature fluctuations on small angular scales. Although both effects are interconnected, they are distinct.

The key formulae derived in the instantaneous recombination approximation are easily modified for the case of delayed recombination. Namely, for the damping scale of radiation inhomogeneities one has to use (9.62) and the finite thickness leads to the appearance of the overall factor

$$\exp\left[-2(\sigma k \eta_r)^2\right]$$

in the integrand of (9.38) for the multipole moments C_l.

9.7 Anisotropies on small angular scales

For large l or small angular scales, the main contribution to C_l comes from perturbations with angular size $\theta \sim \pi/l$ on today's sky. The multipole moment $l \sim 200$ corresponds to the sound horizon scale at recombination. Hence, the perturbations responsible for the fluctuations with $l > 200$ have wavenumbers $k > \eta_r^{-1}$, that is, they entered the horizon before recombination. These perturbations undergo evolution, causing a significant modification of the primordial spectrum.

In Section 7.4.2, we found the transfer function relating the initial spectrum of gravitational potential fluctuations Φ_k^0 to the spectra of Φ and δ_γ at recombination in two limiting cases: namely, for the perturbations that entered the horizon well before equality, (7.143), and perturbations that enter well after equality, (7.134). However, for realistic values of the cosmological parameters, neither of these limits applies directly to the most interesting band of multipole moments corresponding to the first few acoustic peaks. The approximation (7.143) is valid only for modes which undergo at least one oscillation before equality ($k\eta_{eq} > 2\sqrt{3}\pi \sim 10$), and approximation (7.134) is legitimate for modes which enter the horizon after the radiation density has become negligible compared to the matter density. If $\Omega_m h_{75}^2 \simeq 0.3$, then it follows from (9.61) that $z_{eq}/z_r \simeq 4$, and the radiation still constitutes about 20% of the energy density at recombination. Hence, the asymptotics (7.134) and (7.143) are poor approximations for perturbations which enter the horizon *near* equality. These are precisely the modes responsible for the fluctuations in the region of first few acoustic peaks. Since the precise positions and shapes of these peaks provide valuable information about cosmological models, it is worth improving the approximation for the source function $(\Phi_k + \delta_k/4)_r$ in this region.

9.7.1 Transfer functions

If the speed of sound changes slowly one can use after matter–radiation equality the WKB solution (7.127), derived in Section 7.4, for the subhorizon modes. Since the gravitational potential no longer changes significantly at this time, we find that for a given amplitude of the gravitational potential Φ_k^0 of the primordial spectrum the source function at recombination is

$$\left(\Phi_k + \frac{\delta_k}{4}\right)_r \simeq \left[T_p\left(1 - \frac{1}{3c_s^2}\right) + T_o\sqrt{c_s}\cos\left(k\int_0^{\eta_r} c_s d\eta\right) e^{-(k/k_D)^2}\right]\Phi_k^0 \quad (9.63)$$

and

$$\left(\delta_k'\right)_r \simeq -4T_o k c_s^{3/2} \sin\left(k\int_0^{\eta_r} c_s d\eta\right) e^{-(k/k_D)^2}\Phi_k^0, \quad (9.64)$$

where the transfer functions T_p and T_o correspond to the constants of integration in the WKB solution and depend on whether the perturbation entered the horizon before or after equality. They were calculated in Section 7.4.2 in two limiting cases, namely, for the perturbations with $k\eta_{eq} \ll 1$, which entered the horizon *long enough* after equality, see (7.134),

$$T_p \to \frac{9}{10}; \quad T_o \to \frac{9}{10} \times 3^{-3/4} \simeq 0.4, \quad (9.65)$$

and for the perturbations with $k\eta_{eq} \gg 1$, which entered the horizon *well before* equality, see (7.143),

$$T_p \to \frac{\ln(0.15k\eta_{eq})}{(0.27k\eta_{eq})^2} \to 0; \quad T_o \to \frac{3^{5/4}}{2} \simeq 1.97. \quad (9.66)$$

Note that the transfer functions change very significantly. In particular, T_p is negligible for the perturbations which entered the horizon during the radiation-dominated stage and close to unity for the perturbations which entered the horizon long after that. The physical reason for such behavior is obvious. As we found in Section 6.4.3, for the subhorizon modes the gravitational instability in the cold dark matter component is suppressed during radiation-dominated stage and the gravitational potential decays; for the perturbation which enters the horizon when cold matter already dominates the amplitude of inhomogeneity in the cold component grows and the potential does not change. T_o, which defines the amplitude of the sound wave, is about 5 times greater for modes that enter the horizon well before equality than for those that enter the horizon long afterwards. This effect is due to the gravitational field of the radiation; it is significant when the modes with large $k\eta_{eq}$ enter the horizon and boosts the amplitude of the resulting sound wave compared to the case when the contribution of radiation to the gravitational potential is negligible.

It is clear that for those perturbations which entered the horizon near equality, the appropriate values of the transfer functions should lie somewhere between their asymptotic values. As we mentioned above these perturbations with $k\eta_{eq} \sim O(1)$ determine the amplitude of the temperature fluctuations in the region of the first

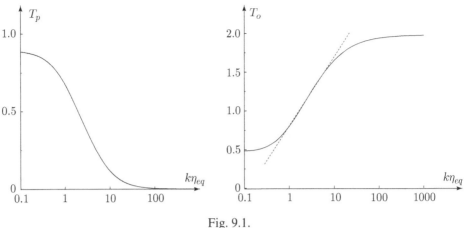

Fig. 9.1.

few acoustic peaks and therefore are most interesting. Unfortunately, the transfer functions in the intermediate region between two asymptotics can be calculated only numerically. In general, T_p and T_o should depend on k and η_{eq}, which by dimensional analysis must enter in the combination $k\eta_{eq}$, and the baryon density. To simplify the analysis, we will restrict ourselves to the case where the baryon density is small compared with the total density of dark matter, $\Omega_b \ll \Omega_m$, a practical limit since this condition is satisfied by the real universe. With this assumption, we can neglect the baryon contribution to the gravitational potential compared with the contribution from cold dark matter, and the Ω_b-dependence of the transfer functions can be ignored. The result of the numerical calculation of T_p and T_o in the limit of negligible baryon density is presented in Figure 9.1.

For intermediate range scales $10 > k\eta_{eq} > 1$, which give the leading contributions to the first acoustic peaks of the microwave background anisotropy, one can approximate T_p by

$$T_p \simeq 0.25 \ln\left(\frac{14}{k\eta_{eq}}\right), \tag{9.67}$$

and T_o by

$$T_o \simeq 0.36 \ln(5.6 k\eta_{eq}). \tag{9.68}$$

The transfer functions are monotonic and approach their asymptotic values given by (9.65) and (9.66) in the appropriate limits.

9.7.2 Multipole moments

To calculate the multipole moments C_l, we have to substitute (9.63) and (9.64) into (9.38) with an extra $\exp(-2(\sigma k \eta_r)^2)$ factor inside the integrand to account for the finite thickness effect. The resulting integral expressions are rather complicated, but can be very much simplified for $l \gg 1$. We first remove the derivatives of the spherical Bessel function in (9.38) using the identity

$$\left(\frac{dj_l(y)}{dy}\right)^2 = \left[1 - \frac{l(l+1)}{y^2}\right] j_l^2(y) + \frac{1}{2y} \frac{d^2(y j_l^2(y))}{dy^2}, \tag{9.69}$$

which can be verified using the Bessel function equation. Substituting (9.69) into (9.38) and integrating by parts, we obtain

$$C_l = \frac{2}{\pi} \int \left[\left|\Phi + \frac{\delta}{4}\right|^2 k^2 + \frac{9|\delta'|^2}{16}\left(1 - \frac{l(l+1)}{(k\eta_0)^2}\right)\right]$$

$$\times (1 + \Delta) e^{-2(\sigma k \eta_r)^2} j_l^2(k\eta_0)\, dk, \tag{9.70}$$

where Δ denotes corrections of order η_r/η_0 and $(k\eta_0)^{-1}$, which were estimated taking into account (9.63) and (9.64) for the source functions. Recall that $\eta_r/\eta_0 \lesssim z_r^{-1/2} \sim 1/30$ is small. As $l \to \infty$, we can approximate the Bessel functions as

$$j_l(y) \to \begin{cases} 0, & y < \nu, \\ \dfrac{1}{y^{\frac{1}{2}}(y^2 - \nu^2)^{\frac{1}{4}}} \cos\left[\sqrt{y^2 - \nu^2} - \nu \arccos\left(\dfrac{\nu}{y}\right) - \dfrac{\pi}{4}\right], & y > \nu, \end{cases} \tag{9.71}$$

where $\nu/y \neq 1$ is held fixed and $\nu \equiv l + 1/2$. Only those modes for which $y = k\eta_0 > l$ contribute to the integral and therefore the corrections of order $(k\eta_0)^{-1} < l^{-1} \ll 1$ can also be neglected. Using the approximation (9.71) in the integrand of (9.70), and bearing in mind that the argument of $j_l^2(k\eta_0)$ changes with k much more rapidly than the argument in the oscillating part of the source functions (9.63) and (9.64), we can replace the cosine squared coming from (9.71) by its average value $1/2$. The result is

$$C_l \simeq \frac{1}{16\pi} \int_{l\eta_0^{-1}}^{\infty} \left[\frac{|4\Phi + \delta|^2 k^2}{(k\eta_0)\sqrt{(k\eta_0)^2 - l^2}} + \frac{9\sqrt{(k\eta_0)^2 - l^2}}{(k\eta_0)^3} |\delta'|^2\right] e^{-2(\sigma k \eta_r)^2} dk, \tag{9.72}$$

where for large l we have set $l + 1 \approx l$.

Let us consider the case of a scale-invariant spectrum of initial density perturbations, $|(\Phi_k^0)^2 k^3| = B$, where B is constant. Substituting (9.63) and (9.64) into (9.72) and changing the integration variable to $x \equiv k\eta_0/l$, we get a sum of integrals with "oscillating"(O) and "nonoscillating" functions (N) in the integrands, so that

$$l(l+1)C_l \simeq \frac{B}{\pi}(O+N). \tag{9.73}$$

Because of the cross-term that arises when the expression in (9.63) is squared in the integrand of (9.72), the oscillating contribution to $l(l+1)C_l$ can be written as a sum of two integrals:

$$O = O_1 + O_2, \tag{9.74}$$

where

$$O_1 = 2\sqrt{c_s}\left(1 - \frac{1}{3c_s^2}\right)\int_1^\infty \frac{T_p T_o e^{\left(-\frac{1}{2}\left(l_f^{-2}+l_s^{-2}\right)^2 l^2 x^2\right)}}{x^2\sqrt{x^2-1}} \cos(l\varrho x)\, dx \tag{9.75}$$

and

$$O_2 = \frac{c_s}{2}\int_1^\infty T_o^2 \frac{(1-9c_s^2)x^2 + 9c_s^2}{x^4\sqrt{x^2-1}} e^{-(l/l_S)^2 x^2} \cos(2l\varrho x)\, dx. \tag{9.76}$$

Note that the periods of the cosines entering O_1 and O_2 differ by a factor of 2. As we will soon see, the acoustic peaks and valleys in the spectrum for $l(l+1)C_l$ result from the constructive and deconstructive interference of these two terms. The parameter

$$\varrho \equiv \frac{1}{\eta_0}\int_0^{\eta_r} c_s(\eta)\, d\eta \tag{9.77}$$

determines the locations of the peaks.

The scales l_f and l_S, which characterize the damping due to the finite thickness and Silk dissipation effects, are equal to

$$l_f^{-2} \equiv 2\sigma^2\left(\frac{\eta_r}{\eta_0}\right)^2; \quad l_S^{-2} \equiv 2\left(\sigma^2 + (k_D\eta)_r^{-2}\right)\left(\frac{\eta_r}{\eta_0}\right)^2, \tag{9.78}$$

where σ is given in (9.60) and $k_D\eta_r$ is given roughly by (9.62).

Likewise, the nonoscillating contribution to $l(l+1)C_l$ is a sum of three integrals:

$$N = N_1 + N_2 + N_3, \tag{9.79}$$

where

$$N_1 = \left(1 - \frac{1}{3c_s^2}\right)^2 \int_1^\infty \frac{T_p^2 e^{-(l/l_f)^2 x^2}}{x^2\sqrt{x^2-1}}\, dx \tag{9.80}$$

is proportional to the baryon density and vanishes in the absence of baryons where $c_s^2 \to 1/3$. The other two integrals are

$$N_2 = \frac{c_s}{2} \int_1^\infty \frac{T_o^2 e^{-(l/l_S)^2 x^2}}{x^2 \sqrt{x^2 - 1}} dx, \qquad (9.81)$$

and

$$N_3 = \frac{9c_s^3}{2} \int_1^\infty T_o^2 \frac{\sqrt{x^2 - 1}}{x^4} e^{-(l/l_S)^2 x^2} dx. \qquad (9.82)$$

The microwave background anisotropy is a powerful cosmological probe because the parameters which determine the spectrum $l(l+1)C_l$, namely, c_s, l_f, l_S, ϱ and the transfer functions T_o and T_p, can all be directly related to the basic cosmological parameters Ω_b, Ω_m, Ω_Λ, the dark energy equation of state w, and the Hubble constant h_{75}. Before we proceed with the calculation of the integrals determining the anisotropies, we explore these relations, making the simplifying assumption that the dark energy is the vacuum energy density, so that $w = -1$.

9.7.3 Parameters

The speed of sound c_s at recombination depends only on the baryon density, which determines how much c_s differs from its value for a purely relativistic gas of photons, $c_s = 1/\sqrt{3}$. If we define the baryon density parameter

$$\xi \equiv \frac{1}{3c_s^2} - 1 = \frac{3}{4}\left(\frac{\varepsilon_b}{\varepsilon_\gamma}\right)_r \simeq 17(\Omega_b h_{75}^2), \qquad (9.83)$$

then the speed of sound is

$$c_s^2 = \frac{1}{3(1+\xi)}.$$

For baryon density $\Omega_b h_{75}^2 \simeq 0.035$, one has $\xi \simeq 0.6$. The physical reason for the dependence of c_s on the baryon density is clear. The baryons interacting with radiation make the sound waves more "heavy" and therefore reduce their speed.

The damping scales l_f and l_S, given in (9.78), each depend on the ratio η_r/η_0. To calculate this ratio we introduce a supplementary moment of time $\eta_0 \gg \eta_x \gg \eta_r$, so that at η_x the radiation energy density is *already* negligible and the cosmological term is *still* small compared with the cold matter energy density. Then we can separately determine η_x/η_0 and η_r/η_x using the exact solutions (1.108) and (1.81) respectively.

Problem 9.9 Show that $\eta_x/\eta_0 \simeq I_\Lambda z_x^{-1/2}$, where

$$I_\Lambda \equiv 3\left(\frac{\Omega_\Lambda}{\Omega_m}\right)^{1/6}\left[\int_0^y \frac{dx}{(\sinh x)^{2/3}}\right]^{-1} \tag{9.84}$$

and $y \equiv \sinh^{-1}(\Omega_\Lambda/\Omega_m)^{1/2}$. In a flat universe $\Omega_\Lambda = 1 - \Omega_m$, and the numerical fitting formula

$$I_\Lambda \simeq \Omega_m^{-0.09} \tag{9.85}$$

approximates (9.84) to an accuracy better than 1% over the interval $0.1 < \Omega_m < 1$.

Verify that η_x/η_r is equal to

$$\frac{\eta_r}{\eta_x} \simeq \left(\frac{z_x}{z_{eq}}\right)^{1/2}\left[\left(1 + \frac{z_{eq}}{z_r}\right)^{1/2} - 1\right]. \tag{9.86}$$

Combining the relations from Problem 9.9 we obtain

$$\frac{\eta_r}{\eta_0} = \frac{1}{\sqrt{z_r}}\left[\left(1 + \frac{z_r}{z_{eq}}\right)^{1/2} - \left(\frac{z_r}{z_{eq}}\right)^{1/2}\right] I_\Lambda. \tag{9.87}$$

Using this result together with (9.60) for σ, (9.78) becomes

$$l_f \simeq 1530\left(1 + \frac{z_r}{z_{eq}}\right)^{1/2} I_\Lambda^{-1}, \tag{9.88}$$

where we recall from (9.61) that

$$\frac{z_r}{z_{eq}} \simeq 7.8 \times 10^{-2}\left(\Omega_m h_{75}^2\right)^{-1} \tag{9.89}$$

for three neutrino species.

The result is that the finite thickness damping coefficient l_f depends only weakly on the cosmological term and $\Omega_m h_{75}^2$. For $\Omega_m h_{75}^2 \simeq 0.3$ and $\Omega_\Lambda h_{75}^2 \simeq 0.7$, we have $l_f \simeq 1580$, whereas for $\Omega_m h_{75}^2 \simeq 1$ and $\Omega_\Lambda h_{75}^2 \simeq 0$, we find $l_f \simeq 1600$.

The scale l_S describing the combination of finite thickness and Silk damping effects can be calculated in a similar way.

Problem 9.10 Using the estimate (9.62) for the Silk dissipation scale, show that

$$l_S \simeq 0.7 l_f \left\{\frac{1 + 0.56\xi}{1+\xi} + \frac{0.8}{\xi(1+\xi)}\frac{\left(\Omega_m h_{75}^2\right)^{1/2}}{\left[1 + \left(1 + z_{eq}/z_r\right)^{-1/2}\right]^2}\right\}^{-1/2}. \tag{9.90}$$

This estimate is not very reliable because the imperfect fluid approximation breaks down when the visibility function is near maximum. Nevertheless, comparison with exact numerical calculations shows that the discrepancy is less than 10% with numerical l_S being slightly smaller than in (9.90). In contrast to l_f, the scale l_S does depend on the baryon density, characterized by ξ. However, this dependence is very strong only for $\xi \ll 1$, when the second term inside the parenthesis in (9.90) dominates. For $\xi = 0.6$, we find $l_S \simeq 1100$ if $\Omega_m h_{75}^2 \simeq 0.3$ and $l_S \simeq 980$ for $\Omega_m h_{75}^2 \simeq 1$.

The parameter ρ, which determines the location of the acoustic peaks, can be calculated by substituting the expression for $c_s(\eta)$

$$c_s(\eta) = \frac{1}{\sqrt{3}} \left[1 + \xi \left(\frac{a(\eta)}{a(\eta_r)} \right) \right]^{-1/2}, \tag{9.91}$$

with $a(\eta)$ given by (1.81) into (9.77) and then integrating.

Problem 9.11 Verify that

$$\varrho \simeq \frac{I_\Lambda}{\sqrt{3 z_r \xi}} \ln \left[\frac{\sqrt{(1 + z_r/z_{eq})\xi} + \sqrt{(1+\xi)}}{1 + \sqrt{\xi(z_r/z_{eq})}} \right]. \tag{9.92}$$

Although it is clear that ϱ depends on both baryon and matter densities, it is not apparent from this expression how ϱ varies when we change their values. For this reason, it is useful to find a fitting formula for ϱ. Verify that the numerical fit

$$\varrho \simeq 0.014 (1 + 0.13\xi)^{-1} \left(\Omega_m h_{75}^2 \right)^{1/4} I_\Lambda \tag{9.93}$$

reproduces the exact result (9.92) to within 7% everywhere in the region $0 < \xi < 5$, $0.1 < \Omega_m h_{75} < 1$, whereas the function ϱ itself varies by roughly a factor of 3. Combining (9.93) with the numerical fit for I_Λ in (9.85), we obtain

$$\varrho \simeq 0.014 (1 + 0.13\xi)^{-1} \left(\Omega_m h_{75}^{3.1} \right)^{0.16}. \tag{9.94}$$

Note the unusual combination $\Omega_m h_{75}^{3.1}$. We will see later that because of this we can hope to determine Ω_m and h_{75} separately by combining the measurements of the location of the peaks with the measurements of other features of the microwave spectrum which depend only on $\Omega_m h_{75}^2$.

The parameter ϱ characterizes the angular size of the sound horizon at recombination on today's sky. The size of the sound horizon drops as the baryon density increases. For a given physical size of the sound horizon, its angular scale on today's sky should of course also depend on the evolution of the universe after recombination. This is why the parameter ϱ and the location of the acoustic peaks in a flat universe are sensitive to the cosmological constant.

The transfer functions T_p and T_o depend only on

$$k\eta_{eq} = \frac{\eta_{eq}}{\eta_0} lx \simeq 0.72 (\Omega_m h_{75}^2)^{-1/2} I_\Lambda l_{200} x, \tag{9.95}$$

where $l_{200} \equiv l/200$ and $x \equiv k\eta_0/l$.

Problem 9.12 Verify that

$$\frac{\eta_{eq}}{\eta_0} = (\sqrt{2}-1) \frac{I_\Lambda}{\sqrt{z_{eq}}} \simeq 3.57 \times 10^{-3} (\Omega_m h_{75}^2)^{-1/2} I_\Lambda. \tag{9.96}$$

As we will see, for the most interesting range $1000 > l > 200$, the dominant contribution to the integrals in (9.73) comes from x close to unity for which $10 > k\eta_{eq} > 1$. Therefore, we can use for the transfer functions the approximations (9.67) and (9.68), which can be rewritten in terms of x and cosmological parameters as

$$T_p(x) \simeq 0.74 - 0.25 (P + \ln x) \tag{9.97}$$

and

$$T_o(x) \simeq 0.5 + 0.36 (P + \ln x), \tag{9.98}$$

where the function

$$P(l, \Omega_m, h_{75}) \equiv \ln\left(\frac{I_\Lambda l_{200}}{\sqrt{\Omega_m h_{75}^2}}\right) \tag{9.99}$$

tells us how the transfer function determining the fluctuations in multipole l scales depends on the cosmological term and cold matter energy density. The physical reason for the dependence of the transfer functions on the matter density is explained in Section 9.7.1.

9.7.4 Calculating the spectrum

We will now proceed to calculate the multipole spectrum $l(l+1) C_l$. The main contribution to the integrals O_1 and O_2 in (9.75) and (9.76) arises in the vicinity of the singular point $x = 1$.

9.7 Anisotropies on small angular scales

Problem 9.13 Using the stationary (saddle) point method, verify that for a slowly varying function $f(x)$

$$\int_1^\infty \frac{f(x)\cos(bx)}{\sqrt{x-1}}dx$$
$$\approx \frac{f(1)}{(1+B^2)^{1/4}}\sqrt{\frac{\pi}{b}}\cos\left(b+\frac{\pi}{4}+\frac{1}{2}\arcsin\frac{B}{\sqrt{1+B^2}}\right), \quad (9.100)$$

where

$$B \equiv \left(\frac{d\ln f}{bdx}\right)_{x=1}.$$

For large b we can set $B \approx 0$ and the above formula becomes

$$\int_1^\infty \frac{f(x)\cos(bx)}{\sqrt{x-1}}dx \approx f(1)\sqrt{\frac{\pi}{b}}\cos\left(b+\frac{\pi}{4}\right) \quad (9.101)$$

(*Hint* Make the substitution $x = y^2 + 1$ in (9.100).)

Using (9.101) to estimate the integrals in (9.75) and (9.76), we obtain

$$O \simeq \sqrt{\frac{\pi}{\varrho l}}(\mathcal{A}_1 \cos(l\varrho+\pi/4)+\mathcal{A}_2\cos(2l\varrho+\pi/4))e^{-(l/l_S)^2}, \quad (9.102)$$

with the coefficients

$$\mathcal{A}_1 \simeq 0.1\xi\frac{((P-0.78)^2 - 4.3)}{(1+\xi)^{1/4}}e^{\left(\frac{1}{2}(l_S^{-2}-l_f^{-2})l^2\right)},$$

$$\mathcal{A}_2 \simeq 0.14\frac{(0.5+0.36P)^2}{(1+\xi)^{1/2}}, \quad (9.103)$$

which are slowly varying functions of l. In deriving this expression we used (9.97) and (9.98) for the transfer functions, which are valid in the range of multipoles $200 < l < 1000$. For $l > 1000$ the fluctuations are strongly suppressed and this effect is roughly taken into account by the exponential factor in (9.102). However, the expected accuracy in this region is not as good as for $200 < l < 1000$.

We note that the contribution of the Doppler term to O is equal to zero in this approximation. A precise numerical evaluation reveals that the Doppler contribution to the oscillating integrals is small, only a few percent or less of the total, for multipoles $l > 200$.

Substituting (9.97) in (9.80) for the nonoscillating contribution N_1, we obtain

$$N_1 \simeq \xi^2\left[(0.74-0.25P)^2 I_0 - (0.37-0.125P) I_1 + (0.25)^2 I_2\right], \quad (9.104)$$

where the integrals

$$I_m(l/l_f) \equiv \int_1^\infty \frac{(\ln x)^m}{x^2\sqrt{x^2-1}} e^{-(l/l_f)^2 x^2} dx \tag{9.105}$$

can be calculated in terms of hypergeometric functions. However, the resulting expressions are not very transparent and therefore it makes sense to find a numerical fit for them. The final result is

$$N_1 \simeq 0.063\xi^2 \frac{\left[P - 0.22(l/l_f)^{0.3} - 2.6\right]^2}{1 + 0.65(l/l_f)^{1.4}} e^{-(l/l_f)^2}. \tag{9.106}$$

Similarly, we obtain

$$N_2 \simeq \frac{0.037}{(1+\xi)^{1/2}} \frac{\left[P - 0.22(l/l_S)^{0.3} + 1.7\right]^2}{1 + 0.65(l/l_S)^{1.4}} e^{-(l/l_S)^2}. \tag{9.107}$$

The Doppler contribution to the nonoscillating part of the spectrum is comparable to N_2 and is equal to

$$N_3 \simeq \frac{0.033}{(1+\xi)^{3/2}} \frac{\left[P - 0.5(l/l_S)^{0.55} + 2.2\right]^2}{1 + 2(l/l_S)^2} e^{-(l/l_S)^2}. \tag{9.108}$$

The numerical fits for N reproduce the exact result for multipoles $200 < l < 1000$ to within a few percent accuracy for a wide range of cosmological parameters.

The ratio of the value of $l(l+1)C_l$ for $l > 200$ to its value for low multipole moments (the flat plateau) is

$$\frac{l(l+1)C_l}{(l(l+1)C_l)_{\text{low }l}} = \frac{100}{9}(O + N_1 + N_2 + N_3), \tag{9.109}$$

where O, N_1, N_2, N_3 are given by (9.102), (9.106), (9.107) and (9.108) respectively. The result in the case of the *concordance model* ($\Omega_m = 0.3$, $\Omega_\Lambda = 0.7$, $\Omega_b = 0.04$, $\Omega_{tot} = 1$ and $H = 70$ km s^{-1} Mpc^{-1}) is presented in Figure 9.2, where we have shown the total nonoscillating contribution and the total oscillatory contribution by the dashed lines. Their sum is the solid line.

Our results are in good agreement with numerical calculations for a rather wide range of cosmological parameters around the concordance model. Although numerical codes are more precise, the analytic expressions enable us to understand how the main features in the anisotropy power spectrum arise and how they depend on cosmological parameters.

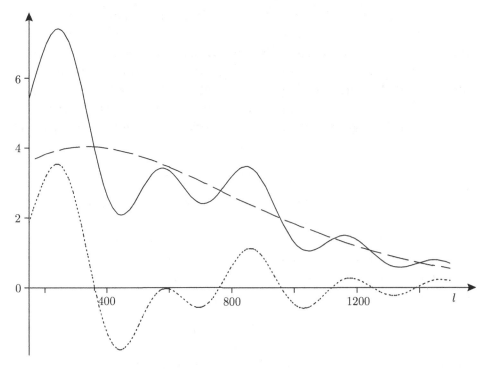

Fig. 9.2.

9.8 Determining cosmic parameters

Assuming a Gaussian, adiabatic spectrum of initial density perturbations, as predicted by inflation, the principal cosmological parameters are: the amplitude B and slope n_S of the primordial spectrum, the baryon density $\xi \equiv 17(\Omega_b h_{75}^2)$, the cold matter density Ω_m, the vacuum density Ω_Λ and the Hubble parameter h_{75}. We shall consider how the spectrum changes as the parameters are varied around the best-fit model, $\Omega_b = 0.04$, $\Omega_m = 0.3$, $\Omega_\Lambda = 0.7$ and $H = 70\,\mathrm{km\,s^{-1}\,Mpc^{-1}}$. Our formulae above are valid for a wide variation in each parameter, more than sufficient to cover the likely values, although some of the expressions break down if used in the limits of very high or very low densities.

Keeping in mind the comments above on the physical origin of parameter dependence of the coefficients in the formulae describing the fluctuations, the reader can easily figure out why the main features of the spectrum depend in one way or another on these parameters. Therefore we omit physical explanations for these dependences in what follows.

The plateau For a nearly scale-invariant spectrum, the anisotropy power spectrum on large angular scales ($l < 30$) is a nearly flat plateau. The amplitude and slope of the plateau can be used to determine the primordial spectral amplitude and spectral index. The accuracy is limited mainly by the cosmic variance and by the fact that C_l for small l is a weighted integral over modes that also include wavelengths smaller than the Hubble scale, the contribution of which depends on other parameters, such as Ω_b, Ω_m, etc. This prevents a determination of the spectral slope to better than 10% accuracy. To improve the accuracy and fix other cosmological parameters, one must go to smaller angular scales and get information about the acoustic peaks and other features of the spectrum.

The location of the peaks and the spatial curvature of the universe The acoustic peaks arise when the oscillating term O, given by (9.102), is superimposed on the "hill" given by the nonoscillating contribution $N(l) = N_1 + N_2 + N_3$ (see Figure 9.2). The peak locations and the heights depend on both contributions. The oscillation peaks in O alone come from the superposition of two cosine terms in (9.102), whose periods differ by a factor of 2. If $|\mathcal{A}_1| \ll \mathcal{A}_2$, the peaks are located at

$$l_n = \pi \varrho^{-1} \left(n - \frac{1}{8}\right), \qquad (9.110)$$

where $n = 1, 2, 3 \ldots$ and ϱ is given by (9.94). The first term on the right hand side in (9.102) has a period twice as large as that of the second term and its amplitude $\mathcal{A}_1$ is negative. Therefore, it interferes constructively for the odd peaks ($n = 1, 3, \ldots$) and destructively for the even peaks ($n = 2, 4, \ldots$). Moreover, because of the relative phase shift of the two cosine terms, their maxima do not coincide and the constructive maxima of their sum lies between the closest maxima of the two individual terms; that is,

$$l_1 \simeq \left(\frac{6 \div 7}{8}\right) \pi \varrho^{-1}, \quad l_3 \simeq \left(2 + \frac{6 \div 7}{8}\right) \pi \varrho^{-1}. \qquad (9.111)$$

Here the notation $6 \div 7$, for instance, denotes a number between 6 and 7. If $|\mathcal{A}_1| \gg \mathcal{A}_2$, the peaks move closer to the lower bounds of the intervals in (9.111).

In the concordance model, where $\xi \simeq 0.6$ and $\Omega_m h_{75}^2 \simeq 0.26$ from (9.94) and (9.111), we find that $l_1 \simeq 225 \div 265$ and $l_3 \simeq 825 \div 865$. The situation is made more complicated by the nonoscillating contribution N. As is clear from Figure 9.2, the hill causes the first peak to move towards the right and the third peak to move towards the left.

The even peaks correspond to the multipoles where two terms in (9.102) destructively interfere. The second peak should be located at

$$l_2 \simeq \left(1 + \frac{6 \div 7}{8}\right) \pi \varrho^{-1} \simeq 525 \div 565 \qquad (9.112)$$

in the current best-fit model. However, for some choices of parameters the destructive interference can annihilate this peak altogether.

The consideration above refers to a spatially flat universe with $\Omega_{tot} = 1$. Let us now consider how the peak locations depend on the values of the fundamental cosmological parameters. If the universe were curved, the angular size of the sound horizon would change and the peaks would shift compared with the flat case. For instance, as follows from (2.73), in a universe without the cosmological constant $l_1 \propto \Omega_{tot}^{-1/2}$. Could we then accurately determine the spatial curvature simply by measuring the location of the first peak? The answer to this question is not as straightforward as it seems at first glance. According to (9.94), the value of ϱ also depends on Ω_m, h_{75} and Ω_b (through ξ), and so it is clear from (9.111) and (9.112) that the peak positions depend on these parameters. Over the range of realistic values, while the sensitivity to these parameters is not as strong as to the spatial curvature, it is nevertheless significant. As an example, if we take a flat universe with the current best-fit values of the cosmological parameters, and then double the baryon density ($\xi \simeq 0.6 \to \xi \simeq 1.2$), the first peak moves to the right by $\Delta l_1 \sim +20$, the second by $\Delta l_2 \sim +40$, and the third by $\Delta l_3 \sim +60$. We note that the locations of the peaks depend on $\xi \propto \Omega_b h_{75}^2$, whereas the dependence on the cold matter density enters through ϱ as $\Omega_m h_{75}^{3.1}$. An increase in Ω_m has the opposite effect on peak locations: if we were to double the cold dark matter density ($\Omega_m h_{75}^{3.1} \simeq 0.3 \to \Omega_m h_{75}^{3.1} \simeq 0.6$), the first peak would move to the left by $\Delta l_1 \sim -20$ and the second and third peaks by $\Delta l_2 \sim -40$ and $\Delta l_3 \sim -60$ respectively. Thus, even keeping the spatial curvature fixed, the first peak can be shifted significantly ($\Delta l_1 \sim 40$) by doubling the baryon density and simultaneously halving the cold matter density. This limits our ability to determine the spatial curvature precisely based on the first peak location only. Fortunately, the parameter degeneracy can be resolved by combining measurements of peak locations with peak heights, as described below.

Acoustic peak heights, the baryon and cold matter densities, and flatness Substituting l_n, given by (9.111) and (9.112), into (9.99) and using (9.92) for ϱ, we see that the factor I_Λ is canceled in the expression for P. Hence, the peak heights predicted by (9.109) depend on the combinations $\Omega_m h_{75}^2$ and $\Omega_b h_{75}^2$ (or ξ). For fixed $\Omega_m h_{75}^2$, an increase in the baryon density increases the height of the the first acoustic peak H_1. For instance, beginning from the current best-fit model, doubling the baryon density increases H_1 by a factor of 1.5, due principally to N_1 (proportional to ξ^2) and O (since $\mathcal{A}_1 \propto \xi$). An increase of the cold matter density for fixed ξ suppresses H_1 since P decreases as $\Omega_m h_{75}^2$ increases. For the cold matter density, the sensitivity comes from the N_2 and N_3 terms. Therefore, playing the various terms off one another, the height of the first peak can be held fixed for certain combinations of changes in the baryon and cold dark matter densities. If the baryon density is made too large, though, an increase in $\Omega_m h_{75}^2$ cannot compensate an increase in $\Omega_b h_{75}^2$ because its effect on H_1 saturates for large values (and moreover $\Omega_m h_{75}^2$ cannot be much greater than unity).

Current observations suggest that H_1 is 6–8 times the amplitude at large angular scales. *Based on the peak height alone, we can be sure that the baryon density is less than* 20% of the critical density. Although much better constraints can be obtained using the full power spectrum and additional data, it is important to appreciate that the peak height alone suffices to rule out a flat, baryon-dominated universe, which was, in essence, the original concept of the big bang universe.

If the height of the first peak is kept fixed, the only freedom left is a simultaneous change of the cold matter and baryon densities, both of which can be either increased or decreased. For instance, we can still keep the height of the first peak unchanged *simultaneously* increasing the baryon and matter density by a factor of about 1.5 from the concordance model densities. However, since the increases of the baryon and cold matter densities have opposing effects on the location of the peak, its net shift will be negligible. This explains why, for a fixed height, the location of the first peak depends sensitively on the spatial curvature only, allowing us to resolve the degeneracy in its determination. The current data on the location of the first peak strongly support the flat universe predicted by inflation.

To break the degeneracy between baryon and matter density altogether, it suffices to consider the second acoustic peak, which results primarily from the destructive interference between oscillatory terms in (9.102), but bearing in mind that they are superimposed on the "hill" due to N. The first term in (9.102) makes a negative contribution to the second peak and has a coefficient that is proportional to ξ. The second term makes a positive contribution to the second peak but slightly decreases as ξ increases. Hence, one can see that the second peak shrinks as the baryon density increases. The two terms nearly cancel altogether when the baryon density is about 8% of the critical density, or about twice the best-fit value. Curiously enough, though, exact numerical calculations show that a tiny second peak survives for $\Omega_m h_{75}^2 \simeq 0.26$ even when the baryon density is made much greater than 8%. This is because an increase in the baryon density also increases N_1, making the hill on which the oscillatory contributions rest much steeper in the vicinity of the second peak. In other words, the appearance of the second peak depends on a delicate cancellation and combination of diverse terms. So, for example, it would be incorrect to conclude that the baryon density is less than 8% of the critical density simply because one observes a second acoustic peak.

However, combining information about the height of the first peak with the fact that the second peak *exists* (ignoring peak locations for the moment) does lead to good limits on both baryon density and cold dark matter density. We initially showed that to keep the first peak height H_1 fixed, the baryon density can be increased only simultaneously with the dark matter density. That is, baryonic and dark matter work in opposite directions in terms of their effect on the first peak. Now we have seen that the second peak decreases as the baryon density increases. For the second peak, it turns out that an increase in cold dark matter density has a similar effect. Hence, the baryon density and the dark matter density work in the same direction in terms of how they alter the height of the second peak. This is because the positive contribution $O_2 \propto T_o^2$ decreases more rapidly than the negative contribution $O_1 \propto T_o T_p$ as $\Omega_m h_{75}^2$ increases. Using both peak heights enables us to determine both the baryon density and the cold dark matter density and thus to resolve the degeneracy in the determination of $\Omega_m h_{75}^2$ and $\Omega_b h_{75}^2$. For instance, assuming that the universe is flat and that cold matter constitutes 100% of the critical density, we can fit the data for the height of the first peak only if $\Omega_b \simeq 0.08$ for $H = 70$ km s^{-1} Mpc^{-1}. However, in this case the second acoustic peak is absent. It reappears only when we simultaneously decrease the cold dark matter and the baryon density. Therefore, based on the data, the combination of the first acoustic peak height with the fact that the second peak *exists* informs us that the cold dark matter density cannot exceed half of the critical density and that the baryon density is less than 8% of the critical density. Although these limit can be improved greatly by an analysis of the full anisotropy power spectrum and other observations, it is important to appreciate that *the height of the first peak together with the existence of the second peak are in themselves convincing evidence of the following key qualitative features of our universe: that the total cold matter density is less than the critical density, that cold dark matter exists and that its density exceeds the baryon density.*

Combining peak height and location If information about the first two peak heights determines the baryon and matter density (of course, in combination with h_{75}), then adding information about the first peak location fixes the spatial curvature precisely. Data strongly suggest that the universe is flat and that the total energy density is equal to the critical density. At the same time, the peak heights suggest that the dark matter and baryon densities are significantly less than the critical density. Hence, some form of dark energy must make up the difference and dominate the density of the universe today.

Hence, combining peak heights and their location we can conclude that dark energy exists. Note that this line of argument is totally independent of the supernova luminosity–redshift test (see Section 2.5.2), which leads to the same conclusion.

Since the heights of the peaks depend on $\Omega_m h_{75}^2$ and their locations on $\Omega_m h_{75}^{3.1}$, we can squeeze out even more information; namely, we can determine the Hubble constant. The dependence of the peak location on h_{75} is modest, so a very accurate measurement of the microwave background is required to obtain a reasonable constraint on h_{75}. For example, if the locations and heights of the peaks are determined to 1% accuracy, then the expected accuracy for the Hubble parameter will be about 7%.

Reconsideration of the spectral tilt Until now we have been assuming that the primordial spectrum of inhomogeneities is scale-invariant with spectral index $n_S = 1$. Inflation predicts that there should be a small deviation from perfect scale invariance, typically $n_S \simeq 0.92$–0.97. The above derivation for the microwave background fluctuations can easily be modified to account for these deviations.

Problem 9.14 Show that the multipoles C_l for a primordial spectrum with tilt n_S are modified by a factor proportional to $l^{n_S - 1}$ compared with a scale-invariant spectrum.

When we include uncertainty in the spectral tilt, then the heights and locations of the first two peaks are insufficient to determine Ω_b, Ω_m and h_{75} separately. Here is where the third acoustic peak comes into play. The height of the third peak is not as sensitive to $\Omega_b h_{75}^2$ and $\Omega_m h_{75}^2$ as the first two peaks, but it is sensitive to the spectral index. Fixing these parameters and the height of the first peak, the ratio of the third peak height to the first, $r \equiv H_3/H_1$, changes by a factor

$$\frac{\Delta r}{r} \sim 1 - \left(\frac{l_3}{l_1}\right)^{1-n_S} \sim (n_S - 1) \ln\left(\frac{l_3}{l_1}\right). \quad (9.113)$$

For instance, if $n_S \simeq 0.95$, the height of the third peak decreases by about 7% compared with the case for $n_S = 1$.

Summary Thus we have seen what a powerful tool the microwave background power spectrum can be. The general shape – a plateau at large angular scales and acoustic peaks at small angular scales – confirms that the spectrum is predominantly nearly scale-invariant and adiabatic. (The higher-order correlation functions should be used to show that the spectrum is also Gaussian.) This supports the basic predictions of the inflationary/big bang paradigm. Then we can proceed to use the quantitative details of the spectrum – the plateau and the heights and locations of the acoustic peaks – to determine the primary cosmological parameters.

Our analysis is valid for a limited parameter set; the inclusion of other physical effects or variants of the best-fit model weakens to some extent the conclusions that can be drawn from measurements of the anisotropy. For example, secondary reionization by early star formation at redshifts $z > 20$ or so reduces the small angular scale power in a way that is difficult to disentangle from tilting the spectral index. The dark energy may comprise quintessence, rather than vacuum energy. In this case, we must introduce a new parameter, the equation of state of the dark energy w (or perhaps a function $w(z)$). Correlated changes in Ω_m, h_{75} and w can produce canceling effects that leave the plateau and the first three peaks virtually unchanged. Thus, the temperature autocorrelation function is a powerful tool, but it is not all-powerful.

To explore the range of possible models fully we need to use all the information the power spectrum offers, in combination with other cosmological observations. For example, the heights, locations and shapes of the peaks also depend on the dissipation scales l_f and l_S, which in turn depend on combinations of the cosmological parameters. For the current best-fit model, for which $l_S \sim 1000$, dissipation does not influence the first peak significantly, but it becomes increasingly important for the higher-order peaks. Hence, using more peaks in the analysis further constrains models.

9.9 Gravitational waves

An important physical effect that we have neglected thus far is that of gravitational waves, one of the basic predictions of inflationary cosmology. As discussed in Section 7.1, to describe gravitational waves we use the metric

$$ds^2 = a^2\left(d\eta^2 - (\delta_{ik} + h_{ik})dx^i dx^k\right). \tag{9.114}$$

The gravitational waves correspond to the traceless, divergence-free part of h_{ik}. They produce perturbations in the microwave background by inducing the redshifts and blueshifts of the photons. Using the equation $p^\alpha p_\alpha = 0$, we can express the zero component of the photon's 4-momentum as

$$p_0 = p^0 = \frac{p}{a^2}\left(1 - \frac{1}{2}h_{ik}l^i l^k\right), \tag{9.115}$$

where as before $p \equiv \left(\Sigma p_i^2\right)^{1/2}$, $l^i \equiv -p_i/p$ and we have kept only the first order terms in metric perturbations. The photon geodesic equations for the metric (9.114) take the following forms:

$$\frac{dx^i}{d\eta} = l^i + O(h), \quad \frac{dp_j}{d\eta} = -\frac{1}{2}p\frac{\partial h_{ik}}{\partial x^j}l^i l^k. \tag{9.116}$$

Taking into account that the distribution function f depends only on the single variable

$$y = \frac{\omega}{T} = \frac{p_0}{T\sqrt{g_{00}}} = \frac{p}{T_0 a}\left(1 - \frac{\delta T}{T} - \frac{1}{2}h_{ik}l^i l^k\right), \tag{9.117}$$

and substituting (9.116) in the Boltzmann equation (9.7), we find that the temperature fluctuations satisfy the equation

$$\left(\frac{\partial}{\partial \eta} + l^j \frac{\partial}{\partial x^j}\right) \frac{\delta T}{T} = -\frac{1}{2} \frac{\partial h_{ik}}{\partial \eta} l^i l^k, \tag{9.118}$$

which has the obvious solution

$$\frac{\delta T(\mathbf{l})}{T} = -\frac{1}{2} \int_{\eta_r}^{\eta_0} \frac{\partial h_{ij}}{\partial \eta} l^i l^j d\eta. \tag{9.119}$$

Note that until now we have not used the fact that h_{ik} is a traceless, divergence-free tensor. Therefore, (9.119) is a general result, which can also be applied when calculating the temperature fluctuations induced by the scalar metric perturbations in the synchronous gauge. For tensor perturbations, h_{ik} satisfies the extra conditions $h_i^i = h_{k,i}^i = 0$ (here we raise and lower the indices with the unit tensor δ_{ik}), which reduce the number of independent components of h_{ik} to two, corresponding to two independent polarizations of the gravitational waves. For random Gaussian fluctuations, the tensor metric perturbations can be written as

$$h_{ik}(\mathbf{x}, \eta) = \int h_{\mathbf{k}}(\eta) e_{ik}(\mathbf{k}) e^{i\mathbf{k}\mathbf{x}} \frac{d^3 k}{(2\pi)^{3/2}}, \tag{9.120}$$

where $e_{ik}(\mathbf{k})$ is the time-independent random polarization tensor. Because of the conditions $e_i^i = e_j^i k_i = 0$, $e_{ik}(\mathbf{k})$ should satisfy

$$\langle e_{ik}(\mathbf{k}) e_{jl}(\mathbf{k}') \rangle = (P_{ij} P_{kl} + P_{il} P_{kj} - P_{ik} P_{jl}) \delta(\mathbf{k} + \mathbf{k}'), \tag{9.121}$$

where

$$P_{ij} \equiv (\delta_{ij} - k_i k_j / k^2), \tag{9.122}$$

is the projection operator. Substituting (9.120) into (9.119) and calculating the tensor contribution to the correlation function of the temperature fluctuations (see the definition in (9.33)), we obtain

$$C^T(\theta) = \frac{1}{4} \int F(\mathbf{l}_1, \mathbf{l}_2, \mathbf{k}) h'_{\mathbf{k}}(\eta) h^{*'}_{\mathbf{k}}(\tilde{\eta}) e^{i\mathbf{k}[\mathbf{l}_1(\eta - \eta_0) - \mathbf{l}_2(\tilde{\eta} - \eta_0)]} d\eta d\tilde{\eta} \frac{d^3 k}{(2\pi)^3}, \tag{9.123}$$

where $\cos\theta = \mathbf{l}_1 \cdot \mathbf{l}_2$. In deriving (9.123) we have averaged over the random polarization with the help of (9.121). The function F entering this expression does not depend on time and is equal to

$$F = 2\left((\mathbf{l}_1 \mathbf{l}_2) - \frac{(\mathbf{l}_1 \mathbf{k})(\mathbf{l}_2 \mathbf{k})}{k^2}\right)^2 - \left(1 - \frac{(\mathbf{l}_1 \mathbf{k})^2}{k^2}\right)\left(1 - \frac{(\mathbf{l}_2 \mathbf{k})^2}{k^2}\right). \tag{9.124}$$

9.9 Gravitational waves

Introducing the new variable $x \equiv k(\eta_0 - \eta)$ instead of η, and noting that

$$\frac{(\mathbf{l}_1\mathbf{k})}{k} = -\frac{\partial}{i\partial x}e^{-i\frac{\mathbf{k}\mathbf{l}_1}{k}x}, \quad \frac{(\mathbf{l}_2\mathbf{k})}{k} = \frac{\partial}{i\partial \tilde{x}}e^{i\frac{\mathbf{k}\mathbf{l}_2}{k}\tilde{x}},$$

after integration over the angular part of $\mathbf{k}$ we can rewrite (9.123) as

$$C^T(\theta) = \frac{1}{4}\int \frac{\partial h_k}{\partial x}\frac{\partial h_k^*}{\partial \tilde{x}}\left[\hat{F}\cdot\frac{\sin(|\mathbf{l}_2\tilde{x} - \mathbf{l}_1 x|)}{|\mathbf{l}_2\tilde{x} - \mathbf{l}_1 x|}\right]dxd\tilde{x}\frac{k^2 dk}{2\pi^2}, \qquad (9.125)$$

where

$$\hat{F} = 2\left(\cos\theta - \frac{\partial}{\partial x}\frac{\partial}{\partial \tilde{x}}\right)^2 - \left(1 + \frac{\partial^2}{\partial x^2}\right)\left(1 + \frac{\partial^2}{\partial \tilde{x}^2}\right). \qquad (9.126)$$

Now we can use formula (9.36) to expand $C^T(\theta)$ as a discrete sum over multipole momenta (see (9.37)). After a lengthy but straightforward calculation, the result for C_l^T can be written in a rather simple form.

Problem 9.15 Substitute (9.36) into (9.125) and use the recurrence relations for $zP(z)$, the Bessel functions equation and the recurrence relations for the spherical Bessel functions to express j_l'', j_{l-2}, j_{l-1}' etc. through j_l, j_l', to verify that

$$C_l^T = \frac{(l-1)l(l+1)(l+2)}{2\pi}\int_0^\infty\left|\int_0^{k(\eta_0 - \eta_r)}\frac{\partial h_k}{\partial x}\frac{j_l(x)}{x^2}dx\right|^2 k^2 dk. \qquad (9.127)$$

The derivative of the metric perturbations takes its maximal value at $k\eta \sim O(1)$ and drops very fast after that. Hence, for those gravitational waves which entered the horizon, the main contribution to the integral over x in (9.127) comes from the relatively narrow region: $k\eta_0 > x > k\eta_0 - O(1)$. For $l \gg 1$ and $k\eta_0 \gg 1$, the function $j_l(x)/x^2$ does not change significantly within this interval and can therefore be approximated by its value at $x_0 = k\eta_0$. As a result, (9.127) simplifies for $l \gg 1$ to

$$C_l^T \simeq \frac{(l-1)l(l+1)(l+2)}{2\pi}\int_0^\infty\left|h_k^2(\eta_r)k^3\right|\frac{j_l^2(x_0)}{x_0^5}dx_0, \qquad (9.128)$$

where $\left|h_k^2(\eta_r)k^3\right|$ should be expressed as a function of $x_0 = k\eta_0$. The gravity waves generated during inflation, which enter the horizon after recombination but well before the present time, have a nearly flat spectrum at $\eta = \eta_r$, so that

$$\left|h_k^2(\eta_r)k^3\right| = B_{gw} \approx \text{const} \qquad (9.129)$$

for $\eta_r^{-1} > k > \eta_0^{-1}$. Taking into account that for $l \gg 1$ the main contribution to (9.128) comes from the perturbations with $k \sim l/\eta_0$, and substituting (9.129) into

(9.128), we can calculate the integral with the help of (9.43). The result, valid for $\eta_0/\eta_r \gg l \gg 1$, reads

$$l(l+1)C_l^T \simeq \frac{2}{15\pi} \frac{l(l+1)}{(l+3)(l-2)} B_{gw} \approx 4.2 \times 10^{-2} B_{gw}. \quad (9.130)$$

(For example, $\eta_0/\eta_r \simeq 55$ for $\Omega_m h_{75}^2 \simeq 0.3$.) This estimate fails when applied to the low multipoles and in particular to the quadrupole ($l = 2$). The quadrupole can be calculated numerically and the result,

$$l(l+1)C_l^T\big|_{l=2} \simeq 4.4 \times 10^{-2} B_{gw}, \quad (9.131)$$

does not differ greatly from the expression in the right hand side of (9.130).

Problem 9.16 Verify that the relative contribution of the tensor and scalar perturbations generated during inflation to the quadrupole is

$$\frac{C_{l=2}^T}{C_{l=2}^S} \simeq 10.4 c_s \left(1 + \frac{p}{\varepsilon}\right), \quad (9.132)$$

where the expression on the right hand side must be estimated at the moment when the perturbations responsible for the quadrupole cross the Hubble scale on inflation. Taking $1 + p/\varepsilon \sim 10^{-2}$ and $c_s = 1$ we find that the gravitational waves should contribute about 10% to the quadrupole component. In k inflation where $c_s \ll 1$ their contribution can be negligible.

As we found in Section 7.3.2, the amplitude of the gravitational waves which entered the horizon is inversely proportional to the scale factor and, for $k > \eta_r^{-1}$, their spectrum at $\eta = \eta_r$ is already significantly modified. For instance, for $k \gg \eta_{eq}^{-1}$,

$$\left|h_k^2(\eta_r) k^3\right| \simeq O(1) B_{gw} \left(\frac{1}{k\eta_{eq}}\right)^2 \left(\frac{z_r}{z_{eq}}\right)^2. \quad (9.133)$$

Substituting this expression into (9.128), we obtain

$$l(l+1)C_l^T \propto B_{gw} \left(\frac{l_{eq}}{l}\right)^2$$

for $l \gg l_{eq}$, where $l_{eq} \equiv \eta_0/\eta_{eq}$ ($l_{eq} \simeq 150$ for $\Omega_m h_{75}^2 \simeq 0.3$). In the intermediate region $55 < l < 150$, the amplitude $l(l+1)C_l^T$ also decays. Note that all results in this section were derived in the approximation of instantaneous recombination, which is valid in the most interesting range of the multipoles.

Problem 9.17 Assuming that $\eta_{eq} \ll \eta_r$ determine how $l(l+1)C_l^T$ depends on l for $\eta_0/\eta_{eq} \gg l \gg \eta_0/\eta_r$.

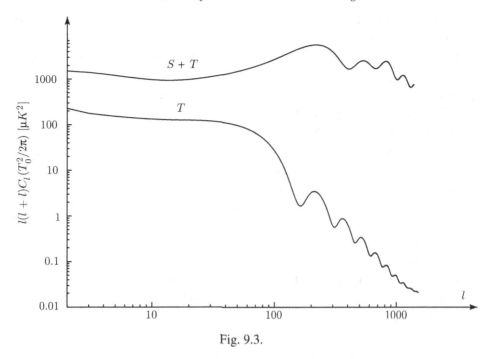

Fig. 9.3.

As with scalar perturbations, the contribution of the tensor perturbations to the CMB power spectrum also consists of a flat plateau at low multi-poles, due to the superhorizon gravitational waves at last scattering. However, for $l > 55$, the amplitude $l(l+1)C_l^T$ decreases quickly. Figure 9.3 was drawn using a precise numerical code showing how the total spectrum is subdivided into scalar and tensor components for the concordance model. Note that the tensor component dies off rapidly where the acoustic peaks appear. Hence, detecting the tensor contribution to the temperature autocorrelation function relies on comparing the height of the plateau to the height of the acoustic peaks. It is difficult to separate this effect from reionization or a spectral tilt. Polarization proves to be the better test for detecting primordial gravitational waves.

9.10 Polarization of the cosmic microwave background

Thus far, we have focused on the temperature fluctuations in the cosmic microwave background, because the temperature autocorrelation provides the single, most powerful test for distinguishing cosmological models and determining cosmological parameters. However, there is more information to be gained by measuring the polarization and its correlation with the temperature fluctuations. In particular, polarization provides the cleanest and most sensitive method of detecting the primordial spectrum of gravitational waves produced by inflation, one of the most challenging predictions to verify.

The polarization of the cosmic microwave background is an inevitable consequence in any model because recombination is not an instantaneous process. The quadrupole anisotropy of the microwave background, which is absent before recombination begins, is produced by both scalar and tensor perturbations as recombination proceeds. In turn, this leads to radiation, scattered off electrons, through Thomson scattering, being linearly polarized. Note that if recombination were instantaneous, no significant polarization would be generated. Hence, measuring the polarization gives us a chance to uncover the subtle details of the recombination history. Note that Thomson scattering on the electrons does not produce any circular polarization.

Just as with the temperature fluctuations, the most useful quantity to compute is the two-point correlation function for polarization. The polarization signal is very weak: it is expected to be only 10% of the total temperature fluctuations on small angular scales, decreasing to much less than 1% on large angular scales. Hence, as difficult as it is to detect the temperature fluctuations, detecting the polarization is an even more extraordinary experimental challenge. Nevertheless, experimentalists are up to the challenge and the prospects seem to be excellent.

9.10.1 Polarization tensor

The electric field $\mathbf{E}$ is always transverse to the direction of propagation of the electromagnetic wave, characterized by the unit vector $\mathbf{n}$. Therefore, this field can be decomposed as $\mathbf{E} = E^a \mathbf{e}_a$, where $a = 1, 2$ and $\mathbf{e}_a$ are two linearly independent basis vectors perpendicular to $\mathbf{n}$ (Figure 9.4). Completely (linearly) polarized light always has vector $\mathbf{E}$ aligned in a definite direction, while in the opposite case of completely unpolarized radiation, all $\mathbf{E}$ directions perpendicular to $\mathbf{n}$ are equally probable. *In the absence of circular polarization,* the radiation polarization properties can be

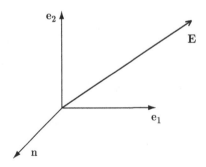

Fig. 9.4.

9.10 Polarization of the cosmic microwave background

characterized entirely by the two-dimensional, second rank symmetric polarization tensor

$$\mathcal{P}_{ab} \equiv \frac{1}{I}\left(\langle E_a E_b \rangle - \frac{1}{2}\langle E_c E^c \rangle g_{ab}\right), \tag{9.134}$$

where the metric tensor $g_{ab} = \mathbf{e}_a \cdot \mathbf{e}_b$ and its inverse are used to raise and lower the indices, e.g. $E_a = g_{ac} E^c$. The brackets represent an average over a time interval much exceeding the typical inverse frequencies of the wave. The scalar product of two three-dimensional vectors $\mathbf{e}_a$ is defined, as usual, with respect to the spatial Euclidean metric. The overall intensity of the radiation is proportional to $I \equiv \langle E_c E^c \rangle$. If the light is polarized, then the brightness temperature of the radiation after it passes through the polarizer depends on its orientation $\mathbf{m} = m^a \mathbf{e}_a$ and the temperature variations are $\delta T(\mathbf{m}) \propto \mathcal{P}_{ab} m^a m^b$. Thus, by measuring this dependence, one can determine the polarization tensor $\mathcal{P}_{ab}$.

Problem 9.18 Calculate the polarization tensor and the fraction of polarization

$$P \equiv -4 \det \left| \mathcal{P}_b^a \right| \tag{9.135}$$

in two extreme cases: completely polarized and completely unpolarized radiation.

Let us assume that $P \neq 0$ and consider the eigenvalue problem for the matrix $\mathcal{P}_b^a$:

$$\mathcal{P}_b^a p_a = \lambda p_b \tag{9.136}$$

with positive eigenvalue λ. Normalizing an eigenvector p_a in such a way that $p^2 \equiv p_c p^c = 2\lambda$, we can express the polarization tensor $\mathcal{P}_{ab}$ through the *polarization vector* p_a as

$$\mathcal{P}_{ab} = p_a p_b - \frac{1}{2} p^2 g_{ab}. \tag{9.137}$$

In fact, one can see that for $\mathcal{P}_{ab}$ given by (9.137), the vector p_a is the solution of (9.136) with $\lambda = p^2/2$. The other independent solution of (9.136) is a vector f_α perpendicular to p_a, that is, $f_a p^a = 0$. Using the orthogonality condition, we immediately obtain from (9.137) that the appropriate eigenvalue is negative and equal to $-p^2/2$, in complete agreement with the fact that the polarization tensor is traceless: $\mathcal{P}_a^a = 0$. The fraction of polarization can be expressed through the magnitude of polarization vector as

$$P \equiv -4 \det \left| \mathcal{P}_b^a \right| = p^4. \tag{9.138}$$

In the orthonormal basis, where $\mathbf{e}_a \cdot \mathbf{e}_b = \delta_{ab}$, one can define the Stokes parameters

$$Q \equiv 2I\mathcal{P}_{11} = -2I\mathcal{P}_{22}, \quad U \equiv -2I\mathcal{P}_{12}. \tag{9.139}$$

Every direction on the sky $\mathbf{n}$ can be entirely characterized by the polar coordinates θ and φ, in terms of which the metric induced on the celestial sphere of unit radius is

$$ds^2 = g_{ab}dx^a dx^b = d\theta^2 + \sin^2\theta d\varphi^2. \tag{9.140}$$

In this case, it is convenient to use as a pair of polarization basis vectors $\mathbf{e}_a$ the coordinate basis vectors $\mathbf{e}_\theta$ and $\mathbf{e}_\varphi$, tangential to the coordinate lines $\varphi = $ const and $\theta = $ const respectively. Considering the appropriate orthonormal vectors $\mathbf{e}_\theta$ and $\hat{\mathbf{e}}_\varphi \equiv \mathbf{e}_\varphi / |\mathbf{e}_\varphi|$, the Stokes parameters are

$$Q_{\theta\theta} \equiv 2I\mathcal{P}_{\theta\theta} = -2I\mathcal{P}_{\hat{\varphi}\hat{\varphi}}, \quad U_{\theta\varphi} \equiv -2I\mathcal{P}_{\theta\hat{\varphi}}.$$

Problem 9.19 Write down in the original basis $\mathbf{e}_\theta$, $\mathbf{e}_\varphi$ the covariant, contravariant and mixed components of the polarization tensor in terms of these Stokes parameters.

The reader may question why we work with the polarization tensor and not simply with the polarization vector. The point is that the polarization tensor multiplied by I, and consequently the Stokes parameters, are additive for incoherent superposition of waves and can easily be calculated. The polarization vector is not additive, but nevertheless the physical interpretation of the polarization pattern is clearest in terms of this vector. In particular, if the radiation is completely polarized, it is aligned along the electric field. Note that only the orientation of p_a, and not its direction, has a physical meaning because the polarization tensor is quadratic in p_a. For partially polarized radiation, the polarization vector points in the direction of the electric field of the waves which dominate in overall flux, and the magnitude of p_a characterizes the excess of the waves with appropriate polarization.

9.10.2 Thomson scattering and polarization

Let us consider the linearly polarized elecromagnetic wave with electric field $\mathbf{E}$ scattered by the electron in the direction $\mathbf{n}$ (see Figure 9.5). After scattering, the wave remains completely polarized and its electric field is

$$\tilde{\mathbf{E}} = A((\mathbf{E} \times \mathbf{n}) \times \mathbf{n}), \tag{9.141}$$

where the coefficient A does not depend on $\mathbf{E}$ and $\mathbf{n}$. Taking into account that the polarization basis vectors $\mathbf{e}_a$ are orthogonal to $\mathbf{n}$, we find that, after scattering, the

9.10 Polarization of the cosmic microwave background

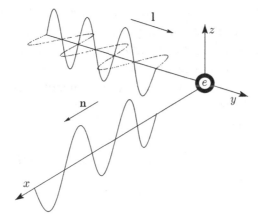

Fig. 9.5.

components of the electric field along vector e_a are

$$\tilde{E}_a = \tilde{\mathbf{E}} \cdot \mathbf{e}_a = A\mathbf{E} \cdot \mathbf{e}_a. \qquad (9.142)$$

If incoming light arriving from direction $\mathbf{l}$ is completely unpolarized, then the resulting polarization tensor can be calculated using (9.142) and averaging over all directions of $\mathbf{E}$ perpendicular to $\mathbf{l}$.

Problem 9.20 Show that in this case

$$\langle \tilde{E}_a \tilde{E}_b \rangle = \tfrac{1}{2} A^2 \langle E^2 \rangle (g_{ab} - (\mathbf{l} \cdot \mathbf{e}_a)(\mathbf{l} \cdot \mathbf{e}_b)), \qquad (9.143)$$

and

$$I = \langle \tilde{E}_a \tilde{E}^a \rangle = \tfrac{1}{2} A^2 \langle E^2 \rangle (1 + (\mathbf{l} \cdot \mathbf{n})^2), \qquad (9.144)$$

where $\langle E^2 \rangle$ is the average of the squared electric field in the incident unpolarized beam. (*Hint* Justify and use the following formula for averaging over directions of the electric field in the incident beam:

$$\langle E^i E^j \rangle = \tfrac{1}{2} \langle E^2 \rangle (\delta^{ij} - l^i l^j), \qquad (9.145)$$

where E^i, l^i ($i = 1, 2, 3$) are the components of the appropriate 3-vectors in some orthonormal basis.)

Write down the polarization tensor and verify that $f_a = (\mathbf{l} \cdot \mathbf{e}_a)$ is an eigenvector of $\mathcal{P}_b^a$ with negative eigenvalue. Show that the polarization vector p^a is the vector perpendicular to f_a with norm

$$p^2 = \frac{1 - (\mathbf{l} \cdot \mathbf{n})^2}{1 + (\mathbf{l} \cdot \mathbf{n})^2}. \qquad (9.146)$$

It follows from (9.146) that the incoming unpolarized radiation scattered at right angles ($\mathbf{l} \cdot \mathbf{n} = 0$) comes out completely polarized in the direction perpendicular to the plane containing the vectors $\mathbf{l}$ and $\mathbf{n}$ as, for instance, the sunlight from the horizon is linearly polarized parallel to the horizon at midday.

Now, if we generalize to the case of an incoming unpolarized radiation field with intensity $J(\mathbf{l}) \propto \langle E^2(\mathbf{l}) \rangle$ and take into account that $\langle \tilde{E}_a \tilde{E}_b \rangle$ and I are additive for the incoherent light, we obtain

$$\mathcal{P}_{ab}(\mathbf{n}) = \frac{\int \left[\frac{1}{2} g_{ab}(1 - (\mathbf{l} \cdot \mathbf{n})^2) - (\mathbf{l} \cdot \mathbf{e}_a)(\mathbf{l} \cdot \mathbf{e}_b) \right] J(\mathbf{l}) d^2 \mathbf{l}}{\int \left[1 + (\mathbf{l} \cdot \mathbf{n})^2 \right] J(\mathbf{l}) d^2 \mathbf{l}} \quad (9.147)$$

If the incoming radiation is isotropic ($J(\mathbf{l}) = $ const), then integrating over $\mathbf{l}$, we find that $\mathcal{P}_{ab}(\mathbf{n}) = 0$, that is, the scattered radiation remains unpolarized. The expression inside the integrand is quadratic in $\mathbf{l}$ and can be expressed in terms of the quadrupole spherical harmonics. Thus, scattering induces polarization in initially unpolarized radiation only if the incident radiation is anisotropic and only the quadrupole component of the anisotropy contributes to the polarization.

Problem 9.21 If the vectors $\mathbf{e}_a$ are orthonormal, then the triplet $\mathbf{e}_1, \mathbf{e}_2$ and $\mathbf{n}$ forms an orthonormal basis in three-dimensional space and one can specify the direction of the incident vector $\mathbf{l}$ by the Euler angles θ and φ. Verify that in this case the polarization tensor components given in (9.147) are

$$\mathcal{P}_{11} = -\mathcal{P}_{22} = \frac{1}{\tilde{I}} \int \sqrt{\frac{3}{40}} \operatorname{Re} Y_{22} J d\Omega,$$
$$\mathcal{P}_{12} = \frac{1}{\tilde{I}} \int \sqrt{\frac{3}{40}} \operatorname{Im} Y_{22} J d\Omega, \quad (9.148)$$

where

$$\tilde{I} = \int \left(Y_{00} - \frac{1}{2\sqrt{5}} Y_{20} \right) J d\Omega, \quad (9.149)$$

and $Y_{lm}(\theta, \varphi)$ are the appropriate spherical functions.

9.10.3 Delayed recombination and polarization

Before recombination begins, the radiation field has only dipole anisotropy (see (9.31)) and therefore polarization cannot be generated. If recombination were instantaneous, then after recombination the photons would propagate without further scattering and no polarization would be generated. Therefore, the background radiation becomes polarized only if recombination is delayed. Taking into account

9.10 Polarization of the cosmic microwave background

that at conformal time $\tilde{\eta}_L$ the probability of last scattering is given by (9.47) and

$$J(\tilde{\eta}_L, \mathbf{l}) \propto (T_0 + \delta T(\tilde{\eta}_L, \mathbf{l}))^4, \qquad (9.150)$$

we obtain (to leading order) from (9.147):

$$\mathcal{P}_{ab}(\mathbf{n}) = 3 \int \left[\frac{1}{2} g_{ab} \left(1 - (\mathbf{l} \cdot \mathbf{n})^2\right) - (\mathbf{l} \cdot \mathbf{e}_a)(\mathbf{l} \cdot \mathbf{e}_b) \right]$$

$$\times \frac{\delta T}{T_0}(\tilde{\eta}_L, \mathbf{l}) \mu'(\tilde{\eta}_L) e^{-\mu(\tilde{\eta}_L)} d\tilde{\eta}_L \frac{d^2 \mathbf{l}}{4\pi}. \qquad (9.151)$$

Hence, the polarization should be proportional to the quadrupole temperature fluctuations generated during the delayed recombination. To calculate $\delta T/T_0(\tilde{\eta}_L, \mathbf{l})$ at the point of scattering $\mathbf{x}$, resulting from the scalar metric perturbations, we can use (9.49) together with (9.48), where one has to replace η_0 by $\tilde{\eta}_L$ and integrate over the time interval $\tilde{\eta}_L > \eta_L > 0$, that is,

$$\frac{\delta T}{T_0}(\tilde{\eta}_L, \mathbf{l}) = \int \int_0^{\tilde{\eta}_L} \left(\Phi + \frac{\delta}{4} - \frac{3\delta'}{4k^2} \frac{\partial}{\partial \tilde{\eta}_L} \right)_{\eta_L} e^{i\mathbf{k} \cdot [\mathbf{x} + \mathbf{l}(\eta_L - \tilde{\eta}_L)]}$$

$$\times \mu'(\eta_L) e^{-\mu(\eta_L)} d\eta_L \frac{d^3 k}{(2\pi)^{3/2}}. \qquad (9.152)$$

Because we will be content with only a rough estimate of the expected polarization we note that the visibility function $\mu'(\tilde{\eta}_L) e^{-\mu(\tilde{\eta}_L)}$ in (9.151) has a sharp maximum at $\tilde{\eta}_L = \eta_r$ corresponding to $z_r \simeq 1050$. Thus the polarization should be about the quadrupole temperature fluctuation at this time. As we have seen, the main contribution to the quadrupole component comes from perturbations with the scales comparable to the horizon scale, that is, with $k\eta_r \sim 1$. One can get an idea of the amplitude of this component by noting that the quadrupole is proportional to terms quadratic in $\mathbf{l}$ arising from the expansion of

$$\exp[i\mathbf{k} \cdot \mathbf{l}(\eta_L - \tilde{\eta}_L)] \sim \exp[i\mathbf{k} \cdot \mathbf{l}(\eta_L - \eta_r)]$$

in powers of $\mathbf{l}$ in (9.152) (note that the higher multipoles also contain terms quadratic in $\mathbf{l}$). Because the visibility function has a sharp peak of width $\Delta \eta \sim \sigma \eta_r$ (see (9.55), (9.57) and (9.60)), we can estimate $\mathbf{k} \cdot \mathbf{l}(\eta_L - \eta_r)$ as σ for $k\eta_r \sim 1$. Therefore, the quadrupole components at η_r, and hence the expected polarization should be about $O(1)\sigma \sim 10^{-2}$–10^{-1} times the temperature fluctuations observed today on angular scales corresponding to the recombination horizon. Polarization, then, is proportional to the duration of recombination and vanishes if recombination is instantaneous. Numerical calculations show that polarization never exceeds 10% of the temperature fluctuations on any angular scales.

9.10.4 E and B polarization modes and correlation functions

To analyze the field of temperature fluctuations we computed the temperature autocorrelation function. The polarization induced at the last scattering surface is characterized by the tensor *field* $\mathcal{P}_{ab}(\mathbf{n})$ on the celestial sphere. The induced polarization is correlated at different points on the sphere and, as with the temperature fluctuation field, can be characterized by the correlation functions $\langle \mathcal{P}_{ab}(\mathbf{n}_1) \mathcal{P}_{cd}(\mathbf{n}_2) \rangle$.

The symmetric traceless tensor $\mathcal{P}_{ab}(\mathbf{n})$ has two independent components and therefore, instead of $\mathcal{P}_{ab}(\mathbf{n})$ itself, it is more convenient to consider two independent scalar functions built out of the polarization tensor:

$$E(\mathbf{n}) \equiv \mathcal{P}_{ab}{}^{;ab}, \qquad B(\mathbf{n}) \equiv \mathcal{P}_a^{b;ac} \epsilon_{cb}, \qquad (9.153)$$

where ; denotes the covariant derivative on the two-dimensional sphere with metric (9.140) and

$$\epsilon_{cb} \equiv \sqrt{g} \begin{pmatrix} 0 & 1 \\ -1 & 0 \end{pmatrix} \qquad (9.154)$$

is the two-dimensional skew-symmetric Levi–Civita "tensor." It behaves as a tensor only under coordinate transformations with positive Jacobian. Under reflections, ϵ_{cb} changes sign. Therefore, only the E mode of polarization is a scalar, while the B mode is a pseudo-scalar, reminiscent of an electric (E) and magnetic (B) field respectively.

The most important thing is that the B mode *is not generated by scalar perturbations*. To prove this let us consider the polarization induced by inhomogeneity with wavenumber $\mathbf{k}$. We use the particular spherical coordinate system where the z-axis determining the Euler angle θ coincides with the direction $\mathbf{k}$. In this coordinate system, the direction of observation $\mathbf{n}$ is characterized by the polar angles θ and φ, and $\mathbf{k} \cdot \mathbf{n} = k \cos \theta$. For every $\mathbf{n}$ we can use the orthogonal coordinate vectors $\mathbf{e}_\theta(\mathbf{n})$ and $\mathbf{e}_\varphi(\mathbf{n})$, tangential to the coordinate lines on the celestial sphere, as the polarization basis. As is clear from (9.152), the $\mathbf{l}$-dependence of the temperature of the incident radiation appears only in the combination $\mathbf{k} \cdot \mathbf{l}$. Therefore, from (9.151) we may easily infer that the nondiagonal component of the polarization tensor should be proportional to

$$\mathcal{P}_{\theta\varphi}(\theta, \varphi) \propto (\mathbf{k} \cdot \mathbf{e}_\theta)(\mathbf{k} \cdot \mathbf{e}_\varphi). \qquad (9.155)$$

This component vanishes because in our coordinate system the vector $\mathbf{e}_\varphi(\mathbf{n})$ is transverse to $\mathbf{k}$ at every point on the sphere. The diagonal components of $\mathcal{P}$ can depend only on $\mathbf{k} \cdot \mathbf{e}_\theta$, $\mathbf{k} \cdot \mathbf{n}$ and the metric (all of which are φ-independent) and

9.10 Polarization of the cosmic microwave background

therefore, the general form of the polarization tensor is

$$\mathcal{P}_{ab}(\theta, \varphi) = \begin{pmatrix} Q(\theta) & 0 \\ 0 & -Q(\theta)\sin^2\theta \end{pmatrix}, \tag{9.156}$$

where we have taken into account that $\mathcal{P}_a^a = 0$.

Problem 9.22 Calculate E and B for the polarization tensor given by (9.156) and verify that $B = 0$ in this case.

Because $B(\theta, \varphi)$ is a pseudo-scalar function, it does not depend on the coordinate system used to calculate it and vanishes for every mode of the density perturbations. Thus density perturbations generate only E mode polarization, which describes the component of the polarization with even parity (the scalar perturbation with given $\mathbf{k}$ is symmetric with respect to rotations around $\mathbf{k}$ and therefore has no handedness). It is easy to see that for $\mathcal{P}_{ab}$ given by (9.156) the appropriate polarization vectors are proportional to $\mathbf{e}_\theta$ at every point where $Q(\theta) > 0$ and proportional to $\mathbf{e}_\varphi$ for $Q(\theta) < 0$. Therefore, in terms of polarization patterns, the E mode produces arrangements of polarization vectors which are oriented radially or tangentially to the circles with respect to the density perturbations, respecting their axial symmetry, as illustrated in Figure 9.6.

In contrast with scalar perturbations, gravitational waves also generate B mode polarization. To see this, let us consider a gravitational wave with wavenumber $\mathbf{k}$ in a coordinate system where the z-axis is aligned along $\mathbf{k}$. Taking into account the general structure of the temperature fluctuations induced by gravitational waves (see (9.119), (9.120)), we can infer from (9.151) that the nondiagonal component of the polarization tensor is proportional to

$$\mathcal{P}_{\theta\varphi} \propto e_{ik}(\mathbf{e}_\theta)^i (\mathbf{e}_\varphi)^k, \tag{9.157}$$

where e_{ik} is the polarization tensor of the gravitational waves.

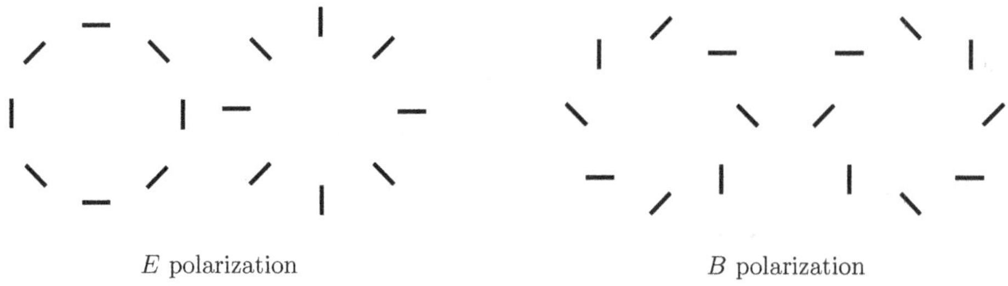

E polarization B polarization

Fig. 9.6.

Problem 9.23 Using (9.121), verify that after averaging over the random polarizations e_{ik}, the component $\langle \mathcal{P}^2_{\theta\varphi}\rangle$ does not vanish and depends only on θ. Calculate the B mode polarization in the case of nondiagonal $\mathcal{P}_{ab}(\theta)$ and show that it is generically different from zero.

The polarization vector p_a induced by the gravitational wave is a linear combination of $\mathbf{e}_\theta$ and $\mathbf{e}_\varphi$. Therefore, the polarization vectors are oriented in circulating patterns, as illustrated in Figure 9.6. In this case the B mode, which has odd parity (handedness), does not vanish. This is due to the fact that the gravitational wave is not symmetric with respect to the rotations around $\mathbf{k}$. Hence, the gravitational waves present at recombination can be detected indirectly via the B mode of the CMB polarization.

To characterize the polarization field on today's sky one can use the appropriate correlation functions, for instance,

$$C^{ET}(\theta) \equiv \left\langle E(\mathbf{n}_1) \frac{\delta T}{T}(\mathbf{n}_2) \right\rangle, \tag{9.158}$$

where the averaging is performed over all directions on the sky satisfying the condition $\mathbf{n}_1 \cdot \mathbf{n}_2 = \cos\theta$. The other correlation functions are C^{BT}, C^{EE}, C^{BB} and C^{EB}. As in the case of temperature fluctuations, the polarizations $E(\mathbf{n})$ and $B(\mathbf{n})$ can be expanded in terms of the scalar spherical harmonics:

$$E = \sum_{l,m} \tilde{a}^E_{lm} Y_{lm}(\theta,\phi), \quad B = \sum_{l,m} \tilde{a}^B_{lm} Y_{lm}(\theta,\phi). \tag{9.159}$$

Since we directly measure the polarization tensor itself, however, it is not very practical to take second derivatives of the experimental data to calculate the coefficients $\tilde{a}_{lm}$. Instead we note that

$$\tilde{a}^E_{lm} = \int E(\mathbf{n}) Y^*_{lm}(\mathbf{n}) d^2\mathbf{n} = \int \mathcal{P}_{ab}{}^{;ab} Y^*_{lm} d^2\mathbf{n} = \frac{1}{N_l} \int \mathcal{P}_{ab} Y^{E*(ab)}_{lm} d^2\mathbf{n}, \tag{9.160}$$

where

$$N_l \equiv \sqrt{\frac{2(l-2)!}{(l+2)!}}$$

and

$$Y^E_{lm(ab)} \equiv N_l\left(Y_{lm;ab} - \tfrac{1}{2} g_{ab} Y_{lm}{}^{;c}{}_c\right)$$

are the E-type tensor harmonics which obey orthogonality relations analogous to the scalar spherical harmonics:

$$\int Y^{E*(ab)}_{lm} Y^E_{l'm'(ab)} d^2\mathbf{n} = \delta_{ll'}\delta_{mm'}. \tag{9.161}$$

In deriving (9.160) we integrated twice by parts and took into account that the polarization tensor is traceless.

Likewise, we have

$$\tilde{a}_{lm}^B = \frac{1}{N_l} \int \mathcal{P}_{ab} Y_{lm}^{B*(ab)} d^2\mathbf{n}, \qquad (9.162)$$

in terms of the normalized B-type tensor harmonics

$$Y_{lm(ab)}^B \equiv \frac{N_l}{2} \left(Y_{lm;ac} \epsilon_b^c + Y_{lm;cb} \epsilon_a^c \right). \qquad (9.163)$$

Note that the E- and B-type tensor harmonics only exist for $l > 1$ and, taken together, form a complete orthonormal basis for second rank tensors on the sphere. Therefore, the polarization tensor can be expanded as

$$\mathcal{P}_{ab} = \sum_{lm} \left[a_{lm}^E Y_{lm(ab)}^E + a_{lm}^B Y_{lm(ab)}^B \right], \qquad (9.164)$$

where, as follows from (9.160) and (9.162), $a_{lm}^{E,B} = N_l \tilde{a}_{lm}^{E,B}$. Thus, instead of first calculating the second derivative of the polarization tensor and then expanding it in terms of the scalar spherical harmonics, we can simply expand the polarization tensor itself in terms of the tensor harmonics. Then, in addition to the usual $C_l \equiv \langle a_{lm}^* a_{lm} \rangle$ characterizing the temperature fluctuations, the polarization of the CMB fluctuations can be described in terms of the sequence of multipoles:

$$\begin{aligned} C_l^{BT} &= \langle a_{lm}^{B*} a_{lm} \rangle, \ C_l^{ET} = \langle a_{lm}^{E*} a_{lm} \rangle, \ C_l^{EE} = \langle a_{lm}^{E*} a_{lm}^E \rangle, \\ C_l^{BB} &= \langle a_{lm}^{B*} a_{lm}^B \rangle, \ C_l^{EB} = \langle a_{lm}^{E*} a_{lm}^B \rangle. \end{aligned} \qquad (9.165)$$

Problem 9.24 Find the explicit expressions for the tensor spherical harmonics in terms of the usual scalar spherical harmonics.

Although the tensor harmonics are technically more complicated than the scalar harmonics, the point is that, given the orthogonality relations, the analysis of the polarization correlation functions is exactly parallel to that of the correlation function for the temperature fluctuations. In Figure 9.7 we present the numerical result for the concordance model. In the models without reionization the l-dependence of $l(l+1) C_l^{EE}$ and $l(l+1) C_l^{BB}$ can easily be understood if we take into account that the polarization is proportional to the quadrupole component of the temperature fluctuations at recombination. In turn, this quadrupole component is mainly due to the perturbations which, at this time, have scales of the order of the horizon and smaller. Therefore, the correlation functions $l(l+1) C_l^{EE}$ and $l(l+1) C_l^{BB}$ drop off for $l < 100$, corresponding to the superhorizon scales at recombination. We have seen above that the amplitude of the gravitational waves, and their contribution to the quadrupole component of the temperature fluctuations at recombination,

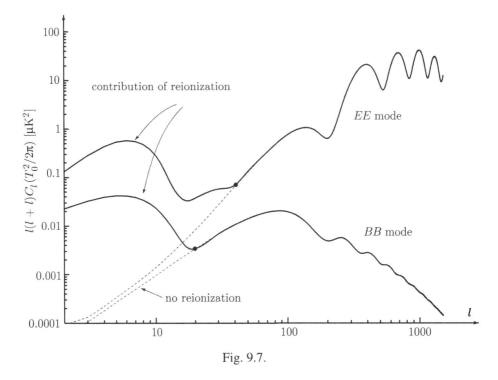

Fig. 9.7.

decreases on subhorizon scales. Hence, they do not contribute to the correlation of the B-type polarization on angular scales corresponding to $l > 100$. As a result, the function $l(l+1) C_l^{BB}$ reaches its maximum at $l \sim 100$. In contrast with B-polarization, due to scalar perturbations on subhorizon scales, there are substantial correlations of E-polarization for $l > 100$.

The correlation function C^{ET} is the easiest to measure since it entails a cross-correlation between the temperature fluctuation amplitude, which is large, and the largest (E mode) polarization component. Measured at $l > 50$, C^{ET} would supply us with information about the history of the recombination.

The B mode polarization is an especially important object of microwave background measurement, since it is the most decisive and probably the only realistic way of detecting the nearly scale-invariant spectrum of gravitational waves predicted by inflation. The cross-polarization multipoles C^{BT} are the easiest moments to detect if one is searching for signs of gravitational waves. The technological challenge of detecting the B mode is considerable. We have already noted that the polarization signal is small, but the B mode polarization component itself is a small fraction of the total polarization in typical inflationary models, as shown in Figure 9.7. In addition, there are foregrounds to consider. For example, the lensing of the microwave background by foreground sources distorts the background polarization pattern in such a way that purely E mode polarization appears to have a B mode component. Nevertheless, present projections suggest that, during the next decade, the most likely range of gravitational wave amplitudes predicted by inflationary cosmology can be fully explored.

9.11 Reionization

At late stages, when nonlinear structure begins to form in the universe, neutral hydrogen can be reionized. In fact, analyzing the spectra of the most distant quasars, one can conclude that most of the intergalactic hydrogen is ionized at $z \simeq 5$. If it were not, the spectra would be significantly corrupted by the absorptions lines of the intergalactic neutral hydrogen. After reionization, the CMB photons can scatter on the free electrons and therefore late reionization affects the resulting CMB fluctuations.

Let us first establish how reionization influences the spectrum of the temperature fluctuations. The probability that the photon has avoided scatterings and propagated freely from time t until the present time t_0 is equal to

$$P(t) = \lim_{\Delta t \to 0} \left(1 - \frac{\Delta t}{\tau(t)}\right) \cdots \left(1 - \frac{\Delta t}{\tau(t_0)}\right)$$
$$= \exp\left(-\int_t^{t_0} \frac{dt}{\tau(t)}\right) \equiv \exp(-\mu(t)), \qquad (9.166)$$

where $\tau = (\sigma_T X n_t)^{-1}$ is the mean free time for Thomson scattering, n_t is the total number of all electrons and X is the ionization fraction. The optical depth $\mu(t) = \mu(z)$ entering (9.166) can be rewritten as an integral over the redshift parameter

$$\mu(z) = \int_t^{t_0} \frac{dt}{\tau(t)} = \sigma_T \int_0^z \frac{X n_t(z)}{H(z)(1+z)} dz. \qquad (9.167)$$

Let us assume instantaneous reionization at redshift $z_r \gg z_{ion} \gg 1$ and calculate the optical depth $\mu(z_{ion})$ in a flat universe. The total number of all (free and bound) electrons is equal to

$$n_t(z) \simeq 0.88 \times \frac{\varepsilon_b(z)}{m_b} = 0.88 \times \frac{3H_0^2}{8\pi m_b} \Omega_b (1+z)^3, \qquad (9.168)$$

where the factor 0.88 accounts for the fact that about 12% of all baryons are neutrons. At high z, which give the main contribution to the integral (9.167), we can neglect the cosmological constant in comparison with the cold matter and use the following expression for the Hubble parameter: $H(z) \simeq H_0 \Omega_m^{1/2} (1+z)^{3/2}$. Substituting it with (9.168) into (9.167), and assuming that $X \simeq 1$ at $z < z_{ion}$, we obtain

$$\mu(z_{ion}) \simeq 0.03 \frac{\Omega_b h_{75}^2}{\sqrt{\Omega_m h_{75}^2}} z_{ion}^{3/2}. \qquad (9.169)$$

In the concordance model, where $\Omega_b h_{75}^2 \simeq 0.04$ and $\Omega_m h_{75}^2 \simeq 0.3$, the reionization at $z_{ion} \simeq 20$ implies an optical depth $\mu \simeq 0.2$. If reionization happens at $z_{ion} \simeq 5$, then $\mu \simeq 0.02$, and in this case the overall effect of reionization on the CMB fluctuations does not exceed 2% or so of the total fluctuations.

As a result of reionization, the fraction

$$1 - \exp(-\mu(z_{ion}))$$

of all photons will be rescattered on the electrons. The remaining fraction $\exp(-\mu(z_{ion}))$ will not be influenced and will give the usual contribution to the fluctuations. For instance, in the model with $z_{ion} \simeq 20$, about 80% of all photons will not be influenced by reionization. The contribution of the 20% of the photons that are rescattered to the angular power spectrum depends on the multipole l. After rescattering, the photon changes its direction of propagation and then the position from which it appears to emanate can be any point that is remote from the original scattering point at a distance not exceeding the horizon scale at this time. As a result, the contribution of the rescattered photons is smeared out within the angular scales corresponding to the reionization horizon scale and does not give rise to temperature fluctuations. For those l which correspond to superhorizon scales the fluctuations will not be influenced because of causality. Of course, at $z < z_{ion}$, the horizon scale continues to grow but the optical depth decreases and the fraction of the photons which are rescattered drops; therefore we neglect this effect. The angular size of the horizon at z_{ion}, corresponding to the conformal time η_{ion}, can easily be found from (2.69) if we take into account that the physical size of the reionization horizon is equal to $a(\eta_{ion})\eta_{ion}$ and $\chi_{em} = \eta_0 - \eta_{ion} \simeq \eta_0$. Then, in a flat universe, we get $\Delta\theta_{ion} \simeq \eta_{ion}/\eta_0$ and the appropriate multipole moment is equal to

$$l_{ion} \simeq \pi/\Delta\theta_{ion} = \pi \eta_0/\eta_{ion} \simeq \pi z_{ion}^{1/2} \Omega_m^{0.09}, \qquad (9.170)$$

where for the ratio η_0/η_{ion} we have used the results of Problem 9.9. Thus, in the model with reionization the observed temperature fluctuations are

$$C_l^{obs} = \begin{cases} C_l, & l \ll l_{ion}, \\ \exp(-\mu) C_l, & l \gg l_{ion}, \end{cases} \qquad (9.171)$$

where the intrinsic temperature fluctuations C_l have been calculated in the previous sections. If $\Omega_m \simeq 0.3$ and $z_{ion} \simeq 20$, we have $l_{ion} \simeq 12$. In this case the amplitudes

9.11 Reionization

of the higher multipoles are suppressed by about 20% compared with their original value, while the lower multipoles are untouched. As we have already mentioned, this can imitate the tilting of the spectral index to a certain extent and lead to an extra degeneracy and further cosmic confusion.

This degeneracy can be easily resolved if we consider the influence of the reionization on polarization spectra. In fact, reionization leads to distinct features in these spectra. The temperature fluctuations in the scattered fraction of photons are not completely washed out on the scales corresponding to the reionization horizon. As a result, there is a net contribution of the rescattered photons to the total temperature fluctuations for the multipoles $\sim l_{ion}$, and it is polarized. It is obvious that the extra polarization induced by reionization is proportional to the fraction of the rescattered photons, $1 - \exp(-\mu(z_{ion}))$, and to the quadrupole anisotropy of the rescattered photons at the beginning of reionization. Because this quadrupole anisotropy is mostly due to the perturbations with scales of order of the reionization horizon, an extra contribution to the polarization correlation functions should have a local maximum at $l \sim l_{ion}$. This explains the behavior of the correlation functions in Figure 9.7 where the results for the polarization in the concordance model with $z_{ion} \simeq 20$ are presented. We would like to stress that because of the presence of the long-wavelength gravitational waves, both E and B modes of polarization will be generated. Thus, we see how measuring the polarization at low multipoles can reveal details of the reionization history and help us to resolve the degeneracy problem.

Bibliography

The literature on the subjects covered in the book runs to many thousands of papers and to document all the important contributions accurately is obviously a task which goes beyond the scope of a textbook. Therefore, I decided to restrict the bibliography to those articles whose results have been explicitly incorporated into the unified account given in this book. Among them are, chiefly, pioneering papers, where the ideas discussed are presented in their modern form. I have also included those papers whose results were directly used in the book. Finally, because the book is devoted mostly to theoretical ideas, I decided to skip altogether references to experimental (observational) papers.

For the convenience of the reader, the full titles of the papers are given, together with a brief mention of the main ideas discussed. In some cases, short quotations from the original papers are given in italics.

Expanding universe (Chapters 1 and 2)

Einstein, A. Kosmologische Betrachtungen zur allgemeinen Relativitaetstheorie. *Sitzungbericht der Berlinische Akademie*, **1** (1917), 142. Introduction of the cosmological constant. Original static Einstein universe with positive curvature (see Problem 1.22).

De Sitter, W. On Einsteins's theory of gravitation and its astronomical consequences. *Monthly Notices of Royal Astronomical Society*, **78** (1917), 3. The original treatment of the de Sitter universe in "static" coordinates (see section 1.3.6).

Friedmann, A. On the curvature of space. *Zeitschrift für Physik*, **10** (1922), 377; On the possibility of a world with constant negative curvature. *Zeitschrift für Physik*, **21** (1924), 326. Discovery of nonstatic solutions for the universe. The papers contain the consideration of closed and open universes, respectively. *"The available data are not sufficient to make numerical estimates and to arrive at a definite conclusion about the features of our universe... Setting $\Lambda = 0$ and taking M to be $5 \cdot 10^{21}$ solar masses, we obtain for the period of the universe* 10 *billion years."* (1922). The expansion of the universe was discovered by Hubble in 1929.

Einstein, A., de Sitter, W. On the relation between the expansion and the mean density of the universe. *Proceedings of the National Academy of Science*, **18** (1932), 213. Discussion of the flat expanding universe with $k = 0$, $\Lambda = 0$ and $p = 0$, which, from the point of view of authors, is a preferable description of the real universe.

McCrea, W., Milne, E. Newtonian universes and the curvature of space. *Quarterly Journal of Mathematics*, **5** (1934), 73. Newtonian treatment of an expanding, matter-dominated universe (see Section 1.2).

Milne, E., A Newtonian expanding universe. *Quarterly Journal of Mathematics*, **5** (1934), 64. For some reason, Milne was uncomfortable with General Relativity and the idea of curved spacetime. Therefore, he suggests an expanding cloud of dust in Minkowski spacetime as an alternative to the expanding curved spacetime (Section 1.3.5).

Penrose, R. Conformal treatment of infinity. *Relativity, Groups and Topology*, eds. C. and B. DeWitt, (1964) p. 563, New York: Gordon and Breach. Describes how ordinary topologically trivial asymptotically flat four-dimensional spacetime can be embedded (in a non-obvious way) in a compact extension.

Carter, B. The complete analytic extension of the Reissner–Nordstrom metric in the special case $e^2 = m^2$. *Physics Letters,* **21** (1966), 23; Complete analytic extension of the symmetry axis of Kerr's solution of Einstein's equations. *Physical Review,* **141** (1966), 1242. The systematic use of conformal diagrams is introduced for geometries with nontrivial global structure.

Hot universe and nucleosynthesis (Chapter 3)

Gamov, G. Expanding universe and the origin of elements. *Physical Review*, **70** (1946), 572; The origin of elements and the separation of galaxies. *Physical Review*, **74** (1948), 505. The hot universe is proposed to solve the nucleosynthesis problem.

Doroshkevich, A., Novikov, I. Mean density of radiation in the metagalaxy and certain problems in relativistic cosmology. *Soviet Physics–Doklady*, **9** (1964), 11. "*Measurements in the region of frequencies $10^9 - 5 \times 10^{10}$ cps are extremely important for experimental checking of Gamov theory. ... According to the Gamov theory, at present time it should be possible to observe equilibrium Planck radiation with a temperature of 1–10 K.*" The paper was not noticed by experimentalists and the cosmic background radiation was discovered accidentally the same year by A. Penzias and R. Wilson.

Hayashi, C. Proton–neutron concentration ratio in the expanding universe at the stages preceding the formation of the elements. *Progress in Theoretical Physics*, **5** (1950), 224. The role of weak interactions in keeping the protons and neutrons in chemical equilibrium is noted and the freeze-out concentration of the neutrons calculated.

Alpher, R., Herman, R. Remarks on the evolution of the expanding universe. *Physical Review,* **75** (1949), 1089. Estimate of the expected temperature of a hot universe. Alpher, R., Follin, J., Herman, R. Physical conditions in the initial stages of the expanding universe. *Physical Review*, **92** (1953), 1347. The calculation of the abundances of the light elements beginning with correct initial conditions for the neutron-to-proton ratio.

Wagoner R., Fowler W., Hoyle F. On the synthesis of elements at very high temperatures. *Astrophysical Journal*, **148** (1967), 3. Contains the modern calculations of the element abundances. The computer programs used today to calculate the primordial abundances are based on the (modified) Wagoner code.

Shvartsman, V. Density of relict particles with zero rest mass in the universe. *JETP Letters*, **9** (1969), 184. The influence of extra relativistic species on primordial nucleosynthesis is noted and it is pointed out that one can obtain bounds on the number of relativistic species present at the epoch of nucleosynthesis.

Zel'dovich, Ya., Kurt, V., Sunyaev, R. Recombination of hydrogen in the hot model of the universe. *ZhETF*, **55** (1968), 278 (translation in *Soviet Physics JETP*, **28** (1969), 146); Peebles, P.J.E. Recombination of the primeval plasma. *Astrophysical Journal*,

153 (1968), 1. Nonequilibrium hydrogen recombination is considered. The roles of Lyman-alpha quanta and two-quanta decay of the $2S$ level are noted.

Particle physics and early universe (Chapter 4)

Yang, C., Mills, R. Conservation of isotopic spin and isotopic gauge invariance. *Physical Review*, **96** (1954), 191. The first non-Abelian gauge theory based on the $SU(2)$ group of isotopic spin conservation is constructed.

Gell-Mann, M. A schematic model of baryons and mesons. *Physics Letters*, **8** (1964), 214; Zweig, G. CERN Preprints TH 401 and TH 412 (1964) (unpublished). The quark model is proposed.

Greenberg, O. Spin and unitary spin independence in a paraquark model of baryons and mesons. *Physical Review Letters*, **13** (1964), 598; Han, M., Nambu, Y., Three triplet model with double $SU(3)$ symmetry. *Physical Review B*, **139** (1965), 1006; Bardeen, W., Fritzsch, H., Gell-Mann, M. Light cone current algebra, π^0 decay, and e^+e^- annihilation. In *Scale and Conformal Symmetry in Hadron Physics*, ed. Gatto, R. (1973) p. 139, New York: Wiley. It is found from baryon systematics and from the rate of neutral pion decay into two photons that quarks of each flavor must come in three colors.

Stuckelberg, E., Petermann, A. The normalization group in quantum theory. *Helvetica Physica Acta*, **24** (1951), 317; La normalisation des constantes dans la theorie des quanta. *Helvetica Physica Acta*, **26** (1953), 499; Gell-Mann, M., Low, F. Quantum electrodynamics at small distances. *Physical Review*, **95** (1954), 1300. The renormalization group method is proposed.

Gross, D., Wilczek, F. Ultraviolet behavior of non-Abelian gauge theories. *Physical Review Letters*, **30** (1973), 1343; Politzer, H. Reliable perturbative results for strong interactions? *Physical Review Letters*, **30** (1973), 1346. The asymptotic freedom of the strong interaction is discovered using the renormalization group method. The asymptotic freedom and its physical implications in the $\lambda\varphi^4$ theory, with negative λ, are discussed in the earliar papers by: Symanzik K. A field theory with computable large-momenta behavior. *Lettere al Nuovo Cimento*, **6** (1973), 77; and Parisi, G. Deep inelastic scattering in a field theory with computable large-momenta behavior. *Lettere al Nuovo Cimento*, **7** (1973), 84.

Chodos, A., Jaffe, R., Johnson, K., Thorn, C., Weisskopf, V. A new extended model of hadrons. *Physical Review D*, **9** (1974), 3471. The bag model is proposed (see Section 4.2.2).

Glashow, S. Partial symmetries of weak interactions. *Nuclear Physics*, **22** (1961), 579; Salam, A., Ward, J. Electromagnetic and weak interactions. *Physics Letters*, **13** (1964), 168. The $SU(2) \times U(1)$ group structure is discussed in relation to electromagnetic and weak interactions.

Higgs, P. Broken symmetries, massless particles and gauge fields. *Physics Letters*, **12** (1964), 132; Broken symmetries and the masses of gauge bosons. *Physics Letters*, **13** (1964), 508; Englert, F., Brout, R. Broken symmetry and the mass of gauge vector mesons. *Physical Review Letters*, **13** (1964), 321; Guralnik, G., Hagen, C., Kibble, T. Global conservation laws and massless particles. *Physical Review Letters*, **13** (1964), 585. The mechanism of the generation of the mass of gauge bosons via interaction with a classical scalar field is discovered.

Weinberg, S. A model of leptons. *Physical Review Letters*, **19** (1967), 1264; Salam, A. Weak and electromagnetic interactions. In *Elementary Particle Theory*, Proceedings of the

8th Nobel Symposium, Svartholm N., ed. (1968), p. 367, Stockholm: Almqvist and Wiksell. The standard theory of electroweak interactions with spontaneously broken symmetry is discovered in its final form.

't Hooft, G. Renormalization of massless Yang–Mills fields. *Nuclear Physics*, **B33** (1971), 173; 't Hooft, G., Veltman, M. Regularization and renormalization of gauge fields. *Nuclear Physics*, **B44** (1972), 189. Proof of the renormalizability of the electroweak theory.

Gell-Mann, M., Levy, M. The axial vector current in beta decay. *Nuovo Cimento*, **16** (1960), 705; Cabibbo, N. Unitary symmetry and leptonic decays. *Physical Review Letters*, **10** (1963), 531. The mixing of two flavors is discussed. In this case it is characterized by one parameter – the Cabibbo angle.

Kobayashi, M., Maskawa, K. CP violation in the renormalizable theory of weak interactions. *Progress of Theoretical Physics,* **49** (1973), 652. It is found in the case of three quark generations that quark mixing generically leads to CP violation. At present, this is the leading explanation of experimentally discovered CP violation.

Kirzhnits, D. Weinberg model in the hot universe. *JETP Letters*, **15** (1972), 529; Kirzhnits, D., Linde, A., Macroscopic consequences of the Weinberg model. *Physics Letters*, **42B** (1972), 471. It is found that, in the early universe at high temperatures, symmetry is restored and the gauge bosons and fermions become massless.

Coleman, S., Weinberg, E. Radiative corrections as the origin of spontaneous symmetry breaking. *Physical Review D*, **7** (1973), 1888. The one-loop quantum corrections to the effective potential are calculated (Section 4.4).

Linde, A. dynamical symmetry restoration and constraints on masses and coupling constants in gauge theories. *JETP Letters,* **23B** (1976), 64; Weinberg, S. Mass of the Higgs boson. *Physical Review Letters*, **36** (1976), 294. The Linde–Weinberg bound on the mass of the Higgs boson is found (Section 4.4.2).

Coleman, S. The fate of the false vacuum, 1: Semiclassical theory. *Physical Review D*, **15** (1977), 2929. The theory of false vacuum decay via bubble nucleation is developed (Section 4.5.2).

Belavin, A., Polyakov, A., Schwartz, A., Tyupkin, Yu. Pseudoparticle solutions of the Yang–Mills equations. *Physics Letters*, **59B** (1975), 85. Instanton solutions in non–Abelian Yang–Mills theories are found.

Bell, J., Jackiw, R. A PCAP puzzle: $\pi^0 \to \gamma\gamma$ in the σ-model. *Nuovo Cimento*, **60A** (1969), 47; Adler, S. Axial-vector vertex in spinor electrodynamics. *Physical Review*, **117** (1969), 2426. Chiral anomaly is discovered. 't Hooft, G. Symmetry breaking through Bell–Jackiw anomalies. *Physical Review Letters*, **37** (1976), 8. The anomalous nonconservation of the chiral current in instanton transitions is noted.

Manton, N. Topology in the Weinberg–Salam theory. *Physical Review D*, **28** (1983), 2019; Klinkhamer, F., Manton, N. A saddle point solution in the Weinberg–Salam theory. *Physical Review D*, **30** (1984), 2212. The role of the sphaleron in transitions between topologically different vacua is discussed.

Kuzmin, V., Rubakov, V., Shaposhnikov, M. On the anomalous electroweak baryon number nonconservation in the early universe. *Physics Letters*, **155B** (1985), 36. It is found that, in the early universe at temperatures above the symmetry restoration scale, transitions between topologically different vacua are not suppressed and, as a result, fermion and baryon numbers are strongly violated.

Gol'fand, Yu., Likhtman, E. Extension of the algebra of Poincare group generators and violation of P invariance. *JETP Letters,* **13** (1971), 323; Volkov, D., Akulov, V. Is the neutrino a Goldstone particle. *Physics Letters*, **46B** (1973), 10. The supersymmetric extension of the Poincare algebra is found.

Wess, J., Zumino, B. Supergauge transformations in four dimensions. *Nuclear Physics*, **B70** (1974), 39. The first supersymmetric model of particle interactions is proposed.

Sakharov, A. Violation of CP invariance, C asymmetry, and baryon asymmetry of the universe. *Soviet Physics, JETP Letters*, **5** (1967), 32. The conditions for the generation of baryon asymmetry in the universe are formulated.

Minkowski, P. Mu to E gamma at a rate of one out of 1-billion muon decays? *Physics Letters*, **B67** (1977), 421; Yanagida, T. In *Workshop on Unified Theories*, KEK report 79-18 (1979), p. 95; Gell-Mann, M., Ramond, P., Slansky, R. Complex spinors and unified theories. In *Supergravity*, eds. van Nieuwenhuizen, P., Freedman, D., (1979) p. 315; Mohapatra, R., Senjanovic, G. Neutrino mass and spontaneous parity nonconservation. *Physical Review Letters*, **44** (1980), 912. The seesaw mechanism is invented (see Section 4.6.2).

Fukugita, M., Yanagida, T. Baryogenesis without grand unification. *Physics Letters*, **B174** (1986), 45. Baryogenesis via leptogenesis is proposed.

Affleck, I., Dine, M., A new mechanism for baryogenesis. *Nuclear Physics*, **B249** (1985), 361. Baryogenesis scenario in supersymmetric models is proposed (see Section 4.6.2).

Peccei, R., Quinn, H. CP conservation in the presence of instantons. *Physical Review Letters*, **38** (1977), 1440. A global $U(1)$ symmetry is proposed to solve the strong CP violation problem.

Weinberg, S. A new light boson? *Physical Review Letters*, **40** (1978), 223; Wilczek, F. *Physical Review Letters*, **40** (1978), 279. It is noted that the breaking of the Peccei–Quinn symmetry leads to a new scalar particle – the axion.

Nielsen, H., Olesen, P. Vortex line models for dual strings. *Nuclear Physics*, **B61** (1973), 45. The string solution in theories with broken symmetry is found.

't Hooft, G. Magnetic monopoles in unified gauge theories. *Nuclear Physics*, **B79** (1974), 276; Polyakov, A. Particle spectrum in the quantum field theory. *JETP Letters*, **20** (1974), 194. The magnetic monopole in gauge theories with broken symmetry is found.

Zel'dovich, Ya., Kobzarev, I., Okun, L. Cosmological consequences of the spontaneous breakdown of discrete symmetry. *Soviet Physics JETP*, **40** (1974), 1; Kibble, T., Topology of cosmic domains and strings. *Journal of Physics*, **A9** (1976), 1387. The production of topological defects in the early universe is discussed (see Section 4.6.3). The subsequent evolution of topological defects is reviewed in Vilenkin, A. Cosmic strings and domain walls. *Physics Report*, **121** (1985), 263.

Inflation (Chapters 5 and 8)

Starobinsky, A. A new type of isotropic cosmological model without singularity. *Physics Letters*, **91B** (1980), 99. The first successful realization of cosmic acceleration with a graceful exit to a Friedmann universe in a higher-derivative gravity theory is proposed. The author wants to solve the singularity problem by assuming that the universe has spent an infinite time in a nonsingular de Sitter state before exiting it to produce the Friedmann universe. In *"... models with the initial superdense de Sitter state... such a large amount of relic gravitational waves is generated... that... the very existence of this state can be experimentally verified in the near future."*

Starobinsky, A. Relict gravitational radiation spectrum and initial state of the universe. *JETP Letters*, **30** (1979), 682. The spectrum of gravitational waves produced during cosmic acceleration is calculated.

Mukhanov, V., Chibisov, G. Quantum fluctuations and a nonsingular universe. *JETP Letters*, **33** (1981), 532. (See also: Mukhanov, V., Chibisov, G. The vacuum energy and large scale structure of the universe. *Soviet Physics JETP*, **56** (1982), 258.) It is shown that the stage of cosmic acceleration considered in Starobinsky (1980) (see above) does not solve the singularity problem because quantum fluctuations make its duration finite. The graceful exit to a Friedmann stage due to the quantum fluctuations is calculated. The red-tilted logarithmic spectrum of initial inhomogeneities produced from initial quantum fluctuations during cosmic acceleration is discovered: *"... models in which the de Sitter stage exists only as an intermediate stage in the evolution are attractive because fluctuations of the metric sufficient for the galaxy formation can occur."*

Guth, A. The inflationary universe: a possible solution to the horizon and flatness problems. *Physical Review D*, **23** (1981), 347. It is noted that the stage of cosmic acceleration, which the author calls inflation, can solve the horizon and flatness problems. It is pointed out that inflation can also solve the monopole problem. No working model with a graceful exit to the Freedman stage is presented: *"... random formation of bubbles of the new phase seems to lead to a much too inhomogeneous universe."*

Linde, A., A new inflationary scenario: a possible solution of the horizon, flatness, homogeneity, isotropy, and primordial monopole problems. *Physics Letters*, **108B** (1982), 389. The new inflationary scenario with a graceful exit based on *"improved Coleman–Weinberg theory"* for the scalar field is proposed.

Albrecht, A., Steinhardt, P. Cosmology for grand unified theories with radiatively induced symmetry breaking. *Physical Review Letters*, **48** (1982), 1220. Confirms the conclusion of Linde (1982) (see above).

Linde, A. Chaotic inflation. *Physics Letters*, **129B** (1983), 177. The generic character of inflationary expansion is discovered for a broad class of scalar field potentials, which must simply satisfy the slow-roll conditions. *"... inflation occurs for all reasonable potentials $V(\varphi)$. This suggests that inflation is not a peculiar phenomenon..., but that it is a natural and maybe even inevitable consequence of the chaotic initial conditions in the very early universe."*

Whitt, B. Fourth order gravity as general relativity plus matter. *Physics Letters*, **B145** (1984), 176. The conformal equivalence between Einstein theory with a scalar field and a higher-derivative gravity is established.

Mukhanov, V. Gravitational instability of the universe filled with a scalar field. *JETP Letters*, **41** (1985), 493; Quantum theory of gauge invariant cosmological perturbations. *Soviet Physics JETP*, **67** (1988), 1297. The self-consistent theory of quantum cosmological perturbations in generic inflationary models is developed[†].

Mukhanov, B., Feldman, H., Brandenberger, R. Theory of cosmological perturbations. *Physics Report*, **215** (1992), 203. This paper contains the derivation of the action for cosmological perturbations in different models from first principles. Explicit formulae in higher-derivative gravity and for cases of nonzero spatial curvature can be found here. (See also Garriga, J., Mukhanov, V. Perturbations in k-inflation. *Physics Letters*, **458B** (1999), 219.)

Damour, T., Mukhanov, V. Inflation without slow-roll. *Physical Review Letters*, **80** (1998), 3440. Fast oscillation inflation in the case of a convex potential is discussed (see Section 4.5.2).

Armendariz-Picon, C., Damour, T., Mukhanov, V. k-Inflation. *Physics Letters*, **458B** (1999), 209. Inflation based on a nontrivial kinetic term for the scalar field is discussed (Section 5.6).

[†] The papers by Hawking, S. *Phys. Lett.*, **115B** (1982), 295; Starobinsky, A. *Phys. Lett.*, **117B** (1982), 175; Guth, A., Pi, S. *Phys. Rev. Lett.*, **49** (1982), 1110; Bardeen, J., Steinhardt, P., Turner, M. *Phys. Rev. D*, **28** (1983), 679 are devoted to perturbations in the new inflationary scenario. However, bearing in mind the considerations of Chapter 8 and solving Problems 8.4, 8.5, 8.7 and 8.8, the reader can easily find out that none of the above papers contains a consistent derivation of the result.

Kofman, L., Linde, A., Starobinsky, A. Reheating after inflation. *Physical Review Letters*, **73** (1994), 3195; Toward the theory of reheating after inflation. *Physical Review D*, **56** (1997), 3258. The self-consistent theory of preheating and reheating after inflation is developed with special stress on the role of broad parametric resonance. The presentation in Section 5.5 follows the main line of these papers.

Everett, H. "Relative state" formulation of quantum mechanics. *Reviews of Modern Physics*, **29** (1957), 454. (See also: *The Many-Worlds Interpretation of Quantum Mechanics*, eds. De Witt, B., Graham, N. (1973), (Princeton, NJ: Princeton University Press.) This remarkable paper is of great interest for those who want to pursue questions related to the interpretation of the state vector of cosmological perturbations, mentioned at the end of Section 8.3.3.

Vilenkin, A. Birth of inflationary universes. *Physical Review D*, **27** (1983), 2848. The eternal self-reproduction regime is found for the new inflationary scenario.

Linde, A. Eternally existing self-reproducing chaotic inflationary universe. *Physics Letters*, **175B** (1986), 395. It is pointed out that self-reproduction naturally arises in chaotic inflation and this generically leads to eternal inflation and a nontrivial global structure of the universe.

Gravitational instability (Chapters 6 and 7)

Jeans, J. *Phil. Trans.*, **129**, (1902), 44; *Astronomy and Cosmogony* (1928), Cambridge: Cambridge University Press. The Newtonian theory of gravitational instability in non-expanding media is developed.

Bonnor, W. *Monthly Notices of the Royal Astronomical Society*, **117** (1957), 104. The Newtonian theory of cosmological perturbations in an expanding matter-dominated universe is developed.

Tolman, R. *Relativity, Thermodynamics and Cosmology* (1934), Oxford: Oxford University Press. The exact spherically symmetric solution for a cloud of dust is found within General Relativity (see Section 6.4.1).

Zel'dovich, Ya. Gravitational instability: an approximate theory for large density perturbations. *Astronomy and Astrophysics*, **5** (1970), 84 . It is discovered that gravitational collapse generically leads to anisotropic structures and the exact nonlinear solution for a one-dimensional collapsing cloud of dust is found (see Section 6.4.2).

Shandarin, S., Zel'dovich, Ya. Topology of the large scale structure of the universe. *Comments on Astrophysics*, **10** (1983), 33; Bond, J. R., Kofman, L., Pogosian, D. How filaments are woven into the cosmic web. *Nature*, **380** (1996), 603. The general picture of the large-scale structure of the universe is developed (Section 6.4.3).

Lifshitz, E. About gravitational stability of expanding world. *Journal of Physics USSR* **10** (1946), 166. The gravitational instability theory of the expanding universe is developed in the synchronous coordinate system.

Gerlach, U., Sengupta, U. Relativistic equations for aspherical gravitational collapse. *Physical Review D*, **18** (1978), 1789. The gauge-invariant gravitational potentials Φ and Ψ used in Chapter 7 are introduced and the equations for these variables are derived.

Bardeen, J. Gauge-invariant cosmological perturbations. *Physical Review D*, **22** (1980), 1882. The solutions for the gauge-invariant variables in concrete models for the evolution of the universe are found.

Chibisov, G., Mukhanov, V. Theory of relativistic potential: cosmological perturbations. Preprint LEBEDEV-83-154 (1983) (unpublished; most of the results of this paper are included in Mukhanov, Feldman and Brandenburger (1992) (see above)). The long-wavelength solutions discussed in Section 7.3 are derived.

Sakharov, A. *Soviet Physics JETP*, **49** (1965), 345. It is found that the spectrum of adiabatic perturbations is ultimately modulated by a periodic function.

CMB fluctuations (Chapter 9)

Sachs, R., Wolfe, A. Perturbation of a cosmological model and angular variations of the microwave background. *Astrophysical Journal,* **147** (1967), 73. The influence of the gravitational potential on the temperature fluctuations is calculated.

Silk, J. Cosmic black-body radiation and galaxy formation. *Astrophysical Journal,* **151** (1968), 459. The radiative dissipation of the fluctuations on small scales is found. The initial conditions for the temperature fluctuations on the last scattering surface (at recombination) are discussed.

Sunyaev, R., Zel'dovich, Ya. Small-scale fluctuations of relic radiation. *Astrophysics and Space Science*, **7** (1970), 3. The fluctuations of background radiation temperature are calculated in a baryon–radiation universe. It is pointed out that "... *a distinct periodic dependence of the spectral density of perturbations on wavelength is peculiar to adiabatic perturbations.*" The approximate formula describing nonequilibrium recombination (see (3.202)) is derived.

Peebles, P.J.E., Yu, J. Primeval adiabatic perturbations in an expanding universe. *Astrophysical Journal*, **162** (1970), 815. The CMB fluctuation spectrum in a baryon–radiation universe is calculated.

Bond, J. R., Efstathiou, G. The statistic of cosmic background radiation fluctuations. *Monthly Notices of the Royal Astronomical Society*, **226** (1987), 655. The modern unified treatment of the CMB fluctuations on all angular scales in cold dark matter models.

Seljak, U., Zaldarriaga, M. A line of sight integration approach to cosmic microwave background anisotropies. *Astrophysical Journal*, **469** (1996), 437. A method of integration of equations for CMB fluctuations is proposed and used to write the CMB-FAST computer code, which is widely used at present.

Index

Affleck–Dine scenario, 215
age of the universe, 8
asymptotic freedom, 141, 146
axions, 204

baryogenesis, 210
  in GUTs, 211
  via leptogenesis, 213
baryon asymmetry, 73, 199, 201, 211
baryon–radiation plasma
  influence on CMB, 365
baryon-to-entropy ratio, 90
baryon-to-photon ratio, 4, 70, 105, 271
  observed value of, 119
baryons, 4, 138
bolometric magnitude, 64
Boltzmann equation, 359
Bose–Einstein distribution, 78
broad resonance, 249, 254

chemical equilibrium, 92
chemical potential, 78
  of bosons, 85
  of electrons, 93
  of fermions, 86, 88
  of neutrinos, 92
  of protons, 93
chiral anomaly, 196
Christoffel symbols, 20
Coleman–Weinberg potential, 172, 259
collision time, 96
color singlets, 139, 140
comoving observers, 7
concordance model, 367, 384
conformal diagrams, 42
continuity equation, 9
coordinates
  comoving, 20
  Lagrangian vs. Eulerian, 272, 279
cosmic coincidence problem, 71
cosmic mean, 365
cosmic microwave background, 4

acoustic peaks, 63, 378, 381, 383
  height of, 387
  location of, 386
  angular scales, 356
  bispectrum, 365
  correlation function, 365
  Doppler peaks, 365
  finite thickness effect, 369, 378
  Gaussian distribution of, 365
  implications for cosmology, 389
  large-angle anisotropy, 368
  last scattering, 72, 356
  multipoles C_l, 366
  plateau, 385
  polarization, 365, 395
    E and B modes, 402, 406
    magnitude of, 396, 401
    mechanism of, 396, 398
    multipoles, 405
    spectrum of, 404
  rest frame of, 360
  small-angle anisotropy, 374
  spectral tilt, 390
  temperature of, 69
  thermal spectrum, 129
  transfer functions, 375, 382
  values of multipoles C_l, 377
  visibility function, 370, 401
cosmic strings, 217, 219
  global vs. local, 220
cosmic variance, 366, 369
cosmological constant Λ, 20
cosmological constant problem, 203
cosmological parameter Ω, 11, 23
cosmological principle, 3
CP violation, 162, 164
CPT invariance, 165
critical density, 11
curvature scale, 39

dark energy, 65, 70, 355
  existence of, 389

dark matter, 70, 355
  candidate particles, 204
  cold relics, 205
  existence of, 389
  hot relics, 204
  nonthermal relics, 207
de Sitter universe, 29, 233, 261
deceleration parameter, 12
delayed recombination, effect on
  CMB, 372, 373
  CMB polarization, 400
  Silk damping, 372
determining cosmological parameters, 367, 379
deuterium abundance, 104, 114
domain walls, 217, 218
dust, 9

effective potential
  one-loop contribution, 168
  thermal contribution, 169
Einstein equations, 20
  linearized, 297
electroweak theory, 150
  fermion interactions, 158
  Higgs mechanism, 154
  phase transitions in, 176, 199
  topological transitions in, 194, 196
energy–momentum tensor, 21
  for imperfect fluid, 311
  for perfect fluid, 21
  for scalar field, 21
equation of state, 21
  discontinuous change in, 305
  for oscillating field, 242
  for scalar field, 235
  ultra-hard, 236
event horizon, 40, 327

Fermi four-fermion interaction, 150
Fermi–Dirac distribution, 78
fermion number violation, 197
Feynman diagrams, 134
flatness problem, 228
free streaming effect, 318, 362
Friedmann equations, 23, 58, 233

gauge symmetry
  global vs. local, 133
  spontaneously broken, 154
Gaussian random fields, 323
  correlation function, 324
gluons, 139
Grand Unification of particle physics, 199
gravitational waves, 348
  effect on CMB, 391
  effect on CMB polarization, 404
  evolution of, 351
  power spectrum of, 349, 351
  quantization of, 348

hadrons, 138
helium-4 abundance, 110

Higgs mechanism, 154
  in electroweak theory, 154
homogeneity, 14
homogeneity problem, 227
homotopy groups, 219
horizon problem, 227
Hubble expansion, 5, 28, 56
Hubble horizon, 39, 327
hybrid topological defects, 225
hydrogen ionization fraction, 123, 127

imperfect fluid approximation, 311
inflation, 73
  attractor solution, 236, 238
  chaotic inflation, 260
  definition of, 230
  different scenarios, 256
  graceful exit, 233, 239
  in higher-derivative gravity, 257
  k inflation, 259
  minimum e-folds, 234
  new inflation, 259
  old inflation, 259
  predictions of, 354
  slow-roll approximation, 241, 329
  with kinetic term, 259
inflaton, 235
initial velocities problem, 228
instantaneous recombination, 357
instantons, 180
  for topological transitions, 194
  in field theory, 185
  thin wall approximation, 188
isotropy, 14

Jeans length, 269

Kobayashi–Maskawa matrix, 162, 164, 215

large-scale structure, 288
lepton era, 89
leptons, 151
Linde–Weinberg bound, 172, 178
Liouville's theorem, 358
Lobachevski space, 16
local equilibrium, 74
Lyman-α photons, 124

Majorana mass term, 213
matter–radiation equality, 72
Mathieu equation, 247
mesons, 138
Milne universe, 27
MIT bag model, 147
monopole problem, 223
monopoles, 217, 221
  local, 222

narrow resonance, 245
  condition for, 248
neutralino, 206
neutrino, 151
  left- and right-handed, 151

neutrino masses, 151
neutron abundance, 115
neutron freeze-out, 102, 103
neutron-to-proton ratio, 94
Newtonian cosmology, 10, 24

optical depth, 370, 407
optical horizon, 39

particle horizon, 38, 327
peculiar velocities, 7, 20
   redshift of, 57
perfect fluid approximation, 266
perturbations
   action for, 340
   adiabatic, 269, 273
   approximate conservation law, 304, 339
   conformal-Newtonian gauge, 295
   decoherence of, 348
   during inflation, 333, 338
   evolution equations, 298, 299
   fictitious modes, 289, 296
   gauge transformation of, 293
   gauge-invariant variables, 294, 297
   generated by inflation, 330, 334
      spectrum of, 345
   in expanding universe, 274
   in inflaton field, 335
   initial state, 343
   long-wavelength modes, 303, 315, 329, 337
   longitudinal gauge, 295
   nonlinear evolution, 279
   of a perfect fluid, 299
   of baryon–radiation plasma, 310, 313
   of dark matter, 312
   of entropy, 270, 306, 315
   on cosmological constant background, 277
   on dust background, 300
   on radiation background, 278
   on sub-Planckian scales, 328, 333
   on ultra-relativistic background, 301
   one-dimensional solution, 283
   quantization of, 341
   scalar modes, 269, 273, 299
   scalar vs. vector vs. tensor modes, 291
   self-similar growth, 276
   short-wavelength modes, 305, 316, 327, 337
   spectral tilt, 346, 355
   spherically symmetric, 282
   sub vs. supercurvature modes, 314
   synchronous gauge, 295
   tensor modes, 309
   transfer functions, 318
   vector modes, 270, 275, 309
phase volume, 357
photon decoupling, 130
polarization tensor, 397
polarization vector, 397
primordial neutrinos, 70
   decoupling of, 73, 96

primordial nucleosynthesis, 73, 98
   overview, 107, 112

quantum chromodynamics, 138
   θ term, 209
quantum tunneling amplitude, 181
quark–gluon plasma, 147, 149
quarks, 138
   colors of, 138
   confinement of, 140
   flavors of, 138
quintessence, 65

recombination, 120
   delayed, 369
   of helium vs. hydrogen, 72, 120
   speed of sound at, 379
redshift parameter, 58
reheating, 243, 245
reionization, 407
   effect on CMB, 408
renormalizable theories, 142
renormalization group equation, 144
Ricci tensor, 20

Sachs–Wolfe effect, 362, 365
Saha formula, 121
Sakharov conditions, 211
seesaw mechanism, 214
self-reproducing universe, 260, 352
self-reproduction scale, 354
shear viscosity coefficient, 311, 314
Silk damping, 311, 317, 372, 378
spaces of constant curvature, 17
sphalerons, 180, 183
   in field theory, 187
standard candles, 7
standard model of particle physics, 131, 199
standard rulers, 7
Stokes parameters, 398
strong energy condition, 22, 233
structure formation, 72
   by inflation, 333
supergravity, 202
supersymmetry, 201

tensor spherical harmonics, 405
textures, 224
thermal history of the universe, 72
thermodynamical integrals, 82
transfer functions
   of CMB, 375, 382
   of primordial perturbations, 315

weak energy condition, 22
weak interactions, 151
Weakly interacting massive particles, 203, 206

Zel'dovich approximation, 286
Zel'dovich pancake, 285